Universitext

T0190481

Christian Grossmann   Hans-Görg Roos

Numerical Treatment of Partial Differential Equations

Translated and revised by Martin Stynes

 Springer

Prof. Dr. Christian Grossmann
Prof. Dr. Hans-Görg Roos

Institute of Numerical Mathematics
Department of Mathematics
Technical University of Dresden
D-01062 Dresden, Germany

e-mail: Christian.Grossmann@tu-dresden.de
        Hans-Goerg.Roos@tu-dresden.de

Prof. Dr. Martin Stynes

School of Mathematical Sciences
Aras na Laoi
University College Cork
Cork, Ireland

e-mail: m.stynes@ucc.ie

Mathematics Subject Classification (2000): 65N, 65F

Translation and revision of the 3rd edition of "Numerische Behandlung Partieller Differentialgleichungen" Published by Teubner, 2005.

Library of Congress Control Number: 2007931595

ISBN 978-3-540-71582-5 Springer Berlin Heidelberg New York

Springer is a part of Springer Science+Business Media
springer.com
© Springer-Verlag Berlin Heidelberg 2007

Cover design: WMXDesign, Heidelberg
Typesetting by the authors and SPi using a Springer LaTeX macro package

Printed on acid-free paper   SPIN: 11830948   46/2244/SPi 5 4 3 2 1 0

Preface

Many well-known models in the natural sciences and engineering, and today even in economics, depend on partial differential equations. Thus the efficient numerical solution of such equations plays an ever-increasing role in state-of-the-art technology. This demand and the computational power available from current computer hardware have together stimulated the rapid development of numerical methods for partial differential equations—a development that encompasses convergence analyses and implementational aspects of software packages.

In 1988 we started work on the first German edition of our book, which appeared in 1992. Our aim was to give students a textbook that contained the basic concepts and ideas behind most numerical methods for partial differential equations. The success of this first edition and the second edition in 1994 encouraged us, ten years later, to write an almost completely new version, taking into account comments from colleagues and students and drawing on the enormous progress made in the numerical analysis of partial differential equations in recent times. The present English version slightly improves the third German edition of 2005: we have corrected some minor errors and added additional material and references.

Our main motivation is to give mathematics students and mathematically-inclined engineers and scientists a textbook that contains all the basic discretization techniques for the fundamental types of partial differential equations; one in which the reader can find analytical tools, properties of discretization techniques and advice on algorithmic aspects. Nevertheless, we acknowledge that in fewer then 600 pages it is impossible to deal comprehensively with all these topics, so we have made some subjective choices of material. Our book is mainly concerned with finite element methods (Chapters 4 and 5), but we also discuss finite difference methods (Chapter 2) and finite volume techniques. Chapter 8 presents the basic tools needed to solve the discrete problems generated by numerical methods, while Chapter 6 (singularly perturbed problems) and Chapter 7 (variational inequalities and optimal control) are special topics that reflect the research interests of the authors.

As well as the above in-depth presentations, there are passing references to spectral methods, meshless discretizations, boundary element methods, higher-order equations and systems, hyperbolic conservation laws, wavelets and applications in fluid mechanics and solid mechanics.

Our book sets out not only to introduce the reader to the rich and fascinating world of numerical methods for partial differential equation, but also to include recent research developments. For instance, we present detailed introductions to a posteriori error estimation, the discontinuous Galerkin method and optimal control with partial differential equations; these areas receive a great deal of attention in the current research literature yet are rarely discussed in introductory textbooks. Many relevant references are given to encourage the reader to discover the seminal original sources amidst the torrent of current research papers on the numerical solution of partial differential equations.

A large portion of Chapters 1–5 constitutes the material for a two-semester course that has been presented several times to students in the third and fourth year of their undergraduate studies at the Technical University of Dresden.

We gratefully acknowledge those colleagues who improved the book by their comments, suggestions and discussions. In particular we thank A. Felgenhauer, S. Franz, T. Linß, B. Mulansky, A. Noack, E. Pfeifer, H. Pfeifer, H.-P. Scheffler and F. Tröltzsch.

We are much obliged to our colleague and long standing friend Martin Stynes for his skill and patience in translating and mathematically revising this English edition.

Dresden, June 2007

Contents

Notation

Often used symbols:

$a(\cdot,\cdot)$	bilinear form
D^+, D^-, D^0	difference quotients
D^α	derivative of order $|\alpha|$ with respect to the multi-index α
I	identity
I_h^H, I_H^h	restriction- and prolongation operator
$J(\cdot)$	functional
$L\, L^*$	differential operator and its adjoint
L_h	difference operator
$O(\cdot), o(\cdot)$	Landau symbols
P_l	the set of all polynomials of degree l
Q_l	the set of all polynomial which are the product of polynomials of degree l with respect to every variable
$\mathbb{R}, \mathbb{N}$	real and natural numbers, respectively
V, V^*	Banach space and its dual
$dim\, V$	dimension of V
V_h	finite-dimensional finite element space
$\|\cdot\|_V$	norm on V
$(\cdot,\cdot)$	scalar product in V, if V is Hilbert space
$f(v)$ or $\langle f, v \rangle$	value of the functional $f \in V^*$ applied to $v \in V$
$\|f\|_*$	norm of the linear functional f
$\rightarrow \quad \rightharpoonup$	strong and weak convergence, respectively
$\oplus$	direct sum
$\mathcal{L}(U, V)$	space of continuous linear mappings of U in V
$\mathcal{L}(V)$	space of continuous linear mappings of V in V
$U \hookrightarrow V$	continuous embedding of U in V
$Z^\perp$	orthogonal complement of Z with respect to the scalar product in a Hilbert space V

Ω	given domain in space
$\partial\Omega = \Gamma$	boundary of Ω
$int\,\Omega$	interior of Ω
$meas\,\Omega$	measure of Ω
n	outer unit normal vector with respect to $\partial\Omega$
$\dfrac{\partial}{\partial n}$	directional derivative with respect to n
$\omega_h,\ \Omega_h$	set of mesh points
$C^l(\Omega),\ C^{l,\alpha}(\Omega)$	space of differentiable and Hölder differentiable functions, respectively
$L_p(\Omega)$	space of functions which are integrable to the power $p\ (1 \le p \le \infty)$
$\|\cdot\|_\infty$	norm in $L_\infty(\Omega)$
$\mathcal{D}(\Omega)$	infinitely often differentiable functions with compact support in Ω
$W_p^l(\Omega)$	Sobolev space
$H^l(\Omega),\ H_0^l(\Omega)$	Sobolev spaces for $p = 2$
$H(div;\Omega)$	special Sobolev space
TV	space of functions with finite total variation
$\|\cdot\|_l$	norm in the Sobolev space H^l
$|\cdot|_l$	semi-norm in the Sobolev space H^l
$t,\ T$	time with $t \in (0,T)$
$Q = \Omega \times (0,T)$	given domain for time-depending problems
$L_2(0,T;X)$	quadratically integrable functions with values in the Banach space X
$W_2^1(0,T;V,H)$	special Sobolev space for time-depending problems
$supp\,v$	support of a function v
∇ or $grad$	gradient
div	divergence
$\triangle$	Laplacian
$\triangle_h$	discrete Laplacian
$h_i,\ h$	discretization parameters with respect to space
$\tau_j,\ \tau$	discretization parameters with respect to time
$det(A)$	determinant of the matrix A
$cond(A)$	condition of the matrix A
$\rho(A)$	spectral radius of the matrix A
$\lambda_i(A)$	eigenvalues of the matrix A
$diag(a_i)$	diagonal matrix with elements a_i
$span\{\varphi_i\}$	linear hull of the elements φ_i
$conv\{\varphi_i\}$	convex hull of the elements φ_i
Π	projection operator
Π_h	interpolation or projection operator which maps onto the finite element space

1

Partial Differential Equations: Basics and Explicit Representation of Solutions

1.1 Classification and Correctness

A partial differential equation is an equation that contains partial derivatives of an unknown function $u : \overline{\Omega} \to \mathbb{R}$ and that is used to define that unknown function. Here Ω denotes an open subset of $\mathbb{R}^d$ with $d \geq 2$ (in the case $d = 1$ one has a so-called ordinary differential equation).

In which applications do partial differential equations arise? Which principles in the modelling process of some application lead us to a differential equation?

As an example let us study the behaviour of the temperature $T = T(x, t)$ of a substance, where t denotes time, in a domain free of sources and sinks where no convection in present. If the temperature is non-uniformly distributed, some energy flux $J = J(x, t)$ tries to balance the temperature. Fourier's law tells us that J is proportional to the gradient of the temperature:

$$J = -\sigma \operatorname{grad} T.$$

Here σ is a material-dependent constant. The principle of conservation of energy in every subdomain $\widetilde{\Omega} \subset \Omega$ leads to the conservation equation

$$\frac{d}{dt} \int\limits_{\widetilde{\Omega}} \gamma \varrho T = - \int\limits_{\partial \widetilde{\Omega}} n \cdot J = \int\limits_{\partial \widetilde{\Omega}} \sigma n \cdot \operatorname{grad} T. \qquad (1.1)$$

Here $\partial \widetilde{\Omega}$ is the boundary of $\widetilde{\Omega}$, γ the heat capacity, ϱ the density and n the outer unit normal vector on $\partial \widetilde{\Omega}$. Using Gauss's integral theorem, one concludes from the validity of (1.1) on each arbitrary $\widetilde{\Omega} \subset \Omega$ that

$$\gamma \varrho \frac{\partial T}{\partial t} = \operatorname{div} (\sigma \operatorname{grad} T). \qquad (1.2)$$

This resulting equation — the heat equation — is one of the basic equations of mathematical physics and plays a fundamental role in many applications.

How does one classify partial differential equations?

In linear partial differential equations the equation depends in a linear manner on the unknown function and its derivatives; any equation that is not of this type is called nonlinear. Naturally, nonlinear equations are in general more complicated than linear equations. *We usually restrict ourselves in this book to linear differential equations and discuss only a few examples of nonlinear phenomena.*

The second essential attribute of a differential equation is its order, which is the same as the order of the highest derivative that occurs in the equation. Three basic equations of mathematical physics, namely

the Poisson equation $-\Delta u = f$,

the heat equation $u_t - \Delta u = f$ and

the wave equation $u_{tt} - \Delta u = f$

are second-order equations. *Second-order equations play a central role in this book.* The bi-harmonic equation (or the plate equation)

$$\Delta \Delta u = 0$$

is a linear equation of the fourth order. Because this equation is important in structural mechanics we shall also discuss fourth-order problems from time to time.

Often one has to handle not just a single equation but a system of equations for several unknown functions. In certain cases there is no difference between numerical methods for single equations and for systems of equations. But sometimes systems do have special properties that are taken into account also in the discretization process. For instance, it is necessary to know the properties of the Stokes system, which plays a fundamental role in fluid mechanics, so we shall discuss its peculiarities.

Next we discuss linear second-order differential operators of the form

$$Lu := \sum_{i,j=1}^{n} a_{ij}(x) \frac{\partial^2 u}{\partial x_i \partial x_j} \qquad \text{with} \qquad a_{ij} = a_{ji}. \tag{1.3}$$

In the two-dimensional case ($n = 2$) we denote the independent variables by x_1, x_2 or by x, y; in the three-dimensional case ($n = 3$) we similarly use x_1, x_2, x_3 or x, y, z. If, however, one of the variables represents time, it is standard to use t. The operator L generates a quadratic form Σ that is defined by

$$\Sigma(\xi) := \sum_{i,j}^{n} a_{ij}(x) \xi_i \xi_j \quad .$$

Moreover, the properties of Σ depend on the eigenvalues of the matrix

$$A := \begin{bmatrix} a_{11} & a_{12} & \cdots & \cdots \\ & \cdots & & \\ a_{n1} & \cdots & & a_{nn} \end{bmatrix}.$$

The differential operator (1.3) is *elliptic* at a given point if all eigenvalues of A are non-zero and have the same sign. The *parabolic* case is characterized by one zero eigenvalue with all other eigenvalues having the same sign. In the *hyperbolic* case, however, the matrix A is invertible but the sign of one eigenvalue is different from the signs of all the other eigenvalues.

Example: The Tricomi equation

$$x_2 \frac{\partial^2 u}{\partial x_1^2} + \frac{\partial^2 u}{\partial x_2^2} = 0$$

is elliptic for $x_2 > 0$, parabolic on $x_2 = 0$ and hyperbolic for $x_2 < 0$.

In the case of equations with constant coefficients, however, the type of the equation does not change from point to point: the equation is uniformly elliptic, uniformly parabolic or uniformly hyperbolic. The most important examples are the following:

Poisson equation   — elliptic,

heat equation   — parabolic,

wave equation   — hyperbolic.

Why it is important to know this classification?

To describe a practical problem, in general it does not suffice to find the corresponding partial differential equation; usually additional conditions complete the description of the problem. The formulation of these supplementary conditions requires some thought. In many cases one gets a well-posed problem, but sometimes an ill-posed problem is generated.

What is a well-posed problem? Let us assume that we study a problem in an abstract framework of the form

$$Au = f \ .$$

The operator A denotes a mapping $A : V \to W$ for some Banach spaces V and W. The problem is well-posed if small changes of f lead to small changes in the solution u; this is reasonable if we take into consideration that f could contain information from measurements.

Example 1.1. (an ill-posed problem of Hadamard for an elliptic equation)

$$\Delta u = 0 \quad \text{in} \quad (-\infty, \infty) \times (0, \delta)$$

$$u|_{y=0} = \varphi(x), \quad \frac{\partial u}{\partial y}\Big|_{y=0} = 0 \quad \text{with} \quad \varphi(x) = \frac{\cos x}{n} \ .$$

The solution of this problem is $u(x,y) = \dfrac{\cos nx \, \cosh ny}{n}$. For large n the function φ is small, but u is not. □

Example 1.1 shows that if we add initial conditions to an elliptic equation, it is possible to get an ill-posed problem.

Example 1.2. (an ill-posed problem for a hyperbolic equation)

Consider the equation

$$\frac{\partial^2 u}{\partial x_1 \, \partial x_2} = 0 \quad \text{in} \quad \Omega = (0,1)^2$$

with the additional boundary conditions

$$u|_{x_1=0} = \varphi_1(x_2), \quad u|_{x_2=0} = \psi_1(x_1), \quad u|_{x_1=1} = \varphi_2(x_2), \quad u|_{x_2=1} = \psi_2(x_1).$$

Assume the following compatibility conditions:

$$\varphi_1(0) = \psi_1(0), \quad \varphi_2(0) = \psi_1(1), \quad \varphi_1(1) = \psi_2(0), \quad \varphi_2(1) = \psi_2(1).$$

Integrating the differential equation twice we obtain

$$u(x_1, x_2) = F_1(x_1) + F_2(x_2)$$

with arbitrary functions F_1 and F_2. Now, for instance, it is possible to satisfy the first two boundary conditions but not all four boundary conditions. In the space of continuous functions the problem is ill-posed. □

Example 1.2 shows that boundary conditions for a hyperbolic equation can result in an ill-posed problem. What is the reason for this behaviour ?
    The first-order characteristic equation

$$\sum_{i,j} a_{ij} \frac{\partial w}{\partial x_i} \frac{\partial w}{\partial x_j} = 0$$

is associated with the differential operator (1.3). A surface (in the two-dimensional case a curve)

$$w(x_1, x_2, \ldots, x_n) = C$$

that satisfies this equation is called a *characteristic* of (1.3). For two independent variables one has the following situation: in the elliptic case there are no real characteristics, in the parabolic case there is exactly one characteristic through any given point, and in the hyperbolic case there are two characteristics through each point.
    It is a well-known fundamental fact that it is impossible on a characteristic Γ to prescribe arbitrary initial conditions

$$u|_\Gamma = \varphi \qquad\qquad \frac{\partial u}{\partial \lambda}\Big|_\Gamma = \psi$$

(λ is a non-tangential direction with respect to Γ): the initial conditions must satisfy a certain additional relation.

For instance, for the parabolic equation

$$u_t - u_{xx} = 0$$

the characteristics are lines satisfying the equation $t = \text{constant}$. Therefore, it makes sense to pose only one initial condition $u|_{t=0} = \varphi(x)$, because the differential equation yields immediately $u_t|_{t=0} = \varphi''(x)$.

For the hyperbolic equation

$$u_{tt} - u_{xx} = 0$$

the characteristics are $x + t = \text{constant}$ and $x - t = \text{constant}$. Therefore, $t = \text{constant}$ is not a characteristic and the initial conditions $u|_{t=0} = \varphi$ and $u_t|_{t=0} = \psi$ yield a well-posed problem.

Let us summarize: *The answer to the question of whether or not a problem for a partial differential equation is well posed depends on both the character of the equation and the type of supplementary conditions.*
The following is a rough summary of well-posed problems for second-order partial differential equations:

elliptic equation	plus	boundary conditions
parabolic equation	plus	boundary conditions with respect to space
	plus	initial condition with respect to time
hyperbolic equation	plus	boundary conditions with respect to space
	plus	two initial conditions with respect to time

1.2 Fourier's Method and Integral Transforms

In only a few cases is it possible to exactly solve problems for partial differential equations. *Therefore, numerical methods are in practice the only possible way of solving most problems.* Nevertheless we are interested in finding an explicit solution representation in simple cases, because for instance in problems where the solution is known one can then ascertain the power of a numerical method. This is the main motivation that leads us in this chapter to outline some methods for solving partial differential equations exactly.

First let us consider the following problem for the heat equation:

$$u_t - \Delta u = f \qquad \text{in} \quad \Omega \times (0, T)$$
$$u|_{t=0} = \varphi(x). \tag{2.1}$$

Here f and φ are given functions that are at least square-integrable; additionally one has boundary conditions on $\partial\Omega$ that for simplicity we take to be homogeneous. In most cases such boundary conditions belong to one of the classes

a) Dirichlet conditions or boundary conditions of the first kind :

$$u|_{\partial\Omega} = 0$$

b) Neumann conditions or boundary conditions of the second kind :

$$\frac{\partial u}{\partial n}|_{\partial\Omega} = 0$$

c) Robin conditions or boundary conditions of the third kind :

$$\frac{\partial u}{\partial n} + \alpha u|_{\partial\Omega} = 0.$$

In case c) we assume that $\alpha > 0$.

Let $\{u_n\}_{n=1}^{\infty}$ be a system of eigenfunctions of the elliptic operator associated with (2.1) that are orthogonal with respect to the scalar product in $L_2(\Omega)$ and that satisfy the boundary conditions on $\partial\Omega$ that supplement (2.1). In (2.1) the associated elliptic operator is $-\Delta$, so we have

$$-\Delta u_n = \lambda_n u_n$$

for some eigenvalue λ_n. Typically, properties of the differential operator guarantee that $\lambda_n \geq 0$ for all n.

Denote by $(\cdot\,,\,\cdot)$ the L_2 scalar product on Ω. Now we expand f in terms of the eigenfunctions $\{u_n\}$:

$$f(x,t) = \sum_{n=1}^{\infty} f_n(t)u_n(x), \quad f_n(t) = (f, u_n).$$

The ansatz

$$u(x,t) = \sum_{n=1}^{\infty} c_n(t)u_n(x)$$

with the currently unknown functions $c_n(\cdot)$ yields

$$c_n'(t) + \lambda_n c_n(t) = f_n(t), \qquad c_n(0) = \varphi_n.$$

Here the φ_n are the Fourier coefficients of the Fourier series of φ with respect to the system $\{u_n\}$. The solution of that system of ordinary differential equations results in the solution representation

$$u(x,t) = \sum_{n=1}^{\infty} \varphi_n u_n(x)e^{-\lambda_n t} + \sum_{n=1}^{\infty} u_n(x)\int_0^t e^{-\lambda_n(t-\tau)}f_n(\tau)d\tau. \qquad (2.2)$$

Remark 1.3. Analogously, one can solve the Poisson equation

$$-\Delta u = f$$

with homogeneous boundary conditions, assuming each $\lambda_n > 0$ (this holds true except in the case of Neumann boundary conditions), in the form

$$u(x) = \sum_{n=1}^{\infty} \frac{f_n}{\lambda_n} u_n(x) \, . \tag{2.3}$$

Then for the wave equation with initial conditions

$$u_{tt} - \Delta u = f, \qquad u|_{t=0} = \varphi(x), \qquad u_t|_{t=0} = \psi(x)$$

and homogeneous boundary conditions, one gets

$$u(x,t) = \sum_{n=1}^{\infty} \left(\varphi_n u_n(x) \cos(\sqrt{\lambda_n}\, t) + \frac{\psi_n}{\sqrt{\lambda_n}} u_n(x) \sin(\sqrt{\lambda_n}\, t) \right)$$
$$+ \sum_{n=1}^{\infty} \frac{1}{\sqrt{\lambda_n}} u_n(x) \int_0^t \sin \sqrt{\lambda_n}(t-\tau) f_n(\tau) d\tau \tag{2.4}$$

where φ_n and ψ_n are the Fourier coefficients of φ and ψ.   □

Unfortunately, the eigenfunctions required are seldom explicitly available. For $-\Delta$, in the one-dimensional case with $\Omega = (0,1)$ one has, for instance,

for $u(0) = u(1) = 0$: $\lambda_n = \pi^2 n^2$, $n = 1, 2, \ldots$; $u_n(x) = \sqrt{2} \sin \pi n x$

for $u'(0) = u'(1) = 0$: $\lambda_0 = 0$; $u_0(x) \equiv 1$

$\lambda_n = \pi^2 n^2$, $n = 1, 2, \ldots$; $u_n(x) = \sqrt{2} \cos \pi n x$

for $u(0) = 0$,
   $u'(1) + \alpha u(1) = 0$: λ_n solves $\alpha \tan \xi + \xi = 0$; $u_n(x) = d_n \sin \lambda_n x$.

(In the last case the d_n are chosen in such a way that the L_2 norm of each eigenfunction u_n equals 1.)

When the space dimension is greater than 1, eigenvalues and eigenfunctions for the Laplacian are known only for simple geometries — essentially only for the cases of a ball and $\Omega = (0,1)^n$. For a rectangle

$$\Omega = \{(x,y) : 0 < x < a, \, 0 < y < b\}$$

for example, for homogeneous Dirichlet conditions one gets

$$\lambda_{m,n} = \left(\frac{m\pi}{a}\right)^2 + \left(\frac{n\pi}{b}\right)^2, \quad u_{m,n} = c \sin \frac{m\pi x}{a} \sin \frac{n\pi y}{b} \quad \text{with} \quad c = \frac{2}{\sqrt{ab}} \, .$$

It is possible to study carefully the convergence behaviour of the Fourier expansions (2.2), (2.3) and (2.4) and hence to deduce conditions guaranteeing that the expansions represent classical or generalized solutions (cf. Chapter 3).

To summarize, we state: *The applicability of Fourier's method to problems in several space dimensions requires a domain with simple geometry and a simple underlying elliptic operator — usually an operator with constant coefficients, after perhaps some coordinate transformation.*

Because Fourier integrals are a certain limit of Fourier expansions, it is natural in the case $\Omega = \mathbb{R}^m$ to replace Fourier expansions by the Fourier transform. From time to time it is also appropriate to use the Laplace transform to get solution representations. Moreover, it can be useful to apply some integral transform only with respect to a subset of the independent variables; for instance, only with respect to time in a parabolic problem. Nevertheless, the applicability of integral transforms to the derivation of an explicit solution representation requires the given problem to have a simple structure, just as we described for Fourier's method.

As an example we study the so-called Cauchy problem for the heat equation:

$$u_t - \Delta u = 0 \quad \text{for} \quad x \in \mathbb{R}^m, \quad u|_{t=0} = \varphi(x). \tag{2.5}$$

For $g \in C_0^\infty(\mathbb{R}^m)$ the Fourier transform $\hat{g}$ is defined by

$$\hat{g}(\xi) = \frac{1}{(2\pi)^{m/2}} \int\limits_{\mathbb{R}^m} e^{-ix\cdot\xi} g(x)\, dx\,.$$

Based on the property

$$\left(\widehat{\frac{\partial^k g}{\partial x_j^k}} \right)(\xi) = (i\xi_j)^k \hat{g}(\xi),$$

the Fourier transform of (2.5) leads to the ordinary differential equation

$$\frac{\partial \hat{u}}{\partial t} + |\xi|^2 \hat{u} = 0 \quad \text{with} \quad \hat{u}|_{t=0} = \hat{\varphi}\,,$$

whose solution is $\hat{u}(\xi, t) = \hat{\varphi}(\xi)\, e^{-|\xi|^2 t}$.

The backward transformation requires some elementary manipulation and leads, finally, to the well known *Poisson's formula*

$$u(x,t) = \frac{1}{(4\pi t)^{m/2}} \int\limits_{\mathbb{R}^m} e^{-\frac{|x-y|^2}{4t}} \varphi(y)\, dy\,. \tag{2.6}$$

Analogously to the convergence behaviour of Fourier series, a detailed analytical study (see [Mic78]) yields conditions under which Poisson's formula represents a classical or a generalized solution. Poisson's formula leads to the conclusion that heat spreads with infinite speed. Figure 1.1, for instance, shows the temperature distribution for $t = 0.0001, 0.01, 1$ if the initial distribution is given by

$$\varphi(x) = \begin{cases} 1 & \text{if } 0 < x < 1 \\ 0 & \text{otherwise.} \end{cases}$$

Although the speed of propagation predicted by the heat equation is not exactly correct, in many practical situations this equation models the heat propagation process with sufficient precision.

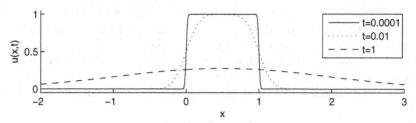

Figure 1.1   Solution at $t = 0.0001$, $t = 0.01$ and $t = 1$

1.3 Maximum Principle, Fundamental Solution, Green's Function and Domains of Dependency

1.3.1 Elliptic Boundary Value Problems

In this section we consider linear elliptic differential operators

$$Lu(x) := -\sum a_{ij}(x)\frac{\partial^2 u}{\partial x_i \partial x_j} + \sum b_i(x)\frac{\partial u}{\partial x_i} + c(x)\,u$$

in some bounded domain Ω. It is assumed that we have:

(1) symmetry and ellipticity: $a_{ij} = a_{ji}$,   and some $\lambda > 0$ exists such that

$$\sum a_{ij}\xi_i\xi_j \geq \lambda \sum \xi_i^2 \qquad \text{for all } x \in \Omega \text{ and all } \xi \in \mathbb{R}^n \qquad (3.1)$$

(2) all coefficients a_{ij}, b_i, c are bounded.

For elliptic differential operators of this type, maximum principles are valid; see [PW67], [GT83] for a detailed discussion.

Theorem 1.4 (boundary maximum principle). *Let $c \equiv 0$ and $u \in C^2(\Omega) \cap C(\bar{\Omega})$. Then*

$$Lu(x) \leq 0 \quad \forall x \in \Omega \qquad \Longrightarrow \qquad \max_{x \in \bar{\Omega}} u(x) \leq \max_{x \in \partial\Omega} u(x).$$

The following comparison principle provides an efficient tool for obtaining a priori bounds:

Theorem 1.5 (comparison principle). *Let $c \geq 0$ and $v, w \in C^2(\Omega) \cap C(\bar{\Omega})$. Then*

$$\left.\begin{array}{l} Lv(x) \leq Lw(x) \; \forall x \in \Omega, \\ v(x) \leq w(x) \quad \forall x \in \partial\Omega \end{array}\right\} \qquad \Longrightarrow \qquad v(x) \leq w(x) \quad \forall x \in \bar{\Omega}.$$

If v, w are a pair of functions with the properties above and if w coincides with the exact solution of a corresponding boundary value problem then v is called a lower solution of this problem. Upper solutions are defined analogously.

Given continuous functions f and g we now consider the classical formulation of the Dirichlet problem ("weak" solutions will be discussed in Chapter 3):

Find a function $u \in C^2(\Omega) \cap C(\bar{\Omega})$ — a so-called classical solution — such that

$$Lu = f \text{ in } \quad \Omega$$
$$u = g \text{ on } \quad \partial\Omega \,. \tag{3.2}$$

Theorem 1.5 immediately implies that *problem (3.2) has at most one classical solution.*

What can we say about the existence of classical solutions?

Unfortunately the existence of a classical solution is not automatically guaranteed. Obstacles to its existence are

- *the boundary Γ of the underlying domain is insufficiently smooth*
- *nonsmooth data of the problem*
- *boundary points at which the type of boundary condition changes.*

To discuss these difficulties let us first give a precise characterization of the smoothness properties of the boundary of a domain.

Definition 1.6. A bounded domain Ω *belongs to the class* $C^{m,\alpha}$ (briefly, $\partial\Omega \in C^{m,\alpha}$) if a finite number of open balls K_i exist with:
(i) $\cup K_i \supset \partial\Omega$, $K_i \cap \partial\Omega \neq 0$;
(ii) There exists some function $y = f^{(i)}(x)$ that belongs to $C^{m,\alpha}(\bar{K}_i)$ and maps the ball K_i in a one-to-one way onto a domain in $\mathbb{R}^n$ where the image of the set $\partial\Omega \cap \bar{K}_i$ lies in the hyperplane $y_n = 0$ and $\Omega \cap K_i$ is mapped into a simple domain in the halfspace $\{y : y_n > 0\}$. The functional determinant

$$\frac{\partial(f_1^{(i)}(x), ..., f_n^{(i)}(x))}{\partial(x_1, ..., x_n)}$$

does not vanish for $x \in \bar{K}_i$.

This slightly complicated definition describes precisely what is meant by the assumption that the boundary is "sufficiently smooth". The larger m is, the smoother is the boundary. Domains that belong to the class $C^{0,1}$ are rather important; they are often called *domains with regular boundary* or *Lipschitz domains*. For example, bounded convex domains have regular boundaries. An important fact is that for domains with a regular boundary (i.e., for Lipschitz domains) a uniquely defined outer normal vector exists at almost all boundary points and functions $v_i \in C^1(\Omega) \cap C(\bar{\Omega})$ on such domains satisfy Gauss's theorem

$$\int_\Omega \sum_i \frac{\partial v_i}{\partial x_i} \, d\Omega = \int_\Gamma \sum_i v_i \, n_i \, d\Gamma.$$

Theorem 1.7. *Let $c \geq 0$ and let the domain Ω have a regular boundary. Assume also that the data of the boundary value problem (3.2) is smooth (at least α-Hölder continuous). Then (3.2) has a unique solution.*

This theorem is a special case of a general result given in [91] for problems on "feasible" domains.

To develop a classical convergence analysis for finite difference methods, rather strong smoothness properties of the solution ($u \in C^{m,\alpha}(\bar{\Omega})$ with $m \geq 2$) are required. But even in the simplest cases such smoothness properties may fail to hold true as the following example shows:

Example 1.8. Consider the boundary value problem

$$-\triangle u = 0 \quad \text{in} \quad \Omega = (0,1) \times (0,1),$$
$$u = x^2 \quad \text{on} \quad \Gamma.$$

By the above theorem this problem has a unique classical solution u. But this solution cannot belong to $C^2(\bar{\Omega})$ since the boundary conditions imply that $u_{xx}(0,0) = 2$ and $u_{yy}(0,0) = 0$, which in the case where $u \in C^2(\bar{\Omega})$ contradicts the differential equation.   □

Example 1.9. In the domain

$$\Omega = \{(x,y) \,|\, x^2 + y^2 < 1, \, x < 0 \text{ or } y > 0\},$$

which has a re-entrant corner, the function $u(r,\varphi) = r^{2/3} \sin((2\varphi)/3)$ satisfies Laplace's equation $-\triangle u = 0$ and the (continuous) boundary condition

$$u = \sin((2\varphi)/3) \text{ for } r = 1, \, 0 \leq \varphi \leq 3\pi/2$$
$$u = 0 \quad \text{elsewhere on } \partial\Omega.$$

Despite these nice conditions the first-order derivatives of the solution are not bounded, i.e. $u \notin C^1(\bar{\Omega})$.   □

The phenomenon seen in Example 1.9 can be described in a more general form in the following way. Let Γ_i and Γ_j be smooth arcs forming parts of the boundary of a two-dimensional domain. Let r describe the distance from the corner where Γ_i and Γ_j meet and let $\alpha\pi$, where $0 < \alpha < 2$, denote the angle between these two arcs. Then near that corner one must expect the following behaviour:

$$u \in C^1 \quad \text{for} \quad \alpha \leq 1,$$
$$u \in C^{1/\alpha}, \quad u_x \text{ and } u_y = O(r^{\frac{1}{\alpha}-1}) \quad \text{for } \alpha > 1.$$

In Example 1.9 we have $\alpha = 3/2$. Later, in connection with finite element methods, we shall study the regularity of weak solutions in domains with corners.

In the case of Dirichlet boundary conditions for $\alpha \leq 1$ the solution has the C^1 property. But if the Dirichlet conditions meet Neumann or Robin conditions at some corner then this property holds true only for $\alpha < 1/2$. Further details on corner singularities can be found in e.g. [121].

If the domain has a smooth boundary and if additionally all coefficients of the differential operator are smooth then the solution of the elliptic boundary value problem is smooth. As a special case we have

Theorem 1.10. *In (3.2) let $c \geq 0$ and $g \equiv 0$. Let the domain Ω belong to the class $C^{2,\alpha}$ and let the data of problem (3.2) be sufficiently smooth (at least $C^\alpha(\bar{\Omega})$ for some $\alpha > 0$). Then the problem has a unique solution $u \in C^{2,\alpha}(\bar{\Omega})$.*

This theorem is a special case of a more general existence theorem given in [1]. If $\partial\Omega \notin C^{2,\alpha}$ but solutions with $C^{2,\alpha}(\bar{\Omega})$ smoothness are wanted, then additional requirements — *compatibility conditions* — must be satisfied. For the problem

$$Lu = f \quad \text{in} \quad \Omega = (0,1) \times (0,1),$$
$$u = 0 \quad \text{on} \quad \Gamma$$

these are assumptions such as $f(0,0) = f(1,0) = f(0,1) = f(1,1) = 0$. Detailed results regarding the behaviour of solutions of elliptic problems on domains with corners can be found in [Gri85].

After this brief discussion of existence theory we return to the question of representation of solutions. As one would expect, the task of finding an explicit formula for the solution for (3.2) is difficult because the differential equation and the boundary conditions must both be fulfilled. In the case where no boundary conditions are given ($\Omega = \mathbb{R}^n$) and the differential operator L has sufficiently smooth coefficients, the theory of distributions provides the helpful concept of a *fundamental solution*. A fundamental solution K is a distribution with the property

$$LK = \delta,$$

where δ denotes Dirac's δ-function (more precisely, δ-distribution). In the case where L has constant coefficients, moreover $L(S * K) = S$, where $S * K$ stands for the convolution of the distributions S and K. For regular distributions s and k one has

$$(s * k)(x) = \int s(x-y)k(y)\, dy = \int s(y)k(x-y)\, dy.$$

For several differential operators with constant coefficients that are of practical significance, the associated fundamental solutions are known. As a rule they are regular distributions, i.e. they can be represented by locally integrable functions.

For the rest of this section we consider the example of the Laplace operator

$$Lu := -\Delta u = -\sum_{i=1}^{n} \frac{\partial^2 u}{\partial x_i^2} .$$ (3.3)

For this the fundamental solution has the explicit form

$$K(x) = \begin{cases} -\dfrac{1}{2\pi} \ln |x| & \text{for } n = 2, \\[2ex] \dfrac{1}{(n-2)|w_n||x|^{n-2}} & \text{for } n \geq 3, \end{cases}$$

here $|w_n|$ denotes the measure of the unit sphere in $\mathbb{R}^n$.

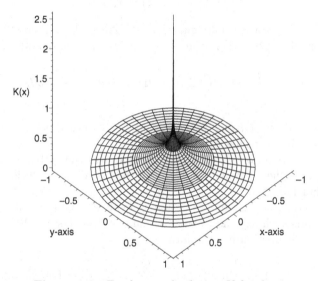

Figure 1.2 Fundamental solution K for $d = 2$

If the right-hand side f of (3.2) is integrable and has compact support then with $K(x, \xi) := K(x - \xi) = K(\xi - x)$ we obtain the representation

$$u(\xi) = \int_{\mathbb{R}^n} K(x, \xi) \, f(x) \, dx$$ (3.4)

for the solution of $Lu = f$ on $\mathbb{R}^n$.

Is there a representation similar to (3.4) for the solution u of $-\Delta u = f$ in bounded domains? To answer this question we first observe (cf. [Hac03a]) that

Theorem 1.11. *Let $\Omega \subset \mathbb{R}^n$ be bounded and have a smooth boundary Γ. Assume that $u \in C^2(\bar{\Omega})$. Then for arbitrary $\xi \in \Omega$ one has*

$$
\begin{aligned}
u(\xi) = & \int_\Omega K(x, \xi)(-\Delta u(x)) \, d\Omega \\
& + \int_\Gamma \left(u(x) \frac{\partial K(x, \xi)}{\partial n_x} - K(x, \xi) \frac{\partial u(x)}{\partial n_x} \right) d\Gamma .
\end{aligned}
\tag{3.5}
$$

This is the fundamental relation underpinning the definition of potentials and the conversion of boundary value problems into related boundary integral equations.

If in (3.5) we replace K by

$$
K(x, \xi) = G(x, \xi) - w_\xi(x) ,
$$

where $w_\xi(x)$ denotes a harmonic function (i.e. $\Delta w = 0$) that satisfies $w_\xi(x) = -K(x, \xi)$ for all $x \in \partial\Omega$, then from (3.5) follows the representation

$$
u(\xi) = \int_\Omega G(x, \xi)(-\Delta u(x)) \, d\Omega + \int_{\partial\Omega} u \frac{\partial G(x, \xi)}{\partial n_x} \, d\Omega .
\tag{3.6}
$$

Here G is called the *Green's function* and G satisfies the same differential equation as K but vanishes on $\partial\Omega$.

The Green's function G essentially incorporates the global effect of local perturbations in the right-hand side f or in the boundary condition $g := u|_{\partial\Omega}$ upon the solution.

Unfortunately, the Green's function is explicitly known only for rather simple domains such as a half-plane, orthant or ball. For the ball $|x| < a$ in $\mathbb{R}^n$, the Green's function has the form

$$
G(x, \xi) =
\begin{cases}
\dfrac{1}{2\pi} \left[\ln|x - \xi| - \ln \dfrac{|\xi|}{a} |x - \xi^*| \right] & \text{for} \quad n = 2, \\[2ex]
K(x, \xi) - \left(\dfrac{a}{|\xi|} \right)^{n-2} K(x, \xi^*) & \text{for} \quad n > 2,
\end{cases}
$$

where $\xi^* := a^2 \xi / |\xi|^2$. In particular this implies for the solution u of

$$
\Delta u = 0 \quad \text{in} \quad |x| < a
$$

$$
u = g \quad \text{on} \quad \partial\Omega
$$

the well-known *Poisson integral formula*

$$
u(\xi) = \frac{a^2 - |\xi|^2}{a|w_n|} \int_{|x|=a} \frac{g(x)}{|x - \xi|^n} \, d\Omega .
$$

While the existence of the Green's function can be guaranteed under rather weak assumptions, this does not imply that the representation (3.6) yields a classical solution $u \in C^2(\bar{\Omega})$ even in the case when the boundary $\partial\Omega$ is smooth, the boundary data g is smooth and the right-hand side $f \in C(\bar{\Omega})$. A sufficient condition for $u \in C^2(\bar{\Omega})$ is the α-Hölder continuity of f, i.e. $f \in C^\alpha(\bar{\Omega})$.

Let us finally remark that Green's functions can be defined, similarly to the case of the Laplace operator, for more general differential operators. Moreover for boundary conditions of Neumann type a representation like (3.6) is possible; here Green's functions of the second kind occur (see [Hac03a]).

1.3.2 Parabolic Equations and Initial-Boundary Value Problems

As a typical example of a parabolic problem, we analyze the following initial-boundary value problem of heat conduction:

$$u_t - \Delta u = f \quad \text{in} \ \ Q = \Omega \times (0, T)$$
$$u = 0 \quad \text{on} \ \ \partial\Omega \times (0, T) \tag{3.7}$$
$$u = g \ \ \text{for} \ \ t = 0, \ \ x \in \Omega.$$

Here Ω is a bounded domain and ∂Q_p denotes the "parabolic" boundary of Q, i.e.

$$\partial Q_p = \{(x, t) \in \bar{Q} : x \in \partial\Omega \text{ or } t = 0\}.$$

The proofs of the following two theorems can be found in e.g. [PW67]:

Theorem 1.12. *(boundary maximum principle)*   Let $u \in C^{2,1}(Q) \cap C(\bar{Q})$. *Then*

$$u_t - \Delta u \leq 0 \quad \text{in } Q \qquad \Longrightarrow \qquad \max_{(x,t) \in \bar{Q}} u(x,t) = \max_{(x,t) \in \partial Q_p} u(x,t).$$

Similarly to elliptic problems, one also has

Theorem 1.13. *(comparison principle)*   Let $v, w \in C^{2,1}(Q) \cap C(\bar{Q})$. *Then*

$$\left. \begin{array}{rl} v_t - \Delta v \leq w_t - \Delta w \ \text{in} & Q \\ v \leq w & \text{on} \ \ \partial\Omega \times (0, T) \\ v \leq w & \text{for} \ \ t = 0 \end{array} \right\} \quad \Longrightarrow \quad v \leq w \ \ \text{on} \ \ \bar{Q}.$$

This theorem immediately implies the uniqueness of classical solutions of (3.7).

The fundamental solution of the heat conduction operator is

$$K(x, t) = \begin{cases} \dfrac{1}{2^n \pi^{n/2} t^{n/2}} e^{-\frac{|x|^2}{4t}} & \text{for} \ \ t > 0, \ x \in \mathbb{R}^n \\[2mm] 0 & \text{for} \ \ t \leq 0, \ x \in \mathbb{R}^n. \end{cases} \tag{3.8}$$

From this we immediately see that the solution of

$$u_t - \Delta u = f \quad \text{in} \quad \mathbb{R}^n \times (0, T) \tag{3.9}$$
$$u = 0 \quad \text{for} \quad t = 0.$$

has the representation

$$u(x, t) = \int_0^t \int_{\mathbb{R}^n} K(x - y, t - s) f(y, s) \, dy \, ds. \tag{3.10}$$

Now in the case of a homogeneous initial-value problem

$$u_t - \Delta u = 0 \quad \text{in} \quad \mathbb{R}^n \times (0, T) \tag{3.11}$$
$$u = g \quad \text{for} \quad t = 0$$

we already have a representation of its solution that was obtained via Fourier transforms in Section 1.2), namely

$$u(x, t) = \int_{\mathbb{R}^n} K(x - y, t) g(y) \, dy. \tag{3.12}$$

Indeed, we can prove

Theorem 1.14. *For the solutions of (3.9) and (3.11) we have the following:*

a) *Let f be sufficiently smooth and let its derivatives be bounded in $\mathbb{R}^n \times (0, T)$ for all $T > 0$. Then (3.10) defines a classical solution of problem (3.9): $u \in C^{2,1}(\mathbb{R}^n \times (0, \infty)) \cap C(\mathbb{R}^n \times [0, \infty))$.*

b) *If g is continuous and bounded then (3.12) defines a classical solution of (3.11). Further we even have $u \in C^\infty$ for $t > 0$ (the "smoothing effect").*

In part b) of this theorem the boundedness condition for g may be replaced by

$$|g(x)| \le M e^{\alpha |x|^2}.$$

Moreover, in the case of functions g with compact support the modulus of the solution u of the homogeneous initial-value problem can be estimated by

$$|u(x, t)| \le \frac{1}{(4\pi t)^{n/2}} e^{-\frac{\text{dist}|x, K|^2}{4t}} \int_K |g(y)| \, dy,$$

where $K := \text{supp} \, g$. Hence, $|u|$ tends exponentially to zero as $t \to \infty$.

Nevertheless, the integrals that occur in (3.10) or in (3.12) can rarely be evaluated explicitly. In the special case of spatially one-dimensional problems

$$u_t - u_{xx} = 0, \quad u|_{t=0} = \begin{cases} 1 & x < 0 \\ 0 & x > 0 \end{cases}$$

with the error function

$$\mathrm{erf}(x) = \frac{2}{\sqrt{\pi}} \int_0^x e^{-t^2} dt,$$

substitution in (3.10) yields the representation

$$u(x,t) = \frac{1}{2} \left(1 - \mathrm{erf}(x/2\sqrt{t}) \right).$$

Further let us point to a connection between the representations (3.10) and (3.12) and the *Duhamel principle*.
If $z(x,t,s)$ is a solution of the homogeneous initial-value problem

$$z_t - z_{xx} = 0, \quad z|_{t=s} = f(x,s),$$

then

$$u(x,t) = \int_0^t z(x,t,s)\, ds$$

is a solution of the inhomogeneous problem

$$u_t - u_{xx} = f(x,t), \quad u|_{t=0} = 0.$$

Analogously to the case of elliptic boundary value problems, for heat conduction in $Q = \Omega \times (0,T)$ with a bounded domain Ω there also exist representations of the solution in terms of Green's functions or heat conduction kernels. As a rule, these are rather complicated and we do not provide a general description. In the special case where

$$u_t - u_{xx} = f(x,t), \quad u|_{x=0} = u|_{x=l} = 0, \quad u|_{t=0} = 0,$$

we can express the solution as

$$u(x,t) = \int_0^t \int_0^l f(\xi,\tau)\, G\left(x,\xi,t-\tau\right) d\xi\, d\tau$$

with

$$G(x,\xi,t) = \frac{1}{2l} \left[\vartheta_3\left(\frac{x-\xi}{2l}, \frac{t}{l^2} \right) - \vartheta_3\left(\frac{x+\xi}{2l}, \frac{t}{l^2} \right) \right].$$

Here ϑ_3 denotes the classical Theta-function

$$\vartheta_3(z,\tau) = \frac{1}{\sqrt{-i\tau}} \sum_{n=-\infty}^{\infty} \exp\left[-i\pi(z+n)^2/\tau \right].$$

An "elementary" representation of the solution is available only in rare cases, e.g. for

$$u_t - u_{xx} = 0 \quad \text{in} \quad x > 0,\, t > 0 \quad \text{with} \quad u|_{t=0} = 0, \quad u|_{x=0} = h(t)\,.$$

In this case one obtains

$$u(x,t) = \frac{1}{2\sqrt{\pi}} \int\limits_0^t \frac{x}{(t-\tau)^{3/2}}\, e^{\frac{-x^2}{4(t-\tau)}}\, h(\tau)\, d\tau\,.$$

One possibility numerical treatment of elliptic boundary value problems is based on the fact that they can be considered as the stationary limit case of parabolic initial-boundary value problems. This application of parabolic problems rests upon:

Theorem 1.15. *Let Ω be a bounded domain with smooth boundary. Furthermore, let f and g be continuous functions. Then as $t \to \infty$ the solution of the initial-boundary value problem*

$$u_t - \Delta u = 0 \quad \text{in} \quad \Omega \times (0, T)$$

$$u = g \ \text{ for } \ x \in \partial\Omega$$

$$u = f \ \text{ for } \ t = 0$$

converges uniformly in $\bar{\Omega}$ to the solution of the elliptic boundary value problem

$$\Delta u = 0 \quad \text{in} \quad \Omega$$

$$u = g \ \text{ on } \ \partial\Omega\,.$$

1.3.3 Hyperbolic Initial and Initial-Boundary Value Problems

The properties of hyperbolic problems differ significantly from those of elliptic and parabolic problems. One particular phenomenon is that, unlike in elliptic and parabolic problems, the spatial dimension plays an important role. In the present section we discuss one-, two- and three-dimensional cases. For all three the fundamental solution of the wave equation

$$u_{tt} - c^2 \Delta u = 0$$

is known, but for $n = 3$ the fundamental solution is a singular distribution [Tri92]. Here we avoid this approach and concentrate upon "classical" representations of solutions.

As a typical hyperbolic problem we consider the following initial-value problem for the wave equation:

$$u_{tt} - c^2 \Delta u = f \qquad \text{in} \qquad \mathbb{R}^n \times (0, T)$$

$$u|_{t=0} = g\,, \qquad u_t|_{t=0} = h\,.$$

Duhamel's principle can also be applied here, i.e. the representation of the solution of the homogeneous problem can be used to construct a solution of the inhomogeneous one. Indeed, if $z(x, t, s)$ denotes a solution of the homogeneous problem

$$z_{tt} - c^2 \Delta z = 0, \quad z|_{t=s} = 0, \quad z_t|_{t=s} = f(x, s),$$

then

$$u(x, t) = \int_0^t z(x, t, s) \, ds$$

is a solution of the inhomogeneous problem

$$u_{tt} - c^2 \Delta u = f(x, t), \quad u|_{t=0} = 0, \quad u_t|_{t=0} = 0.$$

Hence, we can concentrate upon the homogeneous problem

$$u_{tt} - c^2 \Delta u = 0 \quad \text{in} \quad \mathbb{R}^n \times (0, T)$$
$$u|_{t=0} = g, \quad u_t|_{t=0} = h. \tag{3.13}$$

In the one-dimensional case the representation $u = F(x + ct) + G(x - ct)$ of the general solution of the homogeneous equation immediately implies d'Alembert's formula:

$$u(x, t) = \frac{1}{2}(g(x + ct) + g(x - ct)) + \frac{1}{2c} \int_{x-ct}^{x+ct} h(\xi) \, d\xi. \tag{3.14}$$

Several important conclusions can be derived from this formula, e.g.

(a) If $g \in C^2$ and $h \in C^1$, then (3.14) defines a solution that belongs to C^2, but there is no smoothing effect like that for parabolic problems.
(b) The solution at an arbitrary point (x, t) is influenced exclusively by values of g and of h in the interval $[x - ct, x + ct]$ — this is the region of dependence. Conversely, information given at a point ξ on the x-axis influences the solution at the time $t = t_0$ only in the interval $[\xi - ct_0, \xi + ct_0]$. In other words, perturbations at the point ξ influence the solution at another point $x = x^*$ only after the time $t^* = |x^* - \xi|/C$ has elapsed (finite speed of error propagation).

For the two- and three-dimensional case there exist formulas similar to d'Alembert's, which are often called *Kirchhoff's formulas* . In particular in the two-dimensional case one has:

$$u(x_1, x_2, t) = \frac{1}{4\pi} \frac{\partial}{\partial t} \left(2t \int_{|\xi|<1} \frac{g(x_1 + ct\xi_1, x_2 + ct\xi_2)}{\sqrt{1 - |\xi|^2}} \right)$$

$$+ \frac{t}{4\pi} \left(2 \int_{|\xi|<1} \frac{h}{\sqrt{1 - |\xi|^2}} \right).$$

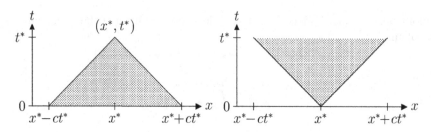

Figure 1.3   Domain of dependency   Domain of influence

Hence, the domain of dependency for the point (x, t) is the disc $\{x + ct\xi$ with $|\xi| \le 1\}$ and the domain of influence forms a cone with the vertex at the point considered. If $g \in C^2$, $h \in C^1$ then the formula guarantees only $u \in C^1$.

For the three-dimensional case the situation is again slightly different. Here we have (with surface integrals of the first kind)

$$u(x, t) = \frac{1}{4\pi} \frac{\partial}{\partial t} \left(t \int_{|\xi|=1} g(x + ct\xi) \, dS_\xi \right) + \frac{t}{4\pi} \int_{|\xi|=1} h(x + ct\xi) \, dS_\xi \,.$$

Now, the domain of dependency and the domain of influence are the surface of a ball and the surface of a cone, respectively. This represents *Huygens' principle*, which states e.g. that a perturbation in a point ξ is visible at another point x^* exactly at the time $t^* = |x^* - c|/c$, but not later ("sharp signals").

If in a bounded domain boundary conditions are also given then we note: *in initial-boundary value problems the solution along characteristics may have discontinuities.* To illustrate this we consider the solution of

$$u_{tt} - c^2 u_{xx} = 0$$

in a parallelogram that is bounded by characteristics.
From the representation

$$u(x, t) = F(x + ct) + G(x - ct) \quad \text{and} \quad F(A) = F(D), \; F(B) = F(C)$$

as well as from $G(A) = G(B)$ and $G(C) = G(D)$ we obtain immediately

$$u(A) + u(C) = u(B) + u(D) \,. \tag{3.15}$$

For the initial boundary value problem

$$u_{tt} - u_{xx} = 0 \quad \text{in} \quad 0 < x < \pi,$$

$$u|_{t=0} = 1, \quad u_t|_{t=0} = 0; \quad u|_{x=0} = 0, \; u|_{x=\pi} = 0$$

now d'Alembert's formula yields $u \equiv 1$ in Q_1 (see Fig. 1.4). From (3.15) it follows that $u \equiv 0$ in Q_2 and in Q_3. Using (3.15) again leads to $u \equiv -1$ in Q_4. By the way, the same result is obtained by the representation

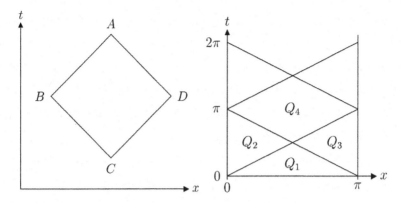

Figure 1.4 Bounding by characteristics   Inclusion of boundary conditions

$$u(x,t) = \frac{4}{\pi} \sum_{n=0}^{\infty} \frac{\sin(2n+1)\,x \cos(2n+1)\,t}{(2n+1)}$$

of the solution that was analyzed in a more general form in Section 1.2 (compare (2.4)).

The discontinuity observed is caused by data incompatibility; conditions of the type

$$u|_{t=0} = g(x); \qquad u|_{x=0} = \alpha(t)$$

are compatible with each other only if $g(0) = \alpha(0)$. To obtain a C^2-solution moreover for the condition $u_t|_{t=0} = h(x)$ one must require

$$\alpha'(0) = h(0) \quad \text{and} \quad \alpha''(0) = c^2 g''(0).$$

Smooth solutions of hyperbolic initial-boundary value problems can be expected only if additional compatibility conditions are satisfied.

2

Finite Difference Methods

2.1 Basic Concepts of the Method of Finite Differences: Grid Functions and Difference Operators

In Collatz's famous 1950 monograph *Numerical Treatment of Differential Equations*, the following statement can be found: *The finite difference method is a procedure generally applicable to boundary value problems. It is easily set up and on a coarse mesh it generally supplies, after a relatively short calculation, an overview of the solution function that is often sufficient in practice. In particular there are classes of partial differential equations where the finite difference method is the only practical method, and where other procedures are able to handle the boundary conditions only with difficulty or not at all.*

Even today, when finite element methods are widely dominant in the numerical solution of partial differential equations and their applications, this high opinion of the finite difference method remains valid. In particular it is straightforward to extend its basic idea from the one-dimensional case to higher dimensions, provided that the geometry of the underlying domain is not too complicated. The classical theoretical foundation of the method of finite differences—which rests on consistency estimates via Taylor's formula and the derivation of elementary stability bounds—is relatively easy, but it has the disadvantage of making excessive assumptions about the smoothness of the desired solution. Non-classical approaches to difference methods, as can be found for example in the standard book [Sam01] published 1977 in Russian (which unfortunately for many years was not widely known in the West) enable a weakening of the smoothness assumptions. These ideas also appear in a slightly concealed form in [Hac03a] and [Hei87]. When we discuss the convergence analysis of finite volume methods in Section 2.5, we shall briefly sketch how to weaken the assumptions in the analysis of finite difference methods.

When a finite difference method (FDM) is used to treat numerically a partial differential equation, the differentiable solution is approximated by some grid function, i.e., by a function that is defined only at a finite number of so-called grid points that lie in the underlying domain and its boundary. Each derivative that appears in the partial differential equation has to be replaced by a suitable divided difference of function values at the chosen grid points. Such approximations of derivatives by difference formulas can be generated in various ways, e.g., by a Taylor expansion, or local balancing equations, or by an appropriate interpretation of finite difference methods as specific finite element methods (see Chapter 4). The first two of these three approaches are generally known in the literature as finite difference methods (in the original sense) and finite volume methods, respectively. The related basic ideas for finite volume methods applied to elliptic differential equations are given in Section 2.5. The convergence of finite difference methods for parabolic and first-order hyperbolic problems is analysed in Sections 2.6 and 2.3, respectively.

As an introduction to finite difference methods, consider the following example. We are interested in computing an approximation to a sufficiently smooth function u that for given f satisfies Poisson's equation in the unit square and vanishes on its boundary:

$$-\Delta u = f \quad \text{in} \quad \Omega := (0,1)^2 \subset \mathbb{R}^2,$$
$$u = 0 \quad \text{on} \quad \Gamma := \partial\Omega. \tag{1.1}$$

Finite difference methods provide values $u_{i,j}$ that approximate the desired function values $u(x_{i,j})$ at a finite number of points, i.e., at the grid points $\{x_{i,j}\}$. Let the grid points in our example be

$$x_{i,j} = (i\,h,\ j\,h)^T \in \mathbb{R}^2,\ i,\ j = 0, 1, \ldots, N.$$

Here $h := 1/N$, with $N \in \mathbb{N}$, is the mesh size of the grid.

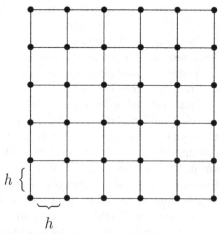

Figure 2.1   Grid for the discretization

At grid points lying on the boundary Γ the given function values (which here are homogeneous) can be immediately taken as the point values of the grid functions. All derivatives in problem (1.1) have however to be approximated by difference quotients. From e.g. Taylor's theorem we obtain

$$\frac{\partial^2 u}{\partial x_1^2}(x_{i,j}) \approx \frac{1}{h^2} \left(u(x_{i-1,j}) - 2u(x_{i,j}) + u(x_{i+1,j}) \right),$$

$$\frac{\partial^2 u}{\partial x_2^2}(x_{i,j}) \approx \frac{1}{h^2} \left(u(x_{i,j-1}) - 2u(x_{i,j}) + u(x_{i,j+1}) \right).$$

If these formulas are used to replace the partial derivatives at the inner grid points and the given boundary values are taken into account, then an approximate description of the original boundary value problem (1.1) is given by the system of linear equations

$$\begin{aligned} 4u_{i,j} - u_{i-1,j} - u_{i+1,j} \\ - u_{i,j-1} - u_{i,j+1} = h^2 f(x_{i,j}), \quad i, j = 1, \dots, N-1. \end{aligned} \tag{1.2}$$

$$u_{0,j} = u_{N,j} = u_{i,0} = u_{i,N} = 0,$$

For any $N \in \mathbb{N}$ this linear system has a unique solution $u_{i,j}$. Under certain smoothness assumptions on the desired solution u of the original problem one has $u_{i,j} \approx u(x_{i,j})$, as will be shown later.

In the method of finite differences one follows the precepts:

- the domain of the given differential equation must contain a sufficiently large number of test points (grid points);
- all derivatives required at grid points will be replaced by approximating finite differences that use values of the grid function at neighbouring grid points.

In problems defined by partial differential equations, boundary and/or initial conditions have to be satisfied. Unlike initial and boundary value problems in ordinary differential equations, the geometry of the underlying domain now plays an important role. This makes the construction of finite difference methods in domains lying in $\mathbb{R}^n$, with $n \geq 2$, not entirely trivial.

Let us consider the very simple domain $\Omega := (0,1)^n \subset \mathbb{R}^n$. Denote its closure by $\overline{\Omega}$. For the discretization of $\overline{\Omega}$ a set $\overline{\Omega}_h$ of grid points has to be selected, e.g., we may chose an equidistant grid that is defined by the points of intersection obtained when one translates the coordinate axes through consecutive equidistant steps with step size $h := 1/N$. Here $N \in \mathbb{N}$ denotes the number of shifted grid lines in each coordinate direction. In the present case we obtain

$$\overline{\Omega}_h := \left\{ \begin{pmatrix} x_1 \\ \vdots \\ x_n \end{pmatrix} \in \mathbb{R}^n : \begin{array}{l} x_1 = i_1\,h, \ \dots, \ x_n = i_n\,h, \\ i_1, \dots, i_n = 0, 1, \dots, N \end{array} \right\} \tag{1.3}$$

as the set of all *grid points*. We distinguish between those grid points lying in the domain Ω and those at the boundary Γ by setting

$$\Omega_h := \overline{\Omega}_h \cap \Omega \quad \text{and} \quad \Gamma_h := \overline{\Omega}_h \cap \Gamma. \tag{1.4}$$

Unlike the continuous problem, whose solution u is defined on all of $\overline{\Omega}$, the discretization leads to a *discrete solution* $u_h : \overline{\Omega}_h \to \mathbb{R}$ that is defined only at a finite number of grid points. Such mappings $\overline{\Omega}_h \to \mathbb{R}$ are called *grid functions*. To deal properly with grid functions we introduce the discrete function spaces

$$U_h := \{ u_h : \overline{\Omega}_h \to \mathbb{R} \}, \quad U_h^0 := \{ u_h \in U_h : u_h|_{\Gamma_h} = 0 \},$$

$$V_h := \{ v_h : \Omega_h \to \mathbb{R} \}.$$

To shorten the writing of formulas for difference quotients, let us define the following *difference operators* where the discretization step size is $h > 0$:

$$(D_j^+ u)(x) := \frac{1}{h} \left(u(x + h\, e^j) - u(x) \right) \quad \text{—forward difference quotient}$$

$$(D_j^- u)(x) := \frac{1}{h} \left(u(x) - u(x - h\, e^j) \right) \quad \text{—backward difference quotient}$$

$$D_j^0 := \frac{1}{2} \left(D_j^+ + D_j^- \right) \qquad \text{—central difference quotient.}$$

Here e^j denotes the unit vector in the positive direction of the j-th coordinate axis. Analogously, we shall also use notation such as D_x^+, D_y^+, D_t^+ etc. when independent variables such as x, y, $t, \ldots$ are present. For grids that are generated by grid lines parallel to the coordinate axes we can easily express difference quotient approximations of partial derivatives in terms of these difference operators.

Next we turn to the spaces of grid functions and introduce some norms that are commonly used in these spaces—which are isomorphic to finite-dimensional Euclidean spaces. The space U_h^0 of grid functions that vanish on the discrete boundary Γ_h will be equipped with an appropriate norm $\|\cdot\|_h$. For the convergence analysis of finite difference methods let us define the following norms on U_h^0:

$$\|u_h\|_{0,h}^2 := h^n \sum_{x_h \in \Omega_h} |u_h(x_h)|^2 \quad \forall u_h \in U_h^0 \tag{1.5}$$

—the *discrete L_2 norm*;

$$\|u_h\|_{1,h}^2 := h^n \sum_{x_h \in \Omega_h} \sum_{j=1}^{n} |[D_j^+ u_h](x_h)|^2 \quad \forall u_h \in U_h^0 \tag{1.6}$$

—the *discrete H^1 norm*; and

$$\|u_h\|_{\infty,h} := \max_{x_h \in \overline{\Omega}_h} |u_h(x_h)| \quad \forall u_h \in U_h^0 \tag{1.7}$$

—the *discrete maximum norm.*

Finally, we introduce the *discrete scalar product* in U_h^0:

$$(u_h, v_h)_h := h^n \sum_{x_h \in \Omega_h} u_h(x_h) \, v_h(x_h) \quad \forall u_h, v_h \in U_h^0. \tag{1.8}$$

It is clear that

$$\|u_h\|_{0,h}^2 = (u_h, u_h)_h \quad \text{and} \quad \|u_h\|_{1,h}^2 = \sum_{j=1}^n (D_j^+ u_h, D_j^+ u_h)_h \quad \forall u_h \in U_h^0.$$

The definitions of the norms $\|\cdot\|_{0,h}$ and $\|\cdot\|_{\infty,h}$ and of the scalar product $(\cdot,\cdot)_h$ use points x_h that lie only in Ω_h, so these norms and scalar product can also be applied to functions in the space V_h. They define norms for V_h that we shall call upon later.

In the case of non-equidistant grids the common multiplier h^n must be replaced by a weight $\mu_h(x_h)$ at each grid point. These weights can be defined via appropriate dual subdomains $D_h(x_h)$ related to the grid points x_h by

$$\mu_h(x_h) := \text{meas} \, D_h(x_h) := \int_{D_h(x_h)} dx, \qquad x_h \in \overline{\Omega}_h.$$

In the equidistant case for $\Omega = (0,1)^n \subset \mathbb{R}^n$ we may for instance choose

$$D_h(x_h) = \{ x \in \Omega : \|x - x_h\|_\infty < h/2 \}. \tag{1.9}$$

This yields $\mu_h(x_h) = h^n$, $\forall x_n \in \Omega_h$. Hence the general scalar product

$$(u_h, v_h)_h := \sum_{x_h \in \Omega_h} \mu_h(x_h) \, u_h(x_h) \, v_h(x_h) \quad \forall u_h, v_h \in U_h^0$$

coincides in the equidistant case with (1.8). Later, when presenting finite volume methods, we shall discuss a procedure for generating dual subdomains for arbitrary grids on rather general domains.

Analogously to norms of continuous real functions, we define discrete L_p norms for $p \in [1, \infty)$ by

$$\|u_h\|_{L_p,h} := \left(\sum_{x_h \in \Omega_h} \mu_h(x_h) \, |u_h(x_h)|^p \right)^{1/p}.$$

Now for $\dfrac{1}{p} + \dfrac{1}{q} = 1$ the *discrete Hölder inequality*

$$|(u_h, v_h)_h| \leq \|u_h\|_{L_p,h} \, \|v_h\|_{L_q,h}$$

is valid, even for the extreme case $p = 1$, $q = +\infty$. As a complement to the standard analysis via discrete maximum principles (see Section 2.1), Hölder's inequality provides a tool for the derivation of uniform error estimates, i.e., estimates for the discrete maximum norm $\|\cdot\|_{\infty,h}$.

If we restrict functions $u \in C(\overline{\Omega})$ to grid functions in a trivial way by reusing the same function values, i.e., by setting

$$[r_h u](x_h) = u(x_h), \quad x_h \in \overline{\Omega}_h, \tag{1.10}$$

then for the restriction operator r_h one has

$$\lim_{h \to 0} \|r_h u\|_{0,h} = \|u\|_{L_2(\Omega)} \quad \text{and also} \quad \lim_{h \to 0} \|r_h u\|_{\infty,h} = \|u\|_{L^\infty(\Omega)}.$$

It is well known that in finite-dimensional spaces all norms are equivalent. Nevertheless, the constant multipliers appearing in the equivalence inequalities will in general depend upon the dimension of the space. Hence for our discrete function spaces these constants can depend upon the discretization step size h. In particular for the norms $\| \cdot \|_{0,h}$ and $\| \cdot \|_{\infty,h}$ we have

$$\min_{x_h \in \overline{\Omega}_h} \mu_h(x_h)^{1/2} \|u_h\|_{\infty,h} \leq \|u_h\|_{0,h} \leq \operatorname{meas}(\Omega)^{1/2} \|u_h\|_{\infty,h},$$
$$\forall u_h \in U_h^0. \tag{1.11}$$

Since

$$\min_{x_h \in \overline{\Omega}_h} \mu_h(x_h) = h^n \quad \text{and} \quad \operatorname{meas}(\Omega) = 1,$$

one infers the inequalities

$$h^{n/2} \|u_h\|_{\infty,h} \leq \|u_h\|_{0,h} \leq \|u_h\|_{\infty,h}, \quad \forall u \in U_h. \tag{1.12}$$

Consequently it is to be expected that estimates for the error $\|r_h u - u_h\|_h$ as $h \to 0$ in finite difference methods will depend heavily upon the norm $\| \cdot \|_h$ that is chosen. In linear problems the error is usually measured in the discrete L_2 norm or the discrete maximum norm and, after further analysis, in the discrete H^1 norm. These norms will be used in the sequel.

How can difference approximations be generated? For sufficiently smooth functions $u : \mathbb{R}^n \to \mathbb{R}$, by a Taylor expansion one has

$$u(x + z) = \sum_{k=0}^{m} \frac{1}{k!} \left(\sum_{j=1}^{n} z_j \frac{\partial}{\partial x_j} \right)^k u(x) + R_m(x, z). \tag{1.13}$$

Here the remainder $R_m(x, z)$ can be written in the Lagrange form, namely

$$R_m(x, z) = \frac{1}{(m+1)!} \left(\sum_{j=1}^{n} z_j \frac{\partial}{\partial x_j} \right)^{(m+1)} u(x + \theta z) \quad \text{for some} \quad \theta = \theta(x, z) \in (0, 1).$$

This form is quite simple, but the alternative integral form of the remainder is more suited to some situations. If derivatives are replaced by approximating difference formulas that are derived from (1.13), one can deduce estimates for the consequent error. For example, if $x \in \mathbb{R}^n$ is fixed then a Taylor expansion (1.13) yields

$$\left| \left[\frac{\partial^2 u}{\partial x_j^2} - D_j^- D_j^+ u \right](x) \right| \le \frac{1}{12} \max_{|\xi| \le h} \left| \frac{\partial^4 u}{\partial x_j^4}(x + \xi) \right| h^2. \tag{1.14}$$

In general Taylor's theorem provides a systematic tool for the generation of difference approximations for derivatives, e.g.,

$$\frac{\partial u}{\partial x_j}(x) = \frac{1}{2h} \Big(-3u(x) + 4u(x + he^j) - u(x + 2he^j) \Big) + O(h^2). \tag{1.15}$$

Indeed, from (1.13) we obtain

$$u(x + h\,e^j) = u(x) + \frac{\partial u}{\partial x_j}(x)\,h + \tfrac{1}{2}\frac{\partial^2 u}{\partial x_j^2}(x)\,h^2 + R_1,$$

$$u(x + 2h\,e^j) = u(x) + 2\frac{\partial u}{\partial x_j}(x)\,h + 2\frac{\partial^2 u}{\partial x_j^2}(x)\,h^2 + R_2$$

with remainders $R_1 = O(h^3)$, $R_2 = O(h^3)$. The difference approximation (1.15) follows.

In general, any given partial differential equation problem, including its boundary and/or initial conditions, can be expressed as an abstract operator equation

$$F u = f \tag{1.16}$$

with appropriately chosen function spaces U and V, a mapping $F : U \to V$, and $f \in V$. The related discrete problem can be stated analogously as

$$F_h u_h = f_h \tag{1.17}$$

with $F_h : U_h \to V_h$, $f_h \in V_h$, and discrete spaces U_h, V_h. Let $r_h : U \to U_h$ denote some restriction operator from the elements of U to grid functions.

Definition 2.1. *The value* $\|F_h(r_h u) - f_h\|_{V_h}$ *is called the consistency error relative to* $u \in U$.

The remainder term in Taylor's theorem gives a way of estimating the consistency error, provided that the solution u of (1.16) is smooth enough and f_h forms an appropriate discretization of f.

Definition 2.2. *A discretization of (1.16) is consistent if*

$$\|F_h(r_h u) - f_h\|_{V_h} \to 0 \quad as \quad h \to 0.$$

If in addition the consistency error satisfies the more precise estimate

$$\|F_h(r_h u) - f_h\|_{V_h} = O(h^p) \quad as \quad h \to 0,$$

then the discretization is said to be consistent of order p.

Convergence of a discretization method is defined similarly:

Definition 2.3. *A discretization method is convergent if the error satisfies*

$$\|r_h u - u_h\|_{U_h} \to 0 \quad as \quad h \to 0$$

and convergent of order q if

$$\|r_h u - u_h\|_{U_h} = O(h^q) \quad as \quad h \to 0.$$

Convergence is proved by demonstrating consistency and *stability* of the discretization.

Definition 2.4. *A discretization method is stable if for some constant $S > 0$ one has*

$$\|v_h - w_h\|_{U_h} \leq S \, \|F_h v_h - F_h w_h\|_{V_h} \quad for\ all \quad v_h, w_h \in U_h. \tag{1.18}$$

Stability also guarantees that rounding errors occurring in the problem will not have an excessive effect on the final result. The definitions above imply immediately the following abstract convergence theorem.

Theorem 2.5. *Assume that both the continuous and the discrete problem have unique solutions. If the discretization method is consistent and stable then the method is also convergent. Furthermore, the order of convergence is at least as large as the order of consistency of the method.*

Proof: Let $u_h \in U_h$ be the solution of the discrete problem, i.e., $F_h u_h = f_h$. Then stability implies

$$S^{-1} \|r_h u - u_h\|_{U_h} \leq \|F_h(r_h u) - F_h u_h\|_{V_h} = \|F_h(r_h u) - f_h\|_{V_h}.$$

Hence, invoking consistency, we obtain convergence of the method. Furthermore, the same calculation implies that the order of convergence is at least as large as the order of consistency. ■

Remark 2.6. If the discrete operators F_h are linear, then stability of the discretization is equivalent to the existence of a constant $c > 0$, independent of h, such that $\|F_h^{-1}\| \leq c$.

When the operators F_h are nonlinear, as a rule the validity of the stability inequality (1.18) will be needed only in some neighbourhood of the discrete solution u_h.   □

Let us remind the reader that the properties considered here, which measure the effectiveness of a discretization, depend strongly upon the spaces chosen and their norms. In particular Definition 2.4 could more precisely be called $U_h - V_h$ stability, where the spaces used are named.

The above basic theory is common to all finite difference methods, yet in practice the convergence analysis differs for various specific classes of problems associated with partial differential equations. After a brief introduction to some general convergence analysis we shall consider separately later the most important classes of these problems and make use of their specific properties.

2.2 Illustrative Examples for the Convergence Analysis

Before we analyse difference methods in detail for the cases of hyperbolic, elliptic and parabolic problems we illustrate some of the basic ideas in the convergence analysis of finite difference methods by two simple examples.

For simplicity let us first consider the linear two-point boundary value problem

$$Lu := -u''(x) + \beta\, u'(x) + \gamma\, u(x) = f(x), \quad x \in (0,1), \ u(0) = u(1) = 0 \quad (2.1)$$

with constant coefficients $\beta, \gamma \in \mathbb{R}$ and $\gamma \geq 0$. This problem has a unique classical solution u for any $f \in C[0,1]$. If $f \in C^k[0,1]$ for some $k \in \mathbb{N}$, then the solution has a certain amount of smoothness: $u \in C^{k+2}[0,1]$. Hence, under such an assumption on f, we can derive realistic discretization error estimates from Taylor's theorem . Let us mention here that in the case of partial differential equations the situation is more complicated because the smoothness of the solution u depends not only on the smoothness of the data but also on the geometry of the underlying domain $\Omega \subset \mathbb{R}^n$.

Set $\Omega := (0,1)$. With

$$\Omega_h := \{\, x_j = j\,h,\ j = 1, \ldots, N-1 \}, \qquad \overline{\Omega}_h := \{\, x_j = j\,h,\ j = 0, 1, \ldots, N \}$$

where $h := 1/N$ for some $N \in \mathbb{N}$, we can discretize (2.1) by

$$[L_h u_h](x_h) := \left[\left(-D^- D^+ + \beta\, D^0 + \gamma \right) u_h \right](x_h) = f(x_h),$$
$$x_h \in \Omega_h,\ u_h \in U_h^0. \quad (2.2)$$

From the definitions of the difference quotients this is equivalent to

$$-\frac{1}{h^2}(u_{j-1} - 2u_j + u_{j+1})$$
$$+\frac{\beta}{2h}(u_{j+1} - u_{j-1}) + \gamma\, u_j = f_j, \quad j = 1, \ldots, N-1, \quad (2.3)$$
$$u_0 = u_N = 0,$$

where $u_j := u_h(x_j)$ and $f_j := f(x_j)$.

Define a restriction of the solution u of problem (2.1) to the grid $\overline{\Omega}_h$ by $[r_h u](x_j) := u(x_j)$. Then, using $w_h := u_h - r_h u$ to denote the error in u_h, from (2.3) it follows that

$$-\frac{1}{h^2}(w_{j-1} - 2w_j + w_{j+1})$$
$$+\frac{\beta}{2h}(w_{j+1} - w_{j-1}) + \gamma\, w_j = d_j, \quad j = 1, \ldots, N-1, \quad (2.4)$$
$$w_0 = w_N = 0.$$

Here the defect $d_h : \Omega_h \to \mathbb{R}^{N-1}$ is defined by

$$d_j := f_j - [L_h r_h u](x_j), \quad j = 1, \ldots, N-1.$$

Equivalently one can write (2.4) in the compact form

$$L_h w_h = d_h, \quad w_h \in U_h^0. \tag{2.5}$$

Similarly to r_h, define a trivial restriction operator $q_h : C(\Omega) \to (\Omega_h \to \mathbb{R})$. Then

$$d_h = q_h L u - L_h r_h u, \tag{2.6}$$

and invoking Taylor's theorem we obtain for the consistency error

$$\|d_h\|_{0,h} \le c h^2 \quad \text{and} \quad \|d_h\|_{\infty,h} \le c h^2 \tag{2.7}$$

for some constant $c > 0$, provided that $u \in C^4[0,1]$. That is, in this example the discretization is second-order consistent in both of these norms.

If (as here) the discrete operator is invertible, then from (2.5)—compare the proof of Theorem 2.5—one obtains the error estimate

$$\|w_h\| \le \|L_h^{-1}\| \, \|d_h\|. \tag{2.8}$$

Which norm does one choose and how can the quantity $\|L_h^{-1}\|$ be estimated? For L_2-norm analysis in the case of constant coefficients, Fourier expansions can be applied; this technique will now be sketched. Another L_2-norm tool, which is applicable to the general case, is the method of energy inequalities (see Section 4). Analysis in the maximum norm will be addressed later.

For the sake of simplicity in our introductory example (2.1), set $\beta = 0$. Thus we deal with the two-point boundary value problem

$$-u''(x) + \gamma u(x) = f(x), \quad x \in (0,1), \qquad u(0) = u(1) = 0, \tag{2.9}$$

and consider the associated discretization

$$[L_h u_h]_l := -\frac{1}{h^2}(u_{l-1} - 2u_l + u_{l+1}) + \gamma u_l = f_l, \quad l = 1, \ldots, N-1,$$
$$u_0 = u_N = 0. \tag{2.10}$$

To analyse the discrete problem we shall use eigenfunction expansions. This idea is well known in the context of continuous functions; one manifestation is Fourier series. It can be shown easily that the eigenfunctions required for the expansion of the discrete problem are simply the restriction to the mesh of the eigenfunctions of the continuous problem. In the present case, as the boundary conditions are homogeneous, this restriction yields the grid functions $v_h^j \in U_h^0$, for $j = 1, \ldots, N-1$, defined by

$$[v_h^j]_l := \sqrt{2} \sin(j \pi x_l) \quad \text{where} \quad j = 1, \ldots, N-1, \; x_l \in \overline{\Omega}_h. \tag{2.11}$$

These grid functions, or vectors, form an orthonormal basis of U_h^0, i.e., one has

$$(v_h^j, v_h^k)_h = \delta_{jk}$$

and $v_0^j = v_N^j = 0$. Furthermore,

$$\left[L_h v_h^j \right]_l = \left(\frac{4}{h^2} \sin^2 \left(\frac{j\pi h}{2} \right) + \gamma \right) v_l^j, \quad j, l = 1, \ldots, N-1.$$

That is, each vector $v_h^j \in U_h^0$ is an eigenvector of L_h with the associated eigenvalue

$$\lambda_j = \frac{4}{h^2} \sin^2 \left(\frac{j\pi h}{2} \right) + \gamma, \quad j = 1, \ldots, N-1. \tag{2.12}$$

In (2.5) let us write w_h and d_h in terms of the basis $\{v_h^j\}$, viz.,

$$w_h = \sum_{j=1}^{N-1} \omega_j \, v_h^j, \qquad d_h = \sum_{j=1}^{N-1} \delta_j \, v_h^j,$$

with coefficients $\omega_j, \delta_j \in \mathbb{C}$ for $j = 1, \ldots, N-1$. Using (2.12), equation (2.5) becomes

$$\sum_{j=1}^{N-1} \lambda_j \omega_j \, v_h^j = \sum_{j=1}^{N-1} \delta_j \, v_h^j.$$

But the vectors v_h^j are linearly independent, so this equation implies that

$$\omega_j = \delta_j / \lambda_j, \quad j = 1, \ldots, N-1.$$

By Parseval's equality (which follows from the orthonormality of the eigenvectors) applied to w_h and d_h, and the observation that $0 < \lambda_1 \le \lambda_j$ for all j, one then has

$$\|w_h\|_{0,h}^2 = \sum_{j=1}^{N-1} |\omega_j|^2 = \sum_{j=1}^{N-1} \left| \frac{\delta_j}{\lambda_j} \right|^2 \le \frac{1}{\lambda_1^2} \sum_{j=1}^{N-1} |\delta_j|^2 = \frac{1}{\lambda_1^2} \|d_h\|_{0,h}^2 .$$

That is,

$$\|w_h\|_{0,h} \le \frac{1}{4 \, h^{-2} \sin^2 \left(\frac{\pi h}{2} \right) + \gamma} \|d_h\|_{0,h}. \tag{2.13}$$

Since $\lim_{h \to 0} 4h^{-2} \sin^2(\pi h/2) = \pi^2 > 0$, we have proved stability in the L_2 norm for the discretization in the case $\gamma \ge 0$; to be precise, $\|L_h^{-1}\| \le k$ for any $k > 1/\pi^2$ if h is sufficiently small.

Stability and consistency together imply (by Theorem 2.5) the error bound

$$\|r_h u - u_h\|_{0,h} = \left(h \sum_{l=1}^{N-1} (u_l - u(x_l))^2 \right)^{1/2} \le c h^2 \tag{2.14}$$

for some constant c.

Next we study convergence in the discrete maximum norm. Applying the norm equivalence inequality (1.12) to (2.14), one obtains immediately the estimate

$$\|r_h u - _h\|_{\infty,h} \leq h^{-1/2} \|r_h u - u_h\|_{0,h} \leq c h^{3/2}, \tag{2.15}$$

but in general the exponent of h in this bound is suboptimal.

In the case $\gamma > 0$ the coefficient matrix that represents L_h is strictly diagonally dominant. This implies the maximum norm stability inequality $\|w_h\|_{\infty,h} \leq c\|d_h\|_{\infty,h}$ and hence

$$\|r_h u - u_h\|_{\infty,h} \leq c h^2. \tag{2.16}$$

In the more general case $\gamma \geq 0$, stability in the discrete maximum norm can be derived from discrete comparison principles using the theory of M-matrices; see Section 4.

Our second example is a parabolic problem in one space dimension:

$$\frac{\partial u}{\partial t}(x,t) - \frac{\partial^2 u}{\partial x^2}(x,t) + \gamma u(x,t) = f(x,t), \ x \in (0,1), \ t > 0,$$
$$u(0,t) = u(1,t) = 0, \qquad t > 0, \tag{2.17}$$
$$u(x,0) = u_0(x), \ x \in (0,1).$$

This problem is time-dependent with (2.9) as its stationary limit case. Once again γ denotes a positive constant and f, u_0 are given functions. As in the boundary value problem (2.9) previously studied, the initial-boundary value problem (2.17) possesses a unique solution u that is sufficiently smooth provided that the data f and u_0 are smooth enough and certain compatibility conditions are satisfied at the corners of the domain.

We use a grid that is equidistant in both the time and space directions. On this grid the time derivative $\partial/\partial t$ is approximated by the backward difference quotient D_t^- and the second-order spatial derivative $\partial^2/\partial x^2$ by the symmetric difference quotient $D_x^- D_x^+$. This leads to the discrete problem

$$[(D_t^- - D_x^- D_x^+ + \gamma I)u_{h,\tau}](x_i, t^k) = f_{i,k}, \quad i = 1, \ldots, N-1,$$
$$k = 1, \ldots, M,$$
$$u_{0,k} = u_{N,k} = 0, \qquad k = 1, \ldots, M, \tag{2.18}$$
$$u_{i,0} = u_0(x_i), \ i = 1, \ldots, N-1,$$

which defines the discrete solution $u_{h,\tau} = \{u_{i,k}\}$. Here N, $M \in \mathbb{N}$ are parameters that define the spatial step-size $h := 1/N$ and the temporal step-size $\tau := T/M$. The points $t^k := k\tau$, for $k = 0, 1, \ldots, M$, define the temporal grid. Then the problem (2.18) can be written as

$$\frac{1}{\tau}(u_{i,k} - u_{i,k-1}) - \frac{1}{h^2}(u_{i-1,k} - 2u_{i,k} + u_{i+1,k}) + \gamma u_{i,k} = f_{i,k},$$
$$u_{0,k} = u_{N,k} = 0, \tag{2.19}$$
$$u_{i,0} = u_0(x_i),$$

with the same ranges for the indices as in (2.18). Let us represent the discrete values $u_{i,k} \approx u(x_i, t^k)$ and $f_{i,k} := f(x_i, t^k)$ at a fixed time level t^k by vectors $u^k \in U_h^0$ and f^k, respectively. Then, using the operators L_h and r_h that were defined for the stationary example, one can write (2.19) in the compact form

$$\frac{1}{\tau}(u^k - u^{k-1}) + L_h u^k = f^k, \quad k = 1, \ldots, M,$$
$$u^0 = r_h u_0. \tag{2.20}$$

As the operator L_h is linear, the error $w^k := u^k - r_h u(\cdot, t^k)$ satisfies

$$\frac{1}{\tau}(w^k - w^{k-1}) + L_h w^k = R^k, \quad k = 1, \ldots, M,$$
$$w^0 = 0. \tag{2.21}$$

Here the norms of the defects R^k can be estimated by

$$\|R^k\|_{0,h} \leq c(\tau + h^2), \quad k = 1, \ldots, M,$$

where $c > 0$ is some constant, provided that the solution u of the original problem (2.17) is sufficiently smooth. Thus (2.21) and the triangle inequality yield

$$\|w^k\|_{0,h} \leq \|(I + \tau L_h)^{-1}\|_{0,h} \|w^{k-1}\|_{0,h} + \tau c(\tau + h^2), \quad k = 1, \ldots, M,$$

where $\|\cdot\|_{0,h}$ is as in the stationary problem. But one can repeat the argument leading to the stability bound (2.13), with L_h replaced by $I + \tau L_h$, to get $\|(I + \tau L_h)^{-1}\|_{0,h} \leq 1$. Hence

$$\|w^k\|_{0,h} \leq \|w^{k-1}\|_{0,h} + \tau c(\tau + h^2), \quad k = 1, \ldots, M.$$

Now $w^0 = 0$; consequently an inductive argument yields

$$\|w^k\|_{0,h} \leq k \tau c(\tau + h^2), \quad k = 1, \ldots, M,$$

and since $M\tau = T$, this is the convergence result

$$\|u_{h,\tau} - u\|_{h,\tau} := \max_{k=0,1,\ldots,M} \|u^k - r_h u(\cdot, t^k)\|_{0,h} \leq T c(\tau + h^2). \tag{2.22}$$

The discrete norm $\|\cdot\|_{h,\tau}$ combines the Euclidean norm in the spatial direction with the maximum norm in the temporal direction.

Given sufficient smoothness of u, discrete maximum principles can be applied to derive a uniform estimate of the form

$$\max_{k=0,1,\ldots,M} \max_{i=1,\ldots,N-1} |u_{i,k} - u(x_i, t^k)| \leq c(\tau + h^2).$$

We shall return to this in more detail in Section 6.

2.3 Transportation Problems and Conservation Laws

In this section, first-order partial differential equations of the type

$$u_t(x,t) + \operatorname{div}(F(x,t,u(x,t))) = 0, \quad x \in \mathbb{R}^n, \ t > 0 \tag{3.1}$$

are considered on the entire space $\mathbb{R}^n$. Here $F : \mathbb{R}^n \times \mathbb{R} \times \mathbb{R} \to \mathbb{R}^n$ is some given differentiable mapping. We are also given the initial state $u(\cdot, 0)$, i.e., as well as the differential equation (3.1) the initial condition

$$u(x,0) = u_0(x), \quad x \in \mathbb{R}^n, \tag{3.2}$$

must be satisfied for some given function u_0. The divergence operator in (3.1) is applied only to the spatial variables, i.e.,

$$\operatorname{div} q := \sum_{j=1}^{n} \frac{\partial q_j}{\partial x_j}$$

for smooth vector fields $q : \mathbb{R}^n \to \mathbb{R}^n$. Assuming appropriate differentiability, the differential equation (3.1) is equivalent to

$$u_t(x,t) + \sum_{j=1}^{n} \left(\frac{\partial F_j}{\partial x_j} + \frac{\partial F_j}{\partial u} \frac{\partial u}{\partial x_j} \right) = 0, \quad x \in \mathbb{R}^n, \ t > 0, \tag{3.3}$$

which can be rearranged as

$$u_t(x,t) + v(x,t) \cdot \nabla u(x,t) = f(x,t), \quad x \in \mathbb{R}^n, \ t > 0, \tag{3.4}$$

where $v = (v_1, v_2, \ldots, v_n)$ with

$$v_j(x,t,u) := \frac{\partial F_j}{\partial u}(x,t,u) \text{ for } j = 1,\ldots,n \quad \text{and} \quad f(x,t) := -\sum_{j=1}^{n} \frac{\partial F_j}{\partial x_j}(x,t).$$

Here the gradient operator ∇, like the divergence operator above, is with respect to the spatial variables only.

Consider now the linear case where v in (3.4) is independent of u. Let $v : \mathbb{R}^n \times \mathbb{R} \to \mathbb{R}^n$ be a continuous function that is Lipschitz-continuous with respect to its first argument. Then Picard-Lindelöf's Theorem guarantees that for any $(\hat{x}, t) \in \mathbb{R}^n \times \mathbb{R}$ the initial-value problem

$$x'(s) = v(x(s), s), \quad s \in \mathbb{R}, \quad x(t) = \hat{x}, \tag{3.5}$$

has a unique solution (x, s). By varying the initial condition one obtains a family of non-intersecting curves $(x(s), s)$, $s \in \mathbb{R}_+$, in the (x, t)-plane. These are called the characteristics (more precisely, the characteristic traces) of (3.4). Along each characteristic the solution of the problem (3.1) is completely defined by an initial-value problem for an ordinary differential equation, as we now explain. The chain rule for differentiation applied to the composite function $u(x(\cdot), \cdot)$ yields

$$\frac{d}{ds} u(x(s), s) = u_t(x(s), s) + x'(s) \cdot \nabla u(x(s), s), \quad s \in \mathbb{R}_+.$$

Let $\hat{x} \in \mathbb{R}^n$ be arbitrary but fixed. Recalling the transport equation (3.4) and the initial condition (3.2), it follows that their solution $u(x, t)$ is explicitly defined along the characteristic passing through $(\hat{x}, 0)$ by

$$u(x, t) = u_0(\hat{x}) + \int_0^t f(x(s), s) \, ds.$$

In the case of constant coefficients—i.e., $v \in \mathbb{R}^n$—one gets

$$x(s) = x + (s - t)v, \quad s \in \mathbb{R},$$

for the solution of (3.5). Thus points (x, t) lying on the characteristic passing through $(\hat{x}, 0)$ satisfy $\hat{x} = x - vt$. In particular for homogeneous problems where $f \equiv 0$, the solution of (3.1) is constant along each characteristic:

$$u(x, t) = u_0(x - vt), \quad t \geq 0.$$

Let us mention here that in the case where v is nonlinear in u, one can still define the characteristics but, unlike the linear case, they may intersect. Even for continuous initial data this intersection can cause discontinuities (shocks) in the solution u at some time $t > 0$.

2.3.1 The One-Dimensional Linear Case

Let F have the form $F(x, t, u) = a u$ for some constant a. Thus we are dealing with a homogeneous linear transport equation with constant coefficients. Problem (3.1) becomes

$$u_t + a u_x = 0, \quad x \in \mathbb{R}, \ t > 0, \qquad u(x, 0) = u_0(x), \qquad (3.6)$$

and its exact solution is

$$u(x, t) = u_0(x - at). \qquad (3.7)$$

We now discuss different discretizations of (3.6) on equidistant grids (x_j, t^k) with $x_j = j h$ and $t^k = k \tau$. Without loss of generality assume that $a \geq 0$; the case $a < 0$ can be handled analogously.

A typical *explicit* discretization of (3.6) that uses three neighbouring grid points in the spatial direction has the form

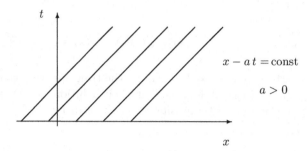

Figure 2.2   Characteristics of the transport equation

$$D_t^+ u = -a \left(\omega D_x^+ + (1 - \omega) D_x^- \right) u \qquad (3.8)$$

for some parameter $\omega \in [0,1]$. For the choices $\omega \in \{1/2, 0, 1\}$ we obtain
the three different stencils of Figure 2.3. Setting $\gamma := (a\tau)/h$, these three
difference schemes are

(a) $u_j^{k+1} = u_j^k + \frac{1}{2}\gamma(u_{j-1}^k - u_{j+1}^k),$

(b) $u_j^{k+1} = (1 - \gamma)u_j^k + \gamma u_{j-1}^k,$ (3.9)

(c) $u_j^{k+1} = (1 + \gamma)u_j^k - \gamma u_{j+1}^k.$

Because one-sided difference quotients are used in the schemes (b) and (c),
these schemes are called *upwind schemes*.

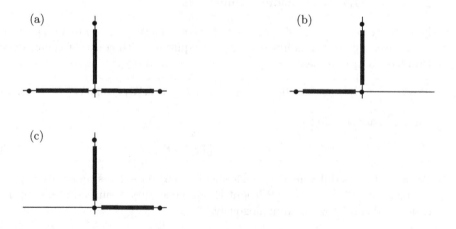

Figure 2.3   Difference stencils of explicit discretizations (3.9)

A comparison of the stencils of Figure 2.3 and the characteristics of Figure 2.2 suggests that scheme (b) is preferable. This preference is also supported by the following argument. The exact solution u of (3.6) satisfies the maximum principle

$$\inf_x u(x,t) \le u(x,t+\tau) \le \sup_x u(x,t), \qquad \forall x \in \mathbb{R}, \quad t \ge 0, \ \tau > 0. \quad (3.10)$$

In terms of the norm $\| \cdot \|_\infty$ on the space of essentially bounded functions defined on $\mathbb{R}$, this principle can be expressed as

$$\|u(\cdot, t+\tau)\|_\infty \le \|u(\cdot, t)\|_\infty.$$

To get the corresponding discrete maximum principle

$$\inf_j u_j^k \le u_j^{k+1} \le \sup_j u_j^k, \qquad \forall j, \ k \in \mathbb{N}, \quad k \ge 0, \quad (3.11)$$

i.e.,

$$\|u^{k+1}\|_{\infty,h} \le \|u^k\|_{\infty,h} \,,$$

one can show that scheme (b) must be chosen for the discretization. For the property (3.11) can never be guaranteed with schemes (a) and (c), while in the case of scheme (b) the *Courant-Friedrichs-Levi (CFL) condition*

$$\gamma = a\frac{\tau}{h} \le 1 \quad (3.12)$$

is sufficient to ensure that this scheme satisfies the discrete maximum principle. The CFL condition also implies the stability of scheme (b) in the maximum norm, as will be shown in the next theorem.

Theorem 2.7. *Let the function u_0 be Lipschitz-continuously differentiable. Then the explicit upwind scheme (b) is consistent of order 1 in the maximum norm. If in addition the CFL condition (3.12) is satisfied then the discrete maximum principle (3.11) holds. On arbitrary finite time intervals $[0, T]$, scheme (b) is stable in the maximum norm and convergent of order 1.*

Proof: First, we study the consistency of the method. For this purpose we do not use the standard Taylor expansion argument; instead we exploit the fact that the solution u has the form (3.7).

The fundamental theorem of integral calculus yields

$$u(x_j, t^{k+1}) - u(x_j, t^k) = \int_{t^k}^{t^{k+1}} u_t(x_j, t)\, dt,$$

$$u(x_j, t^k) - u(x_{j-1}, t^k) = \int_{x_{j-1}}^{x_j} u_x(x, t^k)\, dx.$$

$$(3.13)$$

As the solution u of the differential equation satisfies (3.6), we have also

$$u_t(x_j, t^k) + a\, u_x(x_j, t^k) = 0.$$

This equation and (3.13) imply

$$\frac{u(x_j, t^{k+1}) - u(x_j, t^k)}{\tau} + a\frac{u(x_j, t^k) - u(x_{j-1}, t^k)}{h}$$

$$= \frac{1}{\tau} \int_{t^k}^{t^{k+1}} (u_t(x_j, t) - u_t(x_j, t^k))\, dt + a\frac{1}{h} \int_{x_{j-1}}^{x_j} (u_x(x, t^k) - u_x(x_j, t^k))\, dx.$$

Set $w_j^k = u_j^k - u(x_j, t^k)$ for all j and k. The linearity of the difference operator and the relations $u_x = u_0'$, $u_t = -a\, u_0'$, which follow from the representation (3.7) of the solution, together give

$$\frac{w_j^{k+1} - w_j^k}{\tau} + a\frac{w_j^k - w_{j-1}^k}{h} = -\frac{a}{\tau} \int_{t^k}^{t^{k+1}} [u_0'(x_j - at) - u_0'(x_j - at^k)]\, dt \tag{3.14}$$

$$+ \frac{a}{h} \int_{x_{j-1}}^{x_j} [u_0'(x - at^k) - u_0'(x_j - at^k)]\, dx.$$

The Lipschitz continuity of u_0' now implies the first-order consistency of the difference scheme.

The validity of the discrete maximum principle (3.11) in the case $\gamma \in [0, 1]$ follows immediately for scheme (b) from the triangle inequality. Furthermore, (3.14) implies that

$$w_j^{k+1} = (1 - \gamma)w_j^k + \gamma w_{j-1}^k + \tau r_j^k, \quad j \in \mathbb{Z},\ k = 0, 1, \ldots, M - 1, \tag{3.15}$$

with

$$r_j^k := -\frac{a}{\tau} \int_{t^k}^{t^{k+1}} [u_0'(x_j - at) - u_0'(x_j - at^k)]\, dt \tag{3.16}$$

$$+ \frac{a}{h} \int_{x_{j-1}}^{x_j} [u_0'(x - at^k) - u_0'(x_j - at^k)]\, dx.$$

Here $M \in \mathbb{N}$ denotes the number of time steps in the discretization, i.e., $\tau = T/M$. Now

$$\left| \int_{t^k}^{t^{k+1}} [u_0'(x_j - at) - u_0'(x_j - at^k)]\, dt \right| \le aL \int_{t^k}^{t^{k+1}} (t - t^k)\, dt = \tfrac{1}{2}\, a\, L\, \tau^2,$$

$$\left| \int_{x_{j-1}}^{x_j} [u_0'(x - at^k) - u_0'(x_j - t^k)]\, dx \right| \le L \int_{x_{j-1}}^{x_j} (x_j - x)\, dx = \tfrac{1}{2}\, L\, h^2,$$

where L is the Lipschitz constant of the function u_0'. Consequently

$$|r_j^k| \leq \frac{a}{2} (h + a\tau) L, \qquad j \in \mathbb{Z}, \quad k = 0, 1, \ldots, M - 1. \tag{3.17}$$

Hence, applying the CFL condition (3.12) and the triangle inequality to (3.15), we get

$$\|w^{k+1}\|_{\infty,h} \leq \|w^k\|_{\infty,h} + \tau \frac{a}{2} (h + a\tau) L, \quad k = 0, 1, \ldots, M - 1.$$

But the initial condition here is $\|w^k\|_{\infty,h} = 0$. Thus an inductive argument yields

$$\|w^k\|_{\infty,h} \leq k\tau \frac{a}{2} (h + a\tau) L, \quad k = 1, \ldots, M.$$

We also have

$$\|w\|_{\infty,h,\tau} := \max_{k=0,1,\ldots,M} \|w^k\|_{\infty,h} \leq \frac{aT}{2} (h + a\tau) L. \qquad \blacksquare$$

Remark 2.8. If u_0' is smooth then the solution u is also smooth, and by differentiating the differential equation (3.6) it follows that $u_{tt} = a^2 u_{xx}$. In this case a Taylor expansion can include higher-order terms and yields

$$\frac{u(x, t + \tau) - u(x, t)}{\tau} + a \frac{u(x, t) - u(x - h, t)}{h} =$$

$$= u_t + a u_x + \frac{a\tau - h}{2} a u_{xx} + O(h^2 + \tau^2). \quad \square$$

Remark 2.9. Scheme (b) can also be generated in the following way. If $\gamma \leq 1$ then—see Figure 2.4—the characteristic through (x_j, t^{k+1}) intersects the straight line $t = t^k$ at some point P^* between x_{j-1} and x_j. Since u is constant along each characteristic, it is therefore a good idea to define u_j^{k+1} by a *linear interpolation* of the values u_{j-1}^k and u_j^k. This approach is easily extended to problems with variable coefficients, thereby allowing the discretization to adapt to the local behaviour of the characteristics. $\square$

As an alternative to the maximum-norm stability analysis of Theorem 2.7, we now study the L_2 stability of the difference scheme by means of a Fourier analysis. Unlike Section 2.2, where orthonormal bases of the space of grid functions were used, here we work in a Fourier transform setting that is used for L_2-stability analysis in, e.g., [Str04]. The basic difference from Section 2.2 is that now the spatial grid contains infinitely many grid points. Fourier stability analyses are commonly used in discretizations of time-dependent problems, but the technique is restricted to differential operators with constant coefficients, and boundary conditions in the spatial variable will cause additional difficulties.

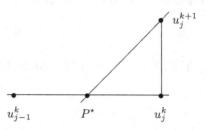

Figure 2.4   Discretization that takes the characteristics into account

Define the complex-valued function φ_j by

$$\varphi_j(x) := \sqrt{\frac{h}{2\pi}}\, e^{ijx}, \quad j \in \mathbb{Z}. \tag{3.18}$$

These functions form a countable basis of the space $L_2(-\pi, \pi)$ that is orthogonal with respect to the complex scalar product (here overline denotes a complex conjugate)

$$\langle v, w \rangle := \int\limits_{-\pi}^{\pi} v(x)\,\overline{w(x)}\, dx \qquad \forall v,\, w \in L_2(-\pi, \pi).$$

More precisely,

$$\langle \varphi_j, \varphi_l \rangle = h\,\delta_{j,l} \quad \forall j,\, l \in \mathbb{Z}, \tag{3.19}$$

using the Kronecker symbol $\delta_{j,l}$. Any grid function $u_h = \{u_j\}_{j \in \mathbb{Z}} \in l^2$ can be mapped in a one-to-one way to a function $\hat{u} \in L_2(-\pi, \pi)$ that is defined by

$$\hat{u}(x) := \sum_{j=-\infty}^{\infty} u_j\,\varphi_j(x).$$

The function $\hat{u} \in L_2(-\pi, \pi)$ is called the Fourier transform of u_h. The representation (3.19) yields the inverse transform formula

$$u_j = \frac{1}{h} \int\limits_{-\pi}^{\pi} \hat{u}(x)\,\overline{\varphi_j(x)}\, dx = \frac{1}{\sqrt{2\pi h}} \int\limits_{-\pi}^{\pi} \hat{u}(x)\, e^{-ijx}\, dx, \qquad j \in \mathbb{Z},$$

that expresses each grid function u_h in terms of its corresponding $\hat{u}$. The orthogonality of the basis and (3.19) also yield Parseval's identity

$$\|u_h\|_{0,h}^2 = h \sum_{j=-\infty}^{\infty} u_j^2 = \langle \hat{u}, \hat{u} \rangle = \|\hat{u}\|_0^2. \tag{3.20}$$

Using this foundation, L_2-norm stability analyses for finite difference methods can be carried out via Fourier transforms to prove bounds like (2.13). The following basic property of the transform facilitates such an analysis.

Let two grid functions v_h, w_h be related to each other by $v_i = w_{i\pm1}$ for all i; then their Fourier transforms are related by

$$\hat{v}(x) = e^{\mp ix}\, \hat{w}(x). \tag{3.21}$$

Consequently the application of a difference operator to a grid function is manifested in the Fourier transform image space as a much simpler operation: multiplication by a (complex) trigonometric polynomial!

Let us now return to (3.6), with t lying in a finite time domain $[0, T]$. For any chosen number $M \in \mathbb{N}$ of time intervals, the discretization time-step size is $\tau = T/M$. Take

$$\|u\|_{h,\tau} := \max_{0 \le k \le M} \|u^k\|_{0,h} \tag{3.22}$$

as the underlying norm. Note that in the present case of an unbounded spatial domain, the Euclidean norm $\|u^k\|_{0,h}$ cannot in general be bounded from above by the maximum norm $\|u^k\|_{\infty,h}$. To overcome this difficulty we shall assume that u_0 has compact support.

Theorem 2.10. *Let the function u_0 be Lipschitz-continuously differentiable and let u_0' have compact support. Then the explicit upwind scheme (b) is consistent of order 1 in the norm $\|\cdot\|_{h,\tau}$ defined by (3.22). If in addition the CFL condition is satisfied, then on finite time domains $[0, T]$ the scheme (b) is stable in the norm (3.22) and as a consequence is convergent of order 1.*

Proof: Set $w_j^k := u_j^k - u(x_j, t^k)$ for all j and k. As already shown in the proof of Theorem 2.7, one has the representation

$$w_j^{k+1} = (1 - \gamma)\, w_j^k + \gamma\, w_{j-1}^k + \tau\, r_j^k \quad j \in \mathbb{Z},\ k = 0, 1, \dots, M - 1, \tag{3.23}$$

with r_j^k defined by (3.16). First, we estimate the grid functions $r^k := \{r_j^k\}_{j\in\mathbb{Z}}$ in the discrete norm $\|\cdot\|_{0,h}$. Since u_0' has compact support there exists some $\rho > 0$ such that

$$u_0'(x) = 0 \qquad \text{for all } x \text{ with } |x| \ge \rho.$$

Together with (3.16) this implies that

$$\|r_k\|_{0,h}^2 = h \sum_{j\in\mathbb{Z}} |r_j^k|^2 \le h \left(\frac{2\rho}{h} + 1\right)\left(\frac{a}{2}(h + a\tau)L\right)^2, \quad k = 0, 1, \dots, M-1.$$

Hence

$$\|r_k\|_{0,h} \le c\,(h + a\tau), \quad k = 0, 1, \dots, M - 1,$$

for some $c > 0$. That is, the scheme is first-order consistent in the norm $\|\cdot\|_{h,\tau}$ of (3.22).

From (3.23) and (3.21), the Fourier transforms $\hat{w}, \hat{r}$ of the grid functions w, r satisfy

$$\hat{w}^{k+1} = ((1 - \gamma) + \gamma e^{ix})\hat{w}^k + \tau \hat{r}^k.$$

Thus

$$\|\hat{w}^{k+1}\|_0 \leq \left(\max_{x \in [-\pi,\pi]} |(1 - \gamma) + \gamma e^{ix}| \right) \|\hat{w}^k\|_0 + \tau \|\hat{r}^k\|_0.$$

The CFL condition implies that $\max_{x \in [-\pi,\pi]} |(1 - \gamma) + \gamma e^{ix}| = 1$, which with (3.20) yields

$$\|w^{k+1}\|_{0,h} \leq \|w^k\|_{0,h} + \tau c(h + a\tau), \quad k = 0, 1, \ldots .$$

Taking into account the initial condition $\|w^k\|_{\infty,h} = 0$, this leads to

$$\|w^k\|_{0,h} \leq k\tau c(h + a\tau), \quad k = 1, \ldots, M.$$

We obtain finally

$$\|w\|_{h,\tau} = \max_{k=0,1,\ldots,M} \|w^k\|_{0,h} \leq Tc(h + a\tau). \quad \blacksquare$$

Remark 2.11. In general, if inequalities of the type

$$\|\hat{w}^{k+1}\|_0 \leq V_F\|\hat{w}^k\|_0 + \tau \|\hat{K}_F\|_0, \quad k = 0, 1, \ldots,$$

are valid, then the *amplification factor* V_F is a critical factor in the stability behaviour of the difference scheme. The amplification factor V_F is always—as in the proof of Theorem 2.10—the maximum modulus of a (complex) trigonometrical polynomial $V(x)$. If on some x-subinterval of $[-\pi, \pi]$ one has $V(x) \geq \delta > 1$ for some δ independent of x and τ, then the method is unstable. The condition $V_F \leq 1$ is obviously sufficient for stability; in fact it can be weakened to

$$V_F \leq 1 + \mu\tau \qquad (3.24)$$

because $\lim_{\tau \to 0}(1 + \tau)^{1/\tau} = e$ is finite. $\quad \square$

Let us continue our study of the upwind scheme (b). If $\gamma > 1$ then

$$|(1 - \gamma) + \gamma e^{i\pi}| = 2\gamma - 1 > 1.$$

The continuity of the function e^{ix} implies that for each $\delta \in (1, 2\gamma - 1)$, there exists $\sigma \in (0, \pi/2)$ such that

$$|(1 - \gamma) + \gamma e^{ix}| \geq \delta \qquad \forall x \in [-\pi + \sigma, \pi - \sigma].$$

Consequently the method is unstable. Thus the CFL sufficient condition $\gamma \leq 1$ is also necessary for stability.

Next, we investigate briefly the two methods (a) and (c). In the case of method (a), the polynomial $V(x)$ of Remark 2.11 satisfies

$$|V(x)| = \left| 1 + \frac{1}{2}\gamma\left(e^{ix} - e^{-ix}\right) \right| = 1 + \gamma \sin x.$$

It follows that there exist $\delta > 1$ and some interval $[\alpha, \beta] \subset [-\pi, \pi]$ such that

$$|V(x)| \geq \delta \quad \forall x \in [\alpha, \beta].$$

As $\gamma > 0$, this unstable situation occurs irrespective of the choices of the step sizes $h > 0$ and $\tau > 0$ in the discretization. The instability of scheme (c) can be shown similarly.

Both these methods also fail to satisfy the discrete maximum principle and are unstable in the maximum norm.

Remark 2.12. A heuristic motivation of the importance of the CFL condition can be given in the following way. When discussing the consistency of scheme (b) we obtained the relation

$$\frac{u(x, t+\tau) - u(x,t)}{\tau} + a\frac{u(x,t) - u(x-h,t)}{h} =$$
$$= u_t + au_x + \frac{a\tau - h}{2}au_{xx} + O(h^2 + \tau^2). \tag{3.25}$$

Given an initial condition at $t = 0$, the well-posedness of the parabolic problem for $t > 0$ requires u_{xx} to have a negative coefficient, viz., $a\tau - h \leq 0$. This is equivalent to the CFL condition.

When $a\tau - h < 0$, the presence of the additional term $\frac{a\tau-h}{2}au_{xx}$ means that, as well as the convection that appeared in the original problem, we have introduced a diffusion term $\frac{a\tau-h}{2}au_{xx}$. This extra term is called *numerical diffusion*; it usually causes the numerical solution to contain an undesirable smearing of any sharp layers present in the original solution.   □

In the above investigations an equidistant spatial grid was assumed, i.e., the spatial grid is $\{x_j\}_{j\in\mathbb{Z}}$ with $x_j - x_{j-1} = h$, $j \in \mathbb{Z}$. We now outline a way of extending the Fourier technique to more general grids. For general grids $\{x_j\}_{j\in\mathbb{Z}}$, set

$$h_j := x_j - x_{j-1} > 0 \quad \text{and} \quad h_{j+1/2} := \frac{1}{2}(h_j + h_{j+1}), \quad j \in \mathbb{Z}.$$

Given a grid function $u_h = (u_j)_{j\in\mathbb{Z}}$, its Euclidean norm $\|\cdot\|_{0,h}$ on this arbitrary grid is defined by

$$\|u_h\|_{0,h}^2 := \sum_{j\in\mathbb{Z}} h_{j+1/2}\, u_j^2.$$

With the aid of modified complex basis functions

$$\varphi_j(x) := \sqrt{\frac{h_{j+1/2}}{2\pi}}\, e^{ijx}, \quad j \in \mathbb{Z}, \tag{3.26}$$

grid functions u_h can be mapped in a one-to-one way to their $\hat{u}$ Fourier transforms, just as in the equidistant case. This transform is defined by

$$\hat{u}(x) := \sum_{j=-\infty}^{\infty} u_j\, \varphi_j(x).$$

The functions φ_j form a countable basis of the space $L_2(-\pi, \pi)$ and they are orthogonal with respect to the complex scalar product

$$\langle v, w \rangle := \int_{-\pi}^{\pi} v(x)\, \overline{w(x)}\, dx \qquad \forall v,\, w \in L_2(-\pi, \pi).$$

Now

$$\langle \varphi_j, \varphi_l \rangle = h_{j+1/2}\, \delta_{j,l} \quad \forall j, l \in \mathbb{Z}, \tag{3.27}$$

so Parseval's identity is valid:

$$\|u_h\|_{0,h}^2 = \sum_{j=-\infty}^{\infty} h_{j+1/2}\, u_j^2 = \langle \hat{u}, \hat{u} \rangle = \|\hat{u}\|_0^2.$$

Using these results our earlier convergence analysis can be extended to non-equidistant grids under weak additional assumptions such as the uniform boundedness of the quotients h_j/h_{j+1} and h_{j+1}/h_j. Nevertheless, it should be noted that in this more general case, quotients of neighbouring step sizes h_j will occur in the modified CFL conditions that are obtained.

Remark 2.13. Instead of three-point approximations, let us consider the general explicit discretization scheme

$$u_j^{k+1} = \sum_{l=-m}^{m} c_l\, u_{j+l}^k \tag{3.28}$$

where $m \geq 1$. This scheme is consistent only if

$$\sum_{l=-m}^{m} c_l = 1 \quad \text{and} \quad \sum_{l=-m}^{m} l\, c_l = -\gamma.$$

The first of these identities implies that we must assume $c_l \geq 0$ for all l to guarantee a discrete maximum principle (3.11) for this general scheme. It turns out that this is impossible for methods whose consistency order is greater than 1. Thus explicit methods of type (3.28) that satisfy the discrete maximum principle can achieve at best first-order convergence. $\square$

Consider now transport problems on a bounded interval, i.e. problems of the type

$$u_t + a\, u_x = 0, \qquad x \in [c, d] \subset \mathbb{R}, \quad t > 0,$$
$$u(\cdot, 0) = u_0(\cdot) \ \text{ on } [c, d], \qquad u(c, \cdot) = g(\cdot), \quad t > 0, \tag{3.29}$$

where $a > 0$. (For bounded spatial intervals, boundary conditions for $t > 0$ must be imposed at one endpoint; as $a > 0$, on recalling the orientation of the characteristics and how the solution u propagates along them, it is clear that the left endpoint should be used so that on each characteristic only one value of u is given a priori.) Now not only explicit schemes but also implicit discretizations are feasible. Similarly to (3.8), we examine those *implicit* discretizations that have three grid points in the spatial direction and are defined by

$$D_t^- u = a\left(\omega D_x^+ + (1 - \omega)\, D_x^-\right) u, \tag{3.30}$$

where $\omega \in [0, 1]$ is some fixed parameter. Set $\gamma = (a\,\tau)/h$. Then for $\omega = 0$ and $\omega = 1$ we obtain the variants

$$\text{(b) } (1 + \gamma)\, u_j^k - \gamma u_{j-1}^k = u_j^{k-1}$$

and $\tag{3.31}$

$$\text{(c) } (1 - \gamma)\, u_j^k + \gamma u_{j+1}^k = u_j^{k-1},$$

respectively. Figure 2.5 shows their difference stencils.

Figure 2.5   Difference stencils of the implicit discretizations (3.31)

The labelling of the implicit methods (b) and (c) corresponds to that of the earlier explicit schemes. Stability for the implicit methods differs from our earlier results for the explicit methods: variant (b) is stable in the L_2-norm for arbitrary step sizes $h, \tau > 0$, but variant (c) is L_2 stable only when $\gamma \geq 1$. Now

$$|1 + \gamma - \gamma e^{\mathrm{i}x}| \geq |1 + \gamma| - |\gamma e^{\mathrm{i}x}| = 1 \quad \forall x \in [-\pi, \pi]$$

and consequently

$$\left| \frac{1}{1 + \gamma - \gamma e^{\mathrm{i}x}} \right| \leq 1 \quad \forall x \in [-\pi, \pi],$$

which is the Fourier L_2-stability condition for (b). For variant (c), under the additional condition $\gamma \geq 1$ we get

$$|1 - \gamma + \gamma e^{-ix}| \geq |\gamma e^{ix}| - |1 - \gamma| = \gamma - (\gamma - 1) = 1 \quad \forall x \in [-\pi, \pi].$$

Thus

$$\left| \frac{1}{1 - \gamma + \gamma e^{-ix}} \right| \leq 1 \quad \forall x \in [-\pi, \pi],$$

which guarantees L_2 stability for scheme (c) if $\gamma \geq 1$.

2.3.2 Properties of Nonlinear Conservation Laws

Let us now consider the problem

$$u_t + f(u)_x = 0, \qquad u(\cdot, 0) = u_0(\cdot), \tag{3.32}$$

where $x \in \mathbb{R}^n$. Throughout this section we assume that $f'' > 0$.

As in the linear case, the characteristics $(x(s), s)$ play an important role in the behaviour of the solution u of problem (3.32). Now

$$\frac{d}{ds} u(x(s), s) = u_t(x(s), s) + \nabla u \cdot x'(s),$$

and we require the characteristics to satisfy

$$x'(s) = f'(u(x(s), s)) \tag{3.33}$$

so that, invoking (3.32), we get

$$\frac{d}{ds} u(x(s), s) = 0.$$

That is, u is constant along each characteristic. It follows from (3.33) that $x'(s)$ is also constant there, i.e., each characteristic is a straight line. These facts combine to yield

$$u(x, t) = u_0(x - f'(u(x, t))t). \tag{3.34}$$

This is the nonlinear form of the solution that we discussed in the previous section. Given sufficient smoothness of f, from the chain rule of differential calculus for composite functions one infers that

$$u_x = u_0' \cdot (1 - f''(u) u_x t) \quad \text{and} \quad u_t = u_0' \cdot (-f'(u) - f''(u) u_t t).$$

Formally solving these equations for u_x and u_t, one gets

$$u_x = \frac{u_0'}{1 + u_0' f''(u)t} \quad \text{and} \quad u_t = -\frac{f'(u)u_0'}{1 + u_0' f''(u)t}. \tag{3.35}$$

Recalling that $f'' > 0$, the following lemma is valid:

Lemma 2.14. *If $u_0' \geq 0$, then (3.32) has a unique classical solution that is implicitly given by (3.34).*

If $u_0' < 0$ then, given two points x_0 and x_1 on the x-axis with $x_0 < x_1$, the characteristics through these two points will intersect at some time $t^* > 0$, where t^* satisfies

$$f'(u_0(x_0))t^* + x_0 = f'(u_0(x_1))t^* + x_1;$$

the existence of a solution t^* follows from $u_0(x_1) < u_0(x_0)$ and $f'' > 0$. See Figure 2.6. Hence, even when the initial function u_0 is continuous, discontinuities of the solution u of (3.32) may occur after a finite time if different values of u propagate to the same point.

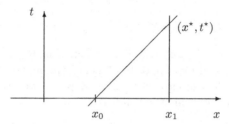

Figure 2.6 Intersecting characteristics

Example 2.15. The special nonlinear conservation law

$$u_t + uu_x = 0$$

is known as *Burgers' equation* and plays an important role in fluid dynamics. Here $f(u) = u^2/2$ and each characteristic has slope $f'(u) = u$. Let us choose the continuous initial condition

$$u(x,0) = \begin{cases} 1 & \text{if } x < 0, \\ 1 - x & \text{if } 0 \le x \le 1, \\ 0 & \text{if } x > 1. \end{cases}$$

Then the solution for $0 \le t \le 1$ is a continuous function given by

$$u(x,t) = \begin{cases} 1 & \text{if } x < t, \\ (1 - x)/(1 - t) & \text{if } t \le x \le 1, \\ 0 & \text{if } x > 1. \end{cases}$$

But for $t > 1$ the solution is discontinuous: $u = 1$ for $x < 1$, $u = 0$ for $x > 1$. □

An inevitable consequence of the possible existence of discontinuous solutions is that our understanding of what is meant by a solution of (3.32) has to be suitably modified. For this purpose we now consider weak formulations and weak solutions. Multiplication of the differential equation by an arbitrary C_0^1 function ϕ (such functions have compact support in $(-\infty, \infty) \times (0, \infty)$) followed by integration by parts yields

$$\int_x \int_t (u\,\phi_t + f(u)\,\phi_x) + \int_x u_0\,\phi(x, 0) = 0 \quad \text{for all} \quad \phi \in C_0^1. \tag{3.36}$$

It turns out that this condition does not characterize weak solutions in a unique way, as will be seen in Example 2.16; something more is needed. Now any discontinuity in the solution travels along a smooth curve $(x(t), t)$. Suppose that such a curve of discontinuity divides the entire region $D := \mathbb{R} \times (0, +\infty)$ into the two disjoint subdomains

$$D_1 := \{\, (x, t) \in D \,:\, x < x(t)\,\}, \qquad D_2 := \{\, (x, t) \in D \,:\, x > x(t)\,\}.$$

Taking ϕ to have support equal to a rectangle R with sides parallel to the coordinate axes and two opposite corners lying a short distance apart on the curve $(x(t), t)$, then integrating (3.36) by parts over R followed by shrinking R to a point, one obtains the condition

$$\frac{dx}{dt} = \frac{f(u_L) - f(u_R)}{u_L - u_R} \qquad (\textit{Rankine-Hugoniot condition}), \tag{3.37}$$

where

$$u_L := \lim_{\varepsilon \to 0+} u(x(t) - \varepsilon, t), \qquad u_R := \lim_{\varepsilon \to 0+} u(x(t) + \varepsilon, t).$$

Example 2.16. Consider again Burgers' equation, but now with a discontinuous initial condition:

$$u_t + u u_x = 0,$$

$$u_0(x) = \begin{cases} 0 \text{ if } x < 0, \\ 1 \text{ if } x > 0. \end{cases} \tag{3.38}$$

The shape of the characteristics (see Figure 2.7) leads us to expect some unusual behaviour in the solution . Let $\alpha \in (0, 1)$ be an arbitrary parameter and set

$$u(x, t) = \begin{cases} 0 \text{ for } x < \alpha t/2, \\ \alpha \text{ for } \alpha t/2 < x < (1 + \alpha)t/2, \\ 1 \text{ for } (1 + \alpha)t/2 < x. \end{cases} \tag{3.39}$$

It is easy to verify that $u(x, t)$ is a weak solution of Burgers' equation, i.e., it satisfies (3.36). But $\alpha \in (0, 1)$ can be chosen freely so there are infinitely many solutions of (3.36). One can also verify by a straightforward calculation that the Rankine-Hugoniot conditions are naturally satisfied along the two curves of discontinuity $x(t) = \alpha t/2$ and $x(t) = (1 + \alpha)t/2$. $\qquad \square$

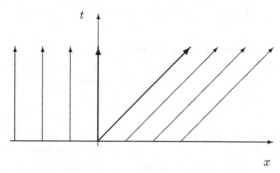

Figure 2.7   Characteristics for problem (3.38)

A physically relevant weak solution can be uniquely defined by an *entropy condition*. To motivate this condition we first consider a classical solution. Let $U(\cdot)$ denote some differentiable convex function and let $F(\cdot)$ be chosen such that

$$F' = U'f'. \qquad (3.40)$$

The function $F(u)$ is called an *entropy flux* and $U(u)$ is called an *entropy function*. (An example is $U(u) = \frac{1}{2}u^2$, $F(u) = \int_0^u sf'(s)ds$.) If u is a classical solution of (3.32), then for any entropy function U and its related entropy flux F one has

$$U(u)_t + F(u)_x = 0. \qquad (3.41)$$

For weak solutions of (3.36), a parabolic regularization and integration by parts (see [Krö97]) yield the *entropy condition*

$$\int_x \int_t [U(u)\Phi_t + F(u)\Phi_x] + \int_x U(u_0)\Phi(x,0) \geq 0, \qquad \forall \Phi \in C_0^1, \ \Phi \geq 0. \quad (3.42)$$

In this equation discontinuous F and U are allowed, but at any point where these functions are differentiable, equation (3.40) must hold. Kruzkov [80] chose (for an arbitrary parameter c) the entropy function and related entropy flux

$$U_c(u) = |u - c|, \qquad (3.43)$$
$$F_c(u) = (f(u) - f(c))\,sign\,(u - c);$$

he then proved

Lemma 2.17. *If $u_0 \in L_1 \cap TV$ then there exists a unique solution of the variational equation (3.36) in the space $L_\infty(0,T; L_1 \cap TV)$ that satisfies the entropy condition (3.42).*

The space TV used in this lemma is the space of locally integrable functions that possess a bounded total variation

$$TV(f) = ||f||_{TV} = \sup_{h \neq 0} \int \frac{|f(x+h) - f(x)|}{|h|}.$$

Remark 2.18. Various other conditions in the literature are also described as entropy conditions. For example, let $(x(t), t)$ be a smooth curve of discontinuity of u and set $s = dx/dt$. Then the inequality

$$f'(u_L) > s > f'(u_R) \tag{3.44}$$

is sometimes called an entropy condition. As $f'' > 0$, clearly (3.44) implies $u_L > u_R$. In [Lax72] it is shown that (3.44) is implied by (3.42) and some properties of the higher-dimensional case are given there.   □

If the entropy condition (3.44) is assumed, then one can construct the complete solution of the *Riemann problem*

$$u_t + uu_x = 0, \qquad u(x, 0) = u_0(x) := \begin{cases} u_L \text{ for } x < 0, \\ u_R \text{ for } x > 0, \end{cases} \tag{3.45}$$

with constant u_L and u_R.

In the case $u_L < u_R$ the solution u cannot have discontinuities because this would violate (3.44); it is given by

$$u(x, t) = \begin{cases} u_L & \text{for } x < u_L t, \\ u_L + (u_R - u_L)(x - u_L t)/((u_R - u_L)t) & \text{for } u_L t \leq x \leq u_R t, \\ u_R & \text{for } u_R t < x. \end{cases}$$

As t increases, the distance between the two constant branches u_L, u_R grows

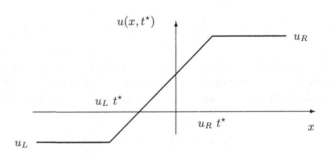

Figure 2.8   Behaviour of the solution of problem (3.45)

(Figure 2.8). In the special case $u_L = 0$, $u_R = 1$ we obtain the entropy solution

$$u(x, t) = \begin{cases} 0 & \text{for } x < 0, \\ x/t & \text{for } 0 \leq x \leq t, \\ 1 & \text{for } t < x. \end{cases}$$

Now we study the case $u_L > u_R$. Then by (3.37) the initial data imply that

$$u(x,t) = \begin{cases} u_L \text{ for } x < \dfrac{f(u_L) - f(u_R)}{u_L - u_R} t, \\[2ex] u_R \text{ for } x > \dfrac{f(u_L) - f(u_R)}{u_L - u_R} t, \end{cases}$$

and in particular for Burgers' equation, where $f(u) = u^2/2$, the solution is

$$u(x,t) = \begin{cases} u_L \text{ for } x < \tfrac{1}{2}(u_L + u_R)t, \\[1ex] u_R \text{ for } x > \tfrac{1}{2}(u_L + u_R)t. \end{cases}$$

If in addition $u_L + u_R = 0$ (e.g., if $u_L = 1$ and $u_R = -1$), then the jump does not move in time. This phenomenon is called a *stationary shock*. If $u_L + u_R \neq 0$ then the shock changes its position as t increases.

Finally, we remark that for piecewise continuous solutions of conservation laws the following properties hold (see [Lax72]):

$$\begin{array}{ll} (a) & \min_x u(x,t) \leq u(x,t+\tau) \leq \max_x u(x,t) \\[1ex] (b) & TV(u(\cdot, t+\tau)) \leq TV(u(\cdot,t)) \\[1ex] (c) & ||u(\cdot,t+\tau))||_1 \leq ||u(\cdot,t)||_1. \end{array} \qquad (3.46)$$

2.3.3 Difference Methods for Nonlinear Conservation Laws

In the previous Section a basic analysis of nonlinear conservation laws was given. Now we shall use this knowledge to discuss and evaluate numerical schemes, i.e., to investigate whether various discretization methods model adequately the essential properties of the underlying continuous problem.

Consider again the problem

$$u_t + f(u)_x = 0, \qquad u(x,0) = u_0(x), \qquad (3.47)$$

under the general assumption $f'' > 0$. If $f(u) = u^2/2$, then $f'' > 0$; thus the Burgers' equation problem

$$\frac{\partial u}{\partial t} + u u_x = 0, \qquad u(x,0) = \begin{cases} -1 \text{ for } x < 0, \\ 1 \text{ for } x \geq 0, \end{cases} \qquad (3.48)$$

is a particular example from the class of problems under consideration.

The obvious generalization to the nonlinear case (3.47) of the explicit methods (3.8) based on three spatial grid points is

$$D_t^+ u + \left(\omega D_x^+ + (1-\omega) D_x^- \right) f(u) = 0, \qquad (3.49)$$

where $\omega \in [0,1]$ denotes some fixed parameter. For example, $\omega = 1/2$ yields the scheme

$$\frac{u_i^{k+1} - u_i^k}{\tau} + \frac{f(u_{i+1}^k) - f(u_{i-1}^k)}{2h} = 0. \qquad (3.50)$$

If any of the methods (3.49) is applied to problem (3.48), it produces the solution

$$u_i^k = \begin{cases} -1 \text{ for } x_i < 0, \\ 1 \text{ for } x_i \geq 0. \end{cases}$$

Thus, taking the limit as the mesh size goes to zero, the computed solution tends to a discontinuous solution that is not the uniquely determined entropy solution that was described in the analysis of the Riemann problem in Section 2.3.2. This simple example shows that the nonlinear case requires careful modifications of methods used for linear problems, if one is to obtain accurate schemes.

A fairly general formula for explicit discretization schemes for the nonlinear conservation law (3.47) is given by

$$\frac{u_i^{k+1} - u_i^k}{\tau} + \frac{1}{h}\left(g(u_{i+1}^k, u_i^k) - g(u_i^k, u_{i-1}^k)\right) = 0, \tag{3.51}$$

or equivalently by

$$u_i^{k+1} = H(u_{i-1}^k, u_i^k, u_{i+1}^k) \tag{3.52}$$

with

$$H(u_{i-1}^k, u_i^k, u_{i+1}^k) := u_i^k - q\left(g(u_{i+1}^k, u_i^k) - g(u_i^k, u_{i-1}^k)\right), \tag{3.53}$$

where $q := \tau/h$. Here $g(\cdot, \cdot)$ denotes some appropriately chosen function that is the *numerical flux*. Further, schemes of type (3.51) are called *conservative*.

To obtain consistency of the scheme (3.51) with (3.47), we assume that

$$g(s, s) = f(s) \qquad \forall s \in \mathbb{R}. \tag{3.54}$$

If g is differentiable then the consistency condition $\partial_1 g + \partial_2 g = f'(u)$ follows automatically from the chain rule of differential calculus.

Some methods that are commonly used will now be described. These methods can be obtained from the general scheme (3.51) by means of properly selected numerical fluxes. In particular, the choice

$$g(v, w) := \frac{1}{2}\left((f(v) + f(w)) - \frac{1}{q}(v - w)\right)$$

in (3.51) yields the well-known *Lax-Friedrichs scheme*

$$u_i^{k+1} = u_i^k - \frac{q}{2}[f(u_{i+1}^k) - f(u_{i-1}^k)] + \frac{u_{i+1}^k - 2u_i^k + u_{i-1}^k}{2}. \tag{3.55}$$

The underlying idea of the Lax-Friedrichs scheme is to add a discretization of the term $h\,u_{xx}/(2q)$ to the basic scheme (3.50). If the solution u is sufficiently smooth, then the stabilization term contributes $O(h/q)$ to the consistency error estimate. Hence, a good recommendation here is to choose $q = 1$.

Another well-known choice for the numerical flux is given by

$$g(v,w) := \begin{cases} f(v) & \text{if } v \geq w \quad \text{and } f(v) \geq f(w), \\ f(w) & \text{if } v \geq w \quad \text{and } f(v) \leq f(w), \\ f(v) & \text{if } v \leq w \quad \text{and } f'(v) \geq 0, \\ f(w) & \text{if } v \leq w \quad \text{and } f'(w) \leq 0, \\ f((f')^{-1}(0)) & \text{otherwise.} \end{cases}$$

With this particular numerical flux the general scheme (3.51) becomes the well-known *Godunov scheme*

$$u_i^{k+1} = u_i^k - q\big(f(u_{i+1/2}^k) - f(u_{i-1/2}^k)\big), \tag{3.56}$$

where the values $u_{i+1/2}^k$ are defined according to the following rules:
Set $f(u_{i+1}^k) - f(u_i^k) = f'(\xi_{i+1/2}^k)(u_{i+1}^k - u_i^k)$. Then we choose

$$u_{i+1/2}^k = u_i^k \qquad \text{if} \quad f'(u_i^k) > 0 \qquad \text{and} \quad f'(\xi_{i+1/2}^k) > 0,$$

$$u_{i+1/2}^k = u_{i+1}^k \qquad \text{if} \quad f'(u_{i+1}^k) < 0 \qquad \text{and} \quad f'(\xi_{i+1/2}^k) < 0,$$

$u_{i+1/2}^k$ is a root of $f'(u) = 0$ in all other cases.

If the numerical flux is specified by

$$g(v,w) := \tfrac{1}{2}[f(v) + f(w) - \int_v^w |f'(s)|ds],$$

then (3.51) yields the *Enquist-Osher scheme*

$$u_i^{k+1} = u_i^k - q\left(\int_{u_{i-1}}^{u_i} f'_+ ds + \int_{u_i}^{u_{i+1}} f'_- ds\right), \tag{3.57}$$

where $f'_+(s) := \max(f'(s), 0), \quad f'_-(s) := \min(f'(s), 0)$.

The scheme (3.52), (3.53) is called *monotone* if H is non-decreasing in each argument. This is a valuable property, for monotone schemes replicate the properties (3.46) of the continuous problem in the discrete one, and convergence to the desired entropy solution can be shown.

Lemma 2.19. *Monotone schemes of the form (3.52), (3.53) possess the property*

$$\min(u_{i-1}^k, u_i^k, u_{i+1}^k) \leq u_i^{k+1} \leq \max(u_{i-1}^k, u_i^k, u_{i+1}^k) \quad \text{for all } i \text{ and } k. \tag{3.58}$$

Proof: Let i and k be arbitrary but fixed. Set $v_j^k = \max\{u_{i-1}^k, u_i^k, u_{i+1}^k\}$ for all j. Using the scheme and the v_j^k to compute all the values v_j^{k+1} on the next time level, monotonicity implies that $v_i^{k+1} \geq u_i^{k+1}$. But since v_j^k is constant as j varies, (3.53) implies that we also have $v_j^{k+1} = v_j^k$ for all j. Thus we have shown that $u_i^{k+1} \leq \max\{u_{i-1}^k, u_i^k, u_{i+1}^k\}$. The other inequality is proved in a similar manner. ∎

The estimate (3.58) is the discrete analog of (3.46a). Now let us consider the properties (3.46b) and (3.46c). Define the discrete norms

$$||u_h||_{1,h} := \sum_i h|u_i| \quad \text{and} \quad ||u_h||_{TV,h} := \sum_i |u_{i+1} - u_i|$$

for those grid functions u_h for which these norms are finite. We say that $u_h \in L_{1,h}$ when $||u_h||_{1,h}$ is finite.

Lemma 2.20. *Let k be arbitrary but fixed. Let u_h^k, $v_h^k \in L_{1,h}$ be given. Suppose that the corresponding u_h^{k+1} and v_h^{k+1} are generated by a monotone scheme of the form (3.52), (3.53). Then $u_h^{k+1}, v_h^{k+1} \in L_{1,h}$ and in fact*

$$
\begin{array}{ll}
(b) & ||u_h^{k+1} - v_h^{k+1}||_{1,h} \le ||u_h^k - v_h^k||_{1,h}, \\
(c) & ||u_h^{k+1}||_{TV,h} \le ||u_h^k||_{TV,h}.
\end{array} \tag{3.59}
$$

Proof: Here we show only that the estimate (b) implies (c). The proof of (b) requires much more effort and for it we refer the reader to [39].

Let $\pi : L_{1,h} \to L_{1,h}$ denote the index-shift operator defined by

$$(\pi u_h)_i := u_{i+1} \quad \text{with} \quad u_h = (u_i)_{i \in \mathbb{Z}}.$$

Let $H : L_{1,h} \to L_{1,h}$ be the operator associated with the discretization scheme, i.e.,

$$(H(u_h))_i := H(u_{i-1}, u_i, u_{i+1}), \quad i \in \mathbb{Z}.$$

Then $\pi H(u_h) = H(\pi u_h)$ and from this property it follows that

$$
\begin{aligned}
h||u_h^{k+1}||_{TV,h} &= \sum h|u_{i+1}^{k+1} - u_i^{k+1}| = \sum h|\pi u_i^{k+1} - u_i^{k+1}| = \\
&= ||\pi u_h^{k+1} - u_h^{k+1}||_{1,h} = ||H(\pi u_h^k) - H(u_h^k)||_{1,h} \\
&\le ||\pi u_h^k - u_h^k||_{1,h} = h||u_h^k||_{TV,h},
\end{aligned}
$$

where the inequality in this calculation is a special case of (b). This completes the proof that (b) implies (c). ∎

Theorem 2.21. *Assume that we are given a monotone and conservative discretization scheme of the form (3.52), (3.53) with some continuous and consistent numerical flux. Then this scheme generates a numerical solution that converges to the entropy solution of the conservation law and its order of convergence is at most one.*

The proof of this theorem basically follows from Lemmas 2.19 and 2.20, but it requires several technicalities. For the special case of the Lax-Friedrichs scheme a nice proof is given in [81], and for the general case see [62].

Remark 2.22. The Lax-Friedrichs, Godunov and Enquist-Osher schemes are monotone when an additional condition is satisfied. In the case of the Lax-Friedrichs scheme,

$$\frac{\partial H}{\partial u_i^k} = 0, \quad \frac{\partial H}{\partial u_{i-1}^k} = \frac{1}{2}(1 + qf'(u_{i-1}^k)), \quad \frac{\partial H}{\partial u_{i+1}^k} = \frac{1}{2}(1 - qf'(u_i^k)).$$

Hence this scheme is monotone if

$$q \max_u |f'(u)| \leq 1 \quad \text{(CFL condition)}. \tag{3.60}$$

In the linear case this condition coincides with the classical CFL condition (3.12).   □

A drawback of monotone schemes is that their convergence order is at most 1. To overcome this limitation one tries to relax the monotonicity condition in some way. One possibility is the following: the scheme (3.52), (3.53) is called *TVNI* (total variation nonincreasing) or *TVD* (total variation diminishing) if the property

$$||u_h^{k+1}||_{TV,h} \leq ||u_h^k||_{TV,h}$$

holds for all k.

Any monotone scheme is also TVNI by Lemma 2.20. The converse is false is general.

For schemes of the form

$$u_i^{k+1} = u_i^k + C_{i+1/2}^+(u_{i+1}^k - u_i^k) - C_{i-1/2}^-(u_i^k - u_{i-1}^k) \tag{3.61}$$

we have (compare [61])

Lemma 2.23. *If $C_{i+1/2}^+ \geq 0$, $C_{i+1/2}^- \geq 0$ and $C_{i+1/2}^- + C_{i+1/2}^+ \leq 1$ for all i, then the scheme (3.61) is TVNI.*

Proof: From

$$u_i^{k+1} = u_i^k + C_{i+1/2}^+(u_{i+1}^k - u_i^k) - C_{i-1/2}^-(u_i^k - u_{i-1}^k)$$

and

$$u_{i+1}^{k+1} = u_{i+1}^k + C_{i+3/2}^+(u_{i+2}^k - u_{i+1}^k) - C_{i+1/2}^-(u_{i+1}^k - u_i^k)$$

it follows that

$$u_{i+1}^{k+1} - u_i^{k+1} = (1 - C_{i+1/2}^- - C_{i+1/2}^+)(u_{i+1}^k - u_i^k)$$
$$+ C_{i-1/2}^-(u_i^k - u_{i-1}^k) + C_{i+3/2}^+(u_{i+2}^k - u_{i+1}^k).$$

Consequently, using the hypotheses of the lemma,

$$|u_{i+1}^{k+1} - u_i^{k+1}| \leq (1 - C_{i+1/2}^- - C_{i+1/2}^+)|u_{i+1}^k - u_i^k|$$
$$+ C_{i-1/2}^-|u_i^k - u_{i-1}^k| + C_{i+3/2}^+|u_{i+2}^k - u_{i+1}^k|.$$

Now a summation over i completes the proof. ■

An important aim in the construction of TVNI schemes is the achievement of a consistency order greater than one. The *Lax-Wendroff scheme* is a well-known second-order scheme that is described by

$$g(v, w) = \frac{1}{2}[f(v) + f(w)] - qf'\left(\frac{v + w}{2}\right)[f(v) - f(w)]. \tag{3.62}$$

But if one applies the Lax-Wendroff scheme to the example

$$u_t + \sin(\pi u)u_x = 0, \qquad u(x, 0) = \begin{cases} -1 \text{ for } x < 0, \\ 1 \text{ for } x \geq 0, \end{cases}$$

then the computed solution—compare the results for (3.8) applied to problem (3.48)—converges to an incorrect solution.

A class of schemes that have a structure like that of (3.62) is given by

$$g(v, w) = \frac{1}{2}\left[f(v) + f(w) - \frac{1}{q}Q(qb)(w - v)\right], \tag{3.63}$$

where

$$b := \begin{cases} [f(w) - f(v)]/(w - v) \text{ for } w \neq v, \\ f'(v) \qquad\qquad\quad \text{for } w = v, \end{cases}$$

and $qb = \lambda$, while $Q(\lambda)$ is some function that has to be chosen. Within this class we can try to find TVNI schemes. If (3.63) is converted to the form (3.61) then

$$C^{\pm} = \frac{1}{2}Q(\lambda) \mp \lambda.$$

Hence, from Lemma 2.23 follows

Lemma 2.24. *Suppose that*

$$|x| \leq Q(x) \leq 1 \quad for \quad 0 \leq |x| \leq \mu < 1. \tag{3.64}$$

Assume that the CFL condition

$$q \max |f'| \leq \mu \leq 1$$

is satisfied. Then the choice (3.63) yields a TVNI scheme.

Unfortunately the choice (3.63) again restricts to one the consistency order of the scheme. Harten [61] analysed the possibility of constructing from a 3-point TVNI scheme a related 5-point TVNI scheme of higher order. This is done by choosing

$$f(v_i) := f(v_i) + qg(v_{i-1}, v_i, v_{i+1})$$

in (3.63), (3.53) with a properly chosen g. See [61] for details.

A further generalization of (3.52), (3.53) is the use of discretizations of the form

$$u_i^{k+1} = H(u_{i-l}^k, u_{i-l+1}^k, \cdots, u_i^k, \cdots, u_{i+l}^k)$$

where $l \geq 1$ and

$$H(u_{i-l}^k, u_{i-l+1}^k, \cdots, u_i^k, \cdots, u_{i+l}^k) =$$
$$= u_i^k - q[g(u_{i+l}^k, \cdots, u_{i-l+1}^k) - g(u_{i+l-1}^k, \cdots, u_{i-l}^k)],$$

where g satisfies the consistency condition

$$g(u, u, \cdots, u) = f(u).$$

In the literature one can find several other approaches to constructing schemes that have the property of convergence to the entropy solution. These include, inter alia, quasi-monotone schemes and MUSCL schemes. For further details and a comprehensive discussion of the discretization of conservation laws, see [Krö97].

Exercise 2.25. Two possible discretizations of the transport equation

$$u_t + bu_x = 0 \quad \text{(constant } b > 0\text{)},$$

with the initial condition $u|_{t=0} = u_0(x)$, are the following:

$$a) \quad \frac{u_i^{k+1} - u_i^{k-1}}{2\tau} + b\frac{u_{i+1}^k - u_{i-1}^k}{2h} = 0,$$

$$b) \quad \frac{u_i^{k+1} - u_i^k}{\tau} + b\frac{u_i^{k+1} - u_{i-1}^{k+1}}{h} = 0.$$

Discuss the advantages and disadvantages of each method.

Exercise 2.26. Analyse the L_2 stability of the following scheme for the discretization of the transport equation of Exercise 2.25 and discuss its consistency:

$$\frac{u_i^{k+1} - u_i^k}{\tau} + \frac{b}{h}\left[\frac{u_i^{k+1} + u_i^k}{2} - \frac{u_{i-1}^{k+1} + u_{i-1}^k}{2}\right] = 0.$$

Exercise 2.27. Consider the multi-level approximation

$$u_i^{k+1} = \sum_{\mu,\nu} \alpha_\nu^\mu u_{i+\mu}^{k+\nu},$$

where $\mu = 0, \pm 1, \pm 2, \cdots, \pm p$ and $\nu = 1, 0, -1, \cdots, -q$, for the discretization of the transport equation.
a) Derive conditions that guarantee consistency of order K.
b) Show that within the class of explicit two-level schemes there does not exist a method that is monotone in the sense that $\alpha_\nu^\mu \geq 0$ for all ν and μ.
c) Construct examples of monotone multi-level schemes and discuss their orders of consistency.

Exercise 2.28. Integration of the conservation law yields the mass balance equation

$$\frac{d}{dt} \int_a^b u(x,t)dx = f(u(a,t)) - f(u(b,t)).$$

Apply an appropriate limit process to this to derive the Rankine-Hugoniot jump condition.

Exercise 2.29. Analyse the limit $\lim_{\varepsilon \to 0} u(x,t,\varepsilon)$ of the parabolic problem

$$u_t - \varepsilon u_{xx} + bu_x = 0 \quad \text{in} \quad (-\infty,\infty) \times (0,\infty),$$
$$u|_{t=0} = u_0(x),$$

for the case of constant $b > 0$.

Exercise 2.30. Consider the parabolic problem

$$u_t - \varepsilon u_{xx} + uu_x = 0$$

with initial condition

$$u(x,0) = \begin{cases} u_L & \text{for} \quad x < 0, \\ u_R & \text{for} \quad x > 0. \end{cases}$$

Investigate whether this problem possesses a solution of the form

$$u(x,t,\varepsilon) = U\left(x - \frac{u_L + u_R}{2}t\right).$$

If such a solution exists, examine its behaviour for small $\varepsilon > 0$.

Exercise 2.31. Consider Burgers' equation

$$u_t + uu_x = 0$$

with the initial condition

$$u(x,0) = \begin{cases} 1 & \text{for} \quad x < 0, \\ 0 & \text{for} \quad x > 0. \end{cases}$$

Discretize this problem by the scheme

$$\frac{u_i^{k+1} - u_i^k}{\tau} + u_i^k \frac{u_i^k - u_{i-1}^k}{h} = 0.$$

a) What is the order of consistency of this method?
b) Determine the limit function to which the numerical solution converges.

Exercise 2.32. Find explicitly the consistency errors of the Lax-Friedrichs and Enquist-Osher schemes for the discretization of the conservation law

$$u_t + f(u)_x = 0.$$

Examine the behaviour of each of these schemes when applied to the example of Exercise 2.31.

2.4 Finite Difference Methods for Elliptic Boundary Value Problems

2.4.1 Elliptic Boundary Value Problems

In this section we deal with boundary value problems for second-order linear differential operators L in n independent variables $x = (x_1, \ldots, x_n)$. These operators take the forms

$$Lu := -\sum_{i,j=1}^{n} a_{ij}(x)\frac{\partial^2 u}{\partial x_i \partial x_j} + \sum_{i=1}^{n} b_i \frac{\partial u}{\partial x_i} + cu \qquad (4.1)$$

and

$$Lu := -\sum_{i=1}^{n} \frac{\partial}{\partial x_i}\left(\sum_{j=1}^{n} a_{ij}(x)\frac{\partial}{\partial x_j}\right)u(x) + \sum_{i=1}^{n} b_i \frac{\partial u}{\partial x_i} + cu. \qquad (4.2)$$

The nature of these differential operators is governed by the properties of their principal parts

$$-\sum_{i,j=1}^{n} a_{ij}\frac{\partial^2}{\partial x_i \partial x_j} \quad \text{and} \quad -\sum_{i=1}^{n} \frac{\partial}{\partial x_i}\left(\sum_{j=1}^{n} a_{ij}(\cdot)\frac{\partial}{\partial x_j}\right).$$

The operator L is said to be *elliptic* in a domain Ω if for each $x \in \Omega$ there exists $\alpha_0 > 0$ such that

$$\sum_{i,j=1}^{n} a_{ij}(x)\xi_i\xi_j \geq \alpha_0 \sum_{i=1}^{n} \xi_i^2 \quad \text{for arbitrary } \xi \in \mathbb{R}^n. \qquad (4.3)$$

The operator L is *uniformly elliptic* in Ω if (4.3) holds for some $\alpha_0 > 0$ that is independent of $x \in \Omega$.

The Laplace operator Δ, which occurs often in practical problems, is clearly uniformly elliptic in the whole space $\mathbb{R}^n$. Throughout Section 2.4 we assume that condition (4.3) holds uniformly and that Ω denotes the domain in which the differential equation holds. Let $\partial\Omega = \Gamma$ denote the boundary of Ω. As a rule, boundary conditions that are appropriate for the elliptic problem will guarantee that the entire problem is well posed.

The problem of finding a function u such that for given L, f, φ, Ω one has

$$\begin{aligned} Lu &= f \quad \text{in} \quad \Omega \\ \text{and} \quad u &= \phi \quad \text{on} \quad \Gamma \end{aligned} \qquad (4.4)$$

is called an *elliptic boundary value problem*. The type of boundary condition in this problem, where u is specified on Γ, is called a boundary condition of the first kind or *Dirichlet* boundary condition.

Let the boundary be differentiable (at least piecewise) and denote by $n(x) \in \mathbb{R}^n$ the outward unit normal at each point $x \in \Gamma$, i.e., a unit vector that is orthogonal to the tangent hyperplane at x. Then, given a function ψ defined on Γ, the equation

$$\frac{\partial u}{\partial n} \equiv n \cdot \nabla u = \psi \quad \text{on} \quad \Gamma$$

defines a boundary condition of the second kind, which is also known as a *Neumann* boundary condition; finally,

$$\frac{\partial u}{\partial n} + \sigma u = \psi \quad (\sigma \neq 0) \qquad \text{on} \quad \Gamma$$

describes a boundary condition of the third kind or *Robin* boundary condition. We shall concentrate on the Dirichlet problem (4.4), but in several places the effects of other types of boundary conditions are considered.

2.4.2 The Classical Approach to Finite Difference Methods

If the main part of the operator is given in divergence form, i.e., if L is of type (4.2), then the discretization should be applied directly—see (4.8); the temptation (when the functions a_{ij} are smooth) to avoid the divergence form by using differentiation to convert L to the form (4.1) should be resisted.

As a model problem for discussing finite difference methods for elliptic boundary value problems, we consider in this section the following linear second-order boundary value problem in divergence form:

$$-\operatorname{div}(A \operatorname{grad} u) = f \text{ in } \Omega := (0,1)^n \subset \mathbb{R}^n,$$
$$u = 0 \text{ on } \Gamma := \partial \Omega. \tag{4.5}$$

Here A is some sufficiently smooth matrix function $A : \overline{\Omega} \to \mathbb{R}^{n \times n}$ with entries A_{ij}, and $f : \Omega \to \mathbb{R}$ denotes a function that is also sufficiently smooth. We assume that A is *uniformly positive definite*, i.e., that a constant $\alpha_0 > 0$ exists such that

$$z^T A(x) z \geq \alpha_0 z^T z, \qquad \forall x \in \overline{\Omega}, \quad \forall z \in \mathbb{R}^n. \tag{4.6}$$

In the particular case $A \equiv I$, the problem (4.5) is simply Poisson's equation with homogeneous Dirichlet boundary conditions. Despite the simplicity of the differential operator, boundary conditions and geometry of the domain Ω in (4.5), this model problem is adequate for an exposition of the fundamental properties of the finite difference method.

To discretize (4.5) the closure $\overline{\Omega}$ of the domain has to be approximated by a set $\overline{\Omega}_h$ that contains a finite number of grid points. For this purpose we use the grid (1.3) of Section 2.1, viz., the set of points

$$\overline{\Omega}_h := \left\{ \begin{pmatrix} x_1 \\ \vdots \\ x_n \end{pmatrix} \in \mathbb{R}^n : x_1 = i_1 h, \ \ldots, \ x_n = i_n h, \ 0 \leq i_j \leq N \text{ for all } j \right\},$$

with $h = 1/N$. In this context recall our earlier notation

$$\Omega_h := \overline{\Omega}_h \cap \Omega, \qquad \Gamma_h := \overline{\Omega}_h \cap \Gamma,$$

with the related spaces of grid functions

$$U_h := \{\, u_h : \overline{\Omega}_h \to \mathbb{R} \,\}, \quad U_h^0 := \{\, u_h \in U_h : u_h|_{\Gamma_h} = 0 \,\},$$
$$V_h := \{\, v_h : \Omega_h \to \mathbb{R} \,\}.$$

The differential operator

$$L\,u := -\operatorname{div}\,(A\operatorname{grad} u) \tag{4.7}$$

of problem (4.5) can be correctly discretized by the difference operator $L_h :$ $U_h \to V_h$ that is defined by

$$L_h\, u_h := -\sum_{i=1}^{n} D_i^- \Big(\sum_{j=1}^{n} A_{ij}\, D_j^+\, u_h\Big). \tag{4.8}$$

Thus the continuous problem (4.5) is described approximately by the finite-dimensional problem

$$u_h \in U_h^0 : \qquad L_h\, u_h = f_h \tag{4.9}$$

where the discrete right-hand side $f_h \in V_h$ is defined by $f_h(x_h) := f(x_h)$, $x_h \in \Omega_h$. The Taylor expansion (1.13) leads to the consistency estimate

$$\left| \left[\frac{\partial^2 u}{\partial x_j^2} - D_j^- D_j^+ u \right](x_h) \right| \le \frac{1}{12} \left\| \frac{\partial^4 u}{\partial x_j^4} \right\|_{C(\overline{\Omega})} h^2 \quad \forall x_h \in \Omega_h,$$

provided that u is sufficiently smooth. In the particular case of Poisson's equation, where $A = I$, with I the $n \times n$ unit matrix, one has $L_h = -\sum_{j=1}^{n} D_j^- D_j^+$. This leads to the estimate

$$|\,[(L - L_h)\, u](x_h)| \le \frac{n}{12} \|u\|_{C^4(\overline{\Omega})}\, h^2 \quad \forall x_h \in \Omega_h. \tag{4.10}$$

More precisely, the right-hand side of (4.10) has the form

$$\frac{h^2}{12} \sum_{j=1}^{n} \max_{\Omega} \left| \frac{\partial^4 u}{\partial x_j^4} \right|.$$

The general case of (4.5) can be handled similarly, but one then obtains only first-order consistency, i.e.,

$$|\,[(L - L_h)\, u](x_h)| \le c\, \|A\|\, \|u\|_{C^3(\overline{\Omega})}\, h \quad \forall x_h \in \Omega_h. \tag{4.11}$$

In the case of Poisson's equation, the notation $\Delta_h := -L_h$ is often used to indicate the *discrete Laplace operator*.

When $n = 2$ with row-wise numbering of the discrete solution $u_h = (u_{i,j})_{i,j=1}^{N-1}$, i.e.,

$$u_h = (u_{1,1}, u_{2,1}, \ldots, u_{N-1,1}, u_{1,2}, \ldots, u_{N-1,2}, \ldots, u_{N-1,N-1})^T \in \mathbb{R}^{(N-1)^2},$$

the operator $-\Delta_h$ can be written as the block tridiagonal matrix

$$-\Delta_h := \frac{1}{h^2} \begin{bmatrix} T & -I & & & 0 \\ -I & T & -I & & \\ & -I & \ddots & \ddots & -I \\ & & \ddots & \ddots & \\ 0 & & & -I & T \end{bmatrix} \tag{4.12}$$

where

$$T := \begin{bmatrix} 4 & -1 & & 0 \\ -1 & \ddots & \ddots & \\ & \ddots & \ddots & -1 \\ 0 & & -1 & 4 \end{bmatrix}.$$

Here the homogeneous boundary values have already been eliminated. In the discretization of partial differential equations by finite difference methods, the discrete problems generated are typically of very high dimension but they do possess a special structure. In particular, linear problems lead to a discrete system of linear equations whose system matrix is rather sparse. For example, the matrix (4.12) that represents the discrete operator $-\Delta_h$ has at most five nonzero entries in each row. In the case of nonlinear problems, the Jacobi matrices are sparse. The specific structure of the discrete systems that are generated by finite difference methods means one should use appropriate methods (see Chapter 8) for their efficient numerical solution.

The discrete solution u_h is defined only at each grid point. Consequently the quality of the approximation of u_h to the desired solution u can be measured naturally only on the discrete set $\overline{\Omega}_h$. It is possible to extend the discrete solution u_h (e.g., by interpolation) to a function $\Pi_h u_h$ defined on $\overline{\Omega}$ and to compare this extension $\Pi_h u_h$ with the solution u on $\overline{\Omega}$. This approach is natural in finite element methods but is less common in finite difference analyses. In our convergence analyses of the finite difference method we shall concentrate on errors at mesh points and sketch only briefly the error analysis of $\Pi_h u_h$ on $\overline{\Omega}$.

As in previous sections, to obtain an error estimate one starts from

$$L_h e_h = d_h \tag{4.13}$$

with $e_h := u_h - r_h u$, where $r_h u$ denotes the restriction of u to a grid function. The consistency error $d_h := f_h - L_h r_h u$ can, as we have seen, be estimated by

$$\|d_h\|_{\infty,h} \le c\|u\|_{C^3(\overline{\Omega})}\, h \text{ or in some cases } \|d_h\|_{\infty,h} \le c\|u\|_{C^4(\overline{\Omega})}\, h^2. \quad (4.14)$$

Next, a specific technique is needed to bound the L_2 convergence error $\|e_h\|_{0,h}$ in terms of the L_2 consistency error $\|d_h\|_{0,h}$; in the same way L_2 stability will be proved. This technique was originally introduced by the Russian school and is usually called the *method of energy inequalities*. (Recall that the Fourier technique is not easily applied to problems with variable coefficients.)

To begin, we have the following technical result:

Lemma 2.33. *For any* $w_h, v_h \in U_h^0$ *one has*

$$(D_j^- w_h, v_h)_h = -(w_h, D_j^+ v_h)_h.$$

That is, the operator $-D_j^+$ *is the adjoint of* D_j^- *with respect to the inner product* $(\cdot,\cdot)_h$.

Proof: The definition of D_j^- gives

$$(D_j^- w_h, v_h)_h = h^n \sum_{x_h \in \Omega_h} \frac{1}{h}\left(w_h(x_h) - w_h(x_h - h\,e^j)\right) v_h(x_h)$$

$$= h^{n-1} \sum_{x_h \in \Omega_h} \left(w_h(x_h) v_h(x_h) - w_h(x_h - h\,e^j)\, v_h(x_h)\right).$$

Recalling that $w_h(x_h) = v_h(x_h) = 0$ for all $x_h \in \Gamma_h$, after a rearrangement we obtain

$$(D_j^- w_h, v_h)_h = h^{n-1} \sum_{x_h \in \Omega_h} w_h(x_h)\left(v_h(x_h) - v_h(x_h + h\,e^j)\right)$$

$$= h^n \sum_{x_h \in \Omega_h} w_h(x_h) \frac{1}{h}\left(v_h(x_h) - v_h(x_h + h\,e^j)\right).$$

This completes the proof. ∎

Remark 2.34. Lemma 2.33 can be interpreted as a *discrete Green's formula*, i.e., as an analogue for grid functions of a standard Green's formula.   □

As an immediate consequence of Lemma 2.33, the operator L_h of (4.7) satisfies

$$(L_h w_h, v_h)_h = \sum_{i,j=1}^{n} (A_{ij}(x_h)D_j^+ w_h, D_i^+ v_h)_h \quad \forall w_h, v_h \in U_h^0. \quad (4.15)$$

But $A(\cdot)$ is assumed to be uniformly positive definite on $\overline{\Omega}$, so it follows that

$$(L_h w_h, w_h)_h \geq \alpha_0 \sum_{j=1}^{n} (D_j^+ w_h, D_j^+ w_h)_h = \alpha_0 \|w_h\|_{1,h}^2 \quad \forall w_h \in U_h^0. \quad (4.16)$$

We now prove a *discrete Friedrichs' inequality* that relates the norms $\|\cdot\|_{0,h}$ and $\|\cdot\|_{1,h}$.

Lemma 2.35. *There is a constant c, which depends only upon the domain Ω, such that*

$$\|w_h\|_{0,h} \leq c\|w_h\|_{1,h} \quad \forall w_h \in U_h^0. \quad (4.17)$$

Proof: Let $x_h \in \Omega_h$ and $j \in \{1, \ldots, n\}$ be arbitrary but fixed. Since $w_h = 0$ on Γ_h, there exists $\hat{l} = \hat{l}(x_h) \leq N - 1$ with

$$|w_h(x_h)| = \left| \sum_{l=0}^{\hat{l}} \left(w_h(x_h + (l+1)\,h\,e^j) - w_h(x_h + l\,h\,e^j) \right) \right|.$$

From the Cauchy-Schwarz inequality we obtain

$$|w_h(x_h)| \leq \sum_{l=0}^{\hat{l}} |w_h(x_h + (l+1)\,h\,e^j) - w_h(x_h + l\,h\,e^j)|$$

$$= h \sum_{l=0}^{\hat{l}} |D_j^+ w_h(x_h + l\,h\,e^j)|$$

$$\leq h\sqrt{N} \left(\sum_{l=0}^{\hat{l}} |D_j^+ w_h(x_h + l\,h\,e^j)|^2 \right)^{1/2}.$$

As $h = N^{-1}$ this leads to

$$\|w_h\|_{0,h}^2 = h^n \sum_{x_h \in \Omega_h} |w_h(x_h)|^2$$

$$\leq h^{n+2} N \sum_{x_h \in \Omega_h} \sum_{l=0}^{\hat{l}(x_h)} |D_j^+ w_h(x_h + l\,h\,e^j)|^2$$

$$\leq h^{n+2} N^2 \sum_{x_h \in \Omega_h} |D_j^+ w_h(x_h)|^2$$

$$\leq h^n \sum_{j=1}^{n} \sum_{x_h \in \Omega_h} |D_j^+ w_h(x_h)|^2 = \|w_h\|_{1,h}^2.$$

Thus the lemma is true, with $c = 1$ in the domain Ω currently under consideration. ∎

Lemma 2.35 implies that $\|\cdot\|_{1,h}$ defines a norm on U_h^0. This property and the positive definiteness assumption (4.16) guarantee the invertibility of the matrix that represents L_h, and the discrete system of linear equations (4.9) has a unique solution u^h for each $f_h \in V_h$.

Theorem 2.36. *Let the solution u of the model problem (4.5) satisfy the regularity assumption $u \in C^3(\overline{\Omega})$. Then there exist positive constants c_1 and c_2 such that*

$$\|u_h - r_h u\|_{0,h} \le c_1 \|u_h - r_h u\|_{1,h} \le c_2 \|u\|_{C^3(\overline{\Omega})} h.$$

In the particular case $A = I$ suppose that $u \in C^4(\overline{\Omega})$. Then there exists a positive constant $\tilde{c}_2 > 0$ such that the stronger estimate

$$\|u_h - r_h u\|_{0,h} \le c_1 \|u_h - r_h u\|_{1,h} \le \tilde{c}_2 \|u\|_{C^4(\overline{\Omega})} h^2$$

is valid.

Proof: The inequality (4.16) and the Cauchy-Schwarz inequality imply that

$$\alpha_0 \|e_h\|_{1,h}^2 \le (L_h e_h, e_h)_h = (d_h, e_h)_h \le \|d_h\|_{0,h} \|e_h\|_{0,h}. \qquad (4.18)$$

Lemma 2.35 then converts this to

$$\alpha_0 \|e_h\|_{1,h}^2 \le \|e_h\|_{1,h} \|d_h\|_{0,h}.$$

The assertions of the theorem for $\|u_h - r_h u\|_{1,h}$ now follow immediately and, by Lemma 2.35, are also true for $\|u_h - r_h u\|_{0,h}$. ■

Remark 2.37. (a second-order discretization)
It is natural to ask: why should one use the discretization

$$-\sum_{i=1}^{n} D_i^- \left(\sum_{j=1}^{n} A_{ij} D_j^+ u_h \right)$$

instead of a

$$-\sum_{i=1}^{n} D_i^+ \left(\sum_{j=1}^{n} A_{ij} D_j^- u_h \right)?$$

In fact there is no fundamental advantage to either form: in the general case both forms produce stable first-order discretizations. Second-order consistency can be achieved by the combination

$$L_h^{new} u_h := -\frac{1}{2} \left(\sum_{i=1}^{n} D_i^- \left(\sum_{j=1}^{n} A_{ij} D_j^+ u_h \right) + \sum_{i=1}^{n} D_i^+ \left(\sum_{j=1}^{n} A_{ij} D_j^- u_h \right) \right).$$

The stability of L_h^{new} can be verified as in our calculation for L_h. In planar problems, i.e., in the two-dimensional case, both first-order methods are associated with 4-point schemes for the approximation of the mixed derivative u_{xy} but the second-order method has a 7-point stencil. □

Remark 2.38. Convergence estimates in the discrete L_2 norm $\|\cdot\|_{0,h}$ can also be derived from the spectrum of the matrix L_h. If $A = I$ and $n = 2$, a complete orthogonal system of eigenfunctions of L_h in the space U_h^0 is given by the vectors $v_h^{k,l} \in U_h^0$, $k,l = 1, 2, \ldots, N - 1$, whose components are

$$[v_h^{k,l}]_{i,j} = \sin\left(\frac{k\pi i}{N}\right) \sin\left(\frac{l\pi j}{N}\right), \quad i,j,k,l = 1, 2, \ldots, N - 1. \quad (4.19)$$

The eigenvalue associated with $v_h^{k,l}$ is

$$\lambda_{k,l}^h = \frac{2}{h^2}\left(\sin^2\left(\frac{k\pi}{2N}\right) + \sin^2\left(\frac{l\pi}{2N}\right)\right), \quad k,l = 1, 2, \ldots, N - 1. \quad (4.20)$$

With the smallest eigenvalue this immediately leads to the estimate

$$\frac{4}{h^2}\sin^2\left(\frac{\pi}{2N}\right)\|v_h\|_{0,h} \leq \|L_h v_h\|_{0,h} \quad \forall v_h \in U_h^0.$$

As $N \geq 2$ we obtain

$$\pi^2 \|v_h\|_{0,h} \leq \|L_h v_h\|_{0,h} \quad \forall v_h \in U_h^0.$$

Thus in the special case $A = I$ and $n = 2$, this technique gives a precise bound, unlike the general bound of Theorem 2.36. But on more general domains or non-equidistant meshes, the eigenvalues and eigenvectors of L_h are not usually known; in these cases generalizations of Lemmas 2.33 and 2.35 lead to crude lower bounds for the eigenvalues of L_h. $\square$

Next, we derive error bounds in the maximum norm. While the inequalities (1.11) allow one to deduce easily L_∞ error bounds from the L_2 bounds $\|u_h - r_h u\|_{0,h}$, this approach leads however to a reduction in the estimated order of convergence and is not sharp enough. Alternatively, the use of the discrete Sobolev inequality [Hei87] allows to derive almost optimal L_∞ rates starting from the error estimates in the discrete H^1 norm.

Instead, discrete comparison principles, which are valid for many discretizations of elliptic boundary value problems, yield bounds for $\|u_h - r_h u\|_{\infty,h}$ that are sharp with respect to the order of convergence. To simplify the representation here, we consider only the case $A = I$, i.e., the discrete Laplace operator. Then a discrete boundary maximum principle and a discrete comparison principle are valid; the situation is completely analogous to the continuous case.

Lemma 2.39. *(discrete boundary maximum principle) Let v_h be a grid function. Then*

$$-[\Delta_h v_h](x_h) \leq 0 \quad \forall x_h \in \Omega_h \implies \max_{x_h \in \overline{\Omega}_h} v_h(x_h) \leq \max_{x_h \in \Gamma_h} v_h(x_h). \quad (4.21)$$

Proof: There exists at least one point $\tilde{x}_h \in \overline{\Omega}_h$ such that

$$v_h(\tilde{x}_h) \geq v_h(x_h) \quad \forall x_h \in \overline{\Omega}_h.$$

Assume that $-[\Delta_h v_h](x_h) \leq 0 \; \forall x_h \in \Omega_h$ but suppose that the conclusion of (4.21) does not hold. Then $\tilde{x}_h \in \Omega_h$ can be chosen such that for at least one immediately neighbouring point $x_h^* \in \overline{\Omega}_h$ one has $v_h(\tilde{x}_h) > v_h(\tilde{x}_h^*)$. But

$$-[\Delta_h v_h](x_h) = \frac{1}{h^2}\left[2n\, v_h(x_h) - \sum_{j=1}^{n} \left(v_h(x_h + h\,e^j) + v_h(x_h - h\,e^j)\right)\right]$$

then gives $-[\Delta_h v_h](\tilde{x}_h) > 0$, contradicting the hypothesis of (4.21). Thus the implication (4.21) is valid. ∎

Lemma 2.40. *(discrete comparison principle) Let v_h and w_h be grid functions. Then*

$$\left.\begin{array}{ll} -[\Delta_h v_h](x_h) \leq -[\Delta_h w_h](x_h) \; \forall x_h \in \Omega_h \\ v_h(x_h) \leq w_h(x_h) \qquad\qquad \forall x_h \in \Gamma_h \end{array}\right\} \implies v_h(x_h) \leq w_h(x_h) \; \forall x_h \in \overline{\Omega}_h.$$

Proof: By linearity of the operator L_h, the statement of the lemma is equivalent to

$$\left.\begin{array}{ll} -[\Delta_h z_h](x_h) \leq 0 \; \forall x_h \in \Omega_h \\ z_h(x_h) \leq 0 \; \forall x_h \in \Gamma_h \end{array}\right\} \implies z_h(x_h) \leq 0 \; \forall x_h \in \overline{\Omega}_h \qquad (4.22)$$

for $z_h := v_h - w_h$. The validity of (4.22) is immediate from Lemma 2.39. ∎

Theorem 2.41. *Let u be the solution of the Poisson problem with homogeneous Dirichlet conditions in the n-dimensional unit cube, i.e., problem (4.5) with $A = I$. Assume that $u \in C^4(\overline{\Omega})$. Then the maximum norm error of the finite difference solution (4.9) satisfies the bound*

$$\|u_h - r_h u\|_{\infty,h} \leq \frac{n}{96}\, \|u\|_{C^4(\overline{\Omega})}\, h^2. \qquad (4.23)$$

Proof: The argument uses an appropriate comparison function and the discrete comparison principle of Lemma 2.40. Let us define $v_h \in U_h$ by

$$v_h(x_h) := \sigma\, x_1\,(1 - x_1), \quad \text{for each } x_h = (x_1, \ldots, x_n) \in \overline{\Omega}_h,$$

where $\sigma \in \mathbb{R}$ is some parameter that is not yet specified. Then

$$-[\Delta_h v_h](x_h) = 2\sigma \quad \forall x_h \in \Omega_h. \qquad (4.24)$$

Choose $\sigma = \frac{n}{24}\|u\|_{C^4(\overline{\Omega})}\, h^2$. Then (4.10) and (4.24) imply that

$$-[\Delta_h(u_h - r_h u)](x_h) \leq -[\Delta_h v_h](x_h) \quad \forall x_h \in \Omega_h.$$

As $\sigma \geq 0$, we also have $u_h - r_h u \leq v_h$ on Γ_h. Thus we can apply Lemma 2.40, obtaining

$$[u_h - r_h u](x_h) \leq v_h(x_h) \leq \frac{1}{4}\sigma \quad \forall x_h \in \overline{\Omega}_h,$$

since $\max_{\Omega_h} v_h(x_h) = \sigma/4$. In a similar way, using $-v_h$, one can show that

$$[u_h - r_h u](x_h) \leq v_h(x_h) \geq -\frac{1}{4}\sigma \quad \forall x_h \in \overline{\Omega}_h.$$

Recalling that $\sigma > 0$, the proof is complete. ■

The proof of Theorem 2.41 depends on a discrete comparison principle and the construction of some appropriate comparison function v_h. Using these tools it was shown that $\|L_h^{-1}\|_{\infty,h} \leq c$ for a constant $c > 0$ that is independent of the mesh size h of the discretization. This technique is related to the theory of M-matrices, which we now describe.

The following definitions depend on the distribution of negative and positive entries in a matrix $A = (a_{ij})$; below, matrix inequalities are understood component-wise.

Definition 2.42. (a) The matrix A is called an L_0-matrix if $a_{ij} \leq 0$ for $i \neq j$.
(b) The matrix A is called an L-matrix if $a_{ii} > 0$ for all i and $a_{ij} \leq 0$
    for $i \neq j$.
(c) An L_0-matrix A that is invertible and satisfies $A^{-1} \geq 0$ is called an
    M-matrix.

Furthermore, a matrix A is *inverse monotone* if

$$A x \leq A y \qquad \Longrightarrow \qquad x \leq y.$$

This property is equivalent to:

$$A^{-1} \text{ exists with } A^{-1} \geq 0.$$

Thus one could define an M-matrix as an inverse monotone L_0-matrix. One can impose other conditions on L_0 or L-matrices that will imply they are M-matrices.

When M-matrices A appear in the discretization of differential equations, one generally needs to estimate $\|A^{-1}\|$, where $\|\cdot\|$ is the matrix norm induced by the discrete maximum norm. In this connection the following M-criterion can often be applied.

Theorem 2.43. *(M-criterion) Let A be an L_0-Matrix. Then A is inverse monotone if and only if there exists a vector $e > 0$ such that $Ae > 0$. Furthermore, in this case one has*

$$\|A^{-1}\| \leq \frac{\|e\|}{\min_k (Ae)_k}. \tag{4.25}$$

Proof:

($\Rightarrow$) If A is inverse monotone then one can choose $e = A^{-1}(1, 1, \cdots, 1)^T$.

($\Leftarrow$) Let $e > 0$ be some vector with $Ae > 0$, i.e.,

$$\sum_j a_{ij} e_j > 0 \quad \text{for each } i.$$

Now $a_{ij} \leq 0$ for $i \neq j$ implies that $a_{ii} > 0$. Thus the matrix $A_D := diag(a_{ii})$ is invertible. We set

$$P := A_D^{-1}(A_D - A), \quad \text{so } A = A_D(I - P).$$

By construction $P \geq 0$. Also, we have

$$(I - P)e = A_D^{-1} Ae > 0, \quad \text{hence } Pe < e.$$

Define the special norm

$$\|x\|_e := \max_i \frac{|x_i|}{e_i}$$

and let $\|\cdot\|_e$ denote the induced matrix norm. From

$$\|P\|_e = \sup_{\|x\|_e = 1} \|Px\|_e$$

and $P \geq 0$ it follows that $\|P\|_e = \|Pe\|_e$. Now

$$\|Pe\|_e = \max_i \frac{(Pe)_i}{e_i} \, ;$$

recalling that $Pe < e$ we obtain $\|P\|_e < 1$. Hence $(I - P)^{-1}$ exists with

$$(I - P)^{-1} = \sum_{j=0}^{\infty} P^j.$$

Since $A = A_D(I - P)$, we see that A^{-1} exists, and furthermore $P \geq 0$ implies $A^{-1} \geq 0$.

To prove the stability bound (4.25), suppose that $Aw = f$. Then

$$\pm w = \pm A^{-1} f \leq \|f\|_\infty A^{-1}(1, ..., 1)^T.$$

The inequality $Ae \geq \min_k (Ae)_k (1, ..., 1)^T$ yields

$$A^{-1}(1, \cdots, 1)^T \leq \frac{e}{\min_k (Ae)_k} \, .$$

Merging these inequalities, we obtain

$$\|w\|_\infty \leq \frac{\|e\|_\infty}{\min\limits_k (Ae)_k} \|f\|_\infty,$$

which implies (4.25). ∎

In many cases one can find a vector e that satisfies the hypotheses of Theorem 2.43. Such a vector is called a *majoring element* of the matrix A.

Theorem 2.43 uses M-matrices to provide an estimate for $\|A^{-1}\|$. This circle of ideas can also be invoked via strict diagonal dominance or weak diagonal dominance combined with the irreducibility of A, as we now outline.

Definition 2.44. (a) A matrix A is *strictly diagonally dominant* if

$$|a_{ii}| > \sum_{j=1, j\neq i}^{n} |a_{ij}| \quad \text{for all } i$$

and *weakly diagonally dominant* if

$$|a_{ii}| \geq \sum_{j=1, j\neq i}^{n} |a_{ij}| \quad \text{for all } i.$$

(b) A matrix A is *irreducible* if no permutation matrix P exists such that

$$PAP^T = \begin{bmatrix} B_{11} & B_{12} \\ 0 & B_{22} \end{bmatrix}.$$

(c) A matrix A has the *chain property* if for each pair of indices i, j there is a sequence of the form

$$a_{i,i_1}, a_{i_1,i_2}, \cdots, a_{i_m,j}$$

of non-vanishing entries of A.

(d) A matrix A is *irreducibly diagonally dominant* if A is weakly diagonally dominant with strict inequality in at least one row of A, and A is irreducible.

The chain property and irreducibility are equivalent. In [OR70] the following result is proved.

Theorem 2.45. *Let A be an L-matrix. If A is either strictly diagonally dominant or diagonally dominant and irreducible, then A is an M-matrix.*

If A is strictly diagonally dominant then the vector $e = (1, \cdots, 1)^T$ is a majoring element of A and by Theorem 2.43 we have

$$\|A^{-1}\| \leq \frac{1}{\min\limits_k \left(a_{kk} - \sum\limits_{j\neq k} |a_{jk}| \right)}.$$

In proving stability for suitable discretizations of differential equations, either demonstrating strict diagonal dominance or constructing a suitable majoring element will yield a bound on $\|A^{-1}\|$.

As a simple example let us sketch how Theorem 2.41 can be proved in the framework of M-matrices. The discrete operator $-\Delta_h$ corresponds to a matrix A that is an L-matrix, and this matrix is weakly diagonally dominant and irreducible. Theorem 2.45 now tells us that A is an M-matrix. The comparison function v_h constructed in the proof of Theorem 2.41 plays the role of the vector e and is a majoring element with $Ae = (2, 2, \ldots, 2)^T$ and $\|e\| = 1/4$. Now $\|A^{-1}\| \leq 1/8$ by Theorem 2.43, and from (4.10) we obtain the estimate (4.23).

The derivation of consistency results from Taylor's theorem, which is an essential component in the convergence analysis of finite differences methods for second-order differential equations, requires the solution u to have the regularity property $u \in C^{2,\alpha}(\bar{\Omega})$ for some $\alpha \in (0,1]$. One can then show that convergence of order $O(h^\alpha)$ is obtained. A natural question is whether one can also show convergence of order 2 for the standard difference method under weaker assumptions. In fact this is possible if, e.g., in the analysis of the difference method one uses discrete Sobolev spaces and arguments typical of finite element analysis. Another way of weakening the assumptions needed on u is available for certain finite difference methods using techniques from the analysis of finite volume methods.

There are computational experiments that indicate second-order convergence under weak assumptions. Consider the following problem:

$$-\Delta u = 1 \quad \text{in} \quad \Omega = (0,1)^2,$$
$$u = 0 \quad \text{on} \quad \Gamma.$$

The solution u does not lie in $C^{2,\alpha}(\bar{\Omega})$ for any positive α.

In practice, to evaluate the rate of convergence of a numerical method one often computes the *numerical convergence rate*. Suppose that at a given point $P \in \Omega_h$ the value $u(P)$ of the exact solution is known. Assume that

$$|u(P) - u_h(P)| \sim Ch^\beta,$$

for some unknown β that we wish to determine experimentally. Then for the finer step size $h/2$ one has

$$|u(P) - u_{h/2}(P)| \sim C \left(\frac{h}{2} \right)^\beta.$$

Dividing the first equation by the second then taking a logarithm yields the numerical convergence rate

$$\beta = [\ln|u(P) - u_h(P)| - \ln|u(P) - u_{h/2}(P)|]/(\ln 2). \qquad (4.26)$$

This is just a illustration to present the idea. Of course, the numerical convergence rate can be determined analogously for other ratios between two different step sizes. Furthermore, instead of a comparison of function values at some point P, one can instead compare the norms of global errors provided that the exact solution u is known. If the exact solution is not available then one could use some highly accurate numerical approximation instead of the exact solution.

Returning to the example above, let us take $P = (\frac{1}{2}, \frac{1}{2})$. The solution u of the original problem can be written in closed from using the standard method of separation of variables and Fourier series; this yields $u(\frac{1}{2}, \frac{1}{2}) = 0.0736713....$ A comparison of this value with the results computed by the standard finite difference method gives

h	$u_h(P)$	$u(P) - u_h(P)$	β
1/8	0.0727826	$8.89 \cdot 10^{-4}$	-
1/16	0.0734457	$2.26 \cdot 10^{-4}$	1.976
1/32	0.0736147373	$5.66 \cdot 10^{-5}$	1.997
1/64	0.0736571855	$1.41 \cdot 10^{-5}$	2.005

Thus the numerical convergence rate obtained for the chosen test point indicates second-order convergence despite the fact that $u \notin C^{2,\alpha}(\bar{\Omega})$ for any $\alpha > 0$.

2.4.3 Discrete Green's Function

The discretization of a linear elliptic boundary value problem leads to a finite-dimensional problem

$$[L_h u_h](x_h) = f_h(x_h) \quad \forall x_h \in \Omega_h, \qquad u_h = 0 \quad \text{on } \Gamma_h. \tag{4.27}$$

If the original problem is well posed then the matrix of the discrete linear system is invertible. Let us define the *discrete Green's function* $G_h(\cdot, \cdot) : \Omega_h \times \Omega_h \to \mathbb{R}$ by

$$L_h G_h(x_h, \xi_h) = \begin{cases} 1/\mu_h(x_h) & \text{if } x_h = \xi_h, \\ 0 & \text{otherwise}, \end{cases} \tag{4.28}$$

where the $\mu_h(x_h)$ are the weights in the scalar product $(\cdot, \cdot)_h$. Then the solution of the discrete problem (4.27) can be expressed as

$$u_h(\xi_h) = \sum_{x_h \in \Omega_h} \mu_h(x_h) \, G_h(x_h, \xi_h) \, f_h(x_h) \qquad \forall \xi_h \in \Omega_h,$$

or, in terms of the discrete scalar product, as

$$u_h(\xi_h) = \left(G_h(\cdot, \xi_h), f_h \right)_h \qquad \forall \xi_h \in \Omega_h. \tag{4.29}$$

As well as the formal analogy to the continuous case, this representation can lead to stability bounds and consequent convergence results for the discretization method. With some restriction operator r_h, the defect d_h is given by $d_h := L_h r_h u - f_h$, and for the discrete solution u_h we obtain

$$L_h (r_h u - u_h) = d_h$$

since L_h is linear. Now (4.27) and (4.29) yield

$$[r_h u - u_h](\xi_h) = \Big(G_h(\cdot, \xi_h), d_h\Big)_h \qquad \forall \xi_h \in \Omega_h.$$

Hence for any p, $q > 0$ with $\frac{1}{p} + \frac{1}{q} = 1$ we have

$$|[r_h u - u_h](\xi_h)| \leq \|G_h(\cdot, \xi_h)\|_{L_p,h} \|d_h\|_{L_q,h}, \qquad \xi_h \in \Omega_h.$$

From this inequality one immediately obtains the maximum norm error bound

$$\|r_h u - u_h\|_{\infty,h} \leq \max_{\xi_h \in \Omega_h} \|G_h(\cdot, \xi_h)\|_{L_p,h} \|d_h\|_{L_q,h}, \tag{4.30}$$

provided one can bound $\max_{\xi_h \in \Omega_h} \|G_h(\cdot, \xi_h)\|_{L_p,h}$ independently of h.

Remark 2.46. In the one-dimensional case $n = 1$ the discrete Green's function is bounded in the maximum norm, independently of h. Then (4.30) implies that the error in the maximum norm is bounded, up to a constant multiplier, by the L_1 norm of the consistency error. In particular, this shows that second-order convergence is preserved in the maximum norm even if the local order of consistency reduces to one at a finite number of points, e.g., at certain points near the boundary.   □

It can be shown that in the higher-dimensional case $n > 1$ the discrete Green's function is no longer bounded uniformly in h. To gain some insight into its behaviour we consider the discrete Green's function for the discrete Laplace operator with an equidistant grid of step size h on the unit square.

The eigenfunctions and eigenvalues are known in this simple example, so G_h can be explicitly written as

$$G_h(x, \xi) = \sum_{k,l=1,\cdots,N-1} \frac{v_h^{kl}(x) \, v_h^{kl}(\xi)}{\lambda_h^{kl}},$$

where the functions v_h^{kl} are defined in (4.19). From Remark 2.38 we deduce that

$$0 \leq G_h(x, \xi) \leq c \sum_{k,l=1,\cdots,N-1} \frac{1}{k^2 + l^2}.$$

The sum on the right-hand side can be estimated in terms of the integral of $1/r^2$ over a quarter of the disk with centre $(0,0)$ and radius $\sqrt{2}/h$. In polar coordinates this integral equals

$$\frac{\pi}{2} \int_1^{\sqrt{2}/h} \frac{dr}{r} = \frac{\pi}{2} \ln(\sqrt{2}/h).$$

Hence

$$0 \le G_h(x, \xi) \le c \ln \frac{1}{h}. \tag{4.31}$$

In several standard situations the boundedness property in one dimension also holds in one direction in two-dimensional problems, i.e., one has

$$\max_{\xi_h} \|G_h(\cdot, \xi_h)\|_{L_1, h} \le c, \tag{4.32}$$

with c independent of h. This inequality enables us to show that order of convergence in the maximum norm is at least as large as the order of consistency in the maximum norm. We sketch a proof of (4.32) for the simple model case of the discrete Laplace operator in the unit square, with an equidistant grid of step size h in both the x- and y-directions.

Let us set

$$G_h^\Sigma(x_i, \xi_k, \eta_l) = h \sum_{y_j} G_h(x_i, y_j, \xi_k, \eta_l).$$

Since G_h is non-negative, the expression G_h^Σ coincides with the discrete L_1 norm of $G_h(x_i, \cdot, \xi_k, \eta_l)$. Multiplying (4.28) by h and summing over y_j yields a one-dimensional discrete difference equation for G_h^Σ. Invoking the uniform boundedness of one-dimensional discrete Green's functions, we get

$$\max_{x_i, \xi_k, \eta_l} \|G_h(x_i, \cdot, \xi_k, \eta_l)\|_{L_1, h} \le C.$$

This anisotropic bound implies (4.32).

In many papers V.B. Andreev has studied the behaviour of discrete fundamental solutions, which are closely related to discrete Green's functions.

2.4.4 Difference Stencils and Discretization in General Domains

Each difference operator appearing in a finite difference method is often characterized by its *difference stencil*, which is also called a *difference star*. For any grid point, this describes the neighbouring nodes that are included in the discrete operator and the weights that are applied to them there. For example, the stencil in Figure 2.9 describes the standard five-point discretization of the Laplace operator on an equidistant grid. The more general stencil in Figure 2.10 represents the local difference operator

$$h^{-2} \sum_{i,j=-1}^{1} c_{i,j} u_h(x + ih, y + jh).$$

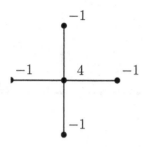

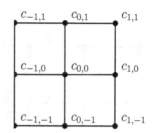

Figure 2.9 Five-point stencil   **Figure 2.10** Nine-point stencil

In general the order of consistency of a difference operator can be increased by including extra grid points in the discretization. From the algorithmic point of view one is naturally in favour of discretizations that include only neighbouring grid points. Difference stencils for such discretizations are said to be *compact*, i.e., for compact stencils that discretize the Laplacian one has $c_{\alpha,\beta} \neq 0$ only for $-1 \leq \alpha, \beta \leq 1$ (see Figure 2.11).

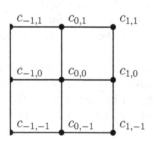

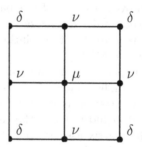

Figure 2.11 Compact nine-point stencil   Consistent nine-point stencil

From a Taylor expansion of the solution u it follows that second-order nine-point discretizations of the Laplace operator whose weights are as in Figure 2.11 must satisfy

$$\mu + 4\nu + 4\delta = 0, \qquad \nu + 2\delta = -1.$$

The choice $\delta = 0$, $\nu = -1$ produces the standard five-point formula, while the choice $\nu = \delta = -1/3$ results in the particular nine-point stencil shown in Figure 2.12. It is straightforward to verify that consistency of order 3 cannot be achieved for any choice of the parameters ν and δ. Hence, third-order discretization methods that use compact nine-point schemes could, if at all, be obtained only if the right hand side f_h includes information from more discretization points.

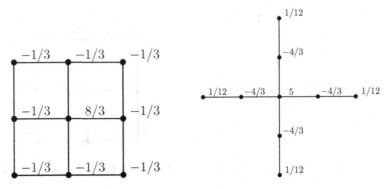

Figure 2.12 Particular nine-point stencil   Non-compact nine-point stencil

If one takes $\nu = -\frac{2}{3}, \delta = -\frac{1}{6}$ and defines the discrete right-hand side f_h by

$$f_h := \frac{1}{12}[f(x - h, y) + f(x + h, y) + f(x, y - h) + f(x, y + h) + 8f(x, y)],$$

then a fourth-order method for Poisson's equation is obtained. This result can be proved using Taylor expansions analogously to the argument for the usual second-order discretization by the standard five-point formula. It does however require that the solution u satisfies the rather restrictive regularity condition $u \in C^6(\bar{\Omega})$.

If non-compact schemes are considered, then the construction of difference stencils of higher order of consistency is possible. One example of a fourth-order discretization is given by the stencil shown in Figure 2.12. But since the stencil includes points that are not immediate neighbours of the central point, the matrix associated with the discrete problem has a wider band of non-zero entries than a matrix associated with a compact scheme. There are also difficulties applying the discretization at grid points close to the boundary Γ_h and as a consequence such formulas are rarely used in practice. Instead, to get higher-order convergence, second-order difference methods are combined with extrapolation or defect correction techniques (see [MS83]).

Let consider now the boundary value problem

$$-\triangle u = f \quad \text{in} \quad \Omega, \tag{4.33}$$
$$u = \varphi \quad \text{on} \quad \Gamma$$

on some arbitrary connected domain Ω with a regular boundary Γ. Similarly to (1.4), we cover the domain with an axi-parallel grid on $\mathbb{R}^2$ and define the set of inner grid points $(x, y) \in \Omega_h \subset \Omega$. Then at each inner grid point (x, y) for which all four neighbours $(x+h, y)$, $(x-h, y)$, $(x, y+h)$, $(x, y-h)$ and the line segments joining them to (x, y) lie in $\overline{\Omega}$, the standard five-point discretization can be used. In other cases the stencil has to be modified. If for example (x, y)

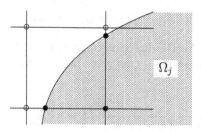

Figure 2.13   Curved boundary

is an inner grid point but $(x - h, y) \notin \Omega$, then there exists some $s \in (0,1]$ such that $(x - sh, y) \in \Gamma$. In this way each inner grid point may generate "left", "right", "upper" and "lower" neighbouring boundary points that do not lie on the original axi-parallel grid but are nevertheless in Γ_h; two such points are shown on Figure 2.13.

An inner grid point is said to be close to the boundary if at least one of its neighbours belongs to Γ_h. We denote the set of all such points by Ω_h^*, and let $\Omega_h' = \Omega_h \setminus \Omega_h^*$ be the set of remaining points (these are inner grid points "far from the boundary"). Note that in case of non-convex domains an inner grid point may be close to the boundary even when $(x \pm h, y)$, $(x, y \pm h)$ all lie in Ω; see Figure 2.14. At grid points that are far from the boundary, the

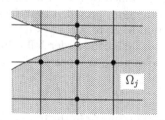

Figure 2.14   Boundary of a non-convex domain

usual five-point discretization of the Laplace operator can be applied. At grid points close to the boundary, the discretization

$$\frac{2}{h_k + h_{k+1}} \left(\frac{u_k - u_{k+1}}{h_{k+1}} + \frac{u_k - u_{k-1}}{h_k} \right)$$

of the second-order derivative $-u''$ on non-equidistant grids will now be applied to each component of $-\Delta u$ in the two-dimensional case. This discretization is however only first-order consistent in the maximum norm, because—unlike the case of equidistant grids—the $O(h_{k+1} - h_k)$ term in the Taylor expansion does not vanish. Denote the four grid points neighbouring (x, y) by

$(x-s_w h, y)$, $(x+s_e h, y)$, $(x, y-s_s h)$, $(x, y+s_n h)$ with $0 < s \leq 1$ in each case,

where w, e, s, n stand for the directions west, east, south and north, according to the direction in the stencil. This gives the following discretization of the Laplace operator:

$$
(\triangle_h u)(x, y) := \frac{2}{h^2} \left[\frac{1}{s_e(s_e + s_w)} u(x + s_e h, y) \right.
$$

$$
+ \frac{1}{s_w(s_e + s_w)} u(x - s_w h, y) + \frac{1}{s_n(s_n + s_s)} u(x, y + s_n h) \qquad (4.34)
$$

$$
+ \frac{1}{s_s(s_n + s_s)} u(x, y - s_n h) - \left. \left(\frac{1}{s_w s_e} + \frac{1}{s_n s_s} \right) u(x, y) \right].
$$

At inner grid points the difference formula (4.34) coincides with the standard five-point stencil. Discretizing (4.33) through (4.34), we obtain

$$
-\triangle_h u_h = f_h \quad \text{in} \quad \Omega_h, \qquad (4.35)
$$
$$
u_h = \varphi_h \quad \text{on} \quad \Gamma_h.
$$

We have already seen that the order of consistency in the maximum norm is only 1 owing to the reduced order at mesh points close to the border. The stability result used in the proof of Theorem 2.41 carries over directly to the current problem. The discretization clearly generates an L_0-matrix and a majoring element is obtained by a restriction of the quadratic function $(x - x_0)(x_0 + d - x)/2$ to the grid, provided that Ω is contained in the strip $(x_0, x_0 + d)$. The M-criterion leads to $\|\triangle_h^{-1}\| \leq d^2/8$.

To prove convergence of the second order following [Sam01], we split the consistency error at grid points into two parts,

$$
c_h = c_h^1 + c_h^2,
$$

here c_h^1 is equal to the consistency error in the interior mesh points close to the border and zero otherwise. Analogously we decompose the error: $e_h = e_h^1 + e_h^2$ with

$$
-\triangle_h e_h^1 = c_h^1, \quad -\triangle_h e_h^2 = c_h^2.
$$

It is easy to estimate e_h^2 because the consistency component c_h^2 is already of the second order:

$$
\|e_h^2\|_\infty \leq \frac{d^2}{8} \|c_h^2\|_\infty.
$$

To estimate e_h^1 we apply a special barrier function (c_h^1 is only of the first order with respect to the maximum norm). Consider, for instance, an interior point close to the border with $s_r = s_o = s_u = 1$ and $s_l < 1$. Then, the barrier function

$$
v_h = \begin{cases} \alpha h^2 & \text{in } \Omega_h \\ 0 & \text{on } \Gamma_h \end{cases}
$$

yields

$$-\Delta_h v_h \geq \alpha(\frac{2}{s_l} - \frac{2}{1+s_l}) \geq \alpha$$

and the discrete comparison principle with an appropriate chosen constant α leads us to

$$\|e_h^1\|_\infty \leq h^2 \|c_h^1\|_\infty.$$

Summarizing, it follows

$$|u(x,y) - u_h(x,y)| \leq h^2 \left(\frac{1}{48} d^2 \|u\|_{C^{3,1}(\bar{\Omega})} + \frac{2}{3} \|u\|_{C^{2,1}(\bar{\Omega})} h \right), \qquad (4.36)$$

i.e., second-order convergence is obtained for our discretization of problem (4.35) in a rather arbitrary domain.

Remark 2.47. In the method we have just described for handling grid points that are close to the boundary, the differential equation is discretized by a modification of the standard difference quotient formula for equidistant grids. An alternative approach at such grid points is to modify the discrete equation by *linear interpolation* of the values at two neighbouring points. This technique also yields a second-order method; see [Hac03a]. □

Exercise 2.48. Consider the boundary value problem

$$-\Delta u(x,y) = 1 \quad \text{in} \quad \Omega = (0,1) \times (0,1),$$
$$u(x,y) = 0 \quad \text{on} \quad \Gamma = \partial\Omega.$$

a) Apply the method of finite differences and determine an approximation for $u(1/2, 1/2)$.
b) Use separation of variables and Fourier series to evaluate $u(1/2, 1/2)$ correct to 6 digits and compare this with the result obtained in part a).

Exercise 2.49. Consider the Dirichlet problem

$$-\Delta u = 0 \quad \text{in} \quad (-1,-1)^2 \setminus (-1,0)^2$$

with the boundary conditions

$$u(x,y) = \begin{cases} 1 - 6x^2 + x^4 & \text{on } y = 1 \text{ and } y = -1, \\ 1 - 6y^2 + y^4 & \text{on } x = 1 \text{ and } x = -1, \\ x^4 & \text{on } y = 0 \text{ with } -1 \leq x \leq 0, \\ y^4 & \text{on } x = 0 \text{ with } -1 \leq y \leq 0. \end{cases}$$

a) Show that the exact solution is symmetric about the straight line $y = x$.
b) Use the finite difference method to find an approximate solution of the problem and compare your computed solution with the exact solution.

Exercise 2.50. Set $\Omega = \{(x, y) : 1 < |x| + |y| < 2.5\}$.
Solve the boundary value problem

$$-\Delta u = 0 \quad \text{in} \quad \Omega$$

with

$$u = \begin{cases} 0 & \text{for} \quad |x| + |y| = 2.5, \\ 1 & \text{for} \quad |x| + |y| = 1, \end{cases}$$

using the method of finite differences with a rectangular grid and step size $h = 1/2$ (make use of the symmetries that occur).

Exercise 2.51. Consider the Dirichlet problem

$$-\Delta u = f \quad \text{in} \quad (0,1)^2,$$
$$u|_\Gamma = 0,$$

and its discretization by the difference scheme

$$\frac{1}{6h^2} \begin{bmatrix} -1 & -4 & -1 \\ -4 & 20 & -4 \\ -1 & -4 & -1 \end{bmatrix} u_h = \frac{1}{12} \begin{bmatrix} 0 & 1 & 0 \\ 1 & 8 & 1 \\ 0 & 1 & 0 \end{bmatrix} f_h$$

on a rectangular grid with step size $h = 1/(N+1)$.
a) Prove that

$$f_{h,1} \leq f_{h,2} \quad \text{and} \quad u_{h,1}|_\Gamma \leq u_{h,2}|_\Gamma \quad \text{together imply that } u_{h,1} \leq u_{h,2}.$$

b) Prove that the scheme is stable in L_∞, i.e., verify that

$$\|u_h\|_\infty \leq C\|f_h\|_\infty,$$

and determine the constant C in this estimate as accurately as possible.
c) Solve the discrete problem that is generated for $N = 1, 3, 5$ in the two cases

$$\alpha) \ f(x, y) = \sin \pi x \sin \pi y,$$
$$\beta) \ f(x, y) = 2(x + y - x^2 - y^2).$$

Hint: it is helpful to exploit the symmetry of the solution.

Exercise 2.52. Show that there does not exist any compact nine-point scheme that approximates the Laplace operator with third-order consistency.

2.4.5 Mixed Derivatives, Fourth-Order Operators and Boundary Conditions of the Second and Third Kinds

Mixed Derivatives

In Section 2.4.2 second-order elliptic equations in divergence form were discussed and in particular the L_2 stability of our discretization was proved.

Next, we investigate the maximum-norm stability properties possessed by discretizations when mixed derivatives are present—up to now we have derived stability results in the maximum norm only for Poisson's equation.

In the two-dimensional case with the simple domain $\Omega := (0,1)^2$, let us consider the linear boundary value problem

$$-\left(a_{11}\frac{\partial^2 u}{\partial x^2} + 2a_{12}\frac{\partial^2 u}{\partial x \partial y} + a_{22}\frac{\partial^2 u}{\partial y^2}\right) + b_1\frac{\partial u}{\partial x} + b_2\frac{\partial u}{\partial y} + cu = f \text{ in } \Omega,$$

$$u = \varphi \text{ on } \partial\Omega. \tag{4.37}$$

All the data are assumed to be continuous (at least). Furthermore, we assume that the main part of the differential operator is elliptic as defined in (4.3) and that $c \geq 0$. For the sake of simplicity in the present section, we work mainly on a rectangular equidistant grid with step size h.

To discretize u_{xx}, u_{yy}, u_x and u_y it is quite common to use the standard second-order central difference quotients that we have met already. But how does one approximate the mixed derivative u_{xy} accurately and stably? An obvious idea would be to first approximate u_x by $(u(x+h,y) - u(x-h,y))/(2h)$, then take a central difference quotient in the y-direction to yield an approximation of u_{xy}. This generates the difference formula

$$\frac{\partial^2 u}{\partial x \partial y} \approx \frac{1}{4h^2}[u(x+h,y+h) - u(x-h,y+h)$$

$$- u(x+h,y-h) + u(x-h,y-h)], \tag{4.38}$$

which corresponds to the difference stencil of Figure 2.15. All the points that

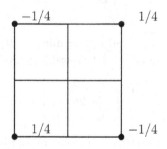

Figure 2.15   A particular four-point stencil for the mixed derivative

are used in this discretization of u_{xy} do not appear in the discretizations of the other derivatives of the differential operator; hence the sign pattern of the weights involved in (4.38) imply that the complete discretization of the differential operator produces a discrete problem whose system matrix cannot be an M-matrix. But it is advantageous to generate M-matrices for the following pair of reasons: the stability analysis is simplified and, if the given second-order elliptic differential operator satisfies a comparison theorem or maximum principle, then the discrete problem also has this property. The latter reason is the main motivation for our desire to have M-matrices in our difference schemes.

It is easy to show by a Taylor expansion that any consistent compact nine-point discretization of u_{xy} must have the stencil

$$\frac{1}{4} \begin{bmatrix} -1-\alpha-\beta+\gamma & 2(\alpha-\gamma) & 1-\alpha+\beta+\gamma \\ 2(\alpha+\beta) & -4\alpha & 2(\alpha-\beta) \\ 1-\alpha-\beta-\gamma & 2(\alpha+\gamma) & -1-\alpha+\beta-\gamma \end{bmatrix}$$

with arbitrary parameters α, β, $\gamma \in \mathbb{R}$. The choice $\alpha = \beta = \gamma = 0$ generates the scheme (4.38). One can verify that no choice of the parameters α, β, γ yields a discretization where all off-diagonal entries have the same sign.

Nevertheless, among the considered stencils with $\beta = \gamma = 0$ are discretizations where all four coefficients c_{11}, $c_{1,-1}$, $c_{-1,-1}$, $c_{-1,1}$ in Figure 2.11 have the same sign. In particular the choices $\alpha = \pm 1$ then give the difference stencils

$$\frac{1}{2} \begin{bmatrix} -1 & 1 & 0 \\ 1 & -2 & 1 \\ 0 & 1 & -1 \end{bmatrix} \quad \text{and} \quad \frac{1}{2} \begin{bmatrix} 0 & -1 & 1 \\ -1 & 2 & -1 \\ 1 & -1 & 0 \end{bmatrix}.$$

The first stencil should be used in the case $-a_{12} > 0$, while the second is suitable when $-a_{12} < 0$. Combining these with the standard discretizations of u_{xx} and u_{yy} yields the following complete discretization of the main part of the differential operator:

$$\frac{1}{h^2} \begin{bmatrix} a_{12}^- & -(a_{22}-|a_{12}|) & -a_{12}^+ \\ -(a_{11}-|a_{12}|) & 2(a_{11}+a_{22}-|a_{12}|) & -(a_{11}-|a_{12}|) \\ -a_{12}^+ & -(a_{22}-|a_{12}|) & a_{12}^- \end{bmatrix}, \qquad (4.39)$$

where $a_{12}^+ := \max\{a_{12},0\} \geq 0$ and $a_{12}^- := \min\{a_{12},0\} \leq 0$. To discretize the lower-order terms $b_1 u_x + b_2 u_y + c$, let us use the approximation

$$\frac{1}{2h} \begin{bmatrix} 0 & b_2 & 0 \\ -b_1 & 2hc & b_1 \\ 0 & -b_2 & 0 \end{bmatrix}. \qquad (4.40)$$

Then the finite difference discretization of (4.37) that is generated is consistent of order 2. Under the assumption that

$$a_{ii} > |a_{12}| + \frac{h}{2}|b_i| \qquad (i = 1, 2)$$

the discretization gives an M-matrix. Consequently a maximum norm convergence analysis can be carried out along the same lines as before.

Theorem 2.53. *If $a_{ii} > |a_{12}| + \frac{h}{2}|b_i|$ for $i = 1, 2$ and $u \in C^{3,1}(\bar{\Omega})$, then for the difference method (4.39),(4.40) we have the error estimate*

$$|u(x,y) - u_h(x,y)| \leq Ch^2 \qquad \text{for all} \quad (x,y) \in \bar{\Omega}_h.$$

Remark 2.54. If $|b|$ is not very large then the condition $a_{ii} > |a_{12}| + \frac{h}{2}|b_i|$ does not usually place a serious restriction on the choice of mesh size h because ellipticity implies already that $a_{11}a_{22} > a_{12}^2$. If however $|b|$ is large—the case of *dominant convection*—then this condition forces a very fine mesh size, which results in a large number of discrete unknowns and cannot be used in practical computations. To avoid this restriction in the convection-dominated case, special discretizations are necessary. □

Remark 2.55. The discussion of the difficulties that arise in the discretization of the mixed derivative u_{xy} underlines strongly the recommendation that differential operators that are given in divergence form, i.e., operators of the type $L = -\mathrm{div}(A\,\mathrm{grad})$, should always be treated directly by (4.8) and not by differentiation and conversion to the form (4.37). Such differentiation should be avoided even for smooth matrix functions A. With a direct discretization of the divergence form of the operator, the positive definiteness of A induces positive definiteness in the corresponding discretized term. □

Fourth-Order Operators

A simple but important example of a fourth-order elliptic boundary value problem is given by

$$\Delta\Delta u = f \text{ in }\quad \Omega,$$
$$u = 0, \quad \frac{\partial u}{\partial n} = 0 \text{ on }\quad \Gamma. \tag{4.41}$$

Problems of this type occur in the modelling of plate bending. The boundary conditions in (4.41) represent the case of a clamped plate because of the prescribed normal derivatives. In the case of a plate that rests on some support but is not clamped, one uses the boundary conditions

$$u = 0, \quad \frac{\partial^2 u}{\partial n^2} = 0 \text{ on }\quad \Gamma, \tag{4.42}$$

i.e., the second-order derivatives in the direction of the normal at the boundary will vanish. Taking into account the fact that in this case the second-order derivatives will also vanish in the tangential direction at the boundary, the boundary conditions can also be written in the form

$$u = 0, \quad \Delta u = 0 \text{ on }\quad \Gamma. \tag{4.43}$$

With these boundary conditions and with the differential operator from (4.41) some boundary maximum principle holds.

The differential operator

$$L := \Delta\Delta$$

of problem (4.41) is called the *biharmonic operator*. In Euclidean coordinates in two dimensions it has the form

$$Lu = \frac{\partial^4 u}{\partial x^4} + 2\frac{\partial^4 u}{\partial x^2 \partial y^2} + \frac{\partial^4 u}{\partial y^4}.$$

Consider an equidistant rectangular grid with mesh size h. If the standard discretization

$$[-\Delta u_h]_{i,j} = \frac{1}{h^2}(4\,u_{i,j} - u_{i-1,j} - u_{i+1,j} - u_{i,j-1} - u_{i,j+1})$$

is applied recursively twice to discretize $L = -\Delta(-\Delta)$, then this yields

$$[\Delta\Delta u_h]_{i,j} = \frac{1}{h^4}\Big(20u_{i,j} - 8(u_{i-1,j} + u_{i+1,j} + u_{i,j-1} + u_{i,j+1})$$

$$+ 2(u_{i-1,j-1} + u_{i-1,j+1} + u_{i+1,j-1} + u_{i+1,j+1})$$

$$+ u_{i-2,j} + u_{i+2,j} + u_{i,j-2} + u_{i,j+2}\Big).$$

This difference stencil is displayed in Figure 2.16.

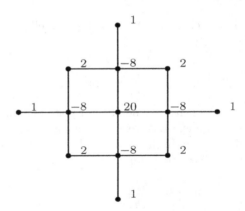

Figure 2.16   Difference stencil for the biharmonic operator

The fourth-order boundary value problem

$$\Delta\Delta u = f \text{ in } \quad \Omega,$$
$$u = 0, \quad \Delta u = 0 \text{ on } \quad \Gamma, \tag{4.44}$$

can be written as a system of two second-order differential equations as follows:

$$-\Delta u = v \text{ in } \quad \Omega,$$
$$-\Delta v = f \text{ in } \quad \Omega, \tag{4.45}$$
$$u = 0, \quad v = 0 \text{ on } \quad \Gamma.$$

If the solution u is sufficiently smooth, then the consistency analysis of the standard discretization of this system is the same as the analysis of the discretization of Poisson's equation. Furthermore, inverse monotonicity guarantees the stability of the standard discretization in the maximum norm and leads directly to a proof that the method is second-order convergent.

Boundary Conditions of the Second and Third Kinds

Consider the boundary condition

$$\nu \cdot \nabla u + \mu u = \varphi \qquad \text{on} \quad \Gamma. \tag{4.46}$$

Here ν is some unit vector that is not tangential to the boundary Γ. In the special case that ν is a normal vector, then for $\mu = 0$ condition (4.46) is simply the classical boundary condition of the second kind, and if $\mu \neq 0$ then it is a boundary condition of the third kind. Boundary conditions of the second and third kinds are often respectively called Neumann and Robin boundary conditions.

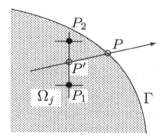

Figure 2.17   Particular boundary condition

Setting $\nu = (\nu_1, \nu_2)$, the straight line

$$y - y^* = \frac{\nu_2}{\nu_1}(x - x^*) \qquad \text{i.e.,} \qquad \nu_2(x - x^*) - \nu_1(y - y^*) = 0,$$

passes through the point $P = (x^*, y^*)$ and is parallel to ν. Let this line have its first intersection with our grid lines at the point P' (see Figure 2.17). Then a first-order approximation of the directional derivative $\nu \cdot \nabla u$ that is required in the boundary condition is given by

$$\frac{\partial u}{\partial \nu} \approx \frac{u(P) - u(P')}{|PP'|}. \tag{4.47}$$

If P' happens to be a grid point then (4.47) can be used in (4.46) and leads directly to a discretization of the boundary condition at P. If P' is not a grid point, let P_1 and P_2 be neighbouring grid points that lie on the grid line through P'. Then the value $u(P')$ in (4.47) can be approximated by linear interpolation through $u(P_1)$ and $u(P_2)$; this will again give a first-order discretization of (4.46) at P.

Exercise 2.56. Show that any compact nine-point stencil that consistently discretizes $\partial^2/(\partial x \partial y)$ must have the form

$$\frac{1}{4}\begin{bmatrix} -1-\alpha-\beta+\gamma & 2(\alpha-\gamma) & 1-\alpha+\beta+\gamma \\ 2(\alpha+\beta) & -4\alpha & 2(\alpha-\beta) \\ 1-\alpha-\beta-\gamma & 2(\alpha+\gamma) & -1-\alpha+\beta-\gamma \end{bmatrix}.$$

Discuss the possibility of selecting the parameters α, β, γ in such a way that the discretization is consistent and all off-diagonal entries have the same sign.

Exercise 2.57. The domain

$$\Omega = \left\{ \begin{pmatrix} x \\ y \end{pmatrix} \in \mathbb{R}^2_+ : 1 < x^2 + y^2 < 4 \right\}$$

is a sector of an annulus. Consider the boundary value problem

$$\Delta u = 0 \quad \text{in} \quad \Omega$$

with boundary conditions

$u(x,0) = x$ for $x \in [1,2]$, $\qquad u(0,y) = y$ for $y \in [1,2]$,
$u(x,y) = 1$ for $x, y > 0$, $x^2 + y^2 = 1$, $u(x,y) = 2$ for $x, y > 0$, $x^2 + y^2 = 4$.

a) Express the Laplace operator in polar coordinates.
b) Write down the difference approximation of this operator that is obtained by the application of the standard discretization on a grid adapted to polar coordinates.
c) Solve the given boundary value problem numerically on the grid that consists of all points that lie on the intersection of the circles

$$x^2 + y^2 = 1, \quad x^2 + y^2 = 2.25, \quad x^2 + y^2 = 4$$

with the straight lines

$$y = 0, \quad y = \frac{1}{3}\sqrt{3}x, \quad y = \sqrt{3}x, \quad x = 0.$$

d) Determine an approximate solution for the potential equation $\Delta u = 0$ using the technique from c), but with the boundary conditions

$u(x,0) = \ln x$ for $x \in [1,2]$, $\qquad u(0,y) = \ln y$ for $y \in [1,2]$,
$u(x,y) = 0$ for $x, y > 0$, $x^2 + y^2 = 1$,
$u(x,y) = \ln 2$ for $x, y > 0$, $x^2 + y^2 = 4$.

The exact solution is $u(x,y) = \frac{1}{2}\ln(x^2 + y^2)$.

2.4.6 Local Grid Refinements

The local consistency error of the discretization of a derivative depends strongly upon the behaviour of the higher-order derivatives that appear in the error terms. Thus it is reasonable to use this information to influence the choice of discretization grid. In certain situations—for example the convection-diffusion problems considered in Chapter 6—one has a priori some knowledge of the principal behaviour of the solution. This enables us to construct a grid suited to the solution; in the case of convection-diffusion, where the solution typically contains layers (narrow regions where the solution changes rapidly), one can avoid excessively large errors using layer-adapted grids, which are discussed in detail in Chapter 6. Alternatively, instead of using a priori information to construct the grid, an idea that is frequently used is to refine the grid adaptively using information obtained in the course of the numerical computation. This approach requires reliable local error indicators or accurate local error estimates. Furthermore, grid refinement techniques must be developed to make proper use of the local error information in constructing a high-quality grid suited to the problem in hand. In Chapter 4 we shall discuss in detail some error indicators and grid refinement strategies in the context of finite element methods.

In what follows here we describe a strategy for grid refinement that uses *hanging knots* in $\mathbb{R}^2$ for the discretization of the operator $-\Delta$. This idea could also be invoked in adaptive finite element discretizations. In contrast to finite element methods, where for example global continuity may be required for the piecewise-defined computed solution, the finite difference method has the advantage that simple difference stencils can be widely used for the discretization.

Consider a plane rectangular grid. Suppose that by means of an error indicator or error estimate we have marked a cell in the grid that must be refined. This is done by adding five new grid points: the mid-points of the cell's four sides and its centre. At the central point a regular stencil can be applied, but this is not the case at the mid-points of the sides—these knots are called *hanging knots*. Here the original grid lines end; by extending these lines as far as the centroids of the neighbouring cells one generates further knots called *slave knots*. Using these slave knots one can apply a standard five-point stencil at the four mid-points that were added to the grid; the function value required at each slave knot is determined by an interpolation of the function values at the vertices of the cell whose centroid is the slave knot. If interpolation formulas with positive weights are used, then the system matrix of the new discrete problem is an M-matrix. Figure 2.18 illustrates this way of refining a grid with hanging knots and slave knots, and also shows how a similar refinement technique can be centred on an existing grid point.

Exercise 2.58. Let the domain Ω and the set of grid points Ω_h be defined by Figure 2.19.

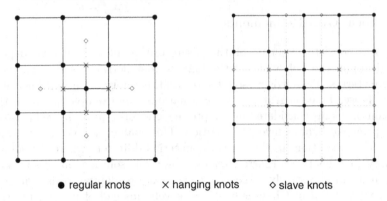

● regular knots   × hanging knots   ◇ slave knots

Figure 2.18 Refinement of a cell   Refinement near a grid point

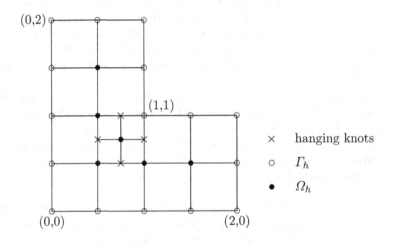

Figure 2.19   Refinement of a rectangle

Use this grid to solve approximately the boundary value problem

$$-\Delta u = 1 \quad \text{in} \quad \Omega,$$
$$u = 0 \quad \text{on} \quad \Gamma = \partial\Omega$$

by the standard finite difference discretization. Here the centroids of neighbouring squares are to be used as slave knots, with the function values at these points determined by bilinear interpolation. Use the symmetries that appear in the problem to reduce the size of the discrete problem generated.

2.5 Finite Volume Methods as Finite Difference Schemes

Finite volume methods form a relatively general class of discretizations for certain types of partial differential equations. These methods start from balance equations over local control volumes, e.g., the conservation of mass in diffusion problems. When these conservation equations are integrated by parts over each control volume, certain terms yield integrals over the boundary of the control volume. For example, mass conservation can be written as a combination of source terms inside the control volume and fluxes across its boundary. Of course the fluxes between neighbouring control volumes are coupled. If this natural coupling of boundary fluxes is included in the discretization, then the local conservation laws satisfied by the continuous problem are guaranteed to hold locally also for the discrete problem. This is an important aspect of finite volume methods that makes them suitable for the numerical treatment of, e.g., problems in fluid dynamics. Another valuable property is that when finite volume methods are applied to elliptic problems that satisfy a boundary maximum principle, they yield discretizations that satisfy a discrete boundary maximum principle (see Lemma 2.60) even on fairly general grids.

Finite volume methods (FVMs) were proposed originally as a means of generating finite difference methods on general grids (see [Hei87]). Today, however, while FVMs can be interpreted as finite difference schemes, their convergence analysis is usually facilitated by the construction of a related finite element method and a study of its convergence properties [Bey98]. It seems that no FVM convergence analysis exists that is completely independent of the finite difference and finite element methods.

As we have already mentioned, FVMs preserve local conservation properties; consequently they play a major role in the discretization of conservation laws. For this important application we refer to [Krö97].

The fundamental idea of the finite volume method can be implemented in various ways in the construction of the control volumes, in the localization of the degrees of freedom (i.e., the free parameters of the method), and in the discretization of the fluxes through the boundaries of the control volumes. There are basically two classes of FVM. First, in *cell-centred* methods each control volume that surrounds a grid point has no vertices of the original triangulation lying on its boundary. The second approach, *vertex-centred* methods, uses vertices of the underlying triangulation as vertices of control volumes.

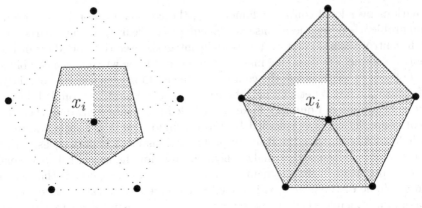

Figure 2.20   cell-centred   vertex-centred

In the present section we discuss a particular cell-centred method that uses Voronoi boxes. The convergence of this method will be analysed in the discrete maximum norm. In Chapter 4 a connection between FVM and finite element methods will be described and the derivation of error estimates in other norms will be sketched.

As a model problem here we consider the elliptic boundary value problem

$$-\Delta u = f \quad \text{in} \quad \Omega, \qquad u = g \quad \text{on} \quad \Gamma := \partial\Omega, \qquad (5.1)$$

where the domain $\Omega \subset \mathbb{R}^n$ is assumed to be a bounded convex polyhedron. Let

$$x_i \in \Omega, \; i = 1, \ldots, N \qquad \text{and} \qquad x_i \in \Gamma, \; i = N+1, \ldots, \overline{N}$$

be the interior discretization points and boundary discretization points, respectively. Define the corresponding index sets

$$J := \{1, \ldots, N\}, \qquad \overline{J} := J \cup \{N+1, \ldots, \overline{N}\}.$$

For each interior grid point x_i define the related subdomain Ω_i (see Figure 2.21) by

$$\Omega_i := \bigcap_{j \in J_i} B_{i,j} \quad \text{with} \quad J_i := \overline{J} \setminus \{i\} \quad \text{and}$$
$$B_{i,j} := \{\, x \in \Omega : |x - x_i| < |x - x_j| \,\}. \qquad (5.2)$$

Then Ω_i is called a *Voronoi box*. By its construction and the properties of Ω, each Ω_i is a convex polyhedron. We shall assume that $\mu_{n-1}(\overline{\Omega}_i \cap \Gamma) = 0$ for all $i \in J$, where μ_{n-1} denotes Lebesgue measure in $\mathbb{R}^{n-1}$. This condition will be satisfied if sufficiently many boundary grid points $x_j \in \Gamma$ are used and if these points are distributed properly over Γ. Let N_i denote the set of indices of essential neighbours of x_i, i.e.,

$$N_i := \{\, j \in J_i : \mu_{n-1}(\overline{\Omega}_j \cap \overline{\Omega}_i) > 0 \,\}.$$

Set

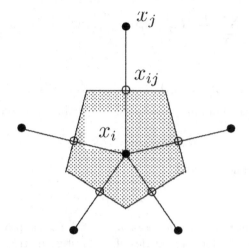

Figure 2.21   Voronoi box

$$\Gamma_{i,j} := \overline{\Omega}_i \cap \overline{\Omega}_j \text{ for } i \in J \text{ and } j \in N_i, \quad \text{and} \quad \Gamma_i = \bigcup_{j \in N_i} \Gamma_{ij} \text{ for } i \in J.$$

Under suitable regularity assumptions the model problem (5.1) has a unique classical solution u, which of course satisfies

$$-\int_{\Omega_i} \Delta u \, d\Omega_i = \int_{\Omega_i} f \, d\Omega_i, \qquad i \in J.$$

Recalling the above partition of the boundary of the subdomain Ω_i, integration by parts (Gauss's Theorem) yields

$$-\sum_{j \in N_i} \int_{\Gamma_{i,j}} \frac{\partial u}{\partial n_{i,j}} \, d\Gamma_{i,j} = \int_{\Omega_i} f \, d\Omega_i, \qquad i \in J, \tag{5.3}$$

where the terms $\dfrac{\partial u}{\partial n_{i,j}}$ are fluxes. Here $n_{i,j}$ denotes the outer unit normal vector on $\Gamma_{i,j}$. The construction (5.2) of the subdomains Ω_i implies that the line segment $[x_i, x_j]$ is bisected by its intersection with $\Gamma_{i,j}$. Denote this intersection point by $x_{i,j} := [x_i, x_j] \cap \Gamma_{i,j}$. Write $|\cdot|$ for the Euclidean norm in $\mathbb{R}^n$. As $n_{i,j} = (x_j - x_i)/|x_j - x_i|$, we obtain

$$\frac{\partial u}{\partial n_{i,j}}(x_{i,j}) = \frac{u(x_j) - u(x_i)}{|x_j - x_i|} + O\left(|x_j - x_i|^2\right), \qquad j \in N_i, \quad i \in J,$$

for sufficiently smooth u. Applying this approximation of the normal derivative to the identity (5.3) leads to the discretization

$$-\sum_{j \in N_i} \frac{m_{i,j}}{d_{i,j}} (u_j - u_i) = \int_{\Omega_i} f \, d\Omega_i, \qquad i \in J, \tag{5.4}$$

where

$$m_{i,j} := \mu_{n-1}(\Gamma_{i,j}) \qquad \text{and} \qquad d_{i,j} := |x_j - x_i|.$$

If in addition the boundary conditions are included by setting

$$u_i = g(x_i), \qquad i \in \overline{J} \setminus J, \tag{5.5}$$

then we obtain a linear system

$$A_h u_h = f_h. \tag{5.6}$$

This system defines the vector $u_h = (u_i)_{i \in J} \in \mathbb{R}^N$ that approximates the true solution $u(x_i)$, $i \in J$. Here the entries of the system matrix $A_h = (a_{i,j})$ and of the right-hand side $f_h \in \mathbb{R}^N$ of (5.6) are defined from (5.4) by

$$a_{i,j} = \begin{cases} \sum_{l \in N_i} \dfrac{m_{i,l}}{d_{i,l}} & \text{if } j = i, \\[2ex] -\dfrac{m_{i,j}}{d_{i,j}} & \text{if } j \in N_i \cap J, \\[2ex] 0 & \text{otherwise,} \end{cases} \qquad \text{and} \quad f_i = \int_{\Omega_i} f \, d\Omega_i + \sum_{l \in N_i \setminus J} \frac{m_{i,l}}{d_{i,l}} g(x_l).$$

Remark 2.59. Consider the two-dimensional case where Poisson's equation is discretized over the unit square using a grid that is equidistant and axiparallel. Then the above FVM generates the usual 5-point stencil, but with a scaling different from the standard finite difference method. Furthermore, integral mean values are used on the right-hand side of (5.6) instead of point values of f. Nevertheless, our example shows that the FVM can be considered as a generalization of the finite difference method to unstructured grids. □

Lemma 2.60. *The discrete problem (5.6) is inverse monotone, i.e., it satisfies the discrete comparison principle*

$$\left.\begin{array}{r} -\sum_{j \in N_i} \dfrac{m_{i,j}}{d_{i,j}} (u_j - u_i) \le 0, \quad i \in J, \\[2ex] u_i \le 0, \quad i \in \overline{J} \setminus J \end{array}\right\} \quad \Longrightarrow \quad u_i \le 0, \quad i \in \overline{J}.$$

Proof: Let $k \in \overline{J}$ be some index with

$$u_k \ge u_j \quad \forall j \in \overline{J}. \tag{5.7}$$

The proof is by contradiction, i.e., we suppose that $u_k > 0$. Then the discrete boundary conditions $u_i \le 0$ for $i \in \overline{J} \setminus J$ imply that $k \in J$. Furthermore, as there is a chain of subdomains linking each subdomain to the boundary Γ on which all $u_i \le 0$, the index $k \in J$ can be chosen such that for at least one

$l \in N_k$ one has $u_k > u_l$. By (5.7), it follows that $\displaystyle\sum_{j \in N_k} \frac{m_{k,j}}{d_{k,j}} u_k > \sum_{j \in N_k} \frac{m_{k,j}}{d_{k,j}} u_j$.

That is,

$$-\sum_{j \in N_k} \frac{m_{k,j}}{d_{k,j}} \left(u_j - u_k\right) > 0$$

which contradicts the hypotheses of the lemma. Hence, our supposition that $u_k > 0$ is false and the lemma is proved. ∎

Remark 2.61. Lemmas 2.60 and 2.39 are proved by means of the same argument. This is a consequence of the fact that the finite volume discretization described above generates an M-matrix. As already mentioned, this is one of the basic advantages of the FVM for elliptic problems in divergence form. □

Invoking Lemma 2.60 with an appropriate comparison function, an error estimate for the FVM in the maximum norm can be derived, provided that the solution u is sufficiently smooth. The next lemma gives us some insight into how such comparison functions might be constructed.

Lemma 2.62. *Define* $v : \mathbb{R}^n \to \mathbb{R}$ *by* $v(x) = -\frac{\alpha}{2}|x|^2 + \beta$ *with parameters* $\alpha, \beta \in \mathbb{R}$. *Then*

$$-\sum_{j \in N_i} \frac{m_{i,j}}{d_{i,j}} \left(v(x_j) - v(x_i)\right) = n\alpha \int_{\Omega_i} d\,\Omega_i, \quad i \in J.$$

Proof: The definition of the function v implies that

$$\nabla v(x) = -\alpha\, x \quad \text{and} \quad -\Delta v(x) = n\alpha \quad \forall x \in \mathbb{R}^n. \tag{5.8}$$

Using Gauss's Theorem we obtain

$$-\sum_{j \in N_i} \int_{\Gamma_{i,j}} \frac{\partial v}{\partial n_{i,j}}\, d\,\Gamma_{i,j} = -\int_{\Omega_i} \Delta v\, d\,\Omega_i = n\alpha \int_{\Omega_i} d\,\Omega_i, \quad i \in J. \tag{5.9}$$

Since v is a quadratic function, its difference quotient over a line segment equals its derivative at the midpoint of the segment. In particular, at the point $x_{i,j} \in \Gamma_{i,j}$ for each i and j, we have

$$-\frac{1}{d_{i,j}}\left(v(x_j) - v(x_i)\right) = -\frac{\partial v}{\partial n_{i,j}}(x_{i,j}) = -\nabla v(x_{i,j})^T n_{i,j}.$$

Recalling (5.8) and the orthogonality of $n_{i,j}$ and $\Gamma_{i,j}$, this formula can be written as

$$-\frac{1}{d_{i,j}}(v(x_j) - v(x_i)) = \alpha\, x_{i,j}^T\, n_{i,j} = \alpha\, x^T n_{i,j} + \alpha\, (x_{i,j} - x)^T n_{i,j}$$

$$= \alpha\, x^T n_{i,j} = -\frac{\partial v}{\partial n_{i,j}}(x) \quad \forall\, x \in \Gamma_{i,j}.$$

Hence

$$-\sum_{j \in N_i} \frac{m_{i,j}}{d_{i,j}}(v(x_j) - v(x_i)) = -\sum_{j \in N_i} \int_{\Gamma_{i,j}} \frac{\partial v}{\partial n_{i,j}}\, d\,\Gamma_{i,j}$$

$$= -\int_{\Omega_i} \Delta v\, d\,\Omega_i = n\alpha \int_{\Omega_i} d\,\Omega_i, \quad i \in J,$$

by (5.9). ∎

Next we investigate the consistency error of the FVM discretization in the maximum norm. Let us define the discretization parameter h by

$$h_i := \left(\max_{j \in N_i} \mu_{n-1}(\Gamma_{i,j}) \right)^{1/(n-1)}, \qquad h := \max_{i \in J} h_i. \tag{5.10}$$

In addition to our previous assumptions, we suppose that for $h \leq h_0$ (some $h_0 > 0$) the family of Voronoi boxes satisfies the following conditions:

- (V1) The number of essential neighbours of each x_i remains uniformly bounded, i.e., $\max_{i \in J}\{\operatorname{card} N_i\} \leq m_*$ for some $m_* \in \mathbb{N}$;
- (V2) Each point $x_{i,j} = [x_i, x_j] \cap \Gamma_{i,j}$ lies at the geometric centre of gravity of $\Gamma_{i,j}$.

The definition of h_i and the assumption (V1) together imply that the diameter of each Ω_i is of order $O(h_i)$. The assumption (V2) is rather restrictive and will hold only for rather regular (uniform) grids; we make it here for the sake of simplicity in the subsequent proofs.

Lemma 2.63. *Assume conditions (V1) and (V2). Let the solution u of the original problem (5.1) belong to $C^4(\overline{\Omega})$. Then there exists some constant $c > 0$ such that*

$$\left| \sum_{j \in N_i} \frac{m_{i,j}}{d_{i,j}}(u(x_j) - u(x_i)) + \int_{\Omega_i} f\, d\Omega_i \right| \leq c\, h_i^{n+1}, \quad i \in J.$$

Proof: From the derivation of our FVM, we must analyse the local discretization errors

$$\sigma_{i,j} := \left| \frac{m_{i,j}}{d_{i,j}}(u(x_j) - u(x_i)) - \int_{\Gamma_{i,j}} \frac{\partial u}{\partial n_{i,j}}\, d\,\Gamma_{i,j} \right|$$

on the components $\Gamma_{i,j}$ of the boundaries of the control volumes. Now $\frac{1}{d_{i,j}}(u(x_j) - u(x_i))$ is a difference quotient that approximates the directional derivative $\frac{\partial u}{\partial n_{i,j}}(x_{i,j})$; since $x_{i,j}$ is the midpoint of the line segment $[x_i, x_j]$ and $u \in C^4(\overline{\Omega})$, there exists $c > 0$ such that

$$\left| \frac{1}{d_{i,j}}(u(x_j) - u(x_i)) - \frac{\partial u}{\partial n_{i,j}}(x_{i,j}) \right| \le c d_{i,j}^2 \le c h_i^2, \quad i \in J, \ j \in N_i, \quad (5.11)$$

where the diameter of Ω_i being $O(h_i)$ implies the second inequality.

To avoid additional indices here and below, c will always denote a generic constant that can take different values at different places in the argument.

By the Cauchy-Schwarz inequality,

$$\left| \sum_{j \in N_i} \frac{m_{i,j}}{d_{i,j}}(u(x_j) - u(x_i)) - \sum_{j \in N_i} m_{i,j} \frac{\partial u}{\partial n_{i,j}}(x_{i,j}) \right|$$

$$\le \left(\sum_{j \in N_i} m_{i,j}^2 \right)^{1/2} \left(\sum_{j \in N_i} \left| \frac{1}{d_{i,j}}(u(x_j) - u(x_i)) - \frac{\partial u}{\partial n_{i,j}}(x_{i,j}) \right|^2 \right)^{1/2}$$

$$\le c h_i^{n+1}, \quad i \in J, \quad (5.12)$$

where we used (V1), the definition (5.10), and (5.11).

For each $i \in J$ define a continuous linear functional T_i on $C^3(\overline{\Omega}_i)$ by

$$T_i u := \sum_{j \in N_i} \left(\int_{\Gamma_{i,j}} \frac{\partial u}{\partial n_{i,j}} - m_{i,j} \frac{\partial u}{\partial n_{i,j}}(x_{i,j}) \right).$$

Then

$$|T_i u| \le c \mu_{n-1}(\Gamma_i) \max_{|\alpha|=1} \max_{x \in \Gamma_i} |[D^\alpha u](x)| \quad (5.13)$$

for some constant c. By assumption (V2) we have

$$\int_{\Gamma_{i,j}} z \, d\Gamma_{i,j} = m_{i,j} z(x_{i,j}) \qquad \text{for all } z \in \mathcal{P}_1, \quad (5.14)$$

where $\mathcal{P}_r$ denotes the space of all polynomials of degree at most r defined on $\overline{\Omega}_i$. It then follows that

$$T_i z = 0 \qquad \text{for all } z \in \mathcal{P}_2.$$

From (5.13), the linearity of the operator T_i, and the triangle inequality we get

$$|T_i u| \leq |T_i(u-z)| + |T_i z| \leq |T_i(u-z)|$$
$$\leq c\,\mu_{n-1}(\Gamma_i) \max_{|\alpha|=1} \max_{x \in \Gamma_i} |[D^\alpha(u-z)](x)| \qquad \text{for all } z \in \mathcal{P}_2. \tag{5.15}$$

Let z_i denote the quadratic Taylor polynomial of u expanded about x_i, i.e.,

$$z_i(x) = u(x_i) + \nabla u(x_i)^T (x-x_i) + \frac{1}{2}(x-x_i)^T H(x_i)(x-x_i) \qquad \text{for } x \in \overline{\Omega}_i,$$

where we introduced the Hessian matrix $H(x) := \left(\dfrac{\partial^2 u}{\partial x_r\, \partial x_s}(x) \right)$. Then there exists $c > 0$ such that

$$|[D^\alpha(u-z_i)](x)| \leq c\,|x-x_i|^2 \qquad \forall x \in \overline{\Omega}_i, \quad |\alpha|=1.$$

This inequality and (5.15) imply that

$$|T_i u| \leq c\,\mu_{n-1}(\Gamma_i)\,h_i^2 \leq c\,h_i^{n+1}, \qquad i \in J, \tag{5.16}$$

for some constant $c > 0$. Invoking (5.12) and (5.16), we obtain

$$\left| \sum_{j \in N_i} \frac{m_{i,j}}{d_{i,j}} (u(x_j) - u(x_i)) + \int_{\Omega_i} f\,d\Omega_i \right|$$

$$\leq \left| \sum_{j \in N_i} \frac{m_{i,j}}{d_{i,j}} (u(x_j) - u(x_i)) - \sum_{j \in N_i} m_{i,j} \frac{\partial u}{\partial n_{i,j}}(x_{i,j}) \right|$$

$$+ \left| \sum_{j \in N_i} m_{i,j} \frac{\partial u}{\partial n_{i,j}}(x_{i,j}) - \int_{\Gamma_i} \frac{\partial u}{\partial n_i}\,d\Gamma_i \right| + \left| \int_{\Omega_i} \Delta u\,d\Omega_i + \int_{\Omega_i} f\,d\Omega_i \right|$$

$$\leq c\,h_i^{n+1}, \qquad i \in J. \qquad \blacksquare$$

Remark 2.64. The central idea in the proof of Lemma 2.63 is the introduction of the operators T_i. This is essentially equivalent to an application of the Bramble-Hilbert Lemma, which is frequently used in the convergence analysis of finite element methods, and which we shall discuss in detail in Chapter 4. □

Theorem 2.65. *Let all the assumptions of Lemma 2.63 be satisfied. Assume also that the division of Ω into subdomains $\{\Omega_i\}_{i \in J}$ is such that a constant $c_1 > 0$ exists for which*

$$\mu_n(\Omega_i) \geq c_1\,h_i^n \quad \text{for all} \quad i \in J. \tag{5.17}$$

Then the error in the solution computed by the FVM can be bounded by

$$\|u_h - r_h u\|_{\infty,h} \leq c\,h. \tag{5.18}$$

Proof: We prove this theorem with the aid of the discrete comparison principle of Lemma 2.60, by using the comparison functions introduced in Lemma 2.62 to bound the consistency error given by Lemma 2.63.

Set $w_h := u_h - r_h u$, with

$$w_i := u_i - u(x_i), \quad i \in \overline{J}.$$

As the Dirichlet boundary conditions are incorporated exactly in the FVM,

$$w_i = u_i - u(x_i) = 0, \quad i \in \overline{J} \setminus J. \tag{5.19}$$

The definition of the FVM and Lemma 2.63 yield

$$\left| \sum_{j \in N_i} \frac{m_{i,j}}{d_{i,j}} (w_j - w_i) \right| = \left| \sum_{j \in N_i} \frac{m_{i,j}}{d_{i,j}} (u_j - u_i) - \sum_{j \in N_i} \frac{m_{i,j}}{d_{i,j}} (u(x_j) - u(x_i)) \right|$$

$$\leq \left| \sum_{j \in N_i} \frac{m_{i,j}}{d_{i,j}} (u_j - u_i) + \int_{\Omega_i} f \, d\Omega_i \right|$$

$$+ \left| \sum_{j \in N_i} \frac{m_{i,j}}{d_{i,j}} (u(x_j) - u(x_i)) + \int_{\Omega_i} f \, d\Omega_i \right|$$

$$= \left| \sum_{j \in N_i} \frac{m_{i,j}}{d_{i,j}} (u(x_j) - u(x_i)) + \int_{\Omega_i} f \, d\Omega_i \right|$$

$$\leq c \, h_i^{n+1}. \tag{5.20}$$

Consider the comparison function $v(x) := -\frac{\alpha}{2} \|x\|^2 + \beta$.
With $z_i := w_i - v(x_i)$, $i \in \overline{J}$, we see from Lemma 2.62 and (5.20) that

$$-\sum_{j \in N_i} \frac{m_{i,j}}{d_{i,j}} (z_j - z_i) \leq c \, h_i^{n+1} - n\alpha \int_{\Omega_i} d\Omega_i, \quad i \in J.$$

Now choose $\alpha = c^* h$ for some sufficiently large c^* and

$$\beta := \frac{\alpha}{2} \max_{i \in \overline{J} \setminus J} \|x_i\|^2.$$

Then, recalling (5.17), we get

$$-\sum_{j \in N_i} \frac{m_{i,j}}{d_{i,j}} (z_j - z_i) \leq 0, \quad i \in J, \quad \text{and } z_i \leq 0, \quad i \in \overline{J} \setminus J.$$

The discrete boundary maximum principle of Lemma 2.60 now tells us that $z_i \leq 0$ for all $i \in \overline{J}$. Hence

$$w_i \leq v_i \leq ch, \qquad i \in \bar{J}.$$

One can prove similarly the existence of some $c > 0$ with

$$w_i \geq -ch, \qquad i \in \bar{J}.$$

This pair of inequalities together complete the proof of the lemma. ∎

Remark 2.66. The assumptions made in the convergence proof for the FVM are rather restrictive. In particular, the point $x_{i,j}$ (which is merely the intersection of the line segment $[x_i, x_j]$ with the component $\Gamma_{i,j}$ of the subdomain boundary Γ_i) is presumed to coincide with the centre of gravity of $\Gamma_{i,j}$. This condition seriously restricts the choice of subdomains Ω_i. We nevertheless made this strong assumption for convenience in the convergence proof, which reveals the structure of the analysis of finite volume methods considered as finite difference methods. An alternative analysis of the FVM that relaxes these assumptions can be achieved by constructing an appropriate non-conforming finite element method that generates the same discrete problem; in this way convergence results for FVM in the H^1 or L_2 norms can be proved under mild assumptions. We will sketch this approach to the FVM in Chapter 4 and refer also, e.g., to [Hac95], [117]. □

Remark 2.67. The finite volume method was introduced for the model problem (5.1), and can be extended easily to other related types of problem. Consider for example the following problem with an inhomogeneous isotropic material:

$$-\mathrm{div}\left(k(\cdot)\,\mathrm{grad}\,u\right) = f \quad \text{in} \quad \Omega, \qquad u = g \quad \text{on} \quad \Gamma := \partial\Omega, \qquad (5.21)$$

where the function $k \in C(\overline{\Omega})$ is given. If the FVM is applied to this problem then, instead of (5.4), the local balance identities

$$-\sum_{j \in N_i} \frac{m_{i,j}\, k(x_{i,j})}{d_{i,j}} (u_j - u_i) = \int_{\Omega_i} f \, d\Omega_i, \qquad i \in J,$$

are obtained. □

Remark 2.68. The key idea of the FVM is to use Gauss's Theorem to convert conservation laws to local boundary integrals of fluxes. This technique can be applied to any differential operator that is given in divergence form. As a further example we consider the conservation law

$$u_t + f(u)_x = 0.$$

Suppose that we use axiparallel boxes

$$\Omega_{i,k} := (x_i - h/2, x_i + h/2) \times (t^k - \tau/2, t^k + \tau/2)$$

centred on the grid points (x_i, t^k). Then the local balance law

$$\int_{\Omega_{i,k}} [u_t + f(u)_x] \, dx \, dt = 0,$$

reformulated via Gauss's Theorem, becomes

$$\int_{x_{i-1/2}}^{x_{i+1/2}} [u(x, t^{k+1/2}) - u(x, t^{k-1/2})] \, dx + \int_{t^{k-1/2}}^{t^{k+1/2}} [f(x_{i+1/2}, t) - f(x_{i-1/2}, t)] \, dt = 0.$$

Now a further approximation of the fluxes through the boundary of the boxes directly yields a difference scheme; cf. (3.51).   □

Treatment of Neumann Boundary Conditions in FVMs

In finite volume methods the boundary integrals that are obtained via Gauss's Theorem allow us to include Neumann conditions directly. Thus no discretization points are needed in those parts of the boundary where Neumann conditions are imposed; the related boundary integrals of fluxes can be evaluated directly from the given boundary data. The following simple example illustrates this treatment of Neumann boundary conditions: find u such that

$$-\Delta u(x, y) = f(x, y), \ (x, y) \in (0, 1)^2,$$
$$u(x, 0) = u(x, 1) = 0, \qquad x \in [0, 1], \tag{5.22}$$
$$-\frac{\partial u}{\partial x}(0, y) = g_1(y), \qquad \frac{\partial u}{\partial x}(1, y) = g_2(y), \quad y \in (0, 1),$$

with given functions f, g_1, g_2. We use double indices to mark the grid points; this enables us to take advantage of the simple structure of the problem. For the discretization step size we set $h := 1/M$ for some $M \in \mathbb{N}$ and choose

$$x_l := -\frac{h}{2} + l \, h, \quad l = 1, \ldots, M \qquad \text{and} \qquad y_m := m \, h, \quad m = 0, 1, \ldots, M.$$

In total there are $\overline{N} = M(M+1)$ grid points. To distinguish between Dirichlet and Neumann data, introduce the index sets

$$J = \{1, \ldots, M\} \times \{1, \ldots, M - 1\} \subset \mathbb{N}^2 \quad \text{and}$$
$$\overline{J} = \{1, \ldots, M\} \times \{0, 1, \ldots, M\} \subset \mathbb{N}^2.$$

For each $(x_i, y_j) \in (0, 1)^2$, the corresponding Voronoi box is

$$\Omega_{i,j} = \left\{ \begin{pmatrix} x \\ y \end{pmatrix} \in \Omega : |x - x_i| < \frac{h}{2}, \ |y - y_j| < \frac{h}{2} \right\}.$$

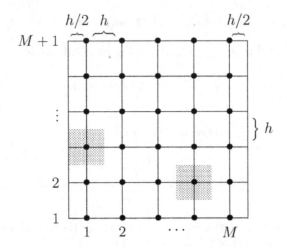

Figure 2.22 Voronoi boxes for Neumann conditions on parts of the boundary

Our finite volume method leads to the following system of linear equations:

$$4u_{i,j} - u_{i-1,j} - u_{i+1,j} - u_{i,j-1} - u_{i,j+1} = \int_{\Omega_{i,j}} f(x,y)\,dx\,dy,$$

$$i = 2,\ldots,M-1,\ j = 1,\ldots,M-1, \tag{5.23}$$

$$3u_{1,j} - u_{2,j} - u_{1,j-1} - u_{1,j+1} = \int_{\Omega_{1,j}} f(x,y)\,dx\,dy + \int_{y_j-h/2}^{y_j+h/2} g_1(y)\,dy,$$

$$j = 1,\ldots,M-1, \tag{5.24}$$

$$3u_{M,j} - u_{M-1,j} - u_{M,j-1} - u_{M,j+1} = \int_{\Omega_{M,j}} f(x,y)\,dx\,dy + \int_{y_j-h/2}^{y_j+h/2} g_2(y)\,dy,$$

$$j = 1,\ldots,M-1, \tag{5.25}$$

and
$$u_{i,j} = 0, \qquad (i,j) \in \overline{J} \backslash J. \tag{5.26}$$

Remark 2.69. In practical implementations of finite volume methods, the integrals that occur must usually be approximated by appropriate quadrature formulas. When this happens our convergence proof still works; the only modification is that now the errors caused by quadrature have to be included in the consistency estimates.   □

Exercise 2.70. Discretize the problem

$$-\Delta u = f \quad \text{in} \ \ (0,1) \times (0,1),$$

with homogeneous Dirichlet boundary conditions, using an equidistant rectangular grid and the finite volume method of this section. Here the points of the grid are used as discretization points. Compare the discrete system that is generated with the classical finite difference method on the same grid.

Exercise 2.71. Apply the FVM to derive a discretization of the Laplace operator over a grid comprising equilateral triangles with side length h. The vertices of the triangles are to be used as grid points. Express the FVM as a finite difference method and analyse the consistency error.

Exercise 2.72. Consider the elliptic boundary value problem

$$- \operatorname{div}\left((1 + x^2 + y^2)\operatorname{grad} u(x,y)\right) = e^{x+y} \text{ in } \Omega := (0,1)^2,$$

$$u(x,y) = 0 \qquad \text{on } \Gamma_D,$$

$$\frac{\partial}{\partial n} u(x,y) = 0 \qquad \text{on } \Gamma_N,$$

with

$$\Gamma_D := \{ (x,y) \in \Gamma : xy = 0 \}, \qquad \Gamma_N := \Gamma \backslash \Gamma_D.$$

Discretize this problem by the finite volume method using an equidistant rectangular grid with step size $h = 1/N$. The Voronoi boxes are

$$\Omega_{ij} := \{ (x,y) \in \Omega : |x - x_i| < h/2, \ |y - y_j| < h/2 \}, \quad i, j = 0, 1, \ldots, N.$$

Determine explicitly the discrete equations generated by the FVM.

2.6 Parabolic Initial-Boundary Value Problems

Let $\Omega \subset \mathbb{R}^n$ be a bounded domain with boundary Γ. Let L denote a differential operator on the spatial variables that is uniformly elliptic. The partial differential equation

$$\frac{\partial}{\partial t} u(x,t) + [Lu](x,t) = f(x,t), \quad x \in \Omega, \ t \in (0,T],$$

with the respective boundary and initial conditions

$$u(x,t) = g(x,t), \quad (x,t) \in \Gamma \times (0,T] \qquad \text{and} \qquad u(x,0) = u_0(x), \quad x \in \overline{\Omega},$$

form a parabolic initial-boundary value problem. Here f, g, u_0 are given data. Once again we shall concentrate on the case of Dirichlet boundary conditions; Neumann or Robin boundary conditions can be treated in a similar way.

When a parabolic problem is treated numerically using some finite difference method, then the temporal derivative and the spatial derivatives have to be approximated by difference quotients. The discretization of the elliptic operator L can be carried out by the techniques described in the preceding sections.

2.6.1 Problems in One Space Dimension

In this section we examine the initial-boundary value problem

$$\frac{\partial u}{\partial t} - \frac{\partial^2 u}{\partial x^2} = f(x,t) \qquad \text{in } (0,1) \times (0,T],$$

$$u(0,t) = g_1(t), \quad u(1,t) = g_2(t), \qquad u(x,0) = u_0(x).$$

(6.1)

Here $\Omega = (0,1)$ and $\Gamma = \{0,1\} \subset \mathbb{R}$. For simplicity we use an equidistant grid in both the spatial and temporal directions. Write this grid as

$$x_i = ih, \ i = 0,1,\ldots,N, \quad \text{and} \quad t^k = k\tau, \ k = 0,1,\ldots,M,$$
$$\text{with} \ \ h := 1/N, \ \tau := T/M.$$

Denote by u_i^k the computed numerical approximation of the value $u(x_i, t^k)$ of the exact solution at the grid point (x_i, t^k).

The derivatives appearing in (6.1) must be replaced by difference approximations. As the differential equation contains only a first-order time derivative, the simplest approximation of this term is a two-point approximation in the t-direction. This leads to so-called *two-layer schemes*. In the x-direction, to discretize the second-order derivative at least three discretization points are required, but these points may lie at either of the two time levels used in the approximation of $\partial u / \partial t$. Thus we obtain schemes lying in the class of *six-point schemes* (see Figure 2.23) for the discretization of (6.1).

If the second-order spatial partial derivative is discretized by

$$D^- D^+ u_i^k := \frac{1}{h^2} \left(u_{i-1}^k - 2u_i^k + u_{i+1}^k \right),$$

then using a freely-chosen parameter $\sigma \in [0,1]$ we can describe a six-point scheme in the general form

$$\frac{u_i^{k+1} - u_i^k}{\tau} = D^- D^+ \left(\sigma u_i^{k+1} + (1-\sigma) u_i^k \right) + \tilde{f}_i^k, \qquad \begin{array}{l} i = 1,\ldots,N-1, \\ k = 1,\ldots,M-1, \end{array}$$

(6.2)

with the discrete initial and boundary conditions

$$u_i^0 = u_0(x_i), \qquad u_0^k = g_1(t^k), \quad u_N^k = g_2(t^k).$$

(6.3)

In (6.2) the quantity $\tilde{f}_i^k$ denotes an appropriate approximation for $f(x_i, t^k)$ that will be written down later.

For certain values of the parameter σ we obtain the following cases that are often used in practice:

- The *explicit (Euler) scheme* for $\sigma = 0$:

$$u_i^{k+1} = (1-2\gamma)u_i^k + \gamma(u_{i-1}^k + u_{i+1}^k) + \tau f(x_i, t^k);$$

(6.4)

- the purely *implicit (Euler)* scheme for $\sigma = 1$:

$$(1 + 2\gamma)u_i^{k+1} - \gamma(u_{i+1}^{k+1} + u_{i-1}^{k+1}) = u_i^k + \tau f(x_i, t^{k+1}); \qquad (6.5)$$

- the *Crank-Nicolson* scheme for $\sigma = \frac{1}{2}$:

$$2(\gamma + 1)u_i^{k+1} - \gamma(u_{i+1}^{k+1} + u_{i-1}^{k+1}) = 2(1 - \gamma)u_i^k + \gamma(u_{i+1}^k + u_{i-1}^k) \\ + 2\tau f(x_i, t^k + \tfrac{\tau}{2}). \qquad (6.6)$$

In these formulas $\gamma := \tau/h^2$. An *explicit* method is obtained only in the case $\sigma = 0$: the values u_i^{k+1} of the numerical solution at the time level $t = t^{k+1}$ are defined directly, i.e., without solving a discrete linear system, by the values at the previous time level $t = t^k$. In the remaining cases (6.5) and (6.6) the methods are *implicit*: a system of discrete linear equations must be solved to compute the numerical solution at the new time level. For the problem and schemes that we are considering, these linear systems have a tridiagonal coefficient matrix, so they can be solved very efficiently by fast Gauss elimination (which is also known as the Thomas algorithm). The different methods that correspond to specific choices of the parameter σ

Figure 2.23   Difference stencils for the schemes (6.4) - (6.6)

possess different consistency and stability properties. Set $Q = \Omega \times (0, T)$, i.e., in the case of problem (6.1) we have $Q = (0, 1) \times (0, T)$.

Lemma 2.73. *The general scheme (6.2), (6.3) has the following order of consistency in the maximum norm:*

(a) $O\,(h^2 + \tau)$ for any $\sigma \in [0, 1]$ and $\tilde{f}_i^k = f(x_i, t^k)$, provided $u \in C^{4,2}(\overline{Q})$
(b) $O\,(h^2 + \tau^2)$ for $\sigma = \frac{1}{2}$ and $\tilde{f}_i^k = f(x_i, t^k + \frac{\tau}{2})$, provided $u \in C^{4,3}(\overline{Q})$.

Here $C^{l,m}(\overline{Q})$ denotes the space of functions that in the domain Q are l times continuously differentiable with respect to x and m times continuously differentiable with respect to t, where all these derivatives can be continuously extended to $\overline{Q}$.

Proof: We prove only part (b) and leave the proof of (a) to the reader. Taylor's theorem yields

$$\frac{u(x, t+\tau) - u(x,t)}{\tau} = u_t + \frac{1}{2}u_{tt}\tau + O(\tau^2).$$

(To simplify the notation we omit the argument (x,t) if no misunderstanding is likely.) Similarly

$$\frac{1}{2}\left(\frac{u(x-h,t+\tau) - 2u(x,t+\tau) + u(x+h,t+\tau)}{h^2}\right.$$
$$\left. + \frac{u(x-h,t) - 2u(x,t) + u(x+h,t)}{h^2}\right) = \frac{1}{2}(2u_{xx} + u_{xxt}\tau) + O(\tau^2 + h^2),$$
$$f(x, t + \frac{\tau}{2}) = f(x,t) + \frac{\tau}{2}f_t + O(\tau^2).$$

Combining these formulas, the consistency error e_{cons} can be written as

$$e_{\text{cons}} = u_t - u_{xx} - f + \frac{1}{2}\tau(u_{tt} - u_{xxt} - f_t) + O(\tau^2 + h^2). \qquad (6.7)$$

But u satisfies the given parabolic differential equation and by differentiating this we obtain $u_{tt} - u_{xxt} = f_t$. Thus several terms in (6.7) cancel, and we see that the Crank-Nicolson scheme has consistency $O(\tau^2 + h^2)$. ∎

Like our earlier discussion in the context of elliptic boundary value problems, it has to be noted that the assumption that $u \in C^{4,2}(\overline{Q})$ is highly restrictive and rarely holds for practical problems since it implies that several compatibility conditions are satisfied at the corners of Q. Nevertheless this does not imply that this class of discretization methods has to be abandoned: the unrealistic assumption arose because we used a simple classical convergence analysis, invoking consistency and stability. We chose this approach so that the reader could easily follow the basic ideas and grasp the principal concepts in finite difference analysis.

We now turn our attention to stability properties of the schemes, starting with stability in the maximum norm. For simplicity we assume homogeneous boundary conditions in (6.1).

Let us write (6.2) in the form

$$-\sigma\gamma u_{i-1}^{k+1} + (2\sigma\gamma + 1)u_i^{k+1} - \sigma\gamma\, u_{i+1}^{k+1} = F_i^k$$

with

$$F_i^k = (1-\sigma)\gamma u_{i-1}^k + (1 - 2(1-\sigma)\gamma)u_i^k + (1-\sigma)\gamma u_{i+1}^k + \tau \tilde{f}_i^k.$$

Strict diagonal dominance gives immediately

$$\max_i |u_i^{k+1}| \le \max_i |F_i^k|.$$

If as well as $0 \leq \sigma \leq 1$ we assume that $1 - 2(1 - \sigma)\gamma \geq 0$, then we obtain

$$\max_i |u_i^{k+1}| \leq \max_i |F_i^k| \leq \max_i |u_i^k| + \tau \max_i |\tilde{f}_i^k|.$$

Applying this inequality iteratively from each time level to the next, we get

$$\max_k \max_i |u_i^{k+1}| \leq \max |u_0(x)| + \tau \sum_{j=0}^{k} \max_i |\tilde{f}_i^j|. \tag{6.8}$$

The estimate (6.8) simply means that the scheme is *stable in the discrete maximum norm*. The assumption $1 - 2(1 - \sigma)\gamma \geq 0$ is equivalent to the inequality

$$(1 - \sigma)\tau/h^2 \leq 1/2. \tag{6.9}$$

This inequality is automatically true if $\sigma = 1$ (the implicit method), but when $\sigma \in [0, 1)$ the condition (6.9) is a restriction on the ratio between the temporal and spatial step sizes. More precisely, it states that the temporal step size has to be much smaller than the spatial step size. A condition of the type (6.9) is essential for the stability of the related method in the maximum norm: numerical experiments show this clearly by producing spurious oscillations if (6.9) is violated. This instability does not occur in the L_2 norm, where the Crank-Nicolson scheme is stable for any step sizes $\tau > 0$ and $h > 0$.

Consistency and stability in the maximum norm lead immediately to the following convergence theorem:

Theorem 2.74. *Choose $\tilde{f}_i^k = f(x_i, t^k)$ for all i, k. Assume that $(1-\sigma)\tau/h^2 \leq 1/2$ in (6.2), (6.3) and $u \in C^{4,2}(\overline{Q})$. Then there exists a constant $C > 0$ such that*

$$\max_{i,k} |u(x_i, t^k) - u_i^k| \leq C(h^2 + \tau).$$

Furthermore, for the Crank-Nicolson scheme (where $\sigma = \frac{1}{2}$), assuming that $\tau/h^2 \leq 1$, one has

$$\max_{i,k} |u(x_i, t^k) - u_i^k| \leq C(h^2 + \tau^2)$$

provided that $u \in C^{4,3}(\overline{Q})$.

A similar stability analysis in the discrete maximum norm can be carried out for more general parabolic problems with variable coefficients of the form

$$c(x, t)\frac{\partial u}{\partial t} - \left[\frac{\partial}{\partial x}\left(a(x, t)\frac{\partial u}{\partial x}\right) + r(x, t)\frac{\partial u}{\partial x} - q(x, t)u\right] = f(x, t)$$

with $a(x, t) \geq \alpha_0 > 0$, $q(x, t) \geq 0$; see, e.g., [Sam01].

We now move on to an examination of stability properties in the L_2 norm.

For simplicity of presentation, assume that $f \equiv 0$ and the boundary conditions are homogeneous in the problem (6.1). Then the general scheme (6.2), (6.3) has the form

$$\frac{u_i^{k+1} - u_i^k}{\tau} = D^- D^+ (\sigma u_i^{k+1} + (1-\sigma)u_i^k), \quad \begin{array}{l} u_0^k = u_N^k = 0, \\ u_i^0 = u_0(x_i). \end{array} \tag{6.10}$$

The L_2 stability of the explicit scheme (i.e., the case $\sigma = 0$) was already investigated in Section 2.2, where we exploited the fact that the system of functions v^j defined by

$$[v^j]_l = \sqrt{2} \sin(j\pi lh), \quad j = 1, \ldots, N-1, \quad l = 0, 1, \ldots, N,$$

forms a complete discrete eigensystem of the operator $L_h = -D^- D^+$ in the space U_h^0. The associated eigenvalues are given by (2.12):

$$\lambda_j = \frac{4}{h^2} \sin^2\left(\frac{j\pi h}{2}\right), \quad j = 1, \ldots, N-1. \tag{6.11}$$

Furthermore, the system $\{v^j\}$ is orthogonal with respect to the scalar product $(\cdot, \cdot)_h$. We express the discrete solutions $u^k = (u_i^k) \in U_h^0$, $k = 0, 1, \ldots M$, in terms of this basis of eigenfunctions:

$$u^k = \sum_{j=1}^{N-1} \omega_j^k v^j, \quad k = 0, 1, \ldots M,$$

where the coefficients $\omega_j^k \in \mathbb{R}$ are uniquely determined. Since the v^j are eigenfunctions of $D^- D^+$, equation (6.10) can be simplified to

$$\frac{\omega_j^{k+1} - \omega_j^k}{\tau} = -\lambda_j (\sigma \omega_j^{k+1} + (1-\sigma)\omega_j^k), \quad \begin{array}{l} j = 1, \ldots, N-1, \\ k = 0, 1, \ldots, M-1. \end{array}$$

Here the $\{\omega_j^0\}$ are the Fourier coefficients of the given function u_0, viz.,

$$\omega_j^0 = (u_0, v^j)_h, \quad j = 1, \ldots, N-1.$$

Thus the coefficients at each time level satisfy

$$\omega_j^{k+1} = q(\tau\lambda_j)\,\omega_j^k, \quad j = 1, \ldots, N-1, \; k = 0, 1, \ldots, M-1, \tag{6.12}$$

where the amplification factor $q(\cdot)$ is defined by

$$q(s) := \frac{1 - (1-\sigma)s}{1 + \sigma s}. \tag{6.13}$$

Our three main schemes have

$$q(s) = 1 - s \quad \text{for } \sigma = 0 \quad \text{(explicit Euler method)},$$

$$q(s) = \frac{1 - \frac{1}{2}s}{1 + \frac{1}{2}s} \quad \text{for } \sigma = \frac{1}{2} \quad \text{(Crank-Nicolson method)}, \tag{6.14}$$

$$q(s) = \frac{1}{1+s} \quad \text{for } \sigma = 1 \quad \text{(purely implicit method)}.$$

A discretization scheme is said to be *stable for all harmonic oscillations* if

$$|q(\tau\lambda_j)| \le 1 \quad \text{for all eigenvalues} \quad \lambda_j. \tag{6.15}$$

For schemes that are stable in this sense, Parseval's equation implies that

$$\|u^k\| \le \|u^0\|, \quad k = 0, 1, \dots, M,$$

and such methods are stable in the discrete L_2 norm. This stability is weaker than stability in the discrete maximum norm, as the next lemma shows.

Lemma 2.75. *The implicit scheme and the Crank-Nicolson scheme are L_2 stable for all values of h and τ. The explicit scheme is L_2 stable if $\tau/h^2 \le 1/2$.*

Proof: Since all eigenvalues λ_j of the operator $-D^-D^+$ are positive by (6.11), the L_2 stability of the implicit and the Crank-Nicolson schemes is immediate from (6.14). For the explicit scheme, for each j we have $|q(\tau\lambda_j)| \le 1$ if and only if $\tau\lambda_j \le 2$. Again invoking (6.11), this condition is satisfied if $\tau/h^2 \le 1/2$. ∎

The above Fourier analysis for the special one-dimensional problem (6.1) is simple and can in principle be extended to much more general cases even when the eigenfunctions and eigenvalues are not explicitly known. Its limitations are that it provides only L_2 stability and is applicable only to linear problems with constant coefficients. Furthermore, to carry out this analysis one must be able to establish bounds on the eigenvalues of the discrete spatial operator L_h associated with L. On the other hand, while the earlier maximum-norm stability analysis via inverse monotonicity seems restrictive at first sight, it can be extended to linear problems with variable coefficients and even to certain nonlinear problems.

2.6.2 Problems in Higher Space Dimensions

For simplicity we take $\Omega = (0,1)^2 \subset \mathbb{R}^2$ and a problem of the form

$$\frac{\partial u}{\partial t} - \Delta u = f \quad \text{in} \quad \Omega \times (0,T],$$
$$u = 0 \quad \text{on} \quad \Gamma \times (0,T],$$
$$u(\cdot, 0) = g(\cdot) \text{ on } \overline{\Omega}.$$

Consider the implicit method

$$\frac{1}{\tau}(u^k - u^{k-1}) - \Delta_h u^k = f^k, \quad k = 1, \dots, M. \tag{6.16}$$

Here Δ_h is some discretization of the Laplace operator Δ on the spatial grid; all discretization methods for elliptic problems that were discussed in

Section 2.4 may be used to generate Δ_h. By solving the system of linear equations (6.16) iteratively, one can compute the grid functions $u^k \in U_h^0$ for $k = 1, \ldots, M$, starting from the discrete initial function u^0. At each time step the linear system that has to be solved is usually of high dimension, but the system matrix is sparse and the locations of its non-zero entries have a certain structure that can be exploited efficiently.

Unlike the above implicit scheme, the explicit method

$$\frac{1}{\tau}(u^k - u^{k-1}) - \Delta_h u^{k-1} = f^{k-1}, \quad k = 1, \ldots, M, \qquad (6.17)$$

gives the new time level solution $u^k \in U_h^0$ directly. No system of equations needs to be solved, but stability requires us to limit the time step size by

$$\tau \leq \frac{1}{2} h^2.$$

The Crank-Nicolson method

$$\frac{1}{\tau}(u^k - u^{k-1}) - \frac{1}{2}(\Delta_h u^k + \Delta_h u^{k-1}) = \frac{1}{2}(f^k + f^{k-1}), \quad k = 1, \ldots, M, \ (6.18)$$

is L_2 stable for any step sizes $h, \tau > 0$. But this is not the case in the L_∞ norm: in particular, if the data of the problem are not smooth and the condition $\tau/h^2 \leq 1$ is violated, then the solution computed by the Crank-Nicolson scheme may contain spurious oscillations—these cannot occur in the original problem because of the boundary maximum principle. This limits the applicability of the Crank-Nicolson method, especially for problems where the true solution u is not sufficiently smooth.

Further methods for the discretization of parabolic problems posed in higher spatial dimensions will be discussed later in the framework of finite element methods.

Semi-discretization (Section 2.6.3) provides a general framework for the construction of difference schemes for higher-dimensional problems; this technique leads to a reduction to problems of lower dimension.

We shall now discuss a modification of the implicit method that generates systems of linear equations which are easier to treat numerically than those generated by the original version of this method. That is, we consider alternating direction implicit (ADI) methods, which are based on an approximation of the implicit scheme for problems posed in domains of spatial dimension greater than one. First, we observe for the standard discretization Δ_h of Δ that

$$I - \tau \Delta_h = I - \tau (D_x^- D_x^+ + D_y^- D_y^+)$$
$$= (I - \tau D_x^- D_x^+)(I - \tau D_y^- D_y^+) - \tau^2 D_x^- D_x^+ D_y^- D_y^+.$$

Now given sufficient smoothness of the solution u, the term $D_x^- D_x^+ D_y^- D_y^+ u$ is bounded. Hence, if we replace (6.16) by the scheme

$$(I - \tau D_x^- D_x^+)(I - \tau D_y^- D_y^+) u^k = u^{k-1} + \tau f^k, \quad k = 1, \ldots, M, \quad (6.19)$$

then this discretization has the same order of consistency as the original implicit method (6.16). Furthermore, since

$$(I - \tau D_x^- D_x^+)^{-1} \geq 0 \quad \text{and} \quad (I - \tau D_y^- D_y^+)^{-1} \geq 0,$$

this scheme (6.19) satisfies the discrete boundary maximum principle. Under smoothness assumptions that preserve the order of consistency, the maximum principle implies the convergence of the method (6.19) in the maximum norm with error $O(h^2 + \tau)$.

An advantage of (6.19) over (6.16) is that the discrete operators $(I - \tau D_x^- D_x^+)$ and $(I - \tau D_y^- D_y^+)$ each generate a tridiagonal matrix. Thus at each time step the linear system that appears in (6.19) can be solved rapidly by two successive applications of fast Gaussian elimination (the Thomas algorithm).

The idea of the ADI splitting can be applied easily also to parabolic problems in higher space dimensions and on general domains and with other types of boundary conditions. The scheme (6.19) can be described equivalently by

$$(I - \tau D_x^- D_x^+) u^{k-1/2} = u^{k-1} + \tau f^k,$$
$$(I - \tau D_y^- D_y^+) u^k = u^{k-1/2}. \qquad k = 1, \ldots, M \qquad (6.20)$$

For this reason such methods are sometimes called half-step schemes in the literature.

Exercise 2.76. The temperature distribution in a rod of length 1 satisfies the heat equation

$$u_t = u_{xx},$$

where t and x are the temporal and spatial coordinates (length), respectively. Assume that the temperature u at the end-points of the rod is time-dependent and given by

$$u(0, t) = u(1, t) = 12 \sin 12\pi t \quad \text{for} \quad t \geq 0,$$

while the initial temperature $u(\cdot, 0)$ should be zero.

Use the explicit Euler method with spatial step size $h = 1/6$ and temporal step size $\tau = 1/72$ to compute approximately the temperature as far as $t = 1/3$.

Exercise 2.77. Boundary conditions of the form

$$u_x(0, t) = f(t)$$

can be handled in the numerical solution of the heat equation

$$u_t = u_{xx}$$

by means of the approximation

$$u(0, t + \tau) \approx u(0, t) + \frac{2\tau}{h^2}[u(h, t) - u(0, t) - hf(t)].$$

Give a justification of this technique and apply it to the discretization of the initial-boundary value problem

$$u_t - u_{xx} = 0 \quad \text{in} \quad (0, 1) \times (0, 1),$$
$$u(x, 0) = 0,$$
$$u_x(0, t) = 0, \quad u(1, t) = 10^6 t,$$

using the explicit Euler scheme. Compute the solution as far as time $t = 0.08$, using (i) the step sizes $h = 0.2$, $\tau = 0.01$ and (ii) $h = 0.1$, $\tau = 0.01$. Compare the two sets of numerical results obtained.

Exercise 2.78. Derive an explicit difference approximation for the partial differential equation

$$\frac{\partial u}{\partial t} = \frac{\partial}{\partial x}\left((1 + x^2)\frac{\partial u}{\partial x}\right).$$

Let the initial and boundary conditions be

$$u(x, 0) = 1000 - |1000x|,$$
$$u(-1, t) = u(1, t) = 0.$$

Solve this initial-boundary value problem approximately in the time interval $0 \le t \le 0.2$ using discretization step sizes $h = 0.4$ and $\tau = 0.04$.

Exercise 2.79. Examine the consistency and stability of the *leapfrog* scheme

$$u_i^{k+1} - u_i^{k-1} - 2\gamma(u_{i-1}^k + u_{i+1}^k) + 4\gamma u_i^k = 0,$$

where $\gamma = \tau/h^2$, for the discretization of the homogeneous heat equation $u_t - u_{xx} = 0$.

Exercise 2.80. A modification of the leapfrog scheme of Exercise 2.79 is the *du Fort-Frankel* scheme

$$u_i^{k+1} - u_i^{k-1} - 2\gamma\left[u_{i-1}^k - (u_i^{k+1} + u_i^{k-1}) + u_{i+1}^k\right] = 0.$$

Analyse its stability and consistency.

Exercise 2.81. Consider the initial-boundary value problem

$$u_t - u_{xx} = x + t \quad \text{in} \quad (0, 1) \times (0, 1),$$
$$u(x, 0) = \phi(x),$$
$$u(0, t) = u(1, t) = 0,$$

where ϕ is some continuous function that satisfies $\phi(0) = \phi(1) = 0$ and $|\phi(x)| \le \alpha$ for some small positive constant α. Show that

$$u_i^k = \alpha \left(-2r - \sqrt{4r^2 + 1} \right)^k \sin \frac{\pi i}{2} \quad \text{with} \quad r = \tau/h^2$$

is a solution of the following difference scheme obtained by discretizing this problem:

$$u_i^{k+1} = u_i^{k-1} + \frac{2\tau}{h^2}(u_{i+1}^k - 2u_i^k + u_{i-1}^k),$$

$$u_0^k = u_N^k = 0, \quad |u_i^0| \le \alpha,$$

for $i = 0, 1, \ldots, N$, where $h = 1/N$ and N is taken to be even. As $h \to 0$ and $\tau \to 0$ with τ/h^2 held fixed, does this solution converge to the solution of the continuous problem?

Exercise 2.82. Prove that the operators that appear in the ADI scheme (6.19) satisfy the discrete Euclidean norm estimates

$$\|(I - \tau D_x^- D_x^+)^{-1}\| \le 1 \quad \text{and} \quad \|(I - \tau D_y^- D_y^+)^{-1}\| \le 1.$$

What do these bounds tell us about the stability of the method in the L_2 norm?

Exercise 2.83. Consider the initial-boundary value problem

$$u_t(x, t) - u_{xx}(x, t) = x + t \quad \text{in } (0, 1) \times (0, 1),$$

$$u(0, t) = u(1, t) = 0 \quad \text{for } t \in [0, 1],$$

$$u(x, 0) = \sin \pi x \text{ for } x \in [0, 1].$$

Show that $0 < u(0.5, 0.5) \le 0.75$.

Exercise 2.84. Analyse the consistency and stability of the scheme

$$\frac{1}{2\tau}(u_{i,k+1} - u_{i,k-1}) - \frac{1}{h^2}(u_{i-1,k} - 2u_{i,k} + u_{i+1,k}) = 0$$

for the discretization of the homogeneous heat equation $u_t - u_{xx} = 0$.

2.6.3 Semi-Discretization

Method of Lines (vertical)

Parabolic initial-boundary value problems can be treated numerically as described previously, where we discretized the space and time derivatives simultaneously. For a better theoretical understanding and more structured approach, however, it is often preferable to separate these two discretizations. In the present section we will first discretize only the spatial derivatives. Since only part of the differential equation is discretized this technique is a type of *semi-discretization.* The version we deal with here is called the *(vertical) method of lines* (MOL).

Let us consider the initial-boundary value problem

$$\frac{\partial u}{\partial t} + L u = f \qquad \text{in } \Omega \times (0, T],$$
$$u = 0 \qquad \text{on } \partial\Omega \times (0, T], \tag{6.21}$$
$$u(\cdot, 0) = u_0(x) \text{ on } \overline{\Omega},$$

where L is some uniformly elliptic differential operator. The domain $\Omega \subset \mathbb{R}^n$ is assumed to be bounded and have a smooth boundary. Similarly to Section 2.4, we could also consider other types of boundary conditions if the boundary is sufficiently regular.

Suppose that the spatial derivatives of (6.21) are discretized by some method—finite differences, finite volumes, finite elements, or any other method. Then this *semi-discretization* generates an initial value problem for a system of ordinary differential equations.

Example 2.85. Consider the initial-boundary value problem

$$u_t - u_{xx} = f \qquad \text{in } (0, 1) \times (0, T),$$
$$u(0, \cdot) = u(1, \cdot) = 0,$$
$$u(\cdot, 0) = u_0(\cdot).$$

To discretize the spatial derivatives we apply the classical difference method on an equidistant grid in the x-direction. Let this grid have step size h. Let $u_i(t)$ denote the resulting computed approximation of $u(x_i, t)$ and set $f_i(t) := f(x_i, t)$. Then the semi-discretization produces the following system of ordinary differential equations:

$$\frac{du_i}{dt} = \frac{u_{i-1} - 2u_i + u_{i+1}}{h^2} + f_i, \quad u_0 = u_N = 0, \quad i = 1, \ldots, N-1,$$

with the initial conditions $u_i(0) = u_0(x_i)$ for $i = 1, \ldots, N-1$. This system defines the unknown functions $u_i(t)$. To finish the numerical solution of the original problem, an ordinary differential equation solver is applied to this system.   □

The properties of any semi-discretization technique depend upon the discretization method used for the spatial derivatives (finite difference method, FEM, ...) and on the numerical method applied to the initial value problem for the system of ordinary differential equations that is generated. If we use the standard finite difference method for the discretization in space (as in the example above), and subsequently choose the explicit Euler or implicit Euler or trapezoidal rule linear one-step method for the integration in time, then we obtain the discretization formulas discussed in the preceding section.

In the chapter on finite element methods for time-dependent problems we shall analyse semi-discretizations based on finite elements in more detail.

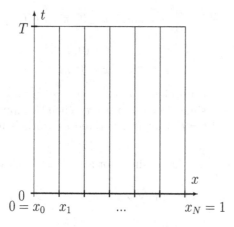

Figure 2.24   Vertical method of lines

A generic property of the systems of ordinary differential equations that are generated by semi-discretization is that they become arbitrarily stiff as $h \to 0$. Consequently their accurate numerical integration requires special techniques.

Rothe's Method (Horizontal Method of Lines)

An alternative semi-discretization is to discretize only the temporal derivatives in the parabolic differential equation. This is called *Rothe's method*. These temporal derivatives are usually discretized by means of the implicit Euler method.

Consider again problem (6.21) and let u^k denote the approximation of the true solution $u(\cdot, t^k)$ at the time level $t = t^k$. Then semi-discretization in time by the implicit Euler method yields the semi-discrete problem

$$\frac{u^k - u^{k-1}}{\tau_k} + Lu^k = f^k \text{ in } \Omega,$$
$$u^k = 0 \text{ on } \Gamma. \qquad k = 1, \ldots, M. \qquad (6.22)$$

Here $\tau_k := t^k - t^{k-1}$ denotes the local time step. Problem (6.22) is a system of elliptic boundary value problems, which can be solved iteratively for $k = 1, \ldots, M$ starting from the given function u^0. See Figure 2.25. When small time steps are used these elliptic problems are singularly perturbed problems of reaction-diffusion type, but under suitable regularity assumptions the solution u^{k-1} at the previous time level provides a very good initial approximation for u^k.

We illustrate Rothe's method for the simple initial-boundary value problem

$$\frac{\partial u}{\partial t} - \frac{\partial^2 u}{\partial x^2} = \sin x, \ x \in (0, \pi), \ t \in (0, T],$$

$$u(0, t) = u(\pi, t) = 0, \qquad t \in (0, T],$$
$$u(x, 0) = 0, \qquad x \in [0, \pi].$$

The exact solution of this problem is $u(x, t) = (1 - e^{-t}) \sin x$. Let us subdivide the time interval $[0, T]$ using a fixed step size τ. Then the approximate solution $u^k(x)$ at the time level $t = t^k$ that is generated by Rothe's method satisfies

$$\frac{u^k(x) - u^{k-1}(x)}{\tau} - \frac{d^2}{dx^2} u^k(x) = \sin x \ \text{ for } x \in (0, \pi), \quad u^k(0) = u^k(\pi) = 0.$$

In this example one can solve analytically to get

$$u^k(x) = \left[1 - \frac{1}{(1 + \tau)^k}\right] \sin x, \quad k = 0, 1, \ldots, M.$$

If we set

$$u^\tau(x, t) := u^{k-1}(x) + \frac{t - t^{k-1}}{\tau}\left(u^k(x)) - u^{k-1}(x)\right) \quad \text{for } t \in [t^{k-1}, t^k],$$

then we obtain a solution $u^\tau(x, t)$ that approximates $u(x, t)$ at each $(x, t) \in [0, \pi] \times [0, T]$. For each fixed time $t^* \in [0, T]$ we have

$$\lim_{\tau \to 0} \frac{1}{(1 + \tau)^{t^*/\tau}} = e^{-t^*}, \quad \text{which implies that} \quad \lim_{\tau \to 0} \|u^\tau - u\|_\infty = 0.$$

That is, the semi-discrete solution converges in the maximum norm to the exact solution as $\tau \to 0$.

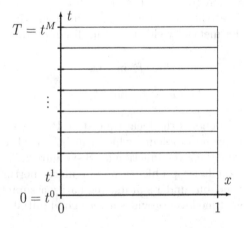

Figure 2.25   Horizontal method of lines

Besides its usefulness as a numerical semi-discretization scheme, Rothe's method is often applied also in existence proofs for time-dependent problems by reducing them to a sequence of stationary problems. Furthermore, other concepts originally developed for stationary problem (e.g., grid generation algorithms) can be transferred via Rothe's method to the time-dependent case. Let us repeat our earlier observation that the elliptic problems that appear in Rothe's method have the main part of the differential operator multiplied by the small parameter τ; thus these stationary problems are singularly perturbed and their analysis requires additional care.

The version of Rothe's method described above can be generalized by replacing the implicit Euler method by more general methods for time discretization. For example, one-step methods can be applied with error control of the discretization error in time (and also in space). In [96] Runge-Kutta methods are studied, including the problem of *order reduction*.

Exercise 2.86. The rotationally symmetric temperature distribution $w(r, t)$ in an infinitely long circular cylinder of radius R satisfies the parabolic differential equation

$$\frac{\partial w}{\partial t}(r, t) = \alpha^2 \frac{1}{r} \frac{\partial}{\partial r} \left(r \frac{\partial}{\partial r} w \right) (r, t) \quad \text{for } r \in (0, R), \, t \in (0, \infty),$$

with the initial and boundary conditions

$$w(r, 0) = h(r), \quad w(R, t) = 0.$$

Here α is a given non-zero constant.

(a) Determine the analytical solution $w(r, t)$ using separation of variables.

(b) Apply the method of (vertical) lines to this problem, using equidistant discretization in space. Determine the system of ordinary differential equations and the initial value problem that are obtained.

Exercise 2.87. Consider the initial-boundary value problem

$$u_t(x, t) - \Delta u(x, t) = 1 \text{ in } \Omega \times (0, T],$$
$$u(x, t) = 0 \text{ in } \Gamma \times (0, T], \quad\quad (6.23)$$
$$u(x, 0) = 0 \text{ for } x \in \bar{\Omega},$$

with $\Omega := (0, 1) \times (0, 1) \subset \mathbb{R}^2$, $\Gamma := \partial\Omega$ and fixed $T > 0$.

(a) Discuss semi-discretization of this problem via Rothe's method, using the step size $\tau := T/M$ in the temporal direction. Determine the boundary value problems that are generated.

(b) Apply the ADI method to (6.23) using equidistant discretization in both spatial directions with step size $h := 1/N$. As step size in time use $\tau := T/M$.

How can the matrix factorizations required be implemented efficiently? Find numerically the approximate solution of (6.23) for $N = 10$ and $M = 5$ with $T = 2$.

Exercise 2.88. Discretize the initial-boundary value problem

$$\frac{\partial u}{\partial t} - \frac{\partial^2 u}{\partial x^2} = \sin x \quad \text{in} \quad (0, \pi/2) \times (0, T),$$

with the conditions

$$u|_{t=0} = 0, \quad u|_{x=0} = 0, \quad u_x|_{x=\pi/2} = 0,$$

using Rothe's method. Compare the computed approximate solution with the exact solution.

2.7 Second-Order Hyperbolic Problems

Let $\Omega \subset \mathbb{R}^n$ be a bounded domain with boundary Γ, and L a second-order differential operator on the spatial derivatives that is uniformly elliptic. The partial differential equation

$$\frac{\partial^2}{\partial t^2} u(x, t) + [Lu](x, t) = f(x, t), \quad x \in \Omega, \ t \in (0, T] \tag{7.1}$$

with the boundary and initial conditions

$$u(x, t) = 0, \qquad\qquad (x, t) \in \Gamma \times (0, T]$$
$$u(x, 0) = p(x), \quad \frac{\partial}{\partial t} u(x, 0) = q(x), \ x \in \overline{\Omega} \tag{7.2}$$

defines a second-order hyperbolic initial-boundary value problem. Here the data f, g, p and q are given.

Let L_h be some discretization of the elliptic differential operator L on a spatial grid Ω_h. Subdivide the time interval $[0, T]$ by the equidistant grid points $t^k := k\tau$, $k = 0, 1, \ldots, M$, where $\tau := T/M$ is the step size in the temporal direction. To approximate the second-order temporal derivative $\frac{\partial^2}{\partial t^2}$ we use the standard discretization $D_t^- D_t^+$, which is second-order consistent. In what follows we consider two basic types of difference methods:

$$D_t^- D_t^+ u^k + L_h u^k = f^k, \qquad k = 1, \ldots, M - 1, \tag{7.3}$$

and

$$D_t^- D_t^+ u^k + \frac{1}{2} L_h \left(u^{k+1} + u^{k-1} \right) = f^k, \qquad k = 1, \ldots, M - 1. \tag{7.4}$$

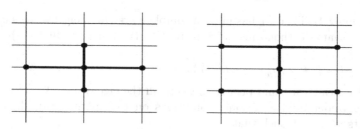

Figure 2.26   Difference stencils related to (7.3), (7.4)

Here $u^k \in U_h^0$ denotes a grid function defined on the spatial grid Ω_h at the discrete time level $t = t^k$. This grid function approximates $u(x, t^k)$, $x \in \Omega_h$. When $n = 1$ (i.e., in one space dimension) and L_h is a three-point scheme, the discretization methods (7.3) and (7.4) are represented by the difference stencils of Figure 2.26. The discretization (7.3) is an explicit scheme: it is equivalent to

$$u^{k+1} = (2I - \tau^2 L_h)\, u^k - u^{k-1} + \tau^2 f^k, \quad k = 1, \ldots, M - 1,$$

so the new grid function u^{k+1} is obtained without solving a system of equations. On the other hand, the discretization (7.4) is an implicit scheme. Here $u^{k+1} \in U_h^0$ is determined by the linear system

$$(I + \tfrac{1}{2}\tau^2 L_h)\, u^{k+1} = 2\, u^k - (I + \tfrac{1}{2}\tau^2 L_h)\, u^{k-1} + \tau^2 f^k, \quad k = 1, \ldots, M - 1.$$

Both methods are used iteratively starting from the two grid functions u^0, $u^1 \in U_h$. While u^0 is given immediately by the initial condition

$$u^0 = p_h, \tag{7.5}$$

the grid function u^1 is determined from a discretization of the second initial condition, e.g., by its approximation $(u^1 - u^0)/\tau = q_h$. Taking (7.5) into account this yields

$$u^1 = p_h + \tau\, q_h. \tag{7.6}$$

Here p_h, $q_h \in U_h$ denote pointwise restrictions of the initial functions p and q to the spatial grid $\overline{\Omega}_h$.

We now analyse the L_2-stability of the methods (7.3) and (7.4). To do this, the discrete solutions $u^k \in U_h$ will be expanded in terms of eigenvectors of the discrete elliptic operator L_h. To ensure that L_h possesses a complete system of eigenvectors, we want L_h to be self-adjoint. Thus assume that $L = -\mathrm{div}(A\,\mathrm{grad})$ for some symmetric positive-definite matrix $A = (a_{ij})$, then define L_h by

$$L_h = -\sum_j D_i^- \sum_j a_{ij} D_j^+$$

(cf. Section 2.4). Then L_h has in U_h^0 a complete system of eigenvectors v_h^j, $j = 1, \ldots, N$, that is orthonormal with respect to the inner product $(\cdot, \cdot)_h$. That is,

$$L_h v_h^j = \lambda_j v_h^j, \qquad (v_h^j, v_h^l)_h = \delta_{jl}, \qquad j, l = 1, \ldots, N. \qquad (7.7)$$

Since A is symmetric and positive definite all the eigenvalues λ_j are real and positive. Under mild additional conditions on the spatial grid, there exist constants c_0, $c_1 > 0$ such that

$$c_0 \le \lambda_j \le c_1 h^{-2}, \quad j = 1, \ldots, N. \qquad (7.8)$$

Here $h > 0$ is an appropriate measure of the fineness of the discretization. In particular, for the one-dimensional case $\Omega = (0, 1) \subset \mathbb{R}$ and two-dimensional case $\Omega = (0, 1)^2 \subset \mathbb{R}^2$ with equidistant grids, the eigenvalues of the discrete Laplace operator are

$$\lambda_j = \frac{4}{h^2} \sin^2 \left(\frac{j \pi h}{2} \right), \quad j = 1, \ldots, N,$$

and

$$\lambda_{j,l} = \frac{4}{h^2} \left[\sin^2 \left(\frac{j \pi h}{2} \right) + \sin^2 \left(\frac{l \pi h}{2} \right) \right], \quad j, l = 1, \ldots, N, \qquad (7.9)$$

respectively, with step size $h = 1/(N+1)$. Hence (7.8) holds in these two cases with $c_0 = \pi^2$, $c_1 = 4$ and $c_0 = 2\pi^2$, $c_1 = 8$, respectively.

Lemma 2.89. *The implicit method (7.4) applied to the homogeneous problem is stable in the discrete L_2 norm for all step sizes h, $\tau > 0$. If for the spatial grid the estimate (7.8) holds with constants c_0, $c_1 > 0$, then the explicit scheme (7.3) is stable in the discrete L_2 norm provided that the temporal and spatial step sizes satisfy $\tau \le 2 c_1^{-1/2} h$.*

Proof: Let $w^k \in U_h^0$ arbitrarily chosen. Since $\{v_h^j\}$ is a basis for U_h^0, there are unique $\xi_j^k \in \mathbb{R}$ such that

$$w^k = \sum_{j=1}^{N} \xi_j^k v_h^j. \qquad (7.10)$$

From (7.7) the homogeneous difference equations associated with (7.3) and (7.4) yield

$$\frac{1}{\tau^2} \left(\xi_j^{k+1} - 2\xi_j^k + \xi_j^{k-1} \right) + \lambda_j \xi_j^k = 0, \qquad j = 1, \ldots, N, \ k = 1, \ldots, M-1$$

and

$$\frac{1}{\tau^2} \left(\xi_j^{k+1} - 2\xi_j^k + \xi_j^{k-1} \right) + \frac{\lambda_j}{2} \left(\xi_j^{k+1} + \xi_j^{k-1} \right) = 0, \qquad \begin{matrix} j = 1, \ldots, N, \\ k = 1, \ldots, M-1, \end{matrix}$$

respectively. Thus for each fixed $j \in \{1, \ldots, N\}$ we have in each case a second-order difference equation with constant coefficients that defines ξ_j^k, $k = 1, \ldots, M$. The related characteristic equations are, writing λ for λ_j,

$$\kappa^2 - (2 - \tau^2 \lambda)\kappa + 1 = 0 \quad \text{and} \quad \left(1 + \frac{\tau^2 \lambda}{2}\right)\kappa^2 - 2\kappa + 1 + \frac{\tau^2 \lambda}{2} = 0,$$

respectively. In the first of these a calculation shows that the characteristic roots κ_1 and κ_2 satisfy

$$|\kappa_{1,2}| \leq 1 \quad \Longleftrightarrow \quad 1 - \frac{\tau^2 \lambda}{2} \geq -1.$$

Here in the extreme case $\tau^2 \lambda = 4$ the characteristic polynomial has a double root $\kappa = 1$. Hence the explicit scheme (7.3) is stable in the L_2 norm if and only if for the time step size τ and for all eigenvalues λ (which depend upon the step size h) one has

$$1 - \frac{\tau^2 \lambda}{2} > -1.$$

Recalling the bounds (7.8) on the eigenvalues, we obtain the sufficient stability condition

$$\tau \leq 2 c_1^{-1/2} h. \tag{7.11}$$

In the case of the implicit scheme, the characteristic equation above has two different conjugate complex zeros κ_1, κ_2 for any step sizes $h, \tau > 0$. These zeros satisfy $|\kappa_{1,2}| = 1$. Hence the implicit method is L_2 stable for any step sizes. ∎

Remark 2.90. Let us draw the reader's attention to the fact that the step size restriction for the explicit scheme for hyperbolic problems is significantly less demanding than the corresponding restriction in the case of parabolic problems (compare Theorem 2.74). This difference is caused by the qualitatively different ways in which the solutions of hyperbolic and parabolic problems develop as time progresses: hyperbolic problems have only a limited speed of propagation while parabolic problems have theoretically an infinite speed of propagation.   □

Stability and consistency together imply convergence. We now study this for a simple two-dimensional model problem. For more general cases see, e.g., [Sam01, Str04] and also the analysis of finite element methods for time-dependent problems that will be given in Chapter 5.

Theorem 2.91. *Let $\Omega = (0,1)^2 \subset \mathbb{R}^2$ with boundary Γ. Suppose that the hyperbolic initial-boundary value problem*

$$u_{tt} - \Delta u = f \text{ in } \Omega \times (0, T],$$

$$u = g \text{ on } \Gamma \times (0, T], \quad u = p, \quad u_t = q \quad \text{on} \quad \overline{\Omega}, \tag{7.12}$$

has a sufficiently smooth solution u. If this problem is discretized on an equidistant spatial grid where the initial conditions are handled by (7.5) and (7.6), then for the implicit scheme (7.4) one has the error estimate

$$\max_{0 \leq k \leq M} \|u^k - r_h u(\cdot, t^k)\|_{2,h} = O(h^2 + \tau^2). \tag{7.13}$$

If the temporal step size and the spatial step size are coupled by the condition $\tau \leq 4\sqrt{2}\,h$, then the estimate (7.13) is also valid for the explicit scheme (7.3).

Proof: We first examine the consistency error. Now $u^0 = r_h p = r_h u(\cdot, t^0)$ so

$$\|u^0 - r_h u(\cdot, t^0)\|_{2,h} = 0.$$

Next, $u(x_h, \tau) = u(x_h, 0) + \tau\, u_t(x_h, 0) + O(\tau^2)$ for all $x_h \in \Omega_h$, $t^1 = \tau$ and (7.6) yield

$$\|u^1 - r_h u(\cdot, t^1)\|_{2,h} = O(\tau^2).$$

Setting $w^k := u^k - r_h u(\cdot, t^k)$ for $k = 0, 1, \ldots, M$, we have

$$\|w^0\|_{2,h} = 0, \qquad \|w^1\|_{2,h} = O(\tau^2). \tag{7.14}$$

For subsequent time levels, a Taylor expansion gives

$$\frac{1}{\tau^2}\left(w^{k+1} - 2w^k + w^{k-1}\right) + \frac{1}{2}L_h(w^{k+1} + w^{k-1}) = O(\tau^2 + h^2)$$

for the implicit scheme (7.4) and

$$\frac{1}{\tau^2}\left(w^{k+1} - 2w^k + w^{k-1}\right) + L_h w^k = O(\tau^2 + h^2)$$

for the explicit scheme (7.3).

The stability of both these numerical methods is dealt with by Lemma 2.89, since one can take $c_1 = 8$ from (7.9). The desired error estimates follow. ∎

Remark 2.92. An alternative way of discretizing the initial condition $u_t(\cdot, 0) = q$ is to use the central difference quotient $\dfrac{u^1 - u^{-1}}{2\tau} = q_h$. In this case the auxiliary grid function u^{-1} can be eliminated by introducing the additional condition

$$D_t^- D_t^+ u^0 + L_h u^0 = f^0,$$

which implies that

$$u^{-1} = 2u^0 - u^1 + \tau^2(f^0 - L_h u^0).$$

Substituting this into the central difference quotient above and setting $u^0 = p_h$, we get

$$u^1 = p_h + \frac{\tau^2}{2}\left(f^0 - L_h\, p_h\right) + \tau\, q_h. \tag{7.15}$$

□

Semi-discretization (the method of lines) can also be applied to second-order hyperbolic problems, as was done for parabolic initial-boundary value problems in Section 2.6.3. The spatial discretization required can be got from finite difference or finite element methods. Using finite differences, we sketch the semi-discretization for the model problem (7.1), (7.2). Let $\Omega_h = \{x_i\}_{i=1}^N$ denote some grid in Ω, with L_h some discretization on Ω_h of the elliptic operator L. Let $u_i : [0, T] \to \mathbb{R}$, $i = 1, \ldots, N$, denote functions

$$u_i(t) \approx u(x_i, t), \quad t \in [0, T],$$

that are associated with the grid points $x_i \in \Omega_h$. In (7.1), (7.2) replace the operator L by L_h and set $f_i := f(x_i, \cdot)$. We then obtain an approximation of the original problem by the following initial-value problem for a system of ordinary differential equations:

$$\left.\begin{aligned}
\ddot{u}_i + [L_h u_h]_i &= f_i, \\
u_i(0) &= p(x_i), \\
\dot{u}_i(0) &= q(x_i),
\end{aligned}\right\} \quad i = 1, \ldots, N, \tag{7.16}$$

where $u_h(t) := (u_1(t), \ldots, u_N(t))$. If the ODE system (7.16) is solved numerically by some suitable integration method (e.g., a BDF method), one has then a complete discretization of the original problem.

Unlike the parabolic case, semi-discretization of second-order hyperbolic problems leads to a system of second-order ordinary differential equations. This could be transformed in the usual way into a first-order system with twice the number of unknown functions then solved by some standard code, but we do not recommend this. Instead one should solve (7.16) using methods that make use of the specific structure of this second-order problem. One such method (compare Exercise 2.93), which could be used for the complete discretization via (7.16), is the *Newmark method* (see [JL01]), which was originally developed in the engineering literature.

If constant time steps $\tau > 0$ are taken, then the Newmark method applied to (7.16) is defined as follows:

$$\begin{aligned}
z_h^{k+1} + L_h u_h^{k+1} &= f_h^{k+1}, \\
v_h^{k+1} &= v_h^k + \tau\left((1 - \gamma)\, z_h^k + \gamma\, z_h^{k+1}\right), \\
u_h^{k+1} &= u_h^k + \tau\, v_h^k + \frac{\tau^2}{2}\left((1 - 2\beta)\, z_h^k + 2\beta\, z_h^{k+1}\right).
\end{aligned} \tag{7.17}$$

Here u_h^k, v_h^k and z_h^k denote the approximations for $u_h(t^k)$, the semi-discrete velocity $\dot{u}_h(t^k)$ and the acceleration $\ddot{u}_h(t^k)$, respectively. The real-valued quantities γ and β are parameters of the method. If $2\beta \geq \gamma \geq \frac{1}{2}$ then the Newmark method is stable for all step sizes τ, $h > 0$ in time and space.

Exercise 2.93. Consider the following initial-boundary problem for the one-dimensional wave equation:

$$
\begin{aligned}
u_{tt} - a^2 u_{xx} &= f, && x \in (0,1), \ t \in (0,T], \\
u(0,t) = u(1,t) &= 0, \\
u(x,0) &= p(x), \\
u_t(x,0) &= q(x).
\end{aligned}
$$

Determine the order of consistency of the discretization

$$
\frac{1}{\tau^2}\left(u^{k+1} - 2u^k + u^{k-1}\right) - \frac{a^2}{12}\left(\Delta_h u^{k+1} + 10\Delta_h u^k + \Delta_h u^{k-1}\right) \tag{7.18}
$$
$$
= \frac{1}{12}\left(f^{k+1} + 10f^k + f^{k-1}\right)
$$

of the wave equation. Here $u^k = (u_i^k) \in \mathbb{R}^{N-1}$,

$$
\left[\Delta_h u^k\right]_i := \frac{1}{h^2}\left(u_{i+1}^k - 2u_i^k + u_{i-1}^k\right), \quad i = 1, \ldots, N-1,
$$

denotes the central difference quotient approximation of u_{xx}, and $h, \tau > 0$ are the spatial and temporal step sizes.

Investigate the L_2 stability of the scheme (7.18).

Exercise 2.94. Analogously to the parabolic case, develop an ADI version of the implicit method (7.4) and analyse its order of consistency.

3

Weak Solutions, Elliptic Problems and Sobolev Spaces

3.1 Introduction

In Chapter 2 we discussed difference methods for the numerical treatment of partial differential equations. The basic idea of these methods was to use information from a discrete set of points to approximate derivatives by difference quotients.

Now we start to discuss a different class of discretization methods: the so-called *ansatz methods*. An ansatz method is characterized by prescribing some approximate solution in a certain form. In general, this is done by determining the coefficients in a linear combination of a set of functions chosen by the numerical analyst. One cannot then expect to get an exact solution of the differential equation in all cases. Thus a possible strategy is to determine the coefficients in a way that approximately satisfies the differential equation (and perhaps some additional conditions).

For instance, one can require that the differential equation be satisfied at a specified discrete set of points; this method is called collocation. It is, however, much more popular to use methods that are based on a weak formulation of the given problem. Methods of this type do not assume that the differential equation holds at every point. Instead, they are based on a related variational problem or variational equation. The linear forms defined by the integrals in the variational formulation require the use of appropriate function spaces to guarantee, for instance, the existence of weak solutions. It turns out that existence theorems for weak solutions are valid under assumptions that are much more realistic than in the corresponding theorems for classical solutions. Moreover, ansatz functions can have much less smoothness than, e.g., functions used in collocation methods where the pointwise validity of the differential equation is required.

As a first simple example let us consider the two-point boundary value problem

$$-u''(x) + b(x)u'(x) + c(x)u(x) = f(x) \quad \text{in} \quad \Omega := (0,1), \quad (1.1)$$

$$u(0) = u(1) = 0 \,. \tag{1.2}$$

Let b, c and f be given continuous functions. Assume that a classical solution exists, i.e., a twice continuously differentiable function u that satisfies (1.1) and (1.2). Then for an arbitrary continuous function v we have

$$\int_\Omega (-u'' + bu' + cu)v \, dx = \int_\Omega fv \, dx. \tag{1.3}$$

The reverse implication is also valid: if a function $u \in C^2(\bar{\Omega})$ satisfies equation (1.3) for all $v \in C(\bar{\Omega})$, then u is a classical solution of the differential equation (1.1).

If $v \in C^1(\bar{\Omega})$, then we can integrate by parts in (1.3) and obtain

$$-u'v \, |_{x=0}^{1} + \int_\Omega u'v' \, dx + \int_\Omega (bu' + cu)v \, dx = \int_\Omega fv \, dx.$$

Under the additional condition $v(0) = v(1) = 0$ this is equivalent to

$$\int_\Omega u'v' \, dx + \int_\Omega (bu' + cu)v \, dx = \int_\Omega fv \, dx. \tag{1.4}$$

Unlike (1.1) or (1.3), equation (1.4) still makes sense if we know only that $u \in C^1(\bar{\Omega})$. But we have not yet specified a topological space in which mappings implicitly defined by a weak form of (1.1) such as (1.4) have desirable properties like continuity, boundedness, etc. It turns out that Sobolev spaces, which generalize L_p spaces to spaces of functions whose generalized derivatives also lie in L_p, are the correct setting in which to examine weak formulations of differential equations. The book [Ada75] presents an excellent general survey of Sobolev spaces. In Section 3.2 we shall give some basic properties of Sobolev spaces that will allow us to analyse discretization methods—at least in standard situations.

But first we explain the relationship of the simple model problem (1.1), (1.2) to variational problems. Assume that $b(x) \equiv 0$ and $c(x) \geq 0$. Define a functional J by

$$J(u) := \frac{1}{2} \int_\Omega (u'^2 + cu^2) \, dx - \int_\Omega fu \, dx. \tag{1.5}$$

Consider now the following problem: Find a function $u \in C^1(\bar{\Omega})$ with $u(0) = u(1) = 0$ such that

$$J(u) \leq J(v) \quad \text{for all} \quad v \in C^1(\bar{\Omega}) \quad \text{with} \quad v(0) = v(1) = 0. \tag{1.6}$$

For such problems a necessary condition for optimality is well known: the first variation $\delta J(u, v)$ must vanish for arbitrarily admissible directions v (see, for instance, [Zei90]). This first variation is defined by $\delta J(u, v) := \Phi'(0)$ where $\Phi(t) := J(u + tv)$ for fixed u, v and real t.

For the functional $J(\cdot)$ defined by (1.5), one has

$$J(u + tv) = \frac{1}{2} \int\limits_{\Omega} [(u' + tv')^2 + c(u + tv)^2] \, dx - \int\limits_{\Omega} f \cdot (u + tv) \, dx \,,$$

and consequently

$$\Phi'(0) = \int\limits_{\Omega} u'v' \, dx + \int\limits_{\Omega} c \, uv \, dx - \int\limits_{\Omega} fv \, dx.$$

Thus in the case $b(x) \equiv 0$, the condition $\delta J(u, v) = 0$ necessary for optimality in (1.6) is equivalent to the *variational equation* (1.4). This equivalence establishes a close connection between boundary value problems and variational problems. The differential equation (1.1) is described as the Euler equation of the variational problem (1.6). The derivation of Euler equations for general variational problems that are related to boundary value problems is discussed in [Zei90]. Later we shall discuss in more detail the role played by the condition $v(0) = v(1) = 0$ in the formulation of (1.4).

Variational problems often appear when modelling applied problems in the natural and technical sciences because in many situations nature follows minimum or maximum laws such as the principle of minimum energy.

Next we consider a simple elliptic model problem in two dimensions. Let $\Omega \subset \mathbb{R}^2$ be a simply connected open set with a (piecewise) smooth boundary Γ. Let $f : \bar{\Omega} \to \mathbb{R}$ be a given function. We seek a twice differentiable function u that satisfies

$$-\Delta u(\xi, \eta) = f(\xi, \eta) \qquad \text{in} \quad \Omega, \tag{1.7}$$

$$u|_{\Gamma} = 0. \tag{1.8}$$

To derive a variational equation, we again take a continuous function v, multiply (1.7) by v and integrate:

$$-\int\limits_{\Omega} \Delta u \, v \, dx = \int\limits_{\Omega} fv \, dx.$$

If $v \in C^1(\bar{\Omega})$ with $v|_{\Gamma} = 0$, the application of an integral theorem (the two-dimensional analogue of integration by parts—see the next section for details) yields

$$\int\limits_{\Omega} \left(\frac{\partial u}{\partial \xi} \frac{\partial v}{\partial \xi} + \frac{\partial u}{\partial \eta} \frac{\partial v}{\partial \eta} \right) dx = \int\limits_{\Omega} fv \, dx. \tag{1.9}$$

This is the variational equation derived from the boundary value problem (1.7), (1.8). In the opposite direction, assuming $u \in C^2(\bar{\Omega})$, we can infer from (1.9) that u satisfies the Poisson equation (1.7).

So far we have said nothing about the existence and uniqueness of solutions in our new variational formulation of boundary value problems, because to deal adequately with these topics it is necessary to work in the framework of Sobolev spaces. In the following sections we introduce these spaces and discuss not only existence and uniqueness of solutions to variational problems but also the numerical approximation of these solutions by means of certain ansatz functions.

3.2 Function Spaces for the Variational Formulation of Boundary Value Problems

In the classical treatment of differential equations, the solution and certain of its derivatives are required to be continuous functions. One therefore works in the spaces $C^k(\bar{\Omega})$ that contain functions with continuous derivatives up to order k on the given domain Ω, or in spaces where these derivatives are Hölder continuous.

When the strong form (e.g. (1.7)) of a differential equation is replaced by a variational formulation, then instead of pointwise differentiability we need only ensure the existence of some integrals that contain the unknown function as certain derivatives. Thus it makes sense to use function spaces that are specially suited to this situation.

We start with some basic facts from functional analysis.

Let U be a linear (vector) space. A mapping $\|\cdot\| : U \to \mathbb{R}$ is called a *norm* if it has the following properties:

i) $\|u\| \geq 0$ for all $u \in U$, $\|u\| = 0$ $\Leftrightarrow$ $u = 0$,

ii) $\|\lambda u\| = |\lambda|\,\|u\|$ for all $u \in U,\ \lambda \in \mathbb{R}$,

iii) $\|u + v\| \leq \|u\| + \|v\|$ for all $u, v \in U$.

A linear space U endowed with a norm is called a *normed space*. A sequence $\{u^k\}$ in a normed space is a *Cauchy sequence* if for each $\varepsilon > 0$ there exists a number $N(\varepsilon)$ such that

$$\|u^k - u^l\| \leq \varepsilon \qquad \text{for all}\quad k, l \geq N(\varepsilon)\,.$$

The next property is of fundamental importance both in existence theorems for solutions of variational problems and in proofs of convergence of numerical methods. A normed space is called *complete* if every Cauchy sequence $\{u^k\} \subset U$ converges in U, i.e., there exists a $u \in U$ with

$$\lim_{k \to \infty} \|u^k - u\| = 0.$$

Equivalently,

$$u = \lim_{k \to \infty} u^k.$$

Complete normed spaces are often called *Banach spaces*.

Let U, V be two normed spaces with norms $\|\cdot\|_U$ and $\|\cdot\|_V$ respectively. A mapping $P : U \to V$ is continuous at $u \in U$ if for any sequence $\{u^k\} \subset U$ converging to u one has

$$\lim_{k \to \infty} Pu^k = Pu,$$

i.e.,

$$\lim_{k \to \infty} \|u^k - u\|_U = 0 \quad \Rightarrow \quad \lim_{k \to \infty} \|Pu^k - Pu\|_V = 0.$$

A mapping is continuous if it is continuous at every point $u \in U$.

A mapping $P : U \to V$ is called *linear* if

$$P(\lambda u + \mu v) = \lambda Pu + \mu Pv \text{ for all } u, v \in U, \quad \lambda, \mu \in \mathbb{R}.$$

A linear mapping is *continuous* if there exists a constant $M \geq 0$ such that

$$\|Pu\| \leq M\|u\| \text{ for all } u \in U.$$

A mapping $f : U \to \mathbb{R}$ is usually called a functional. Consider the set of all continuous linear functionals $f : U \to \mathbb{R}$. These form a normed space with norm defined by

$$\|f\|_* := \sup_{v \neq 0} \frac{|f(v)|}{\|v\|} .$$

This space is in fact a Banach space. It is the *dual space* U^* of U. When $f \in U^*$ and $u \in U$ we shall sometimes write $\langle f, u \rangle$ instead of $f(u)$.

Occasionally it is useful to replace convergence in the normed space by convergence in a weaker sense: if

$$\lim_{k \to \infty} \langle f, u^k \rangle = \langle f, u \rangle \text{ for all } f \in U^*$$

for a sequence $\{u^k\} \subset U$ and $u \in U$, then we say that the sequence $\{u^k\}$ *converges weakly* to u. It is standard to use the notation

$$u^k \rightharpoonup u \quad \text{for} \quad k \to \infty$$

to denote weak convergence. If $u = \lim_{k \to \infty} u^k$ then $u^k \rightharpoonup u$, i.e. convergence implies weak convergence, but the converse is false: a weakly convergent sequence is not necessarily convergent.

It is particularly convenient to work in linear spaces that are endowed with a scalar product. A mapping $(\cdot, \cdot) : U \times U \to \mathbb{R}$ is called a (real-valued) *scalar product* if U has the following properties:

i) $\quad (u, u) \geq 0 \quad$ for all $\quad u \in U, \quad (u, u) = 0 \quad \Leftrightarrow \quad u = 0,$

ii) $\quad (\lambda u, v) = \lambda(u, v) \quad$ for all $\quad u, v \in U, \quad \lambda \in \mathbb{R},$

iii) $\quad (u, v) = (v, u) \quad$ for all $\quad u, v \in U,$

iv) $\quad (u + v, w) = (u, w) + (v, w) \quad$ for all $\quad u, v, w \in U.$

Given a scalar product, one can define an induced norm by $\|u\| := \sqrt{(u, u)}$. But not all norms are induced by related scalar products.

A Banach space in which the norm is induced by a scalar product is called a (real) *Hilbert space*. From the properties of the scalar product one can deduce the useful *Cauchy-Schwarz inequality*:

$$|(u, v)| \leq \|u\| \, \|v\| \qquad \text{for all} \quad u, v \in U.$$

Continuous linear functionals on Hilbert spaces have a relatively simple structure that is important in many applications. It is stated in the next result.

Theorem 3.1 (Riesz). *Let $f : V \to \mathbb{R}$ be a continuous linear functional on a Hilbert space V. Then there exists a unique $w \in V$ such that*

$$(w, v) = f(v) \qquad \text{for all } v \in V.$$

Moreover, one has $\|f\|_* = \|w\|$.

The Lebesgue spaces of integrable functions are the starting point for the construction of the Sobolev spaces. Let $\Omega \subset \mathbb{R}^n$ (for $n = 1, 2, 3$) be a bounded domain (i.e., open and connected) with boundary $\Gamma := \partial\Omega$. Let $p \in [1, +\infty)$. The class of all functions whose p-th power is integrable on Ω is denoted by

$$L_p(\Omega) := \left\{ v : \int_\Omega |v(x)|^p \, dx < +\infty \right\}.$$

Furthermore

$$\|v\|_{L_p(\Omega)} := \left[\int_\Omega |v(x)|^p \, dx \right]^{1/p}$$

is a norm on L_p. It is important to remember that we work with Lebesgue integrals (see, e.g., [Wlo87]), so all functions that differ only on a set of measure zero are identified. It is in this sense that $\|v\| = 0$ implies $v = 0$. Moreover, the space $L_p(\Omega)$ is complete, i.e., is a Banach space.

In the case $p = 2$ the integral

$$(u, v) := \int_\Omega u(x) \, v(x) \, dx$$

defines a scalar product, so $L_2(\Omega)$ is a Hilbert space.

The definition of these spaces can be extended to the case $p = \infty$ with

$$L_\infty(\Omega) := \left\{ v : \operatorname*{ess\,sup}_{x \in \Omega} |v(x)| < +\infty \right\}$$

and associated norm

$$\|v\|_{L_\infty(\Omega)} := \operatorname*{ess\,sup}_{x \in \Omega} |v(x)|.$$

Here ess sup denotes the essential supremum, i.e., the lowest upper bound over Ω excluding subsets of Ω of Lebesgue measure zero.

To treat differential equations, the next step is to introduce derivatives into the definitions of suitable spaces. In extending the Lebesgue spaces to Sobolev spaces one needs generalized derivatives, which we now describe. For the reader familiar with derivatives in the sense of distributions this introduction will be straightforward.

Denote by $cl_V A$ the closure of a subset $A \subset V$ with respect to the topology of the space V. For $v \in C(\bar{\Omega})$ the *support* of v is then defined by

$$supp\, v := cl_{\mathbb{R}^n} \{x \in \Omega : v(x) \neq 0\}.$$

For our bounded domain Ω, set

$$C_0^\infty(\Omega) := \{v \in C^\infty(\Omega) : \quad supp\, v \subset \Omega\,\}.$$

In our further considerations the role of integration by parts in several dimensions is very important. For instance, for arbitrary $u \in C^1(\bar{\Omega})$ and $v \in C_0^\infty(\Omega)$ one has

$$\int_\Omega \frac{\partial u}{\partial x_i} v\, dx = \int_\Gamma uv \cos(n, e^i)\, ds - \int_\Omega u \frac{\partial v}{\partial x_i}\, dx$$

where e^i is the unit vector in the ith coordinate direction and n is the outward-pointing unit vector normal to Γ. Taking into account that $v|_\Gamma = 0$ we get

$$\int_\Omega u \frac{\partial v}{\partial x_i}\, dx = - \int_\Omega \frac{\partial u}{\partial x_i} v\, dx. \tag{2.1}$$

This identity is the starting point for the generalization of standard derivatives on Lebesgue spaces.

First we need more notation. To describe partial derivatives one uses a multi-index $\alpha := (\alpha_1, \ldots, \alpha_n)$ where each α_i is a non-negative integer. Set $|\alpha| = \sum_i \alpha_i$. We introduce

$$D^\alpha u := \frac{\partial^{|\alpha|}}{\partial x_1^{\alpha_1} \cdots x_n^{\alpha_n}} u$$

for the derivative of order $|\alpha|$ with respect to the multi-index α.

Now, recalling (2.1), we say that an integrable function u is in a generalized sense differentiable with respect to the multi-index α if there exists an integrable function w with

$$\int_\Omega u D^\alpha v\, dx = (-1)^{|\alpha|} \int_\Omega wv\, dx \quad \text{for all } v \in C_0^\infty(\Omega). \tag{2.2}$$

The function $D^\alpha u := w$ is called the *generalized derivative* of u with respect to the multi-index α.

Applying this definition to each first-order coordinate derivative, we obtain a generalized gradient ∇u. Furthermore, if for a componentwise integrable vector-valued function $\underline{u}$ there exists a integrable function z with

$$\int_\Omega \underline{u} \, \nabla v \, dx = - \int_\Omega z v \, dx \qquad \text{for all } v \in C_0^\infty(\Omega),$$

then we call z the *generalized divergence* of $\underline{u}$ and we write $\operatorname{div} \underline{u} := z$.

Now we are ready to define the Sobolev spaces. Let l be a non-negative integer. Let $p \in [2, \infty)$. Consider the subspace of all functions from $L_p(\Omega)$ whose generalized derivatives up to order l exist and belong to $L_p(\Omega)$. This subspace is called the *Sobolev space* $W_p^l(\Omega)$ (Sobolev, 1938). The norm in $W_p^l(\Omega)$ is chosen to be

$$\|u\|_{W_p^l(\Omega)} := \left[\int_\Omega \sum_{|\alpha| \le l} |[D^\alpha u](x)|^p \, dx \right]^{1/p}. \tag{2.3}$$

Starting from $L_\infty(\Omega)$, the Sobolev space $W_\infty^l(\Omega)$ is defined analogously.

Today it is known that Sobolev spaces can be defined in several other equivalent ways. For instance, Meyers and Serrin (1964) proved the following (see [Ada75]):

> *For $1 \le p < \infty$ the space $C^\infty(\Omega) \cap W_p^l(\Omega)$ is dense in $W_p^l(\Omega)$.* \tag{2.4}

That is, for these values of p the space $W_p^l(\Omega)$ can be generated by completing the space $C^\infty(\Omega)$ with respect to the norm defined by (2.3). In other words,

$$W_p^l(\Omega) = cl_{W_p^l(\Omega)} \, C^\infty(\Omega).$$

This result makes clear that one can approximate functions in Sobolev spaces by functions that are differentiable in the classical sense. Hence, various desirable properties of Sobolev spaces can be proved by first verifying them for classical functions and then using the above density identity to extend them to Sobolev spaces.

When $p = 2$ the spaces $W_p^l(\Omega)$ are Hilbert spaces with scalar product

$$(u, v) = \int_\Omega \left(\sum_{|\alpha| \le l} D^\alpha u D^\alpha v \right) dx. \tag{2.5}$$

It is standard to use the notation $H^l(\Omega)$ in this case, i.e., $H^l(\Omega) = W_2^l(\Omega)$. In the treatment of second-order elliptic boundary value problems the Sobolev spaces $H^1(\Omega)$ play a fundamental role, while for fourth-order elliptic problems one uses the spaces $H^2(\Omega)$.

If additional boundary conditions come into the game, then additional information concerning certain subspaces of these Sobolev spaces is required. Let us first introduce the spaces

$$\mathring{W}^l_p(\Omega) := cl_{W^l_p(\Omega)} C^\infty_0(\Omega).$$

In the case $p = 2$ these spaces are Hilbert spaces with the same scalar product as in (2.5), and they are denoted by $H^l_0(\Omega)$. When $l = 1$ this space can be considered as a subspace of $H^1(\Omega)$ comprising those functions that vanish (in a certain sense) on the boundary Γ; we shall explain this in detail later in our discussion of traces following following Lemma 3.3. The standard notation for the dual spaces of the Sobolev spaces $H^l_0(\Omega)$ is

$$H^{-l}(\Omega) := \left(H^l_0(\Omega) \right)^*. \tag{2.6}$$

In some types of variational inequalities—for instance, in mixed formulations of numerical methods—we shall also need special spaces of vector-valued functions. As an example we introduce

$$H(div; \Omega) := \{ \underline{u} \in L_2(\Omega)^n \ : \ \text{div}\,\underline{u} \in L_2(\Omega) \} \tag{2.7}$$

with

$$\|\underline{u}\|^2_{div,\Omega} := \|\underline{u}\|^2_{H(div;\Omega)} := \sum_{i=1}^n \|u_i\|^2_{L_2(\Omega)} + \|\text{div}\,\underline{u}\|^2_{L_2(\Omega)}. \tag{2.8}$$

Now we begin to use Sobolev spaces in the weak formulation of boundary value problems. Our first example is the Poisson equation (1.7) with homogeneous Dirichlet boundary conditions (1.8). The *variational problem* related to that example can be stated precisely in the following way:

Find $u \in H^1_0(\Omega)$ such that

$$\int_\Omega \left(\frac{\partial u}{\partial x_1} \frac{\partial v}{\partial x_1} + \frac{\partial u}{\partial x_2} \frac{\partial v}{\partial x_2} \right) dx = \int_\Omega fv\,dx \quad \text{for all } v \in H^1_0(\Omega). \tag{2.9}$$

The derivatives here are generalized derivatives and the choice of spaces is made to ensure the existence of all integrals. Every classical solution of the Dirichlet problem (1.7), (1.8) satisfies the variational equation (2.9), as we already saw in (1.9), using integration by parts. But is a weak solution in the sense of (2.9) also a classical solution? To answer this question we need further properties of Sobolev spaces. In particular we need to investigate the following:

- What classical differentiability properties does the *weak solution* of the variational problem (2.9) possess?
- In what sense does the weak solution satisfy the boundary conditions?

To address these issues one needs theorems on regularity, embedding and traces for Sobolev spaces, which we now discuss.

The validity of embedding and trace theorems depends strongly on the properties of the boundary of the given domain. It is not our aim to discuss here minimal boundary assumptions for these theorems, as this is a delicate task; results in many cases can be found in [Ada75].

Here we assume generally—as already described in Chapter 1.3—that for every point of the boundary $\partial\Omega$ there exists a local coordinate system in which the boundary corresponds to some hypersurface with the domain Ω lying on one side of that surface. The regularity class of the boundary (and domain) is defined by the smoothness of the boundary's parametrization in this coordinate system: we distinguish between Lipschitz, C^k and C^∞ boundaries and domains.

Many other characterizations of boundaries are also possible.

In most practical applications it is sufficient to consider Lipschitz domains. In two dimensions, a polygonal domain is Lipschitz if all its interior angles are less than 2π, i.e., if the domain contains no slits.

Let U, V be normed spaces with norms $\|\cdot\|_U$ and $\|\cdot\|_V$. We say the space U is *continuously embedded* into V if $u \in V$ for all $u \in U$ and moreover there exists a constant $c > 0$ such that

$$\|u\|_V \leq c\|u\|_U \quad \text{for all } u \in U. \tag{2.10}$$

Symbolically, we write $U \hookrightarrow V$ for the continuous embedding of U into V. The constant c in inequality (2.10) is called the embedding constant .

The obvious embedding

$$W_p^l(\Omega) \hookrightarrow L_p(\Omega) \text{ for every integer } l \geq 0$$

is a direct consequence of the definitions of the spaces $W_p^l(\Omega)$ and $L_p(\Omega)$ and their norms. It is more interesting to study the imbedding relations between different Sobolev spaces or between Sobolev spaces and the classical spaces $C^k(\bar\Omega)$ and $C^{k,\beta}(\bar\Omega)$ with $\beta \in (0,1)$. The corresponding norms are

$$\|v\|_{C^k(\bar\Omega)} = \sum_{|\alpha|\leq k} \max_{x\in\bar\Omega} |[D^\alpha v](x)|,$$

$$\|v\|_{C^{k,\beta}(\bar\Omega)} = \|v\|_{C^k(\bar\Omega)} + \sum_{|\alpha|=k} |D^\alpha v|_{C^\beta(\bar\Omega)}$$

with the Hölder seminorm

$$|v|_{C^\beta(\bar\Omega)} = \inf\{c : |v(x) - v(y)| \leq c|x - y|^\beta \text{ for all } x, y \in \bar\Omega\}.$$

Then one has the following important theorem (see [Ada75, Wlo87]):

Theorem 3.2 (Embedding theorem). *Let $\Omega \subset \mathbb{R}^n$ be a bounded domain with Lipschitz boundary. Assume that $0 \leq j \leq k$, $1 \leq p,q < +\infty$ and $0 < \beta < 1$.*

i) For $k - j \geq n\left(\frac{1}{p} - \frac{1}{q}\right)$ one has the continuous embeddings

$$W_p^k(\Omega) \hookrightarrow W_q^j(\Omega), \qquad \mathring{W}_p^k(\Omega) \hookrightarrow \mathring{W}_q^j(\Omega).$$

ii) For $k - j - \beta > \frac{n}{p}$ one has the continuous embeddings

$$W_p^k(\Omega) \hookrightarrow C^{j,\beta}(\bar{\Omega}).$$

Note that the definition of the Hölder spaces $C^{j,\beta}(\bar{\Omega})$ shows that they are continuously embedded into $C^j(\bar{\Omega})$, i.e.,

$$C^{j,\beta}(\bar{\Omega}) \hookrightarrow C^j(\bar{\Omega}).$$

Next we study the behaviour of restrictions of functions $u \in W_p^l(\Omega)$ to the boundary Γ, which is a key step in understanding the treatment of boundary conditions in weak formulations. The following lemma from [Ada75] is the basic tool.

Lemma 3.3 (Trace lemma). *Let Ω be a bounded domain with Lipschitz boundary Γ. Then there exists a constant $c > 0$ such that*

$$\|u\|_{L_p(\Gamma)} \leq c \|u\|_{W_p^1(\Omega)} \text{ for all } u \in C^1(\bar{\Omega}).$$

Lemma 3.3 guarantees the existence of a linear continuous mapping

$$\gamma : W_p^1(\Omega) \to L_p(\Gamma)$$

which is called the *trace mapping*. The image of $W_p^1(\Omega)$ under this mapping is a subspace of $L_p(\Gamma)$ that is a new function space defined on the boundary Γ. For us the case $p = 2$ is particularly important; we then obtain

$$H^{1/2}(\Gamma) := \{\, w \in L_2(\Gamma) \ : \ \text{there exists a } v \in H^1(\Omega) \text{ with } w = \gamma v \,\}.$$

It is possible to define a norm on $H^{1/2}(\Gamma)$ by

$$\|w\|_{H^{1/2}(\Gamma)} = \inf\{\, \|v\|_{H^1(\Omega)} \ : \ v \in H^1(\Omega), \ w = \gamma v \,\}.$$

The space dual to $H^{1/2}(\Gamma)$ is denoted by $H^{-1/2}(\Gamma)$, and its norm is given by

$$\|g\|_{H^{-1/2}(\Gamma)} = \sup_{w \in H^{1/2}(\Gamma)} \frac{|g(w)|}{\|w\|_{H^{1/2}(\Gamma)}}.$$

The relationship between the spaces $H^1(\Omega)$ and $H^{1/2}(\Gamma)$ allows a characterization of the norms in $H^{1/2}(\Gamma)$ and $H^{-1/2}(\Gamma)$ by means of suitably defined variational inequalities; see [BF91].

Taking into account the definition of the spaces $\mathring{W}_p^l(\Omega)$, Lemma 3.3 implies that

$$\gamma u = 0 \text{ for all } u \in \mathring{W}_p^1(\Omega)$$

and

$$\gamma D^\alpha u = 0 \text{ for all } u \in \mathring{W}_p^l(\Omega) \text{ and } |\alpha| \leq l - 1.$$

In handling boundary value problems, we usually need not only the norms $\|\cdot\|_{W_p^l(\Omega)}$ defined by (2.3) but also the associated seminorms

$$|u|_{W_p^s(\Omega)} := \left[\int_\Omega \sum_{|\alpha|=s} |[D^\alpha u](x)|^p \mathrm{d}x \right]^{1/p}.$$

It is clear that these seminorms can be estimated by the foregoing norms:

$$|v|_{W_p^s(\Omega)} \le \|v\|_{W_p^l(\Omega)} \text{ for all } v \in W_p^l(\Omega) \text{ and } 0 \le s \le l. \tag{2.11}$$

Is a converse inequality true (at least for certain v)? Here the following result plays a fundamental role.

Lemma 3.4. *Let $\Omega \subset \mathbb{R}^n$ be a bounded domain. Then there exists a constant $c > 0$ such that*

$$\|v\|_{L_2(\Omega)} \le c|v|_{W_2^1(\Omega)} \quad \text{for all } v \in H_0^1(\Omega). \tag{2.12}$$

Inequality (2.12) is known as the *Friedrichs inequality*. Once again a proof is in [Ada75].

Remark 3.5. The smallest constant c in Friedrichs' inequality can be characterized as the reciprocal of the minimal eigenvalue λ of the problem

$$-\Delta u = \lambda u \text{ on } \Omega, \quad u|_{\partial\Omega=0}.$$

For parallelepipeds Ω the value of this eigenvalue is known. Furthermore, the eigenvalue does not increase in value if the domain is enlarged. Consequently in many cases one can compute satisfactory bounds for the constant in (2.12).

A detailed discussion of the values of constants in many fundamental inequalities related to Sobolev spaces can be found in the book [Mik86]. □

From Lemma 3.4 and (2.11) it follows that

$$c_1\|v\|_{W_2^1(\Omega)} \le |v|_{W_2^1(\Omega)} \le \|v\|_{W_2^1(\Omega)} \quad \text{for all } v \in H_0^1(\Omega) \tag{2.13}$$

for some constant $c_1 > 0$. Therefore the definition

$$\|v\| := |v|_{W_2^1(\Omega)}$$

is a new norm on $H_0^1(\Omega)$, which is equivalent to the H^1 norm. This norm is often used as the natural norm on $H_0^1(\Omega)$. It is induced by the scalar product

$$(u, v) := \int_\Omega \sum_{|\alpha|=1} D^\alpha u D^\alpha v \, dx.$$

Using this scalar product, the unique solvability of the weak formulation of the Poisson equation with homogeneous boundary conditions follows immediately from Riesz's theorem if the linear functional defined by

$$v \mapsto \int_\Omega fv \, dx$$

is continuous on $H_0^1(\Omega)$. This is true when $f \in L_2(\Omega)$, for instance.

Based on the following inequalities it is possible ([GGZ74], Lemma 1.36) to define other norms that are equivalent to $\| \cdot \|_{W_2^1(\Omega)}$:

Lemma 3.6. *Let $\Omega \subset \mathbb{R}^n$ be a bounded Lipschitz domain. Assume also that Ω_1 is a subset of Ω with positive measure and Γ_1 a subset of Γ with positive $(n-1)$-dimensional measure. Then for $u \in H^1(\Omega)$ one has*

$$\|u\|_{L_2(\Omega)}^2 \leq c \left\{ |u|_{1,\Omega}^2 + \left(\int_{\Omega_1} u \right)^2 \right\},$$

$$\|u\|_{L_2(\Omega)}^2 \leq c \left\{ |u|_{1,\Omega}^2 + \left(\int_{\Gamma_1} u \right)^2 \right\}.$$

These types of inequalities are proved for the more general $W^{1,p}(\Omega)$ case in [GGZ74]. In the special case $\Omega_1 = \Omega$ the first inequality is called the *Poincaré inequality*. The second inequality generalizes Friedrichs' inequality.

To simplify the notation, we shall write in future

$$|v|_{l,p,\Omega} := |v|_{W_p^l(\Omega)} \qquad \text{and} \qquad |v|_{l,\Omega} := |v|_{W_2^l(\Omega)}.$$

Next we consider the technique of integration by parts and study its application to the weak formulation of boundary value problems.

Lemma 3.7 (integration by parts). *Let $\Omega \subset \mathbb{R}^n$ be a bounded Lipschitz domain. Then one has*

$$\int_\Omega \frac{\partial u}{\partial x_i} v \, dx = \int_\Gamma uv \cos(n, e^i) \, ds - \int_\Omega u \frac{\partial v}{\partial x_i} \, dx$$

for arbitrary $u, v \in C^1(\bar{\Omega})$. Here n is the outward-pointing unit vector normal to Γ and e^i is the unit vector in the ith coordinate direction.

Hence one obtains *Green's formula*:

$$\int_\Omega \Delta u \, v \, dx = \int_\Gamma \frac{\partial u}{\partial n} v \, ds - \int_\Omega \nabla u \nabla v \, dx \quad \text{for all } u \in H^2(\Omega), v \in H^1(\Omega) \ (2.14)$$

—first apply integration by parts to classical differentiable functions then extend the result to $u \in H^2(\Omega)$ and $v \in H^1(\Omega)$ by a density argument based on (2.4).

Here and subsequently the term $\nabla u \nabla v$ denotes a scalar product of two vectors; it could be written more precisely as $(\nabla u)^T \nabla v$. In general we tend to use the simplified form, returning to the precise form only if the simplified version could lead to confusion.

The validity of Green's formula depends strongly on the geometry of Ω. In general we shall consider only bounded Lipschitz domains so that (2.14) holds. For its validity on more general domains, see [Wlo87].

Now we resume our exploration of Poisson's equation with homogeneous Dirichlet boundary conditions:

$$-\Delta u = f \text{ in } \Omega, \qquad u|_\Gamma = 0. \tag{2.15}$$

Every classical solution u of (2.15) satisfies, as we have seen, the variational equation

$$\int_\Omega \nabla u \nabla v \, dx = \int_\Omega f v \, dx \text{ for all } v \in H_0^1(\Omega). \tag{2.16}$$

If one defines the mapping $a(\cdot, \cdot) : H_0^1(\Omega) \times H_0^1(\Omega) \to \mathbb{R}$ by

$$a(u, v) := \int_\Omega \nabla u \nabla v \, dx,$$

then Lemma 3.4 ensures the existence of a constant $\gamma > 0$ such that

$$a(v, v) \geq \gamma \|v\|_{H_0^1(\Omega)}^2 \text{ for all } u, v \in H_0^1(\Omega).$$

This inequality is of fundamental importance in proving the existence of a unique solution $u \in H_0^1(\Omega)$ of the variational equation (2.16) for each $f \in L_2(\Omega)$. In the next section we shall present a general existence theory for variational equations and discuss conditions sufficient for guaranteeing that weak solutions are also classical solutions.

If one reformulates a boundary value problem as a variational equation in order to define a weak solution, the type of boundary condition plays an important role. To explain this basic fact, we consider the following example:

$$\begin{aligned}
-\Delta u + cu &= f \text{ in } \Omega, \\
u &= g \text{ on } \Gamma_1, \\
\frac{\partial u}{\partial n} + pu &= q \text{ on } \Gamma_2.
\end{aligned} \tag{2.17}$$

Here Γ_1 and Γ_2 are subsets of the boundary with $\Gamma_1 \cap \Gamma_2 = \emptyset$, $\Gamma_1 \cup \Gamma_2 = \Gamma$ and the given functions c, f, g, p, q are continuous (say) with $c \geq 0$ in Ω. As usual, multiply the differential equation by an arbitrary function $v \in H^1(\Omega)$, then integrate over Ω and apply integration by parts to get

$$\int_\Omega (\nabla u \nabla v + c \, uv) \, dx - \int_\Gamma \frac{\partial u}{\partial n} v \, ds = \int_\Omega f v \, dx.$$

Taking into account the boundary conditions for u we have

$$\int_\Omega (\nabla u \nabla v + c\,uv)\,dx + \int_{\Gamma_2} (pu - q)v\,ds - \int_{\Gamma_1} \frac{\partial u}{\partial n} v\,ds = \int_\Omega fv\,dx.$$

On Γ_1 we have no information about the normal derivative of u. Therefore we restrict v to lie in $V := \{\, v \in H^1(\Omega) \;:\; v|_{\Gamma_1} = 0 \,\}$. Then we obtain the variational equation

$$\int_\Omega (\nabla u \nabla v + c\,uv)\,dx + \int_{\Gamma_2} (pu - q)v\,ds = \int_\Omega fv\,dx \qquad \text{for all } v \in V. \quad (2.18)$$

Of course, we require u to satisfy $u \in H^1(\Omega)$ and $u|_{\Gamma_1} = g$. The variational equation (2.18) then defines weak solutions of our example (2.17). If the weak solution has some additional smoothness, then it is also a classical solution:

Theorem 3.8. Let $u \in H^1(\Omega)$ with $u|_{\Gamma_1} = g$ be a solution of the variational equation (2.18). Moreover, let u be smooth: $u \in C^2(\bar{\Omega})$. Then u is a solution of the boundary value problem (2.17).

Proof: Taking $v|_{\Gamma_1} = 0$ into account, Green's formula applied to (2.18) yields

$$\int_\Omega (\Delta u + c\,u)v\,dx + \int_{\Gamma_2} \left(\frac{\partial u}{\partial n} + pu - q \right) v\,ds = \int_\Omega fv\,dx \qquad \text{for all } v \in V. \quad (2.19)$$

Because $H_0^1(\Omega) \subset V$ it follows that

$$\int_\Omega (\Delta u + c\,u)v\,dx = \int_\Omega fv\,dx \qquad \text{for all } v \in H_0^1(\Omega).$$

Hence, using a well-known lemma of de la Vallée-Poussin, one obtains

$$-\Delta u + cu = f \quad \text{in } \Omega.$$

Now (2.19) implies that

$$\int_{\Gamma_2} \left(\frac{\partial u}{\partial n} + pu - q \right) v\,ds = 0 \qquad \text{for all } v \in V.$$

Again we can conclude that

$$\frac{\partial u}{\partial n} + pu = q \qquad \text{on } \Gamma_2. \quad (2.20)$$

The remaining condition $u|_{\Gamma_1} = g$ is already satisfied by hypothesis. This is a so-called *essential boundary condition* that does not affect the variational equation but must be imposed directly on the solution itself. ∎

Remark 3.9. In contrast to the essential boundary condition, the condition (2.20) follows from the variational equation (2.18) so it is not necessary to impose it explicitly on u in the variational formulation of the problem. Observe that the weak form of the boundary value problem was influenced by (2.20). Boundary conditions such as (2.20) are called *natural boundary conditions.* □

Remark 3.10. Let A be a positive definite matrix. Consider the differential equation

$$-\operatorname{div}(A \operatorname{grad} u) = f \qquad \text{in } \Omega.$$

Integration by parts gives

$$-\int_\Omega \operatorname{div}(A \operatorname{grad} u)\, v\, dx = -\int_\Gamma n \cdot (A \operatorname{grad} u)\, v\, ds + \int_\Omega \operatorname{grad} v \cdot (A \operatorname{grad} u)\, dx.$$

In this case the natural boundary conditions contain the so-called *conormal derivative* $n \cdot (A \operatorname{grad} u)$ instead of the normal derivative $\frac{\partial u}{\partial n}$ that we met in the special case of the Laplacian (where A is the identity matrix).   □

As we saw in Chapter 2, maximum principles play an important role in second-order elliptic boundary value problems. Here we mention briefly that even for weak solutions one can have maximum principles. For instance, the following *weak maximum principle* (see [GT83]) holds:

Lemma 3.11. *Let $\Omega \subset \mathbb{R}^n$ be a bounded Lipschitz domain. If $u \in H_0^1(\Omega)$ satisfies the variational inequality*

$$\int_\Omega \nabla u \nabla v\, dx \geq 0 \qquad \text{for all } v \in H_0^1(\Omega) \text{ with } v \geq 0,$$

then $u \geq 0$.

Here $u \geq 0$ and $v \geq 0$ are to be understood in the L_2 sense, i.e., almost everywhere in Ω.

Exercise 3.12. Let $\Omega = \{x \in \mathbb{R}^n : |x_i| < 1, \, i = 1, \ldots, n\}$. Prove:
a) The function defined by $f(x) = |x_1|$ has on Ω the generalized derivatives

$$\frac{\partial f}{\partial x_1} = \operatorname{sign}(x_1), \quad \frac{\partial f}{\partial x_j} = 0 \quad (j \neq 1).$$

b) The function defined by $f(x) = \operatorname{sign}(x_1)$ does not have a generalized derivative $\partial f / \partial x_1$ in L_2.

Exercise 3.13. Prove: If $u : \Omega \to \mathbb{R}$ has the generalized derivatives $v = D^\alpha u \in L_2(\Omega)$ and v the generalized derivatives $w = D^\beta v \in L_2(\Omega)$, then $w = D^{\alpha+\beta} u$.

Exercise 3.14. Let $\Omega \subset \mathbb{R}^n$ be a bounded domain with $0 \in \Omega$. Prove that the function defined by $u(x) = \|x\|_2^\sigma$ has first-order generalized derivatives in $L_2(\Omega)$ if $\sigma = 0$ or $2\sigma + n > 2$.

Exercise 3.15. Let $\Omega = (a, b) \in \mathbb{R}$. Prove that every function $u \in H^1(\Omega)$ is continuous, and moreover u belongs to the Hölder space $C^{1/2}(\Omega)$.

Exercise 3.16. Let $\Omega = \{ (x, y) \in \mathbb{R}^2 : x^2 + y^2 < r_0^2 \}$ with $r_0 < 1$. Determine if the function

$$f(x, y) = \left(\ln \frac{1}{\sqrt{x^2 + y^2}} \right)^k, \quad \text{where } k < 1/2,$$

is continuous in Ω. Is $f \in H^1(\Omega)$?

Exercise 3.17. Consider the space of all continuous functions on the interval $[a, b]$. Prove that the norms

$$\|f\|_1 = \max_{x \in [a,b]} |f(x)| \quad \text{and} \quad \|f\|_2 = \int_a^b |f(x)| dx$$

are not equivalent.

Exercise 3.18. Let $\Omega \subset [a_1, b_1] \times \cdots \times [a_n, b_n]$ be a convex domain. Let $v \in H_0^1(\Omega)$. Prove the Friedrichs inequality

$$\int_\Omega v^2 \le \gamma \int_\Omega |\nabla v|^2 \quad \text{with} \quad \gamma = \sum_{k=1}^n (b_k - a_k)^2 .$$

Exercise 3.19. Let $\Omega \subset \mathbb{R}^n$. Let $u \in H^k(\Omega)$ for some integer k. For which dimensions n does Sobolev's embedding theorem guarantee that (i) u (ii) ∇u is continuous?

Exercise 3.20. a) Let $\Omega = (0, 1)$, $u(x) = x^\alpha$. Use this example to show that it is impossible to improve the continuous embedding $H^1(\Omega) \hookrightarrow C^{1/2}(\overline{\Omega})$ to $H^1(\Omega) \hookrightarrow C^\lambda(\overline{\Omega})$ with $\lambda > 1/2$.

b) Investigate for $\Omega \subset \mathbb{R}^2$ whether or not the embedding $H^1(\Omega) \hookrightarrow L_\infty(\Omega)$ holds.

Exercise 3.21. Let $\Omega \subset \mathbb{R}^n$ with $0 \in \Omega$. Does the mapping

$$g \mapsto \langle f, g \rangle = g(0) \quad \text{for} \quad g \in H_0^1(\Omega)$$

define a continuous linear functional f on $H_0^1(\Omega)$? If yes, determine $\|f\|_*$.

Exercise 3.22. Consider the boundary value problem

$$-u'' = f \text{ on } (0, 1), \quad u(-1) = u(1) = 0$$

with f the δ distribution. How one can define the problem correctly in a weak sense? Determine the exact solution!

Exercise 3.23. Consider the boundary value problem

$$-(a(x)u')' = 0 \text{ on } (0,1), \quad u(-1) = 3, \, u(1) = 0$$

with

$$a(x) = \begin{cases} 1 & \text{for } -1 \le x < 0, \\ 0.5 & \text{for } 0 \le x \le 1. \end{cases}$$

Formulate the related variational equation and solve the problem exactly.

3.3 Variational Equations and Conforming Approximation

In the previous section we described the relationship between an elliptic boundary value problem and its variational formulation in the case of the Poisson equation with homogeneous Dirichlet boundary conditions. Before we present an abstract framework for the analysis of general variational equations, we give weak formulations for some other standard model problems.

Let $\Omega \subset \mathbb{R}^2$ with $\Gamma = \partial\Omega$. We consider, for a given sufficiently smooth function f, the boundary value problem

$$\frac{\partial^4}{\partial x^4} u(x,y) + 2 \frac{\partial^4}{\partial x^2 \, \partial y^2} u(x,y) + \frac{\partial^4}{\partial y^4} u(x,y) = f(x,y) \quad \text{in } \Omega \tag{3.1}$$

$$u|_\Gamma = \frac{\partial}{\partial n} u|_\Gamma = 0.$$

This problem models the behaviour of a horizontally clamped plate under some given load distribution. Thus the differential equation in (3.1) is often called the *plate equation*. In terms of the Laplacian we have equivalently

$$\Delta^2 u = f \quad \text{in } \Omega,$$
$$u|_\Gamma = \frac{\partial}{\partial n} u|_\Gamma = 0.$$

Now we formulate this problem weakly. Taking into account the boundary conditions, we apply Green's formula twice to obtain

$$\int_\Omega \Delta^2 u \, v \, dx = \int_\Gamma \frac{\partial}{\partial n}(\Delta u) \, v \, ds - \int_\Omega \nabla(\Delta u)\nabla v \, dx$$

$$= -\int_\Gamma \Delta u \frac{\partial v}{\partial n} \, ds + \int_\Omega \Delta u \, \Delta v \, dx$$

$$= \int_\Omega \Delta u \, \Delta v \, dx \quad \text{for all } u \in H^4(\Omega), \, v \in H_0^2(\Omega).$$

Therefore, the weak formulation of the given problem (3.1) reads as follows:
Find $u \in H_0^2(\Omega)$ such that

$$\int_\Omega \Delta u \, \Delta v \, dx = \int_\Omega f v \, dx \qquad \text{for all } v \in H_0^2(\Omega). \tag{3.2}$$

With the abbreviations $V := H_0^2(\Omega)$ and

$$a(u,v) := \int_\Omega \Delta u \, \Delta v \, dx, \qquad (f,v) := \int_\Omega f v \, dx \qquad \text{for all } u, v \in V,$$

the variational equation (3.2) can be written in the abstract form

$$a(u,v) = (f,v) \qquad \text{for all } v \in V.$$

Using Friedrichs' inequality one can show that there exists some constant $c > 0$ such that

$$c \, \|v\|_{H^2(\Omega)}^2 \le a(v,v) \qquad \text{for all } v \in H_0^2(\Omega).$$

This property is critical in the general existence theory for the weak solution of (3.2), as we shall see shortly in the Lax-Milgram lemma.

Remark 3.24. Up to this point in the plate problem, we considered the boundary conditions

$$u|_\Gamma = \frac{\partial}{\partial n} u|_\Gamma = 0 \,,$$

which correspond to a clamped plate. Both of these conditions are essential boundary conditions. If instead we study a simply supported plate, whose boundary conditions are

$$u|_\Gamma = 0 \quad \text{and} \quad \Delta u|_\Gamma = \phi \,,$$

then the standard technique produces the weak formulation

$$a(u,v) = (f,v) + \int_\Gamma \phi \, \frac{\partial v}{\partial n}$$

with $u, v \in H^2(\Omega) \cap H_0^1(\Omega)$. This means that the first boundary condition is essential but the second is natural. Of course, in practical applications still other boundary conditions are important and in each case a careful study is required to classify each condition.   □

Our last model problem plays an important role in fluid mechanics. Consider a domain $\Omega \subset \mathbb{R}^n$ for $n = 2$ or 3 and given functions $f_i : \Omega \to \mathbb{R}$, $i = 1, \ldots, n$. We seek solutions of the following system of partial differential equations with unknowns $u_i : \overline{\Omega} \to \mathbb{R}$ for $i = 1, \ldots, n$ and $p : \overline{\Omega} \to \mathbb{R}$:

$$-\Delta u_i + \frac{\partial p}{\partial x_i} = f_i \qquad \text{in } \Omega, \qquad i = 1, \ldots, n,$$

$$\sum_{i=1}^n \frac{\partial u_i}{\partial x_i} = 0, \tag{3.3}$$

$$u_i|_\Gamma = 0, \qquad i = 1, \ldots, n.$$

This is the so-called *Stokes problem*. In fluid mechanics, the quantities u_i denote the components of the velocity field while p represents the pressure.

Let $u = (u_1, .., u_n)$ denote a vector-valued function. Let us choose the function space

$$V = \{ u \in H_0^1(\Omega)^n : \operatorname{div} u = 0 \} \subset H(div; \Omega).$$

Then applying our standard technique (multiplication, integration, integration by parts) and adding all the resulting equations yields the following weak formulation of (3.3):

Find some $u \in V$ with

$$\sum_{i=1}^{n} \int_{\Omega} \nabla u_i \nabla v_i \, dx = \sum_{i=1}^{n} \int_{\Omega} f_i v_i \, dx \qquad \text{for all } v \in V. \tag{3.4}$$

It is remarkable that the pressure p has disappeared: integration by parts in the corresponding term gives 0 because $\operatorname{div} v = 0$. In the theory of mixed methods the pressure can be interpreted as a dual quantity; see Chapter 4.6. Alternative weak formulations of the Stokes problem are also possible.

If we introduce

$$a(u,v) := \sum_{i=1}^{n} \int_{\Omega} \nabla u_i \nabla v_i \, dx \quad \text{and} \quad f(v) := \sum_{i=1}^{n} \int_{\Omega} f_i v_i \, dx \qquad \text{for all } u, v \in V,$$

then the weak formulation (3.4) of the Stokes problem can also be written in the form

$$a(u,v) = f(v) \qquad \text{for all } v \in V. \tag{3.5}$$

Now we are ready to present an abstract theory encompassing (2.16), (3.2) and (3.5). The abstract setting allows us to characterize clearly those properties of variational equations that guarantee the existence of a unique solution. Then in every concrete situation one has only to check these properties.

Let V be a given Hilbert space with scalar product $(\cdot, \cdot)$ and corresponding norm $\| \cdot \|$. Furthermore, let there be given a mapping $a : V \times V \to \mathbb{R}$ with the following properties:

i) for arbitrary $u \in V$, both $a(u, \cdot)$ and $a(\cdot, u)$ define linear functionals on V;

ii) there exists a constant $M > 0$ such that

$$|a(u,v)| \leq M \|u\| \|v\| \qquad \text{for all } u, v \in V;$$

iii) there exists a constant $\gamma > 0$ such that

$$a(u,u) \geq \gamma \|u\|^2 \qquad \text{for all } u \in V.$$

A mapping $a(\cdot, \cdot)$ satisfying i) and ii) is called a *continuous bilinear form on* V. Property ii) guarantees the boundedness of the bilinear form. The essential property iii) is called V-*ellipticity*.

The existence of solutions of variational equations is ensured by the following fundamental result.

Lemma 3.25 (Lax-Milgram). *Let $a(\cdot,\cdot) : V \times V \to \mathbb{R}$ be a continuous, V-elliptic bilinear form. Then for each $f \in V^*$ the variational equation*

$$a(u,v) = f(v) \qquad \text{for all } v \in V \tag{3.6}$$

has a unique solution $u \in V$. Furthermore, the a priori estimate

$$\|u\| \le \frac{1}{\gamma}\|f\|_*. \tag{3.7}$$

is valid.

Proof: First we show that the solution of (3.6) is unique. Suppose that $u \in V$ and $\tilde{u} \in V$ are both solutions. Then the linearity of $a(\cdot,v)$ implies that

$$a(\tilde{u} - u, v) = 0 \qquad \text{for all } v \in V.$$

Choosing $v := \tilde{u} - u$ we get $a(v,v) = 0$, which by V-ellipticity implies that $v = 0$, as desired. Note that V-ellipticity, however, is stronger than the condition "$a(v,v) = 0$ implies $v = 0$".

To prove the existence of a solution to (3.6) we use Banach's fixed-point theorem. Therefore, we need to choose a contractive mapping that has as a fixed point a solution of (3.6).

For each $y \in V$ the assumptions i) and ii) for the bilinear form guarantee that

$$a(y,\cdot) - f \in V^*.$$

Hence, Riesz's theorem ensures the existence of a solution $z \in V$ of

$$(z,v) = (y,v) - r[a(y,v) - f(v)] \qquad \text{for all } v \in V \tag{3.8}$$

for each real $r > 0$. Now we define the mapping $T_r : V \to V$ by

$$T_r y := z$$

and study its properties—especially contractivity. The relation (3.8) implies

$$(T_r y - T_r w, v) = (y - w, v) - r\, a(y - w, v) \qquad \text{for all } v, w \in V. \tag{3.9}$$

Given $p \in V$, by applying Riesz's theorem again we define an auxiliary linear operator $S : V \to V$ by

$$(Sp,v) = a(p,v) \qquad \text{for all } v \in V. \tag{3.10}$$

Property ii) of the bilinear form implies that

$$\|Sp\| \le M \|p\| \qquad \text{for all } p \in V. \tag{3.11}$$

The definition of the operator S means that (3.9) can be rewritten as

$$(T_r y - T_r w, v) = (y - w - rS(y - w), v) \quad \text{for all } v, w \in V.$$

This allows us to investigate whether T_r is contractive:

$$\|T_r y - T_r w\|^2 = (T_r y - T_r w, T_r y - T_r w)$$
$$= (y - w - rS(y - w), y - w - rS(y - w))$$
$$= \|y - w\|^2 - 2r(S(y - w), y - w) + r^2(S(y - w), S(y - w)).$$

By (3.10) and (3.11) this yields

$$\|T_r y - T_r w\|^2 \leq \|y - w\|^2 - 2ra(y - w, y - w) + r^2 M^2 \|y - w\|^2.$$

Finally, invoking the V-ellipticity of $a(\cdot, \cdot)$ we get

$$\|T_r y - T_r w\|^2 \leq (1 - 2r\gamma + r^2 M^2) \|y - w\|^2 \quad \text{for all } y, w \in V.$$

Consequently the operator $T_r : V \to V$ is contractive if $0 < r < 2\gamma/M^2$.

Choose $r = \gamma/M^2$. Now Banach's fixed-point theorem tells us that there exists $u \in V$ with $T_r u = u$. Since $r > 0$, the definition (3.8) of T_r then implies that

$$a(u, v) = f(v) \qquad \text{for all } v \in V. \tag{3.12}$$

The a priori estimate (3.7) is an immediate consequence of the ellipticity of $a(\cdot, \cdot)$: choose $v = u$ in (3.6). ∎

We remark that in the case where $a(\cdot, \cdot)$ is symmetric, the existence of u in the Lax-Milgram lemma follows directly from Riesz's theorem. In the symmetric case, moreover, there is a close relationship between variational equations and variational problems:

Lemma 3.26. *In addition to the assumptions of Lemma 3.25, suppose that $a(\cdot, \cdot)$ is symmetric, i.e.,*

$$a(v, w) = a(w, v) \qquad \text{for all } v, w \in V.$$

Then $u \in V$ is a solution of the variational problem

$$\min_{v \in V} J(v), \text{ where } J(v) := \frac{1}{2} a(v, v) - f(v) \quad \text{for } v \in V, \tag{3.13}$$

if and only if u is a solution of the variational equation (3.6).

Proof: The symmetry of the bilinear form $a(\cdot, \cdot)$ implies that

$$a(w, w) - a(u, u) = a(w + u, w - u)$$
$$= 2a(u, w - u) + a(w - u, w - u) \quad \text{for } u, w \in V. \tag{3.14}$$

First we shall show that (3.6) is a sufficient condition for the optimality of u in the variational problem (3.13). From (3.14) one has

$$J(w) = \tfrac{1}{2}a(w,w) - f(w)$$
$$= \tfrac{1}{2}a(u,u) - f(u) + a(u, w - u) \tag{3.15}$$
$$- f(w - u) + \tfrac{1}{2}a(w - u, w - u).$$

Taking $v := w - u$ in (3.6) and using property iii) of the bilinear form $a(\cdot,\cdot)$ leads to

$$J(w) \geq J(u) \qquad \text{for all } w \in V;$$

that is, u is a solution of the variational problem (3.13).

We prove the converse implication indirectly. Given some $u \in V$, assume that there exists $v \in V$ with

$$a(u,v) \neq f(v).$$

Because V is a linear space we can assume without loss of generality that in fact

$$a(u,v) < f(v). \tag{3.16}$$

Now set $w := u + tv$ with a real parameter $t > 0$.
Definition (3.13) implies, using standard properties of $a(\cdot,\cdot)$ and f, that

$$J(w) = J(u) + t[a(u,v) - f(v)] + t^2 \frac{1}{2} a(v,v).$$

By (3.16) we can choose $t > 0$ in such a way that

$$J(w) < J(u).$$

That is, u cannot be an optimal solution of the variational problem. Consequently (3.6) is a necessary optimality condition for the variational problem (3.13). ∎

Lemma 3.26 can be applied to our familiar example of the Poisson equation with homogeneous Dirichlet boundary conditions:

$$-\Delta u = f \text{ on } \Omega, \quad u|_\Gamma = 0.$$

We proved already in Section 3.1 that the corresponding bilinear form is V-elliptic:

$$a(v,v) \geq \gamma \|v\|_1^2.$$

The boundedness of the bilinear form is obvious. Therefore, the Lax-Milgram lemma tells us that there exists a unique weak solution if we assume only that

$f \in H^{-1}(\Omega)$. Because the bilinear form is symmetric, this weak solution is also a solution of the variational problem

$$\min_{v \in H_0^1(\Omega)} \left[\frac{1}{2} \int_\Omega (\nabla v)^2 - f(v) \right].$$

Next we study the nonsymmetric convection-diffusion problem

$$-\Delta u + b \cdot \nabla u + cu = f \text{ on } \Omega, \quad u|_\Gamma = 0.$$

The associated bilinear form

$$a(u,v) := (\nabla u, \nabla v) + (b \cdot \nabla u + cu, v)$$

is not necessarily H_0^1-elliptic. Integration by parts of the term $(b \cdot \nabla v, v)$ shows that the condition

$$c - \frac{1}{2} \operatorname{div} b \geq 0$$

is sufficient for H_0^1-ellipticity.

Remark 3.27 (Neumann boundary conditions). Consider the boundary value problem

$$-\Delta u + cu = f \text{ on } \Omega, \quad \frac{\partial u}{\partial n}\Big|_\Gamma = 0.$$

If $c(x) \geq c_0 > 0$, then the bilinear form

$$a(u,v) := (\nabla u, \nabla v) + (cu, v)$$

is V-elliptic on $V = H^1(\Omega)$. The Lax-Milgram lemma can now be readily applied to the weak formulation of the problem, which shows that it has a unique solution in $H^1(\Omega)$.

If instead $c = 0$, then any classical solution of the Neumann problem above has the property that adding a constant to the solution yields a new solution. How does one handle the weak formulation in this case? Is it possible to apply the Lax-Milgram lemma?

To deal with this case we set $V = \{v \in H^1(\Omega) : \int_\Gamma v = 0\}$. Then Lemma 3.6 implies that the bilinear form $a(u,v) = (\nabla u, \nabla v)$ is V-elliptic with respect to the space V. It is easy to see that the bilinear form is bounded on $V \times V$. Therefore, surprisingly, our weak formulation for the Neumann problem with $c = 0$ is

$$a(u,v) = (f,v) \quad \text{for all } v \in V, \tag{3.17}$$

and this equation has a unique solution $u \in V$ for each $f \in L_2(\Omega)$.

But in the case $c = 0$ if one wants for smooth u to return from the variational equation (3.17) to the classical formulation of the problem, then (3.17) must be valid for all $v \in H^1(\Omega)$. On choosing $v = 1$, this implies the condition

$$\int_\Omega f = 0.$$

In this way we get the well-known solvability condition for classical solutions, which alternatively follows from the classical formulation by invoking Gauss's integral theorem. A detailed discussion of the consequences for the finite element method applied to this case can be found in [17].   □

Classical solutions of boundary value problems are also weak solutions. For the converse implication, weak solutions must have sufficient smoothness to be classical solutions. We now begin to discuss regularity theorems that give sufficient conditions for additional regularity of weak solutions. Embedding theorems are also useful in deducing smoothness in the classical sense from smoothness in Sobolev spaces.

From [Gri85] we quote

Lemma 3.28. *Let Ω be a domain with C^k boundary. If $f \in H^k(\Omega)$ for some $k \geq 0$, then the solution u of (2.16) has the regularity property*

$$u \in H^{k+2}(\Omega) \cap H_0^1(\Omega).$$

Furthermore, there exists a constant C such that

$$\|u\|_{k+2} \leq C\|f\|_k .$$

A result of this type—where a certain regularity of f yields a higher degree of regularity in u—is called a *shift theorem.*

Corollary 3.29. *Let $\Omega \subset \mathbb{R}^n$ have C^k boundary. Let $f \in H^k(\Omega)$ with $k > \frac{n}{2}$. Then the solution u of (2.16) satisfies*

$$u \in C^2(\bar{\Omega}) \cap H_0^1(\Omega).$$

Thus u is a solution of the boundary value problem (2.15) in the classical sense.

Proof: Lemma 3.28 implies that $u \in H^{k+2}(\Omega)$. Then the continuous embedding $W_2^{k+2}(\Omega) \hookrightarrow C^2(\bar{\Omega})$ for $k > n/2$ yields the result. ■

The assumption of Lemma 3.28 that the domain Ω possesses a C^k boundary is very restrictive, for in many practical examples the domain has corners. Thus it is more realistic to assume only that the boundary is piecewise smooth.

What regularity does the solution have at a corner of the domain? To answer this question, we shall study the Laplace equation in the model domain

$$\Omega = \left\{ \begin{pmatrix} x \\ y \end{pmatrix} \in \mathbb{R}^2 : x = r\cos\varphi,\ y = r\sin\varphi,\ r \in (0,1),\ \varphi \in (0,\omega) \right\} \quad (3.18)$$

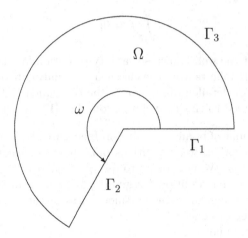

Figure 3.1 Example: corner singularity

for some parameter $\omega \in (0, 2\pi)$; see Figure 3.1. This domain has a piecewise smooth boundary Γ with corners at the points $\begin{pmatrix} 0 \\ 0 \end{pmatrix}$, $\begin{pmatrix} 1 \\ 0 \end{pmatrix}$ and $\begin{pmatrix} \cos \omega \\ \sin \omega \end{pmatrix}$. We decompose Γ into the three smooth pieces

$$\Gamma_1 = \left\{ \begin{pmatrix} x \\ y \end{pmatrix} : x \in [0, 1], \ y = 0 \right\},$$

$$\Gamma_2 = \left\{ \begin{pmatrix} x \\ y \end{pmatrix} : x = r \cos \omega, \ y = r \sin \omega, \ r \in (0, 1) \right\},$$

$$\Gamma_3 = \left\{ \begin{pmatrix} x \\ y \end{pmatrix} : x = \cos \varphi, \ y = \sin \varphi, \ \varphi \in (0, \omega] \right\}.$$

Then $\Gamma = \Gamma_1 \cup \Gamma_2 \cup \Gamma_3$. Now consider the following Dirichlet problem for the Laplacian:

$$\begin{aligned} -\Delta u &= 0 \quad \text{in } \Omega, \\ u|_{\Gamma_1 \cup \Gamma_2} &= 0, \\ u|_{\Gamma_3} &= \sin(\tfrac{\pi}{\omega}\varphi). \end{aligned} \qquad (3.19)$$

The problem has the unique solution

$$u(r, \varphi) = r^{\pi/\omega} \sin\left(\frac{\pi}{\omega}\varphi\right).$$

Consequently $u \in H^2(\Omega)$ if and only if $\omega \in (0, \pi]$. We infer that the solutions of Dirichlet problems in non-convex domains do not in general have the regularity property $u \in H^2(\Omega)$.

Next, consider instead of (3.19) the boundary value problem

$$-\Delta u = 0 \quad \text{in } \Omega,$$
$$u|_{\Gamma_1} = 0,$$
$$\frac{\partial u}{\partial n}\Big|_{\Gamma_2} = 0, \tag{3.20}$$
$$u|_{\Gamma_3} = \sin(\tfrac{\pi}{2\omega}\varphi).$$

Its solution is

$$u(r, \varphi) = r^{\pi/2\omega} \sin\left(\frac{\pi}{2\omega}\varphi\right).$$

For this problem with mixed boundary conditions one has $u \notin H^2(\Omega)$ if $\omega > \pi/2$. In the case $\omega = \pi$, for instance, the solution has a corner singularity of the type $r^{1/2}$.

These examples show clearly that the regularity of the solution of a boundary value problem depends not only on the smoothness of the data but also on the geometry of the domain and on the type of boundary conditions. It is important to remember this dependence to guard against proving convergence results for discretization methods under unrealistic assumptions. Lemma 3.28, for instance, is powerful and elegant but it does treat an ideal situation because of the smoothness of the boundary and the homogeneous Dirichlet boundary conditions.

In the books of Dauge [Dau88] and Grisvard [Gri85, Gri92] the reader will find many detailed results regarding the behaviour of solutions of elliptic boundary value problems in domains with corners. We shall quote only the following theorem, which ensures H^2-regularity for convex domains.

Theorem 3.30. *Let Ω be a convex domain. Set $V = H_0^1(\Omega)$. Let $a(\cdot, \cdot)$ be a V-elliptic bilinear form that is generated by a second-order elliptic differential operator with smooth coefficients. Then for each $f \in L_2(\Omega)$, the solution u of the Dirichlet problem*

$$a(u, v) = (f, v) \quad \text{for all} \quad v \in V$$

lies in the space $H^2(\Omega)$. Furthermore, there exists a constant C such that

$$\|u\|_2 \le C\|f\|_0.$$

A similar result holds for elliptic second-order boundary value problems in convex domains if the boundary conditions are of a different type—but not mixed as the example above has shown us. For fourth-order boundary value problems, however, a convex domain is not sufficient in general to guarantee $u \in H^4(\Omega)$.

Now we start to discuss the approximation of solutions of variational equations.

First we describe *Ritz's method*. It is a technique for approximately solving variational problems such as (3.13). Instead of solving the given problem in the space V, which is in general a infinite-dimensional space, one chooses a finite-dimensional subspace $V_h \subset V$ and solves

$$\min_{v_h \in V_h} J(v_h), \text{ where } J(v_h) = \frac{1}{2} a(v_h, v_h) - f(v_h). \tag{3.21}$$

As V_h is finite-dimensional, it is a closed subspace of V and therefore a Hilbert space endowed with the same scalar product $(\cdot, \cdot)$. Consequently the bilinear form $a(\cdot, \cdot)$ has the same properties on V_h as on V. Thus our abstract theory applies to the problem (3.21). Hence (3.21) has a unique solution $u_h \in V_h$, and u_h satisfies the necessary and sufficient optimality condition

$$a(u_h, v_h) = f(v_h) \qquad \text{for all } v_h \in V_h. \tag{3.22}$$

Ritz's method assumes that the bilinear form $a(\cdot, \cdot)$ is symmetric. Nevertheless in the nonsymmetric case it is an obvious idea to go directly from the variational equation

$$a(u, v) = f(v) \qquad \text{for all } v \in V$$

to its finite-dimensional counterpart (3.22). The discretization of the variational equation by (3.22) is called the *Galerkin method* . Because in the symmetric case the Ritz method and the Galerkin method coincide, we also use the terminology *Ritz-Galerkin method.*

The following result, often called *Cea's lemma*, is the basis for most convergence results for Ritz-Galerkin methods:

Theorem 3.31 (Cea). *Let $a(\cdot, \cdot)$ be a continuous, V-elliptic bilinear form. Then for each $f \in V^*$ the continuous problem (3.6) has a unique solution $u \in V$ and the discrete problem (3.22) has a unique solution $u_h \in V_h$. The error $u - u_h$ satisfies the inequality*

$$\|u - u_h\| \le \frac{M}{\gamma} \inf_{v_h \in V_h} \|u - v_h\|. \tag{3.23}$$

Proof: Existence and uniqueness of u and u_h are immediate consequences of the Lax-Milgram lemma.

As $V_h \subset V$, it follows from (3.6) that

$$a(u, v_h) = f(v_h) \qquad \text{for all } v_h \in V_h.$$

By the linearity of the bilinear form and (3.22) we then get

$$a(u - u_h, v_h) = 0 \qquad \text{for all } v_h \in V_h.$$

This identity and linearity yield

$$a(u - u_h, u - u_h) = a(u - u_h, u - v_h) \qquad \text{for all } v_h \in V_h.$$

The V-ellipticity and boundedness of $a(\cdot, \cdot)$ now imply

$$\gamma \|u - u_h\|^2 \le M \|u - u_h\| \|u - v_h\| \qquad \text{for all } v_h \in V_h.$$

The estimate (3.23) follows since v_h is an arbitrary element of V_h . ■

The property
$$a(u - u_h, v_h) = 0 \qquad \text{for all } v_h \in V_h$$
that we met in the above proof tells us that the error $u - u_h$ is "orthogonal" in a certain sense to the space V_h of ansatz functions. Galerkin used this idea in formulating his method in 1915. We call the property *Galerkin orthogonality*.

Remark 3.32. Cea's lemma relates the discretization error to the best approximation error
$$\inf_{v_h \in V_h} \|u - v_h\|. \tag{3.24}$$
Because the two errors differ only by a fixed multiplicative constant, the Ritz-Galerkin method is described as *quasi-optimal*.

If as (say) $h \to 0$ the best approximation error goes to zero, then it follows that
$$\lim_{h \to 0} \|u - u_h\| = 0.$$

It is often difficult to compute the best approximation error. Then we choose an easily-computed projector $\Pi_h : V \to V_h$, e.g. an interpolation operator, and estimate the approximation error by
$$\inf_{v_h \in V_h} \|u - v_h\| \le \|u - \Pi_h u\|.$$

In Section 4.4 we shall estimate $\|u - \Pi_h u\|$ explicitly for specially chosen spaces V_h used in finite element methods.   □

Remark 3.33. If the bilinear form $a(\cdot, \cdot)$ is symmetric, then instead of (3.23) one can prove that

$$\|u - u_h\| \le \sqrt{\frac{M}{\gamma}} \inf_{v_h \in V_h} \|u - v_h\|.$$

□

Remark 3.34. The assumption that $V_h \subset V$ guarantees that certain properties valid on V remain valid on the finite-dimensional space V_h. If we do not require $V_h \subset V$, then we have to overcome some technical difficulties (see Chapter 4). Methods with $V_h \subset V$ that use the same bilinear form $a(\cdot, \cdot)$ and functional $f(\cdot)$ in both the continuous and discrete problems are called *conforming methods.*   □

Remark 3.35. For the practical implementation of the Galerkin method one needs a suitably chosen space of ansatz functions $V_h \subset V$ and one must compute $a(w, v)$ and $f(v)$ for given $v, w \in V_h$. The exact computation of the integrals involved is often impossible, so quadrature formulas are used. But the introduction of such formulas is equivalent to changing $a(\cdot,\cdot)$ and $f(\cdot)$, so it makes the method nonconforming; see Chapter 4.   □

Remark 3.36. The dimension of V_h is finite. Thus this space has a basis, i.e., a finite number of linearly independent functions $\varphi_i \in V_h$, for $i = 1,\ldots,N$, that span V_h:

$$V_h = \left\{ v \,:\, v(x) = \sum_{i=1}^{N} d_i \varphi_i(x) \right\}.$$

Because $a(\cdot,\cdot)$ and $f(\cdot)$ are linear, the relation (3.22) is equivalent to

$$a(u_h, \varphi_i) = f(\varphi_i), \quad i = 1,\ldots,N.$$

Writing the unknown $u_h \in V_h$ as

$$u_h(x) = \sum_{j=1}^{N} s_j \varphi_j(x), \quad x \in \Omega,$$

the unknown coefficients $s_j \in \mathbb{R}$ $(j = 1,\ldots,N)$ satisfy the linear system of equations

$$\sum_{j=1}^{N} a(\varphi_j, \varphi_i) s_j = f(\varphi_i), \quad i = 1,\ldots,N. \tag{3.25}$$

We call the system (3.25) the *Galerkin equations.* In Chapter 8 we shall discuss its properties in detail, including practical effective methods for its solution. For the moment we remark only that the V-ellipticity of $a(\cdot,\cdot)$ implies that the coefficient matrix of (3.25) is invertible: let $z = (z_1,\ldots,z_N) \in \mathbb{R}^N$ be a solution of the homogeneous system

$$\sum_{j=1}^{N} a(\varphi_j, \varphi_i) z_j = 0, \quad i = 1,\ldots,N. \tag{3.26}$$

Then

$$\sum_{i=1}^{N} \sum_{j=1}^{N} a(\varphi_j, \varphi_i) z_j z_i = 0.$$

By the linearity of $a(\cdot,\cdot)$ this is the same as

$$a\left(\sum_{j=1}^{N} z_j \varphi_j, \sum_{i=1}^{N} z_i \varphi_i \right) = 0,$$

which by V-ellipticity forces

$$\sum_{j=1}^{N} z_j \varphi_j = 0.$$

Because the functions φ_j are linearly independent, we get $z = \mathbf{0}$. That is, the homogeneous system (3.26) has only the trivial solution. Consequently the coefficient matrix of (3.25) is nonsingular.   □

In the derivation of the Galerkin equations (3.25) we used the same basis functions $\{\varphi_i\}_{i=1}^N$ of V_h for both the ansatz and the test functions. This guarantees that the *stiffness matrix* $A_h = (a_{ij}) = (a(\varphi_j, \varphi_i))$ has nice properties; in the case of a symmetric bilinear form the stiffness matrix is symmetric and positive definite.

Alternatively, one can use different spaces V_h and W_h for the ansatz and the test functions, but they must have the same dimension. Let us denote by $\{\varphi_i\}_{i=1}^N$ and $\{\psi_i\}_{i=1}^N$ the basis functions of V_h and W_h, i.e.,

$$V_h = \text{span}\{\varphi_i\}_{i=1}^N, \quad W_h = \text{span}\{\psi_i\}_{i=1}^N.$$

Setting

$$u_h(x) = \sum_{j=1}^N s_j\, \varphi_j(x),$$

the discrete variational equation

$$a(u_h, v_h) = f(v_h) \qquad \text{for all } v_h \in W_h \tag{3.27}$$

is equivalent to

$$\sum_{j=1}^N a(\varphi_j, \psi_i)\, s_j = f(\psi_i), \qquad i = 1, \ldots, N. \tag{3.28}$$

This generalization of the Galerkin method, where the *ansatz functions* differ from the *test functions*, is called the *Petrov-Galerkin method* .

One could choose V_h and W_h with the aim of imposing certain properties on the discrete problem (3.28), but Petrov-Galerkin methods are more usually the result of a weak formulation that is based on different ansatz and test spaces: see the next Section. For instance, they are often used in the treatment of first-order hyperbolic problems and singularly perturbed problems.

Setting $J(v) = \frac{1}{2}a(v, v) - f(v)$, here is a summary of our basic discretizations for variational equations:

Before continuing our study of the properties of the Ritz-Galerkin method, we illustrate it by some simple examples.

Example 3.37. Let us study the two-point boundary value problem

$$-u'' = f \quad \text{in } (0,1),$$
$$u(0) = u(1) = 0. \tag{3.29}$$

We choose $V = H_0^1(0,1)$. As the Dirichlet boundary conditions are homogeneous, integration by parts generates the bilinear form

$$a(u, v) = \int_0^1 u'(x)v'(x)\, dx.$$

(3.30)

Next we choose as ansatz functions

$$\varphi_j(x) = \sin(j\pi x), \quad j = 1, \ldots, N,$$

and set $h := 1/N$. Then $V_h \subset V$ is defined by

$$V_h := \mathrm{span}\{\varphi_j\}_{j=1}^N := \left\{ v \ : \ v(x) = \sum_{j=1}^N c_j \varphi_j(x) \right\}.$$

It is possible to show (see, e.g., [Rek80]) that

$$\lim_{h \to 0} \left[\inf_{v \in V_h} \|u - v\| \right] = 0$$

for any given $u \in V$. The estimate (3.23) proves convergence of the Galerkin method in this case. Now

$$a(\varphi_i, \varphi_j) = \pi^2 \int_0^1 ij \cos(i\pi x) \cos(j\pi x)\, dx$$

$$= \begin{cases} \pi^2 j^2/2 & \text{if } i = j, \\ 0 & \text{if } i \neq j. \end{cases}$$

Setting

$$q_i := \int_0^1 f(x)\varphi_i(x)\, dx,$$

the solution of the Galerkin equations (3.25) is easily seen to be

$$s_j = \frac{2q_j}{\pi^2 j^2}, \quad j = 1, \ldots, N.$$

(3.31)

The Galerkin approximation u_h for the solution of (3.22) is then

$$u_h(x) = \sum_{j=1}^N s_j \sin(j\pi x).$$

Why was it possible to derive an explicit formula for the Galerkin approximation? The reason is that our ansatz functions were the eigenfunctions of the differential operator of (3.29). This is an exceptional situation, since in general the differential operator's eigenfunctions are not known. Consequently, we do not usually have the orthogonality relation

$$a(\varphi_i, \varphi_j) = 0 \quad \text{for } i \neq j.$$

$\square$

Example 3.38. Let us modify problem (3.29) by considering instead the problem

$$-u'' = f \quad \text{in } (0, 1),$$
$$u(0) = u'(1) = 0. \tag{3.32}$$

The condition $u'(1) = 0$ is a natural boundary condition. In the weak formulation we get $a(u, v) = \int_0^1 u'(x)v'(x)\,dx$ as before but now the underlying function space is

$$V = \{\, v \in H^1(0, 1) \,:\, v(0) = 0 \,\}.$$

We choose V_h to be the polynomial subspace

$$V_h = \text{span}\left\{\frac{1}{i}x^i\right\}_{i=1}^{N}. \tag{3.33}$$

The Galerkin method generates a linear system of equations

$$As = b \tag{3.34}$$

for the unknown coefficients $s_i \in \mathbb{R}$, $i = 1, \ldots, N$, in the representation

$$u_h(x) = \sum_{i=1}^{N} \frac{s_i}{i} x^i.$$

The entries in the coefficient matrix $A = (a_{ij})$ are

$$a_{ij} = a(\varphi_j, \varphi_i) = \frac{1}{i + j - 1}, \quad i, j = 1, \ldots, N.$$

This particular matrix A is called the Hilbert matrix. It is well known to be extremely ill-conditioned. For example, when $N = 10$ the condition number $cond(A)$ is approximately 10^{13}.

The example reveals that the choice of the ansatz functions is very important—the Galerkin method with ansatz (3.33) for the boundary value problem (3.37) is impracticable because the linear system generated is numerically unstable and cannot be solved satisfactorily owing to rounding errors. □

Example 3.39. Consider again the boundary value problem (3.29) with $V = H_0^1(0, 1)$. Now we choose the discrete space $V_h = \text{span}\{\varphi_j\}_{j=1}^{N}$ to be the span of the piecewise linear functions

$$\varphi_j(x) = \begin{cases} \dfrac{x - x_{j-1}}{h} & \text{if } x \in (x_{j-1}, x_j], \\ \dfrac{x_{j+1} - x}{h} & \text{if } x \in (x_j, x_{j+1}), \\ 0 & \text{otherwise.} \end{cases} \tag{3.35}$$

where $j = 1, \ldots, N - 1$. Here $\{x_j\}_{j=0}^N$ is an equidistant mesh on the given interval $(0, 1)$, i.e., $x_j = j \cdot h$ for $j = 0, 1, \ldots, N$ with $h = 1/N$. The best approximation error (3.24) from this discrete space will be studied in detail in Chapter 4.

From (3.30) and (3.35) it follows that

$$a(\varphi_i, \varphi_j) = \begin{cases} \frac{2}{h} & \text{if } i = j, \\ -\frac{1}{h} & \text{if } |i - j| = 1, \\ 0 & \text{otherwise.} \end{cases} \tag{3.36}$$

The Galerkin equations (3.23) yield in this case the linear tridiagonal system

$$-s_{i-1} + 2s_i - s_{i+1} = h \int_0^1 f(x)\varphi_i(x)\,dx, \qquad i = 1, \ldots, N - 1, \tag{3.37}$$
$$s_0 = s_N = 0,$$

for the unknown coefficients s_i in the representation $u_h(x) = \sum_{i=1}^{N-1} s_i\varphi_i(x)$ of the approximate solution. Because $\varphi_i(x_j) = \delta_{ij}$, we have the important property $s_i = u_h(x_i)$ for all i.

For smooth f there exists a constant L such that

$$\left| f(x_i) - \frac{1}{h} \int_0^1 f(x)\varphi_i(x)\,dx \right| \leq \frac{2}{3} L h^2 \quad \text{for } i = 1, \ldots, N - 1.$$

This observation reveals the affinity of (3.37) with the standard central difference scheme for the boundary value problem (3.29). More precisely, (3.37) is a difference scheme where each function value $f(x_i)$ is replaced by the integral mean $\frac{1}{h} \int_0^1 f(x)\varphi_i(x)\,dx$. □

We hope that the examples discussed above make clear the importance of choosing a good discrete space V_h in the Galerkin method.

The finite element method, which we shall discuss in great detail in Chapter 4, generalizes the choice of ansatz functions in Example 3.39: one uses ansatz functions—often piecewise polynomials—with a relatively small support

$$supp\,\varphi_i := cl_{\mathbb{R}^n}\{x \in \Omega : \varphi_i(x) \neq 0\},$$

and one aims to ensure that the quantity

$$\sum_{i=1}^N card\{j \in \{1, \ldots, N\} : (supp\,\varphi_i \cap supp\,\varphi_j) \neq \emptyset\}$$

is not too large since it is an upper bound for the number of nonzero elements in the stiffness matrix $A = (a(\varphi_j, \varphi_i))_{i,j=1}^N$ of the Galerkin system (3.25).

Let us go through the details of the method for a simple example in 2D:

Example 3.40. Let $\Omega = (0,1) \times (0,1) \subset \mathbb{R}^2$. Consider the boundary value problem

$$-\Delta u = f \text{ in } \Omega,$$
$$u|_\Gamma = 0. \tag{3.38}$$

We choose $V = H_0^1(\Omega)$ for the weak formulation and decompose Ω into a uniform triangular mesh as in Figure 3.2:

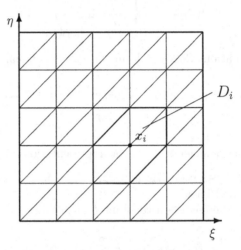

Figure 3.2 uniform triangular mesh

The decomposition of Ω is generated by a uniform mesh of mesh size $h = 1/N$ in each of the coordinate directions ξ and η, then the resulting squares are bisected by drawing diagonals as in Figure 3.2.

Denote the inner mesh points by $x_i = \begin{pmatrix} \xi_i \\ \eta_i \end{pmatrix}$ for $i = 1, \ldots, M$ with $M = (N-1)^2$, and the points on the boundary by x_i for $i = M+1, \ldots, N^2$. Analogously to Example 3.39 we define the piecewise linear ansatz functions $\varphi_i \in C(\overline{\Omega})$ indirectly by the property

$$\varphi_i(x_j) := \delta_{ij}, \qquad i = 1, \ldots, M, \ j = 1, \ldots, N. \tag{3.39}$$

Then for the support of each basis function φ_i we have

$$supp\ \varphi_i = \left\{ \begin{pmatrix} \xi \\ \eta \end{pmatrix} \in \overline{\Omega} : |\xi - \xi_i| + |\eta - \eta_i| + |\xi - \eta - \xi_i + \eta_i| \leq 2h \right\}.$$

Using the bilinear form

$$a(u,v) = \int_\Omega \nabla u \nabla v \, dx,$$

the Galerkin method generates the linear system

$$As = b \qquad (3.40)$$

with stiffness matrix $A = (a_{ij})_{i,j=1}^{M}$ and right-hand side vector $b = (b_i)_{i=1}^{M}$. A direct computation yields

$$a_{ij} = \begin{cases} 4 & \text{if } i = j, \\ -1 & \text{if } |\xi_i - \xi_j| + |\eta_i - \eta_j| = h, \\ 0 & \text{otherwise,} \end{cases}$$

and

$$b_i = \int_{\Omega} f(x)\varphi_i(x)\,dx.$$

The small support of each basis function results in only five nonzero elements in each row of the stiffness matrix. Similarly to Example 3.39, we recognize the affinity of this method with the five-point difference scheme for the boundary value problem (3.38) that appeared in Chapter 2. □

The examples and model problems we have just studied, though they are relatively simple, nevertheless demonstrate some essential features of the Galerkin method:

- It is necessary to choose a discrete space that has good approximation properties and generates linear systems that can be solved efficiently.
- When using piecewise-defined ansatz functions one has to ensure that the discrete space satisfies $V_h \subset V$. As we shall see, this is not a problem for elliptic second-order problems but difficulties can arise with, e.g., fourth-order problems where globally smooth functions are needed and for the Stokes problem where some care is needed to satisfy the divergence condition.
- The computation of the stiffness matrix $A = (a_{ij})_{ij}$ with $a_{ij} = a(\varphi_j, \varphi_i)$ and the vector b of the Galerkin equations both require, in general, the application of numerical integration.

These and other requirements have lead to intensive work on several manifestations of the Galerkin method. The most popular variants are *spectral methods*, where (usually) orthogonal polynomials are used as ansatz functions—for an excellent overview of spectral methods see [QV94]and the recent [CHQZ06]—and the *finite element method* where splines as used as ansatz functions. In Chapter 4 we shall examine the finite element method in detail; as well as presenting the basic facts and techniques, we also discuss advances in the method and its practical implementation.

Exercise 3.41. Approximately solve the boundary value problem

$$Lu := u'' - (1 + x^2)u = 1 \ \text{ on } \ (-1,1), \qquad u(-1) = u(1) = 0,$$

using the ansatz

$$\tilde{u}(x) = c_1\varphi_1(x) + c_2\varphi_2(x) \qquad \text{with} \qquad \varphi_1(x) = 1 - x^2, \quad \varphi_2(x) = 1 - x^4.$$

Determine c_1 and c_2

a) by means of the Ritz-Galerkin technique;

b) using the "Galerkin equations"

$$(L\tilde{u} - 1, \varphi_1) = 0, \quad (L\tilde{u} - 1, \varphi_2) = 0;$$

c) by computing

$$\min_{\tilde{u}} \int_{-1}^{1} \left[(\tilde{u}')^2 + (1 + x^2)\tilde{u}^2 + 2\tilde{u} \right] dx.$$

Exercise 3.42. Consider the boundary value problem

$$-\Delta u(x, y) = \pi^2 \cos \pi x \quad \text{in } \Omega = (0, 1) \times (0, 1),$$
$$\frac{\partial u}{\partial n} = 0 \quad \text{on} \quad \partial \Omega.$$

a) Construct the weak formulation and compute a Ritz-Galerkin approximation $\tilde{u}$ using the basis

$$\varphi_1(x, y) = x - 1/2, \quad \varphi_2(x, y) = (x - 1/2)^3.$$

b) Verify that the problem formulated in a) has a unique solution in

$$W = \left\{ v \in H^1(\Omega) : \int_{\Omega} v = 0 \right\}$$

and that $\tilde{u} \in W$.

c) Verify that the problem formulated in a) does not have a unique classical solution in $C^2(\Omega)$. Determine the solution u in $C^2(\Omega) \cap W$. For the approximation $\tilde{u}$, determine

– the pointwise error at $x = 0.25$
– the defect (i.e., amount by which it is in error) in the differential equation at $x = 0.25$
– the defect in the boundary condition at $x = 0$.

Exercise 3.43. Let $\Omega = \{ (x, y) \in \mathbb{R}^2 : x > 0, y > 0, x + y < 1 \}$. Approximately determine the minimal eigenvalue in the eigenvalue problem

$$\Delta u + \lambda u = 0 \quad \text{in } \Omega, \quad u = 0 \quad \text{on} \quad \partial \Omega$$

by using the ansatz function $\tilde{u}(x, y) = xy(1 - x - y)$ and computing $\tilde{\lambda}$ from the Galerkin orthogonality property

$$\int_{\Omega} (\Delta \tilde{u} + \tilde{\lambda}\tilde{u})\tilde{u} = 0.$$

Exercise 3.44. Let $\Omega \subset \mathbb{R}^N$ be a bounded domain.
a) Verify that

$$\|u\|_{\Omega,c}^2 = \int_\Omega [|grad\, u|^2 + c(x)u^2]dx$$

defines a norm on $V = H_0^1(\Omega)$ if the function $c \in L_\infty(\Omega)$ is nonnegative almost everywhere.
b) Prove the coercivity over V of the bilinear form associated with the Laplacian and discuss the dependence of the coercivity constant on the norm used.

In the remaining sections of this chapter we shall present some generalizations of the earlier theory that include certain nonlinear features.

3.4 Weakening V-ellipticity

In Section 3.3 we investigated elliptic variational equations and used the Lax-Milgram lemma to ensure existence and uniqueness of solutions for both the continuous problem and its conforming Galerkin approximation. The V-ellipticity of the underlying bilinear form $a(\cdot, \cdot)$ was a key ingredient in the proofs of the Lax-Milgram and Cea lemmas.

In the present section we weaken the V-ellipticity assumption. This is important, for example, when analysing finite element methods for first-order hyperbolic problems or mixed finite element methods.

First we study variational equations that satisfy some stability condition and hence derive results similar to the Lax-Milgram lemma.

Let V be a Hilbert space and $a : V \times V \to \mathbb{R}$ a continuous bilinear form. Then there exists a constant $M > 0$ such that

$$|a(u, v)| \leq M \|u\| \|v\| \qquad \text{for all } u, v \in V. \tag{4.1}$$

Now we assume that the variational equation

$$a(u, v) = f(v) \qquad \text{for all } v \in V \tag{4.2}$$

has for each $f \in V^*$ a solution $u \in V$ that satisfies the stability condition

$$\|u\| \leq \sigma \|f\|_* \tag{4.3}$$

for some constant $\sigma > 0$. This stability condition implies uniqueness of the solution of the variational equation (4.2): for if two elements $\tilde{u}, \hat{u} \in V$ are solutions of (4.2), then the linearity of $a(\cdot, \cdot)$ leads to

$$a(\tilde{u} - \hat{u}, v) = 0 \qquad \text{for all } v \in V,$$

and now the estimate (4.3) yields

$$0 \leq \|\tilde{u} - \hat{u}\| \leq (\sigma)(0),$$

whence $\tilde{u} = \hat{u}$.

Consider a conforming Ritz-Galerkin approximation of the problem (4.2). Thus with $V_h \subset V$ we seek $u_h \in V_h$ such that

$$a(u_h, v_h) = f(v_h) \qquad \text{for all } v_h \in V_h. \tag{4.4}$$

Analogously to the continuous problem, we require that for each $f \in V^*$ the discrete problem (4.4) is solvable and that its solution $u_h \in V_h$ satisfies

$$\|u_h\| \leq \sigma_h \|f\|_{*,h} \tag{4.5}$$

for some constant $\sigma_h > 0$. Here we used

$$\|f\|_{*,h} := \sup_{v_h \in V_h} \frac{|f(v_h)|}{\|v_h\|}.$$

Then, similarly to Cea's lemma, we obtain:

Lemma 3.45. *Assume that the bilinear form $a(\cdot, \cdot)$ is continuous on $V \times V$, with M defined in (4.1). Assume that both the continuous problem (4.2) and the discrete problem (4.4) have solutions, and that the solution u_h of the discrete problem satisfies the stability estimate (4.5). Then the error of the Ritz-Galerkin approximation satisfies the inequality*

$$\|u - u_h\| \leq (1 + \sigma_h M) \inf_{v_h \in V_h} \|u - v_h\|.$$

Proof: Since $u \in V$ and $u_h \in V_h$ satisfy (4.2) and (4.4) respectively and $V_h \subset V$, we get

$$a(u - u_h, v_h) = 0 \qquad \text{for all } v_h \in V_h.$$

Hence, for arbitrary $y_h \in V_h$ one has

$$a(u_h - y_h, v_h) = a(u - y_h, v_h) \qquad \text{for all } v_h \in V_h.$$

But $a(u - y_h, \cdot) \in V^*$ so the stability estimate (4.5) implies that

$$\|u_h - y_h\| \leq \sigma_h \|a(u - y_h, \cdot)\|_{*,h}.$$

The continuity of $a(\cdot, \cdot)$ and the property $V_h \subset V$ then lead to

$$\|u_h - y_h\| \leq \sigma_h M \|u - y_h\|.$$

An application of the triangle inequality yields

$$\|u - u_h\| \leq \|u - y_h\| + \|y_h - u_h\| \leq (1 + \sigma_h M) \|u - y_h\|.$$

As $y_h \in V_h$ is arbitrary, the statement of the lemma follows. ∎

Remark 3.46. If for a family of discretizations the variational equations (4.4) are uniformly stable, i.e., there exists a constant $\tilde{\sigma} > 0$ with

$$\sigma_h \leq \tilde{\sigma} \qquad \text{for all } h < h_0,$$

with some $h_0 > 0$ then, like Cea's lemma in the case of V-ellipticity, our Lemma 3.45 guarantees the quasi-optimalitity of the Ritz-Galerkin method. □

In Section 4.6 we shall apply these results to extended variational equations that correspond to so-called mixed formulations. Special conditions there will ensure existence of solutions and the uniform stability of the discrete problem.

Next we consider a different weakening of V-ellipticity. Recall that in Section 3.3 we already met Petrov-Galerkin methods, where it can be useful to choose differing ansatz and test spaces. This is of interest in various situations such as first-order hyperbolic problems, singularly perturbed problems, and error estimates in norms other than the norm on V (e.g. for second-order problems the norm on V is typically an "energy norm", but one might desire an error estimate in the L_∞ norm).

Example 3.47. Let us consider the first-order hyperbolic convection problem

$$b \cdot \nabla u + cu = f \quad \text{in } \Omega, \quad u = 0 \quad \text{on } \Gamma^-.$$

Here the inflow boundary of Ω is defined by $\Gamma^- = \{x \in \Gamma : b \cdot n < 0\}$, where n is as usual an outer-pointing unit vector that is normal to the boundary Γ. Setting

$$W = L_2(\Omega), \quad V = H^1(\Omega)$$

and

$$a(u, v) = -\int_\Omega u \operatorname{div}(bv) + \int_{\Gamma \backslash \Gamma^-} (b \cdot n) uv + \int_\Omega cuv,$$

a standard weak formulation of the problem reads as follows:
Find $u \in W$ such that

$$a(u, v) = f(v) \quad \text{for all } v \in V.$$

It turns out that for this problem it is useful to work with different ansatz and test spaces.   □

Analogously to this example, consider the general problem: Find $u \in W$ such that

$$a(u, v) = (f, v) \quad \text{for all } v \in V, \tag{4.6}$$

where W and V are Hilbert spaces that are not necessarily identical. The following generalization of the Lax-Milgram lemma goes back to Nečas (1962); its proof is similar to our earlier proof of Lax-Milgram. (see also [EG04])

Theorem 3.48. *Let W and V be two Hilbert spaces with norms $\|\cdot\|_W$ and $\|\cdot\|_V$. Assume that the bilinear form $a(\cdot,\cdot)$ on $W \times V$ has the following properties (with constants C and $\gamma > 0$):*

$$|a(w,v)| \leq C\|w\|_W\|v\|_V \quad \text{for all } v \in V, \ w \in W,$$

$$\sup_{v \in V} \frac{a(w,v)}{\|v\|_V} \geq \gamma\|w\|_W \quad \text{for all } w \in W,$$

and

$$\sup_{w \in W} a(w,v) > 0 \quad \text{for all } v \in V.$$

Then (4.6) has for each $f \in V^*$ a unique solution u with

$$\|u\|_W \leq \frac{1}{\gamma}\|f\|_*.$$

Babuška (see [8]) formulated the corresponding generalization of Cea's lemma using the discrete condition

$$\sup_{v_h} \frac{a(w_h,v_h)}{\|v_h\|_{V_h}} \geq \gamma_h\|w_h\|_{W_h} \quad \text{for all } w_h \in W_h, \tag{4.7}$$

for some constant $\gamma_h > 0$. It is important to note that the discrete condition (4.7) does not in general follow from its continuous counterpart. Nevertheless there are several techniques available to investigate its validity—see Chapter 4.6.

Babuška proved the error estimate

$$\|u - u_h\| \leq (1 + C/\gamma_h) \inf_{v_h \in V_h} \|u - v_h\|. \tag{4.8}$$

Recently it was shown in [124] that one can remove the constant 1 from this estimate.

Finally, as a third extension of V-ellipticity, we discuss V-coercivity.

Let $V \subset H^1(\Omega)$ be a space related to the weak formulation of a problem based on a second-order differential operator. We say that a bilinear form $a(\cdot,\cdot)$ is *V-coercive* if there exist constants β and $\gamma > 0$ such that

$$a(v,v) + \beta\|v\|_0^2 \geq \gamma\|v\|_1^2 \quad \text{for all } v \in V.$$

In this situation the operator $A : V \mapsto V^*$ defined by

$$\langle Av, w \rangle := a(v,w),$$

still satisfies the so-called Riesz-Schauder theory. With some further assumptions, one has (see Chapter 8 of [Hac03a]) the following error estimate for the Ritz-Galerkin method:

Theorem 3.49. *Assume that the variational equation*

$$a(u,v) = f(v) \quad \text{for all } v \in V,$$

where the bilinear form is V-coercive, has a solution u. If the bilinear form $a(\cdot,\cdot)$ is moreover continuous and satisfies

$$\inf\big\{\sup\{|a(u,v)| : v \in V_h, \|v\| = 1\} : u \in V_h, \|u\| = 1\big\} = \gamma_h > 0,$$

then the Ritz-Galerkin discrete problem has a solution u_h whose error is bounded by

$$\|u - u_h\| \le (1 + C/\gamma_h) \inf_{w \in V_h} \|u - w\|.$$

In [Hac03a] the validity of the inf-sup condition used in this theorem is discussed.

3.5 Extensions to Nonlinear Boundary Value Problems

In the previous sections we discussed abstract variational equations that treated only linear boundary value problems. Under certain conditions it is possible, however, to extend the technique used in the proof of the Lax-Milgram lemma—the construction of a suitably chosen contractive mapping—to more general differential operators. In this context monotone operators play an essential role; see [GGZ74, Zei90, ET76]. A different approach to proving the existence of solutions of nonlinear boundary value problems is to combine monotone iteration schemes with compactness arguments. To use this technique one needs assumptions that guarantee the monotonicity of the iteration process and carefully chosen starting points for the iteration; see [LLV85].

We now sketch the basic facts of the theory of monotone operators. This will enable us to apply the Galerkin method to some nonlinear elliptic boundary value problems.

Let V be a Hilbert space with scalar product $(\cdot,\cdot)$ and let $B : V \to V$ be an operator with the following properties:

i) There exists a constant $\gamma > 0$ such that

$$(Bu - Bv, u - v) \ge \gamma\|u - v\|^2 \quad \text{for all } u, v \in V.$$

ii) There exists a constant $M > 0$ such that

$$\|Bu - Bv\| \le M\|u - v\| \quad \text{for all } u, v \in V.$$

Property (i) is called *strong monotonicity*, and (ii) *Lipschitz continuity* of the operator B.

Consider the abstract operator equation: find $u \in V$ with

$$Bu = 0. \tag{5.1}$$

This is equivalent to the nonlinear variational equation

$$(Bu, v) = 0 \qquad \text{for all } v \in V. \tag{5.2}$$

The next statement generalizes the Lax-Milgram lemma:

Lemma 3.50. *Assume that B is monotone and Lipschitz continuous. Then equation (5.1) has a unique solution $u \in V$. This solution is a fixed point of the auxiliary operator $T_r : V \to V$ defined by*

$$T_r v := v - rBv, \qquad v \in V,$$

which is contractive when the parameter r lies in $\left(0, \dfrac{2\gamma}{M^2}\right)$.

Proof: As in the proof of the Lax-Milgram lemma we check whether T_r is contractive:

$$
\begin{aligned}
\|T_r y - T_r v\|^2 &= \|y - rBy - [v - rBv]\|^2 \\
&= \|y - v\|^2 - 2r(By - Bv, y - v) + r^2\|By - Bv\|^2 \\
&\leq (1 - 2\gamma r + r^2 M^2)\|y - v\|^2 \qquad \text{for all } y, v \in V.
\end{aligned}
$$

Hence T_r is indeed a contraction mapping for $r \in (0, \frac{2\gamma}{M^2})$. Consequently T_r possesses a unique fixed point $u \in V$, i.e.,

$$u = T_r u = u - rBu.$$

That is, u is a solution of the operator equation (5.1).

Uniqueness of the solution follows immediately from the strong monotonicity property using the same argument as in the proof of the Lax-Milgram lemma. ∎

Next we consider operators $A : V \to V^*$. Here V^* denotes the dual space of V and $\langle \cdot, \cdot \rangle$ the dual pairing, i.e., $\langle l, v \rangle$ denotes the value of the continuous linear functional $l \in V^*$ applied to $v \in V$. We assume that A has the following properties:

i) The operator A is *strongly monotone*, i.e, there exists a constant $\gamma > 0$ such that

$$\langle Au - Av, u - v \rangle \geq \gamma \|u - v\|^2 \qquad \text{for all } u, v \in V.$$

ii) The operator A is *Lipschitz continuous*, i.e., there exists a constant $M > 0$ such that

$$\|Au - Av\|_* \leq M\|u - v\| \qquad \text{for all } u, v \in V.$$

Then the problem

$$Au = f \tag{5.3}$$

has for each $f \in V^*$ a unique solution $u \in V$.

This statement follows immediately from Lemma 3.50 using the auxiliary operator $B : V \to V$ defined by

$$Bv := J(Av - f), \qquad v \in V.$$

Here $J : V^* \to V$ denotes the Riesz operator that maps each continuous linear functional $g \in V^*$ to an element $Jg \in V$ such that

$$\langle g, v \rangle = (Jg, v) \qquad \text{for all } v \in V.$$

Problem (5.3) is equivalent to the nonlinear variational equation

$$\langle Au, v \rangle = \langle f, v \rangle \qquad \text{for all } v \in V. \tag{5.4}$$

Existence and uniqueness of solutions hold not only for the case of a Hilbert space V, as in fact it is sufficient that V be a reflexive Banach space—see [Zei90].

We now discuss two examples of nonlinear elliptic boundary value problems that can be treated with the theory of this section. Note however that some important practical problems cannot be dealt with using monotone and Lipschitz continuous operators; then more sophisticated techniques are necessary. First we present a semi-linear problem and then a special quasi-linear boundary value problem.

Example 3.51. Let $\Omega \subset \mathbb{R}^2$ be a bounded domain with smooth boundary Γ. Consider the weakly nonlinear problem

$$\begin{aligned} -\text{div}\,(M\,\text{grad}\,u) + F(x, u(x)) &= 0 \quad \text{in } \Omega, \\ u|_\Gamma &= 0. \end{aligned} \tag{5.5}$$

Here $M = M(x) = (m_{ij}(x))$ is a matrix-valued function satisfying the estimate

$$\overline{\sigma}\|z\|^2 \ge z^T M(x)z \ge \underline{\sigma}\|z\|^2 \qquad \text{for all } x \in \Omega,\ z \in \mathbb{R}^2, \tag{5.6}$$

for some constants $\overline{\sigma} \ge \underline{\sigma} > 0$. Furthermore, let $F : \overline{\Omega} \times \mathbb{R} \longrightarrow \mathbb{R}$ be a continuous function with the properties

$$\left.\begin{aligned} |F(x, s) - F(x, t)| &\le L\,|s - t| \\ (F(x, s) - F(x, t))(s - t) &\ge 0 \end{aligned}\right\} \qquad \text{for all } x \in \Omega,\ s, t \in \mathbb{R}, \tag{5.7}$$

where L is some constant. Choose $V = H_0^1(\Omega)$. Define a mapping $a(\cdot, \cdot) : V \times V \to \mathbb{R}$ by

$$a(u, v) := \int_\Omega \left[(\nabla u)^T M^T \nabla v + F(x, u(x))v \right] dx, \qquad u, v \in V.$$

For fixed $u \in V$ our assumptions guarantee that $a(u, \cdot) \in V^*$. We define the related operator $A : V \to V^*$ by $Au := a(u, \cdot)$ and study its properties.

First we obtain

$$|\langle Au, v \rangle - \langle Ay, v \rangle|$$

$$= \left| \int_\Omega \left[(\nabla u - \nabla y)^T M^T \nabla v + (F(x, u(x)) - F(x, y(x))) v(x) \right] dx \right|$$

$$\leq \int_\Omega \left(\bar{\sigma} \|\nabla u - \nabla v\| \, \|\nabla v\| + L \, \|u - y\| \, \|v\| \right) dx$$

$$\leq c \|u - y\| \, \|v\|.$$

Hence the operator A is Lipschitz continuous.

Friedrichs' inequality, (5.6) and (5.7) give the following estimates:

$$\langle Au - Av, u - v \rangle = \int_\Omega \nabla(u - v)^T M^T \nabla(u - v) \, dx$$

$$+ \int_\Omega (F(x, u(x)) - F(x, v(x)))(u(x) - v(x)) \, dx$$

$$\geq \underline{\sigma} \int_\Omega \nabla(u - v) \nabla(u - v) \, dx$$

$$\geq \underline{\sigma} \gamma \, \|u - v\|^2 \qquad \text{for all } u, \, v \in V.$$

Thus A is strongly monotone as well and our earlier theory is applicable. □

The next example sketches the analysis of a quasi-linear boundary value problem. This is more difficult to handle than Example 3.51, so we omit the details which can be found in [Zei90].

Example 3.52. Consider the following equation, where a nonlinearity appears in the main part of the differential operator:

$$-\sum_i \frac{\partial}{\partial x_i} \left(\varphi(x, |Du|) \frac{\partial u}{\partial x_i} \right) = f(x) \quad \text{in } \Omega.$$

Assume homogeneous Dirichlet boundary conditions for u, and that φ is a continuous function satisfying the following conditions:

(i) $\varphi(x, t)t - \varphi(x, s)s \geq m(t - s)$ for all $x \in \Omega$, $t \geq s \geq 0$, $m > 0$;

(ii) $|\varphi(x, t)t - \varphi(x, s)s| \leq M|t - s|$ for all $x \in \Omega$, $t, s \geq 0$, $M > 0$.

If, for instance, $\varphi(x, t) = g(t)/t$ and g is differentiable, then both these conditions are satisfied if

$$0 < m \leq g'(t) \leq M.$$

Under these hypotheses one can show that the theory of monotone and Lipschitz continuous operators is applicable, and deduce the existence of weak solutions for this nonlinear boundary value problem. Of course, the conditions (i) and (ii) are fairly restrictive.   □

If one has both strong monotonicity and Lipschitz continuity, then it is not difficult to generalize Cea's lemma:

Lemma 3.53. *Let $A : V \to V^*$ be a strongly monotone, Lipschitz continuous operator. Let $f \in V^*$. If $V_h \subset V$ is a finite-dimensional subspace of V, then there exists a unique $u_h \in V_h$ satisfying the discrete variational equation*

$$\langle Au_h, v_h \rangle = \langle f, v_h \rangle \qquad \text{for all } v_h \in V_h. \tag{5.8}$$

Moreover, the error $u - u_h$ of the Galerkin method satisfies the quasi-optimality estimate

$$\|u - u_h\| \leq \frac{M}{\gamma} \inf_{v_h \in V_h} \|u - v_h\|.$$

Proof: The finite dimensionality of V_h implies that it is a closed subspace of V and therefore it too is a Hilbert space with the same inner product as V. Clearly A is strongly monotone and Lipschitz continuous on V_h. These properties yield existence and uniqueness of a solution $u_h \in V_h$ of the discrete problem (5.8).

From (5.4), (5.8), $V_h \subset V$ and $u_h \in V_h$ we have

$$\langle Au - Au_h, v_h - u_h \rangle = 0 \qquad \text{for all } v_h \in V_h.$$

This identity, strong monotonicity and Lipschitz continuity together imply

$$\begin{aligned}
\gamma \|u - u_h\|^2 &\leq \langle Au - Au_h, u - u_h \rangle \\
&= \langle Au - Au_h, u - v_h \rangle \\
&\leq M \|u - u_h\| \, \|u - v_h\| \qquad \text{for all } v_h \in V_h,
\end{aligned}$$

and the desired result follows. ∎

Unlike the case of linear boundary value problems, the Galerkin equations (5.8) are now a set of nonlinear equations. With the ansatz

$$u_h(x) = \sum_{j=1}^{N} s_j \, \varphi_j(x),$$

the Galerkin equations are equivalent to the nonlinear system

$$\langle A(\sum_{j=1}^{n} s_j \varphi_j), \varphi_i \rangle = \langle f, \varphi_i \rangle, \qquad i = 1, \ldots, N.$$

In principle, this nonlinear system can be solved by standard techniques such as Newton's method; see [Sch78, OR70]. But because the number of unknowns is in general large and the conditioning of the problem is bad, one should take advantage of the special structure of the system.

Moreover, one can systematically use information from discretizations on coarser meshes to obtain good starting points for iterative solution of the system on finer meshes. In [4] a variant of Newton's method exploits properties of the discretization on coarse and finer meshes. In [OR70] Newton's method is combined with other iterative methods suited to the discretization of partial differential equations.

Alternatively, one could deal with a nonlinear problem at the continuous level by some successive linearization technique (such as Newton's method) applied in an infinite-dimensional function space. The application of Newton's method does however demand strong regularity assumptions because differentiation is required.

A further linearization technique at the continuous level is the method of *frozen coefficients*. To explain this technique we consider the boundary value problem

$$-\operatorname{div}\left(D(x, u, \nabla u) \operatorname{grad} u\right) = f \quad \text{in } \Omega, \tag{5.9}$$
$$u|_\Gamma = 0,$$

with a symmetric positive-definite matrix-valued function $D(\cdot, \cdot, \cdot)$. Let $u^0 \in V = H_0^1(\Omega)$ be a suitably chosen starting point for the iteration. Then a sequence $\{u^k\} \subset V$ of approximate solutions for (5.9) is generated, where (given u^k) the function u^{k+1} is a solution of the linear problem

$$\int_\Omega \nabla u^{k+1} D(x, u^k, \nabla u^k) \nabla v \, dx = \int_\Omega f v \, dx \qquad \text{for all } v \in V.$$

This technique is also known as the *secant modulus* method or *Kačanov method*. Its convergence properties are examined in [Neč83] and [Zei90].

4

The Finite Element Method

4.1 A First Example

In Chapter 3 we pointed out that for the Ritz-Galerkin method the choices of the ansatz and test functions are crucial because they strongly influence whether or not the method can be implemented in an effective manner. Classical ansatz functions such as orthogonal polynomials lead to spectral methods, which we shall not discuss; we prefer functions that are defined piecewise, and in particular piecewise polynomials. Then the Ritz-Galerkin method generates discrete problems that have a special structure: the coefficient matrix of the linear system is sparse. This structure allows us to design and apply pertinent numerical techniques that solve the discrete problem efficiently.

The discrete spaces constructed in the finite element method have the following typical features:

- decomposition of the given domain into geometrically simple subdomains using, in general, triangles and quadrilaterals in 2D and tetrahedra and parallelepipeds in 3D;
- definition of local spaces—usually polynomials—on these subdomains;
- fulfilment of inter-subdomain conditions that guarantee certain global properties of the discrete spaces, such as $V_h \subset V$.

Of course these three points are closely related. For instance, with a suitably chosen decomposition and reasonable local spaces one can formulate fairly simple sufficient conditions for the global properties of the discrete spaces, as we shall see.

Let us first consider an introductory example. Denote by $\Omega \subset \mathbb{R}^2$ the triangular domain

$$\Omega = \left\{ \begin{pmatrix} x \\ y \end{pmatrix} : x > 0, \, y > 0, \, x + y < 1 \right\}.$$

Let f be a given continuous function on $\overline{\Omega}$. We consider the following problem: Find $u \in H_0^1(\Omega)$ such that

$$\int_{\Omega} \nabla u \nabla v \, d\Omega = \int_{\Omega} f v \, d\Omega \quad \text{for all } v \in H_0^1(\Omega). \tag{1.1}$$

In Chapter 3 we saw that this is the weak formulation of the Poisson problem

$$\begin{aligned} -\Delta u &= f \ \text{ in } \Omega, \\ u &= 0 \ \text{ on } \Gamma = \partial\Omega, \end{aligned} \tag{1.2}$$

and the starting point for its Ritz-Galerkin discretization.

We decompose Ω into rectangles and triangles, denoting the subdomains by Ω_j for $j = 1, \ldots, m$. It is natural to assume the following conditions for the decomposition $\mathcal{Z} = \{\, \Omega_j \,\}_{j=1}^m$:

$$\overline{\Omega} = \bigcup_{j=1}^{m} \overline{\Omega}_j \quad \text{and} \quad int \ \Omega_i \cap int \ \Omega_j = \emptyset \ \text{ if } i \neq j. \tag{1.3}$$

To be precise, consider the decomposition of Ω given by Figure 4.1 into the quadrilaterals $\Omega_1, \Omega_2, \ldots, \Omega_{10}$ and triangles $\Omega_{11}, \ldots, \Omega_{15}$. These are generated by equidistant lines parallel to the axes at a distance $h = 0.2$ apart.

Next, we choose a space of functions u_h defined by:

i) $u_h \in C(\overline{\Omega})$;

ii) $u|_{\Omega_i}$ is bilinear on Ω_i for $i = 1, \ldots, 10$;

iii) $u|_{\Omega_i}$ is linear on Ω_i for $i = 11, \ldots, 15$.

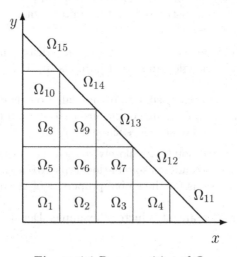

Figure 4.1 Decomposition of Ω

Set $u^i(x, y) := u_h|_{\Omega_i}(x, y)$. Then by ii) and iii) we can write

$$u^i(x, y) = a_i xy + b_i x + c_i y + d_i \quad \text{for} \quad \begin{pmatrix} x \\ y \end{pmatrix} \in \Omega_i, \ i = 1, \ldots, 15 \qquad (1.4)$$

where $a_i = 0$ for $i = 11, \ldots, 15$. The parameters a_i, b_i, c_i, d_i of the ansatz (1.4) are restricted by the boundary condition in (1.2) and the continuity condition i); in the next section we shall prove that i) guarantees that our discrete space is a subspace of $H_0^1(\Omega)$.

Let Ω_h denote the set of all interior mesh points and $\overline{\Omega}_h$ the set of all mesh points of the decomposition of Ω. Define

$$I(p) := \{ i : p \in \overline{\Omega}_i \}, \qquad p \in \overline{\Omega}_h.$$

Then the condition $u_h \in C(\overline{\Omega})$ reduces to

$$u^i(p) = u^j(p) \qquad \text{for} \quad i, j \in I(p), \ p \in \Omega_h, \qquad (1.5)$$

because the functions (1.4) are linear on each edge of each subdomain. Similarly, the boundary condition in (1.2) is satisfied if

$$u^i(p) = 0 \qquad \text{for} \quad i \in I(p), \ p \in \overline{\Omega}_h \setminus \Omega_h. \qquad (1.6)$$

The conditions (1.5) and (1.6) constrain our chosen discrete space and must be satisfied when implementing the Galerkin method.

With this subdomain-oriented approach to the formulation of the discrete problem, the equations (1.4)–(1.6) and the Galerkin equations corresponding to (1.2) form a relatively large system of equations for the unknowns: in the above example one has 55 equations for the 55 parameters a_i, b_i, c_i, d_i, $i = 1, \ldots, 15$. It is possible however to reduce significantly the number of unknowns by replacing the ansatz (1.4) by a better one that is mesh-point oriented.

Denote the interior mesh points by p^j for $j = 1, \ldots, N$ and all mesh points by p^j for $j = 1, \ldots, \overline{N}$. Thus we have

$$\Omega_h = \{p^j\}_{j=1}^N \qquad \text{and} \qquad \overline{\Omega}_h = \{p^j\}_{j=1}^{\overline{N}}.$$

Define a function $\varphi_j \in C(\overline{\Omega})$, which is piecewise linear or bilinear respectively on each subdomain, by

$$\varphi_j(p^k) = \delta_{jk}, \qquad j, k = 1, \ldots, \overline{N}. \qquad (1.7)$$

Now choose the ansatz

$$u_h(x, y) = \sum_{j=1}^{N} u_j \varphi_j(x, y). \qquad (1.8)$$

Then conditions i), ii) and iii) and the homogeneous Dirichlet boundary condition in (1.2) are all satisfied. Using (1.8), the Ritz-Galerkin method yields the linear system

$$\int\limits_\Omega \sum_{j=1}^N u_j \nabla\varphi_i \nabla\varphi_j \, d\Omega \;=\; \int\limits_\Omega f\varphi_i \, d\Omega, \quad i = 1, \ldots, N, \tag{1.9}$$

for the unknown coefficients $\{u_j\}_{j=1}^N$ in (1.8). Setting

$$\begin{aligned}
&A_h := (a_{ij})_{i,j=1}^N, \quad a_{ij} := \int_\Omega \nabla\varphi_i \nabla\varphi_j \, d\Omega, \\
&f_h := (f_i)_{i=1}^N, \qquad f_i := \int_\Omega f\varphi_i \, d\Omega, \quad \text{and } u_h := (u_j)_{j=1}^N,
\end{aligned} \tag{1.10}$$

we can write (1.9) in the matrix form

$$A_h u_h = f_h. \tag{1.11}$$

The matrix A_h is called the *stiffness matrix*. As the functions φ_j are piecewise linear or bilinear, from (1.7) each has support

$$supp\, \varphi_j \;=\; \bigcup_{k \in I(p^j)} \overline{\Omega}_k. \tag{1.12}$$

Hence, by (1.10) we have

$$I(p^i) \cap I(p^j) = \emptyset \qquad \text{implies } a_{ij} = 0. \tag{1.13}$$

If the decomposition of the domain Ω is fine then the method generates a very large matrix A_h, but this matrix is sparse. Consequently not much space is needed to store the matrix A_h and it is possible to solve the discrete problem efficiently using specially tailored techniques—see Chapter 8. These two features are significant advantages of the finite element method over other competing methods. There are further advantages of the finite element method, such as its flexibility with respect to general geometries of the given domain and the powerful and well-developed tools for its analysis, but the two properties first enunciated are the main reasons that the finite element method is the most popular and widely-used discretization method for the numerical solution of partial differential equations.

We resume the study of our concrete example. As in Figure 4.2, number the inner mesh points $p^1, p^2, \ldots, p^6$. Let p^j have coordinates (x_j, y_j) for $j = 1, \ldots, 6$. The basis functions then have the following explicit formulas, which depend on the location of p^j:

$$\varphi_j(x, y) = \begin{cases} \frac{1}{h^2}(h - |x - x_j|)(h - |y - y_j|) & \text{if } \max\{|x - x_j|, |y - y_j|\} \leq h, \\ 0 & \text{otherwise,} \end{cases}$$

in the cases $j = 1, 2, 4$;

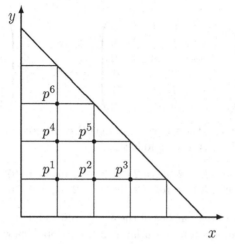

Figure 4.2 Numbering of the interior mesh points

$$\varphi_j(x,y) = \begin{cases} \frac{1}{h^2}(h - |x - x_j|)(h - |y - y_j|) & \text{if } \max\{|x - x_j|, |y - y_j|\} \le h \\ & \text{and } \min\{x - x_j, y - y_j\} \le 0, \\[2mm] \frac{1}{h}(h - (x - x_j) - (y - y_j)) & \text{if } |x - x_j| + |y - y_j| \le h \\ & \text{and } \min\{x - x_j, y - y_j\} \ge 0, \\[2mm] 0 & \text{otherwise,} \end{cases}$$

in the cases $j = 3, 5, 6$.

Figure 4.3 shows such piecewise bilinear and piecewise linear basis functions.

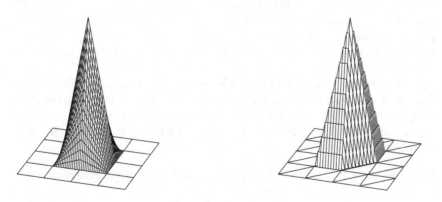

Figure 4.3 Basis functions φ_j

In the special case $f \equiv 1$ we can compute all integrals in (1.10) and obtain

$$A_h = \frac{1}{3} \begin{pmatrix} 8 & -1 & 0 & -1 & -1 & 0 \\ -1 & 8 & -1 & -1 & -1 & 0 \\ 0 & -1 & 9 & 0 & -1 & 0 \\ -1 & -1 & 0 & 8 & -1 & -1 \\ -1 & -1 & -1 & -1 & 9 & -1 \\ 0 & 0 & 0 & -1 & -1 & 9 \end{pmatrix}, \qquad f_h = h^2 \begin{pmatrix} 1 \\ 1 \\ 11/12 \\ 1 \\ 11/12 \\ 11/12 \end{pmatrix}.$$

Then on solving (1.11), the values $u_j = u_h(p^j)$ computed at the interior mesh points p^j for $j = 1, \ldots, 6$ are as follows:

$$u_1 = 2.4821E - 2 \quad u_2 = 2.6818E - 2 \quad u_3 = 1.7972E - 2$$
$$u_4 = 2.6818E - 2 \quad u_5 = 2.4934E - 2 \quad u_6 = 1.7972E - 2.$$

These values are approximations of the exact solution u of the problem. As we do not know u, we postpone until later a discussion of how one can evaluate the quality of the approximation.

4.2 Finite Element Spaces

4.2.1 Local and Global Properties

In the introductory example of Section 4.1 we have explained already the typical features of the finite element method. In particular the ansatz functions are chosen in a piecewise manner, unlike the classical Ritz method. Given a problem whose solution lies in V, if one discretizes it with a conforming Ritz-Galerkin technique then one has to ensure that the discrete space of ansatz functions V_h is a subspace of V.

In practice we often have to handle second-order elliptic boundary problems and V is then a certain subspace of the Sobolev space $H^1(\Omega)$ whose precise specification depends on the given boundary conditions. Sometimes fourth-order elliptic boundary problems arise and V is then a subspace of $H^2(\Omega)$.

The following two lemmata provide us with easily-applied sufficient conditions that guarantee the global property $V_h \subset V$. Assume that the given domain $\Omega \subset R^n$ is decomposed into subdomains Ω_j (for $j = 1, \ldots, m$) having the property (1.3). Moreover we assume that each subdomain Ω_j has a Lipschitz boundary.

Lemma 4.1. *Let $z : \overline{\Omega} \to \mathbb{R}$ be a function that satisfies the condition $z|_{\Omega_j} \in C^1(\overline{\Omega}_j)$ for $j = 1, \ldots, m$. Then*

$$z \in C(\overline{\Omega}) \implies z \in H^1(\Omega).$$

Proof: The hypothesis $z \in C^1(\overline{\Omega}_j)$ implies that there exist constants c_{kj}, for $k = 0, 1, \ldots, n$ and $j = 1, \ldots, m$, such that

$$\left.\begin{array}{ll} |z(x)| \le c_{0j} & \text{for all } x \in \Omega_j \\ |\frac{\partial}{\partial x_k} z(x)| \le c_{kj} & \text{for all } x \in \Omega_j \ , \quad k = 1, \ldots, n \end{array}\right\} \ j = 1, \ldots, m. \quad (2.1)$$

From (1.3) it follows that

$$\int_\Omega z^2(x)\, dx \ = \ \sum_{j=1}^m \int_{\Omega_j} z^2(x)\, dx \ \le \ \sum_{j=1}^m c_{0j}^2 \, \text{meas}(\Omega_j) \ < \ +\infty.$$

That is, $z \in L_2(\Omega)$. For $k = 1, \ldots, n$ define the function w_k piecewise by

$$w_k|_{\Omega_j} \ := \ \frac{\partial}{\partial x_k} z|_{\Omega_j} \ , \quad j = 1, \ldots, m.$$

Analogously to the above calculation for the function z, we conclude from (2.1) that $w_k \in L_2(\Omega)$ for each k. It remains to show that w_k is the generalized derivative of z with respect to x_k, i.e., that

$$\int_\Omega z(x) \frac{\partial}{\partial x_k} \varphi(x)\, dx \ = \ - \int_\Omega w_k(x) \varphi(x)\, dx \quad \text{for all } \varphi \in \mathcal{D}(\Omega), \quad (2.2)$$

where we use the abbreviation $\mathcal{D}(\Omega) := C_0^\infty(\Omega)$. Let us introduce the notation $\Gamma_j := \partial \Omega_j$,

$$\Gamma_{jl} := \overline{\Omega}_j \cap \overline{\Omega}_l \quad \text{and} \quad \Gamma_{j0} := \Gamma_j \backslash \Big\{ \bigcup_{l \ne j,\, l \ne 0} \Gamma_{jl} \Big\}.$$

Integration by parts results in

$$\int_\Omega z(x) \frac{\partial}{\partial x_k} \varphi(x)\, dx = \sum_{j=1}^m \int_{\Omega_j} z(x) \frac{\partial}{\partial x_k} \varphi(x)\, dx$$

$$= \sum_{j=1}^m \left(\int_{\Gamma_j} z(x) \varphi(x) \, \cos(n_j, e_k)\, dx - \int_{\Omega_j} \varphi(x) \frac{\partial}{\partial x_k} z(x)\, dx \right) ;$$

here n_j denotes the outer unit normal vector on Γ_j and e_k is the unit vector associated with x_k. From the decomposition of the boundary of Ω_j we then get

$$\int_\Omega z(x) \frac{\partial}{\partial x_k} \varphi(x)\, dx \ = \ \sum_{j=1}^m \sum_{l \ne j} \int_{\Gamma_{jl}} z(x)\, \varphi(x)\, \cos(n_j, e_k)\, dx$$

$$- \int_\Omega w_k(x) \varphi(x)\, dx \qquad \text{for all } \varphi \in \mathcal{D}(\Omega). \quad (2.3)$$

It now remains only to verify that

$$\sum_{j=1}^{m} \sum_{l \neq j} \int_{\Gamma_{jl}} z(x)\,\varphi(x)\,\cos(n_j, e_k)\,dx = 0. \tag{2.4}$$

First, $\varphi \in \mathcal{D}(\Omega)$ implies that

$$\int_{\Gamma_{j0}} z(x)\,\varphi(x)\,\cos(n_j, e_k)\,dx = 0, \quad j = 1, \ldots, m. \tag{2.5}$$

The remaining terms in (2.4) appear in pairs of the form

$$\int_{\Gamma_{jl}} z(x)\,\varphi(x)\,\cos(n_j, e_k)\,dx + \int_{\Gamma_{lj}} z(x)\,\varphi(x)\,\cos(n_l, e_k)\,dx$$

on adjacent subdomains Ω_j and Ω_l. For the normal vectors we have $n_l = -n_j$ on $\overline{\Omega}_j \cap \overline{\Omega}_l$. Because z is assumed continuous across $\Gamma_{jl} = \Gamma_{lj}$, we infer that

$$\int_{\Gamma_{jl}} z(x)\,\varphi(x)\,\cos(n_j, e_k)\,dx + \int_{\Gamma_{lj}} z(x)\,\varphi(x)\,\cos(n_l, e_k)\,dx$$
$$= \int_{\Gamma_{jl}} z(x)\,\varphi(x)\,\cos(n_j, e_k)\,dx - \int_{\Gamma_{lj}} z(x)\,\varphi(x)\,\cos(n_j, e_k)\,dx = 0$$
$$\text{for all } \varphi \in \mathcal{D}(\Omega).$$

This identity together with (2.5) yields (2.4). Hence (2.3) shows that w_k satisfies the identity

$$\int_{\Omega} z(x)\frac{\partial}{\partial x_k}\varphi(x)\,dx = -\int_{\Omega} w_k(x)\varphi(x)\,dx \quad \text{for all } \varphi \in \mathcal{D}(\Omega).$$

We have proved that w_k is the generalized derivative of z with respect to x_k. As $w_k \in L_2(\Omega)$ it follows that $z \in H^1(\Omega)$. ∎

Lemma 4.2. *Let $z : \overline{\Omega} \to \mathbb{R}$ be a function satisfying the condition $z|_{\Omega_j} \in C^2(\overline{\Omega}_j)$ for $j = 1, \ldots, m$. Then*

$$z \in C^1(\overline{\Omega}) \implies z \in H^2(\Omega).$$

Proof: Set $v_l = \frac{\partial}{\partial x_l} z$ for $l = 1, \ldots, n$. By definition $z \in C^1(\overline{\Omega})$ implies that $v_l \in C^0(\overline{\Omega})$. The functions v_l satisfy the assumptions of Lemma 4.1, so $v_l \in H^1(\Omega)$. That is, there exist $w_{lk} \in L_2(\Omega)$ for $l, k = 1, \ldots, n$ such that

$$\int_{\Omega} v_l(x)\frac{\partial}{\partial x_k}\varphi(x)\,dx = -\int_{\Omega} w_{lk}(x)\varphi(x)\,dx \quad \text{for all } \varphi \in \mathcal{D}(\Omega). \tag{2.6}$$

On the other hand

$$\int_{\Omega} z(x)\frac{\partial}{\partial x_l}\psi(x)\,dx \;=\; -\int_{\Omega} v_l(x)\psi(x)\,dx \qquad \text{for all } \psi \in \mathcal{D}(\Omega). \qquad (2.7)$$

For arbitrary $\varphi \in \mathcal{D}(\Omega)$ one also has $\frac{\partial}{\partial x_k}\varphi \in \mathcal{D}(\Omega)$.
Thus from (2.6) and (2.7) it follows that

$$\int_{\Omega} z(x)\frac{\partial^2}{\partial x_l \partial x_k}\varphi(x)\,dx \;=\; -\int_{\Omega} w_{lk}(x)\varphi(x)\,dx \qquad \text{for all } \varphi \in \mathcal{D}(\Omega),$$
$$l,k = 1,\dots,n.$$

In summary, we have $z \in L_2(\Omega)$, and the generalized derivatives of z up to the second order are also in $L_2(\Omega)$. That means $z \in H^2(\Omega)$. ∎

The construction of the trace of H^1-functions on the boundary immediately implies

$$\left.\begin{array}{c} z \in C(\Omega),\; z|_{\Omega_j} \in C^1(\overline{\Omega}) \\ z|_\Gamma = 0 \end{array}\right\} \quad\Longrightarrow\quad z \in H_0^1(\Omega).$$

Analogously,

$$\left.\begin{array}{c} z \in C^1(\Omega),\; z|_{\Omega_j} \in C^2(\overline{\Omega}) \\ z|_\Gamma = \frac{\partial}{\partial n}z = 0 \end{array}\right\} \quad\Longrightarrow\quad z \in H_0^2(\Omega).$$

The following statement is useful in the construction of conforming approximations of the space $H(div;\Omega)$.

Lemma 4.3. *Let $\underline{z} : \overline{\Omega} \to \mathbb{R}^n$ be a vector-valued mapping. Assume that $\underline{z}_j \in C^1(\overline{\Omega}_j)^n$ for $j = 1,\dots,M$, where $\underline{z}_j := \underline{z}|_{\Omega_j}$.*
If additionally on every interior edge $\Gamma_{jk} := \overline{\Omega}_j \cap \overline{\Omega}_k$ (with unit normal $\underline{n}_{jk}$ pointing from $\overline{\Omega}_j$ to $\overline{\Omega}_k$) one has

$$\underline{z}_j \cdot \underline{n}_{jk} = \underline{z}_k \cdot \underline{n}_{jk}, \qquad j,k = 1,\dots,M, \qquad (2.8)$$

then $\underline{z} \in H(div;\Omega)$.

Proof: Our assumptions immediately give $\underline{z} \in L_2(\Omega)^n$.
We have to verify that a generalized divergence of $\underline{z}$ exists with $div\,\underline{z} \in L_2(\Omega)$. To do that we define piecewise

$$q(x) := (div\,\underline{z}_j)(x) \qquad \text{for all } x \in \Omega_j. \qquad (2.9)$$

It is clear that $q \in L_2(\Omega)$. Using integration by parts, from (2.9) it follows that

$$\int_{\Omega} \varphi q \, dx = \sum_{j=1}^{M} \int_{\Omega_j} \varphi \operatorname{div} \underline{z}_j \, dx$$

$$= \sum_{j=1}^{M} \int_{\Gamma_j} \varphi \underline{z}_j \cdot \underline{n}_j \, dx - \sum_{j=1}^{M} \int_{\Omega_j} \underline{z}_j \cdot \nabla \varphi \, dx \qquad \text{for all } \varphi \in \mathcal{D}(\Omega);$$

here $\underline{n}_j$ denotes the outer unit normal on $\Gamma_j := \partial \Omega_j$. Taking into consideration the decomposition of Γ_j into Γ_{jk}, the relation $\underline{n}_{jk} = -\underline{n}_{kj}$ for $k \neq j$, and $\varphi|_\Gamma = 0$ for $\varphi \in \mathcal{D}(\Omega)$, we obtain

$$\int_{\Omega} \varphi q \, dx = \frac{1}{2} \sum_{j=1}^{M} \sum_{k=1}^{M} \int_{\Gamma_{jk}} \varphi \left(\underline{z}_j - \underline{z}_k \right) \cdot \underline{n}_{jk} \, ds - \int_{\Omega} \underline{z} \cdot \nabla \varphi \, dx \qquad \text{for all } \varphi \in \mathcal{D}(\Omega).$$

The hypothesis (2.8) simplifies this to

$$\int_{\Omega} \varphi q \, dx = - \int_{\Omega} \underline{z} \cdot \nabla \varphi \, dx \qquad \text{for all } \varphi \in \mathcal{D}(\Omega).$$

That is, $q = \operatorname{div} \underline{z}$ as we wanted to prove. ∎

Next we describe simple examples of the construction of finite elements. We concentrate on the two-dimensional case, i.e., $\Omega \subset \mathbb{R}^2$. Since the one-dimensional case is the simplest we also explain some basic facts in 1D; often the three-dimensional case can be handled by using the basic ideas from 2D.

Let $\Omega = (a, b) \subset \mathbb{R}^1$. Decompose Ω into subdomains Ω_j using the mesh $\{x_i\}_{i=0}^{N}$:

$$a = x_0 < x_1 < x_2 < \cdots < x_{N-1} < x_N = b.$$

Set $\Omega_i = (x_{i-1}, x_i)$ for $i = 1, \ldots, N$. Furthermore, we introduce the local step size notation $h_i := x_i - x_{i-1}$ for $i = 1, \ldots, N$.

The space of piecewise linear and globally continuous functions is defined to be $V_h = \operatorname{span}\{\varphi_i\}_{i=0}^{N}$ where the basis functions are

$$\varphi_i(x) = \begin{cases} \dfrac{1}{h_i}(x - x_{i-1}) & \text{for } x \in \Omega_i, \\ \dfrac{1}{h_{i+1}}(x_{i+1} - x) & \text{for } x \in \Omega_{i+1}, \\ 0 & \text{otherwise.} \end{cases} \qquad (2.10)$$

By construction $\varphi_i \in C(\overline{\Omega})$ and $\varphi_i|_{\Omega_j} \in C^1(\overline{\Omega}_j)$. Therefore Lemma 4.1 ensures that $\varphi_i \in H^1(\Omega)$.

Moreover, Lemma 4.1 and (2.10) yield, for $x \neq x_j$, the representation

$$\varphi_i'(x) = \begin{cases} \dfrac{1}{h_i} & \text{for } x \in \Omega_i, \\ -\dfrac{1}{h_{i+1}} & \text{for } x \in \Omega_{i+1}, \\ 0 & \text{otherwise} \end{cases}$$

for the generalized derivative of φ_i. Figure 4.4 shows the graph of the function φ_i. It is clear why it is often called a "hat function".

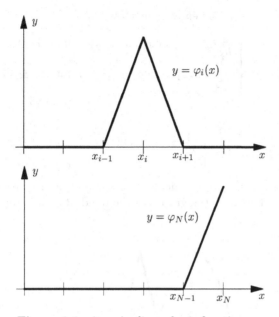

Figure 4.4 piecewise linear basis functions φ_i

The basis functions of (2.10) satisfy $\varphi_i(x_k) = \delta_{ik}$ for $i, k = 0, 1, \ldots, N$. Thus in the representation

$$v(x) = \sum_{i=0}^{N} v_i \varphi_i(x), \tag{2.11}$$

each parameter $v_i \in \mathbb{R}$ is the value of the function v at the mesh point x_i.

We now move on to piecewise quadratic and globally continuous functions defined on the same mesh. Similarly to the linear case, these functions can be described by the representation

$$v(x) = \sum_{i=0}^{N} v_i \psi_i(x) + \sum_{i=1}^{N} v_{i-1/2} \psi_{i-1/2}(x) \tag{2.12}$$

where the globally continuous basis functions ψ_i and $\psi_{i-1/2}$ are defined by the following conditions:

i) $\psi_i|_{\Omega_j}$ and $\psi_{i-1/2}|_{\Omega_j}$ are quadratic polynomials,
ii) $\psi_i(x_k) = \delta_{ik}$ and $\psi_i(x_{k-1/2}) = 0,$

iii) $\psi_{i-1/2}(x_k) = 0$ and $\psi_{i-1/2}(x_{k-1/2}) = \delta_{ik}$.

Here $x_{i-1/2} := \frac{1}{2}(x_{i-1} + x_i)$ denotes the midpoint of the subinterval Ω_i for $i = 1, \ldots, N$.

These basis functions are given explicitly by:

$$\psi_i(x) = \begin{cases} \frac{2}{h_i^2}(x - x_{i-1})(x - x_{i-1/2}) & \text{for } x \in \bar{\Omega}_i, \\ \frac{2}{h_{i+1}^2}(x_{i+1} - x)(x_{i+1/2} - x) & \text{for } x \in \Omega_{i+1}, \\ 0 & \text{otherwise,} \end{cases} \qquad (2.13)$$

and

$$\psi_{i-1/2}(x) = \begin{cases} \frac{4}{h_i^2}(x - x_{i-1})(x_i - x) & \text{for } x \in \Omega_i, \\ 0 & \text{otherwise.} \end{cases} \qquad (2.14)$$

A function of the type $\psi_{i-1/2}$, which vanishes at x_{i-1} and x_i, is called a *bubble function*. Figure 4.5 shows the graphs of the quadratic basis functions ψ_i and $\psi_{i-1/2}$.

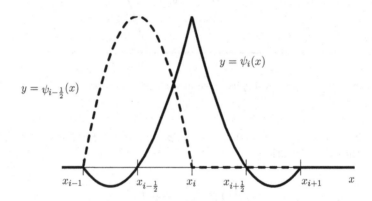

Figure 4.5 The quadratic basis functions ψ_i and $\psi_{i-\frac{1}{2}}$

The ansatz (2.12) with basis functions (2.13) and (2.14) is a continuous function v for any choice of the coefficients v_i and $v_{i-1/2}$. The conditions ii) and iii) imply that the parameters v_i and $v_{i-1/2}$ are the values of v at the points $x_0,\ x_{1/2},\ x_1, \cdots,\ x_{N-1/2},\ x_N$. Consequently this basis is called a *nodal basis*.

It is possible to choose a different basis for the space of piecewise quadratic and globally continuous functions. Let us define

$$v(x) = \sum_{i=0}^{N} v_i \varphi_i(x) + \sum_{i=1}^{N} w_{i-1/2}\psi_{i-1/2}(x). \qquad (2.15)$$

Then it is not difficult to see that

$$\text{span}\{\psi_i\}_{i=0}^N \oplus \text{span}\{\psi_{i-1/2}\}_{i=1}^N = \text{span}\{\varphi_i\}_{i=0}^N \oplus \text{span}\{\psi_{i-1/2}\}_{i=1}^N.$$

The basis in the representation (2.15) of the piecewise quadratic and globally continuous functions is generated from the basis $\{\varphi_i\}_{i=0}^N$ of the piecewise linear and globally continuous functions by adding to it the basis functions $\{\psi_{i-1/2}\}_{i=1}^N$; consequently the basis of (2.15) is a so-called *hierarchical basis*. One can increase the polynomial degree still further and describe spaces of piecewise polynomials of degree p that are globally continuous, but we shall do this later in 2D.

In numerical analysis cubic splines that are globally C^2 are commonly used, but the application of the finite element method to fourth-order differential equations requires only the global C^1 property. Thus we next describe a set of basis functions that are piecewise cubic and globally C^1. Let ζ_i and $\eta_i \in C^1(\overline{\Omega})$ be piecewise cubic polynomials with the following properties:

$$\begin{aligned}
\zeta_i(x_k) &= \delta_{ik}, \quad \zeta_i'(x_k) = 0, \\
\eta_i(x_k) &= 0, \quad \eta_i'(x_k) = \delta_{ik}.
\end{aligned} \tag{2.16}$$

It is easy to calculate these functions explicitly:

$$\eta_i(x) = \begin{cases}
\dfrac{1}{h_i^2}(x - x_i)(x - x_{i-1})^2 & \text{for } x \in \bar{\Omega}_i, \\
\dfrac{1}{h_{i+1}^2}(x - x_i)(x - x_{i+1})^2 & \text{for } x \in \Omega_{i+1}, \\
0 & \text{otherwise,}
\end{cases} \tag{2.17}$$

and

$$\zeta_i(x) = \begin{cases}
\varphi_i(x) - \dfrac{1}{h_i}[\eta_{i-1}(x) + \eta_i(x)] & \text{for } x \in \bar{\Omega}_i, \\
\varphi_i(x) + \dfrac{1}{h_{i+1}}[\eta_i(x) + \eta_{i+1}(x)] & \text{for } x \in \Omega_{i+1}, \\
0 & \text{otherwise,}
\end{cases} \tag{2.18}$$

with φ_i as in (2.10). Figure 4.6 shows the graphs of the basis functions defined by (2.17) and (2.18).

Every piecewise cubic and globally C^1 function has a representation

$$v(x) = \sum_{i=0}^N v_i \zeta_i(x) + \sum_{i=0}^N w_i \eta_i(x)$$

with the *Hermite basis* η_i, ζ_i. The parameters v_i and w_i are now the values of the function v and its derivative v' at the mesh points x_i.

Consider the two-dimensional case where $\Omega \subset \mathbb{R}^2$. It is clear that in contrast to the one-dimensional case the geometry of the given domain Ω now plays an essential role. How should one decompose Ω into simple subdomains?

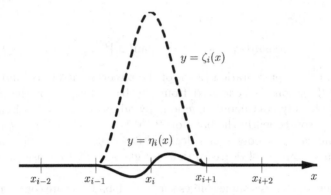

Figure 4.6 Hermite basis functions η_i and ζ_i

In finite element codes there are special algorithms called mesh generators that do this job. For the moment we simply assume that we have such a decomposition of Ω; we shall examine some aspects of mesh generation in Section 4.3.

For simplicity, let Ω be polygonal: that is, the boundary of Ω consists of straight line segments. We decompose Ω in such a way that all its subdomains are triangles or quadrilaterals. These subdomains (which are also called elements) are denoted by Ω_j for $j = 1, \ldots, M$. We assume that the decomposition satisfies (1.3), viz.,

$$\overline{\Omega} = \bigcup_{j=1}^{M} \overline{\Omega}_j, \qquad int\, \Omega_i \cap int\, \Omega_j = \emptyset \ \text{ if } i \neq j. \tag{2.19}$$

We use the terms *decomposition* and *triangulation* interchangeably irrespective of whether triangles or quadrilaterals are used. Figure 4.7 shows an example of a triangulation.

As we shall see, the ansatz functions over each element will be defined by function values and/or values of function derivatives at some points of that element. As we also require global properties of the ansatz functions—such as global continuity—it is more or less necessary that adjacent elements fit together neatly so that the triangulation of the given domain Ω is *admissible*:

A triangulation is admissible if for each pair of subdomains Ω_i and Ω_j exactly one of the following four cases occurs:

- $\Omega_i = \Omega_j$,
- $\overline{\Omega}_i \cap \overline{\Omega}_j$ is a complete edge of both Ω_i and Ω_j,
- $\overline{\Omega}_i \cap \overline{\Omega}_j$ is a nodal point of the triangulation,
- $\overline{\Omega}_i \cap \overline{\Omega}_j = \emptyset$.

Thus we exclude situations where an edge of some subdomain Ω_i is a proper subset of an edge of some other subdomain Ω_j, as in Figure 4.8. In other words, we exclude *hanging nodes*.

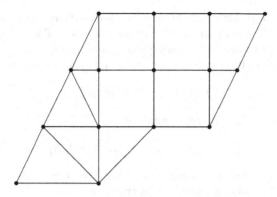

Figure 4.7 Example: admissible triangulation

From now on our triangulations are always assumed to be admissible.

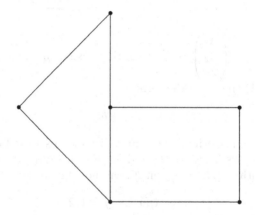

Figure 4.8 Hanging node in a non-admissible triangulation

Let a triangulation $\mathcal{Z}$ of the domain Ω be given by subdomains

$$\Omega_i\,, \quad i = 1,\dots, M,$$

that satisfy (2.19). Now any convex polygon Ω_i^* can be described in the following way. If the vertices of Ω_i^* are p^j, for j in some index set J_i, then $\overline{\Omega}_i^*$ is the convex hull of those vertices:

$$\overline{\Omega}_i^* = conv\{p^j\}_{j \in J_i} := \left\{ x = \sum_{j \in J_i} \lambda_j p^j \,:\, \lambda_j \geq 0,\, \sum_{j \in J_i} \lambda_j = 1 \right\}. \quad (2.20)$$

As previously mentioned, we shall use mainly triangles and convex quadrilaterals in each triangulation; this corresponds to the cases $|J_i| = 3$ and

$|J_i| = 4$. The representation (2.20) is a standard parameterization of a convex domain, which we use often, especially for triangles. Then the local coordinates λ_j (for $j \in J_i$) are called *barycentric coordinates*.

Let us consider the common edge of two adjacent triangles:

$$\overline{\Omega}_i \cap \overline{\Omega}_k = conv\{p^j, p^l\},$$

say. Then this edge has the representation

$$\overline{\Omega}_i \cap \overline{\Omega}_k = \left\{ x = \lambda_j p^j + \lambda_l p^l \; : \; \lambda_j, \, \lambda_l \geq 0, \, \lambda_j + \lambda_l = 1 \right\}.$$

This observation is important when studying the global continuity of ansatz functions that are defined piecewise on triangles.

Now let the element $K \subset \mathcal{Z}$ be a non-degenerate triangle with vertices p^1, p^2, p^3. Each point $x \in \overline{K}$ has unique nonnegative barycentric coordinates λ_1, λ_2 and λ_3 that are the solution of

$$x = \sum_{i=1}^{3} \lambda_i p^i \qquad \text{and} \qquad 1 = \sum_{i=1}^{3} \lambda_i. \tag{2.21}$$

If we introduce $\lambda := \begin{pmatrix} \lambda_1 \\ \lambda_2 \\ \lambda_3 \end{pmatrix} \in \mathbb{R}^3$, then there exists an affine mapping corresponding to (2.21) that is of the form

$$\lambda = Bx + b. \tag{2.22}$$

This way of representing points in K enables us to define local ansatz functions on triangles using λ_1, λ_2 and λ_3. For instance, the affine functions φ_j, $j = 1, 2, 3$, with $\varphi_j(p^k) = \delta_{jk}$ on K, can now be described as follows:

$$\varphi_j(x) = \lambda_j(x), \quad j = 1, 2, 3. \tag{2.23}$$

In the next subsection we specify the local ansatz functions for polynomials of higher degree. It is important that the interpolation points for these polynomials can easily be described in barycentric coordinates. Triangular elements of type (l) use the points

$$p^\alpha = \sum_{j=1}^{3} \frac{\alpha_j}{|\alpha|} p^j \tag{2.24}$$

where $\alpha = (\alpha_1, \alpha_2, \alpha_3)$ is a multi-index with $|\alpha| := \Sigma_{j=1}^3 \alpha_j = l$.

In principle, the case $\Omega \subset \mathbb{R}^n$ for $n \geq 3$ can be handled similarly to the two-dimensional case: if one uses simplices and convex hyper-quadrilaterals for the triangulation, then one has analogous principles. For instance, a simplex $K \subset \mathbb{R}^3$ with vertices p^1, p^2, p^3 and p^4 can be described in barycentric coordinates by

$$K = \left\{ x \in \mathbb{R}^3 \ : \ x = \sum_{i=1}^{4} \lambda_i p^i, \ \lambda_i \geq 0, \ \sum_{i=1}^{4} \lambda_i = 1 \right\}.$$

See, e.g., [Cia78] for more details in the higher-dimensional case.

Nevertheless in the higher-dimensional case additional tools are sometimes necessary. In particular the validity of embedding theorems depends on the dimension; for $n = 4$ the embedding of $H^2(\Omega)$ in $C(\overline{\Omega})$ does not hold, unlike the cases $n = 2, 3$.

4.2.2 Examples of Finite Element Spaces in $\mathbb{R}^2$ and $\mathbb{R}^3$

Now we describe some finite element spaces that are often used in practice in two and three dimensions.

Let $K \subset \mathcal{Z}_h$ denote an element of the triangulation $\mathcal{Z}_h$ of the given domain $\Omega \subset \mathbb{R}^n$ where $n = 2$ or 3. As we have seen, K can be represented as a convex combination of the vertices $p^1, p^2, ..., p^s$:

$$K = \left\{ x \in \mathbb{R}^n \ : \ x = \sum_{i=1}^{s} \lambda_i p^i, \ \lambda_i \geq 0, \ \sum_{i=1}^{s} \lambda_i = 1 \right\}.$$

If K is a regular simplex in $\mathbb{R}^n$, then there is a one-to-one and onto mapping between points $x \in K$ and barycentric coordinates $\lambda_1, \lambda_2, ...,\lambda_{n+1}$.
If K is not a simplex then additional conditions are necessary. For rectangles or quadrilaterals tensor product ideas are often used. For a rectangle with vertices p^1, p^2, p^3 and p^4, a special parameterization is given by

$$\begin{array}{ll} \lambda_1 = (1 - \xi)(1 - \eta), & \lambda_2 = \xi(1 - \eta), \\ \lambda_3 = \xi\eta, & \lambda_4 = (1 - \xi)\eta. \end{array} \qquad (2.25)$$

Here $\xi, \eta \in [0, 1]$ are free parameters. From (2.25) one has immediately

$$\lambda_i \geq 0, \ i = 1, \ldots, 4, \qquad \sum_{i=1}^{4} \lambda_i = 1 \qquad \text{for all } \xi, \eta \in [0, 1].$$

In general, every finite element is characterized by:

- the geometry of the subdomain K,
- the concrete form of the ansatz functions on K
- a set of *unisolvent* linear functionals (which defines the ansatz function uniquely)

The abstract triple (K, P_K, Σ_K) is called a *finite element*. Here P_K is the space of local ansatz functions on K and Σ_K the set of given unisolvent linear functionals. If only function values are used in Σ_K then we call the finite element a *Lagrange element*. If derivatives evaluated at certain points in K are used also, we call this a *Hermite element*. Integral means on K or its edges can also be used as functionals in Σ_K.

In engineering mechanics the linear functionals are also called degrees of freedom. Furthermore, the local ansatz functions are commonly known as *shape functions*.

While a finite element is defined locally, one must also investigate its global properties, often by invoking Lemmas 4.1 and 4.2. Thus it is necessary to study the behaviour of the shape functions on the shared boundary of two adjacent elements.

Let $P_l(K)$ be the set of all polynomials on K of degree at most l. The space P_l is often used as the space of shape functions on triangles or simplices.

For a rectangle $K \subset \mathbb{R}^2$ or parallelepiped $K \subset \mathbb{R}^3$, the set $Q_l(K)$ of polynomials on K plays an important role. A polynomial belongs to Q_l if it is the product of polynomials of degree at most l in each single variable—for a rectangle $K \subset \mathbb{R}^2$ we have $p \in Q_l(K)$ if

$$p(x) = p_l(\xi)q_l(\eta)$$

with polynomials $p_l(\cdot), q_l(\cdot) \in P_l(\mathbb{R}^1)$. It is easy to see that in general

$$P_l(K) \subset Q_l(K) \subset P_{nl}(K).$$

To define an element concretely, we have next to fix its degrees of freedom or, equivalently, to specify its unisolvent set of linear functionals.

As already mentioned, for Lagrange elements on simplices the function values at the interpolation points

$$p^\alpha = \sum_{j=1}^{s} \frac{\alpha_j}{|\alpha|} p^j$$

are important. Here p^1, ..., p^s are the vertices of K.

We now characterize several special elements by their shape functions and degrees of freedom. In the corresponding figures each degree of freedom at a point is symbolized as follows:

● - function value
o - first order derivatives
O - second order derivatives
| - normal derivative

To ensure unisolvence the number of degrees of freedom has to be equal to the dimension of the space of shape functions; this number is denoted by d. For Lagrange elements we also include the shape functions associated with some typical interpolation points p^α.

Triangular Elements

i) linear C^0 element

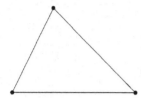

Lagrange shape function for p^α:

$$\Psi_\alpha(\lambda) = \sum_{j=1}^{3} \alpha_j \lambda_j$$

$d = 3$.

ii) discontinuous linear element (Crouzeix-Raviart element)

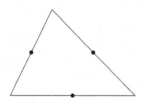

Lagrange shape function for p^α:

$$\Psi_\alpha(\lambda) = \sum_{j=1}^{3} (1 - \alpha_j)\lambda_j$$

$d = 3$.

iii) quadratic C^0 element

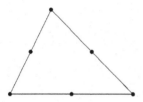

Lagrange shape functions for the points p^{200} and p^{110}:

$$\Psi_{200}(\lambda) = \lambda_1(2\lambda_1 - 1)$$
$$\Psi_{110}(\lambda) = \lambda_1\lambda_2$$

$d = 6$.

iv) cubic C^0 element

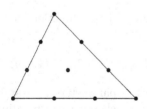

Lagrange shape functions for the points p^{300}, p^{210} and p^{111}:

$$\Psi_{300}(\lambda) = \tfrac{1}{2}\lambda_1(3\lambda_1 - 1)(3\lambda_1 - 2)$$
$$\Psi_{210}(\lambda) = \tfrac{9}{2}\lambda_1\lambda_2(3\lambda_1 - 1)$$
$$\Psi_{111}(\lambda) = 27\lambda_1\lambda_2\lambda_3$$

$d = 10$.

v) cubic C^0 Hermite element

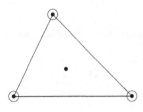

$d = 10.$

vi) quintic C^1 element (Argyris element)

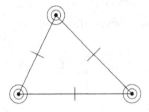

$d = 21.$

vii) reduced quintic C^1 element (Bell element)

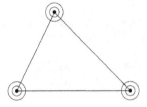

$d = 18.$

Rectangular Elements

viii) bilinear C^0 element

Lagrange shape function for the point p^{1000}:

$$\Psi_{1000}(\lambda) = (1 - \xi)(1 - \eta),$$

where the parameters ξ, η and λ_1, ..., λ_4 are chosen corresponding to (2.25).

$d = 4.$

ix) biquadratic C^0 element

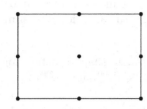

Lagrange shape function for the points p^{4000}, p^{2200} and p^{1111}:

$$\Psi_{4000}(\lambda) = (1 - \xi)(2\xi - 1)(1 - \eta)(2\eta - 1)$$
$$\Psi_{2200}(\lambda) = 4(1 - \xi)\xi(1 - \eta)(1 - 2\eta)$$
$$\Psi_{1111}(\lambda) = 16\xi(1 - \xi)\eta(1 - \eta)$$

$$d = 9.$$

The following figures give an impression of the shape functions viii) and ix).

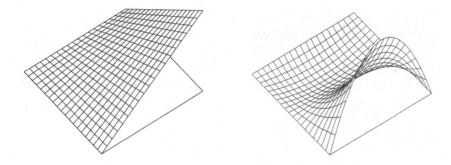

Figure 4.9.a Shape functions for viii) and ix)

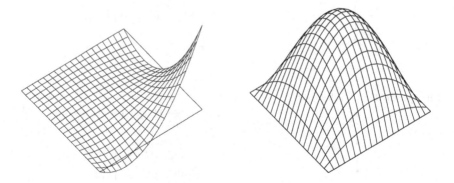

Figure 4.9.b Shape functions for ix)

x) biquadratic serendipity element
(the word "serendipity" was coined in 1754; it means a lucky find and is derived from a tale of three princes of Sri Lanka, which was formerly Serendip)

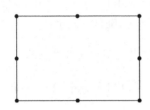

Lagrange shape function for the points p^{4000} and p^{2200}:

$$\tilde{\Psi}_{4000}(\lambda) = \Psi_{4000}(\lambda) - \tfrac{1}{4}\Psi_{1111}(\lambda)$$

$$\tilde{\Psi}_{2200}(\lambda) = \Psi_{2200}(\lambda) + \tfrac{1}{2}\Psi_{1111}(\lambda)$$

—here Ψ_α are the biquadratic shape functions from ix)

$d = 8$.

xi) Bogner-Fox-Schmit C^1 element

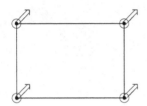

Here 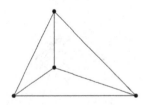 means that one uses the mixed derivative $\dfrac{\partial^2 u}{\partial\xi\partial\eta}$ as a degree of freedom at the corresponding mesh point.

$d = 16$.

Tetrahedral Elements

xii) linear C^0 element

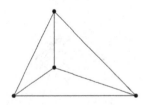

Lagrange shape function for p^{1000}:

$$\Psi_{1000}(\lambda) = \lambda_1,$$

$d = 4$.

xiii) quadratic C^0 element

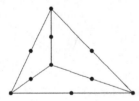

Lagrange shape functions for p^{2000} and p^{1100}:

$$\Psi_{2000}(\lambda) = \lambda_1(2\lambda_1 - 1)$$
$$\Psi_{1100}(\lambda) = 4\lambda_1\lambda_2,$$

$d = 10$.

Elements on a Parallelepiped in 3D

xiv) trilinear C^0 element

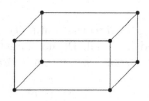

Lagrange shape function for $p^{10000000}$:

$$\Psi_{10000000}(\lambda) = (1 - \xi)(1 - \eta)(1 - \zeta),$$

with a transformation analogous to (2.25).

$d = 8$.

xv) triquadratic C^0 element

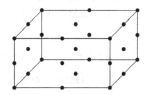

$d = 27$.

xvi) triquadratic serendipity element

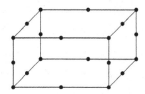

$d = 20$.

For the C^0 elements in each concrete example one must verify the global continuity property—equivalently, the continuity of the shape functions across edges of adjacent elements. Let us take as an example the cubic triangular element of example iv) above:

Lemma 4.4. *Let $\hat{K}$ and $\tilde{K} \subset \mathbb{R}^2$ be two adjacent triangles of a triangulation with a common edge. Then the cubic functions $\hat{u}$ and $\tilde{u}$ defined on $\hat{K}$ and $\tilde{K}$ respectively are identical when restricted to the joint edge. Consequently the piecewise cubic function that is defined by $\hat{u}$ on $\hat{K}$ and $\tilde{u}$ on $\tilde{K}$ is globally continuous on $K := \hat{K} \cup \tilde{K}$.*

Proof: The functions $\hat{u}$ and $\tilde{u}$ are cubic. Hence, the restriction of such a function to an edge yields a cubic polynomial in one variable. But these two cubic polynomials in one variable must coincide at the four interpolation points on the edge common to $\hat{K}$ and $\tilde{K}$, so these polynomials are identical, i.e.,

$$\hat{u}(x) = \tilde{u}(x) \qquad \text{for all } x \in \hat{K} \cap \tilde{K}.$$

Clearly $\hat{u} \in C(\hat{K})$ and $\tilde{u} \in C(\tilde{K})$. Thus the piecewise-defined function

$$u(x) = \begin{cases} \hat{u}(x) & \text{for } x \in \hat{K}, \\ \tilde{u}(x) & \text{for } x \in \tilde{K}, \end{cases}$$

is continuous on K. ∎

In many textbooks on finite elements (e.g., [Cia78]) the continuity of the C^0 elements listed in our overview is verified in detail. Do not forget that the admissibility of the triangulation is an important assumption in drawing these conclusions.

Next we shall motivate the serendipity elements, which are obtained by eliminating a degree of freedom in the interior of the element.

Suppose that, starting from biquadratic elements (i.e., Q_2), we want to eliminate the interior degree of freedom while requiring that the polynomial obtained has at most degree 3. With the biquadratic shape functions of ix) we obtain

$$p(\xi, \eta) = u_{4000}(1 - \xi)(2\xi - 1)(1 - \eta)(2\eta - 1) + u_{0400}\xi(1 - 2\xi)(1 - \eta)(2\eta - 1)$$
$$+ u_{0040}\xi(1 - 2\xi)\eta(1 - 2\eta) + u_{0004}(1 - \xi)(2\xi - 1)\eta(1 - 2\eta)$$
$$+ 4(u_{2200}(1 - \xi)\xi(1 - \eta)(2\eta - 1) + u_{0220}(1 - \xi)(2\xi - 1)\eta(1 - \eta)$$
$$+ u_{0022}(1 - \xi)\xi\eta(2\eta - 1) + u_{2002}(1 - \xi)(2\xi - 1)\eta(1 - \eta)$$
$$+ 16u_{1111}\xi(1 - \xi)\eta(1 - \eta).$$

Gathering the various powers of ξ and η, we get

$$p(\xi, \eta) = u_{4000} + u_{0400} + u_{0040} + u_{0004} + \cdots$$
$$+ \big[16u_{1111} + 4(u_{4000} + u_{0400} + u_{0040} + u_{0004})$$
$$- 8(u_{2200} + u_{0220} + u_{0022} + u_{2002})\big]\xi^2\eta^2.$$

Thus $p|_K \in P_3(K)$ if and only if

$$4u_{1111} + u_{4000} + u_{0400} + u_{0040} + u_{0004} = 2(u_{2200} + u_{0220} + u_{0022} + u_{2002}). \quad (2.26)$$

The shape functions $\tilde{\Psi}_\alpha$ of the serendipity element x) are such that condition (2.26) is satisfied.

The biquadratic serendipity element has, moreover, the following property:

Lemma 4.5. *Let $\tilde{Q}_2(K)$ be the class of all shape functions belonging to the serendipity element* x). *Then*

$$P_2(K) \subset \tilde{Q}_2(K) \subset P_3(K).$$

Proof: The second set inclusion is immediate from the facts that the shape functions $\tilde{\Psi}_\alpha$ of x) lie in $Q_2(K)$ and satisfy (2.26).

Now let u be an arbitrary polynomial in $P_2(K)$. Then

$$u(\xi, \eta) = a_0 + a_{10}(2\xi - 1) + a_{01}(2\eta - 1)$$
$$+ a_{20}(2\xi - 1)^2 + a_{11}(2\xi - 1)(2\eta - 1) + a_{02}(2\eta - 1)^2.$$

where the coefficients a_0, $a_{10}, \cdots$, a_{02} are arbitrary. Hence, for the values u_α of the function $u(\cdot)$ at the points p^α we get

$$
\begin{aligned}
u_{4000} &= a_0 &-a_{10} &-a_{01} &+a_{20} &+a_{11} &+a_{02} \\
u_{0400} &= a_0 &+a_{10} &-a_{01} &+a_{20} &-a_{11} &+a_{02} \\
u_{0040} &= a_0 &+a_{10} &+a_{01} &+a_{20} &+a_{11} &+a_{02} \\
u_{0004} &= a_0 &-a_{10} &+a_{01} &+a_{20} &-a_{11} &+a_{02} \\
u_{2200} &= a_0 & &-a_{01} & & &+a_{02} \\
u_{0220} &= a_0 &+a_{10} & &+a_{20} & & \\
u_{0022} &= a_0 & &+a_{01} & & &+a_{02} \\
u_{2002} &= a_0 &-a_{10} & &+a_{20} & & \\
u_{1111} &= a_0 & & & & &
\end{aligned}
\qquad (2.27)
$$

Because $u \in P_2(K)$ and $P_2(K) \subset Q_2(K)$, the polynomial u will be exactly represented in the space of biquadratic functions ix). We already know that biquadratic functions can be exactly represented in $\tilde{Q}_2(K)$ if and only if (2.26) is valid. But by combining the equations (2.27) one sees easily that (2.26) holds. Thus u will be exactly represented in $\tilde{Q}_2(K)$, i.e., $P_2(K) \subset \tilde{Q}_2(K)$. ∎

Compared with C^0 elements, the construction of C^1 elements is relatively complicated.

For triangular elements and polynomial shape functions, Zenisek [126] proved that the minimal number of degrees of freedom needed to ensure the global C^1 property is $d = 18$. This number is attained by the Bell element; for Argyris's triangle the dimension is slightly larger. Even for rectangular elements the dimension of the space of shape functions is relatively large for C^1 elements; for instance, it is 16 for the Bogner-Fox-Schmit element xi). Of course a finite element method based on shape function spaces of locally high dimension leads to a much larger number of Galerkin equations that must be solved numerically.

Accordingly, in practice often one avoids using C^1 elements and instead embraces nonconforming methods or mixed methods that require only C^0 elements; see Sections 4.5 and 4.6. Here we now discuss the possibility of reducing the number of degrees of freedom by eliminating those degrees of freedom associated with interior nodes. On triangles, well-known elements of this type are the Powell-Sabin element and the Clough-Tocher element; see [Cia78]. In these two elements the local ansatz functions are, respectively, piecewise quadratics and piecewise cubics. For the Clough-Tocher element the dimension of the local space is 12. The verification of global differentiability properties of such *composite elements* on triangulations can be simplified if one uses the so-called Bézier-Bernstein representation of polynomials on triangles.

We describe some principles of composite C^1 elements in a simple one-dimensional example.

Let $K = [x_{i-1}, x_i]$ be a closed subinterval of the decomposition $\mathcal{Z}$ of $\Omega \subset \mathbb{R}^1$. If we prescribe values of the function and its first-order derivative at both endpoints of K, then there exists a unique cubic polynomial interpolating to these. We have already met the corresponding Hermite element.

Instead we choose the intermediate point $x_{i-1/2} := (x_{i-1} + x_i)/2$ and aim to determine a quadratic polynomial on $[x_{i-1}, x_{i-1/2}]$ and a quadratic polynomial on $[x_{i-1/2}, x_i]$ in such a way that the composite function is in $C^1(K)$.

Transform K to the reference interval $K' = [0, 1]$, i.e., represent K in the form

$$K = \{\, x = (1 - \xi)x_{i-1} + \xi x_i \; : \; \xi \in [0, 1] \}.$$

Every globally continuous and piecewise quadratic function on K can now be written as

$$u(x) = \begin{cases} u_{i-1}\varphi(\xi) + h_i u'_{i-1}\psi(\xi) + u_{i-1/2}\sigma(\xi) & \text{for } \xi \in [0, 0.5], \\ u_i\varphi(1 - \xi) + h_i u'_i\psi(1 - \xi) + u_{i-1/2}\sigma(1 - \xi) & \text{for } \xi \in [0.5, 1]. \end{cases} \tag{2.28}$$

Here $u_{i-1/2}$ is the function value at the midpoint $x_{i-1/2}$ while u_{i-1}, u'_{i-1}, u_i and u'_i are the values of the function and its derivatives at the boundary points. Furthermore, $h_i := x_i - x_{i-1}$ and

$$\varphi(\xi) = (1 - 2\xi)(1 + 2\xi), \qquad \psi(\xi) = \xi(1 - 2\xi), \qquad \sigma(\xi) = 4\xi^2. \tag{2.29}$$

These functions are the Hermite-Lagrange shape functions with respect to the interval $[0, 0.5]$: they satisfy

$$\begin{aligned} \varphi(0) &= 1, \; \varphi'(0) = 0, \; \varphi(0.5) = 0, \\ \psi(0) &= 0, \; \psi'(0) = 1, \; \psi(0.5) = 0, \\ \sigma(0) &= 0, \; \sigma'(0) = 0, \; \sigma(0.5) = 1. \end{aligned} \tag{2.30}$$

Now we can evaluate the derivatives at our intermediate point $x_{i-1/2}$. Observe that $\xi = (x - x_{i-1})/h_i$. Then (2.28) and (2.30) yield

$$u'(x_{i-1/2} - 0) = \frac{u_{i-1}}{h_i}\varphi'(0.5) + u'_{i-1}\psi'(0.5) + \frac{u_{i-1/2}}{h_i}\sigma'(0.5),$$

$$u'(x_{i-1/2} + 0) = -\frac{u_i}{h_i}\varphi'(0.5) - u'_i\psi'(0.5) - \frac{u_{i-1/2}}{h_i}\sigma'(0.5).$$

Therefore, our piecewise quadratic function (2.28) is differentiable at $x_{i-1/2}$ for arbitrarily chosen u_{i-1}, u'_{i-1}, u_i, and u'_i if and only if

$$(u_{i-1} + u_i)\varphi'(0.5) + h_i(u'_{i-1} + u'_i)\psi'(0.5) + 2u_{i-1/2}\sigma'(0.5) = 0.$$

Recalling (2.29), this equation produces the condition for $u \in C^1(K)$:

$$u_{i-1/2} = \frac{1}{2}(u_{i-1} + u_i) + \frac{1}{8}h_i(u'_{i-1} + u'_i). \tag{2.31}$$

Therefore, eliminating the artificial degree of freedom $u_{i-1/2}$ by means of (2.31) yields a composite element in $C^1(K)$.

Exercise 4.6. Consider the boundary value problem

$$-(pu')' + qu = f, \quad u(0) = u(1) = 0.$$

Assume that p and q are piecewise constant on the given mesh: $p(x) \equiv p_{i-1/2}$, $q(x) \equiv q_{i-1/2}$ when $x \in (x_{i-1}, x_i)$ for $i = 1, \ldots, N$, where $x_0 = 0$, $x_N = 1$. Compute the coefficient matrix of the linear system that is generated by the application of the finite element" method with linear elements.

Exercise 4.7. Let the differential operator

$$-\frac{d}{dx}\left(k\frac{dT}{dx}\right) + a\frac{dT}{dx} \quad (k, a \text{ constants})$$

be discretized by quadratic finite elements on an equidistant mesh. What approximation of the operator is generated?

Exercise 4.8. Consider a uniform triangular mesh that is derived from a mesh of squares by drawing one diagonal in each square. Now discretize the partial derivative $\dfrac{\partial u(x, y)}{\partial x}$ on that mesh using linear finite elements. What difference stencil is generated?

Exercise 4.9. Let us decompose the domain $\Omega = (0, 1) \times (0, 1)$ into squares of equal size and discretize $-\Delta u$ on this mesh using bilinear elements. Verify that the difference stencil generated is

$$\frac{1}{3}\begin{bmatrix} -1 & -1 & -1 \\ -1 & 8 & -1 \\ -1 & -1 & -1 \end{bmatrix}.$$

Exercise 4.10. If one starts with biquadratic elements and deletes the term $x^2 y^2$, one obtains an element of the serendipity family. Discuss possible choices of basis functions for this element.

Exercise 4.11. Assume that a polygonal domain in 2D is decomposed into triangles in such a way that no obtuse angle appears (we call such a triangulation *weakly acute*). Discretize $-\Delta u$ on this mesh using linear finite elements. Prove that the stiffness matrix for the Dirichlet problem

$$-\Delta u = f \quad \text{in } \Omega, \qquad u|_\Gamma = 0$$

is an M-matrix.

Exercise 4.12. Let us consider the weak formulation of the boundary value problem

$$Lu = -u'' + c(x)u = f(x) \quad \text{in } (0, 1),$$
$$u(0) = 0, \quad u'(1) = \beta,$$

$$(2.32)$$

in $V = \{v \in H^1(0,1) : v(0) = 0\}$, where we assume that $c \in L_\infty(0,1)$ and $f \in V^*$.

a) Verify that the embedding $V \hookrightarrow H = H^* \hookrightarrow V^*$ with $H = L_2(0,1)$ implies

$$\langle Lu, v \rangle = \int_0^1 [u'v' + c(x)uv]dx - \beta v(1).$$

b) Verify that $V_h = \{\varphi_k\}_{k=1}^N$ with

$$\varphi_k(t) = \max\{0, 1 - |Nt - k|\}, \ 0 \le t \le 1, \quad k = 1, \dots, N,$$

where $h = 1/N$, is an N-dimensional subspace of V.

c) Show that every $v_h \in V_h$ can be represented as

$$v_h(x) = \sum_{k=1}^N v_k \varphi_k(x) \quad \text{with} \quad v_k = v_h(kh).$$

d) Compute the quantities

$$a_{ik} = \int_0^1 \varphi_i' \varphi_k' dx \quad \text{and} \quad b_{ik} = \int_0^1 \varphi_i \varphi_k dx.$$

e) Discretize problem (2.32) using the Galerkin method with discrete space $V_h \subset V$. What difference scheme is generated in the case

$$c(x) = c_k \quad \text{when} \quad kh < x \le (k+1)h ?$$

Exercise 4.13. Suppose we are given $d + 1$ points $a_i = (a_{ij})_{j=1}^d$, $i = 1, \dots, d+1$, each lying in the n-dimensional space $\mathbb{R}^d$, and having the property

$$V = V(a_1, \dots, a_{d+1}) = \begin{vmatrix} 1 & a_{11} & a_{12} & \cdots & a_{1d} \\ \vdots & \vdots & \vdots & & \vdots \\ 1 & a_{d+11,1} & a_{d+1,2} & \cdots & a_{d+1,d} \end{vmatrix} \ne 0.$$

a) Verify that the a_i are the vertices of a non-degenerate simplex $S \subset \mathbb{R}^d$.

b) Prove that the equations

$$x = \sum_{i=1}^{d+1} \lambda_i a_i \quad \text{and} \quad \sum_{i=1}^{d+1} \lambda_i = 1$$

define a one-to-one mapping from $x \in \mathbb{R}^d$ onto the numbers $\lambda_i = \lambda_i(x; S)$ — these are the barycentric coordinates of x.

c) Let $l : \mathbb{R}^d \to \mathbb{R}^d$ be an affine mapping. Prove that

$$\lambda_i(x; S) = \lambda_i(l(x); l(S)).$$

d) Define $l = (l_i)_{i=1}^d$ by $l_i(x) = \lambda_i(x; S)$. Prove that l is an affine mapping. Characterize $\hat{S} = l(S)$, the image of the simplex S under l, and formulate an algorithm for computing $\hat{A}$ and $\hat{b}$ in

$$x = \hat{A}l(x) + \hat{b}.$$

e) Use the result of d) to prove that the volume of the simplex generated by the points a_i is $V/d!$. What geometric interpretation can you give the barycentric coordinates?

Exercise 4.14. Let T be a triangle with vertices a_1, a_2, a_3. Given $k \in N$, consider the following set of points:

$$L_k(T) = \left\{ x : x = \sum_{i=1}^3 \lambda_i a_i, \ \sum_{i=1}^3 \lambda_i = 1, \right.$$

$$\left. \lambda_i \in \{0, \tfrac{1}{k}, \cdots, \tfrac{k-1}{k}, 1\} \text{ for } i = 1, 2, 3 \right\}.$$

Now we take polynomials of degree k as shape functions and wish to use the function values at the points of $L_k(T)$ as degrees of freedom.
Prove P_k-unisolvence, perhaps first for $k = 2$ and then for general k.

Exercise 4.15. Let T be a triangle with vertices a_1, a_2, a_3. We want to use the following degrees of freedom:
– at each vertex: function value and directional derivatives along the edges
– at the centre of the triangle: function value.
a) prove P_3-unisolvence for this set of degrees of freedom.
b) construct a basis for this element for an equilateral triangle T.

Exercise 4.16. Consider an admissible triangulation of a polygonal domain Ω. Let X_h be the finite element space associated with the Argyris element. Verify that $X_h \subset C^1(\bar{\Omega})$.

Exercise 4.17. Consider a given triangle T. Denote by a_{ijk} the points of the triangle T with the barycentric coordinates $(\tfrac{i}{3}, \tfrac{j}{3}, \tfrac{k}{3})$ where $i + j + k = 3$ and $i, j, k \in \{0, 1, 2, 3\}$.
Define a functional on P_3 by

$$\varphi(p) = 3 \sum_{i+j+k=3} p(a_{ijk}) - 15p(a_{111}) - 5[p(a_{300}) + p(a_{030}) + p(a_{003})]$$

and define a subspace of P_3:

$$P_3' = \{p \in P_3 : \varphi(p) = 0\}.$$

Prove that $P_2 \subset P_3'$.

4.3 Practical Aspects of the Finite Element Method

4.3.1 Structure of a Finite Element Code

If one uses a finite element method to solve a boundary value problem for a partial differential equation on a domain $\Omega \subset \mathbb{R}^n$, then the precise choice of method depends mainly on the given differential equation but the geometry of Ω and the type of boundary conditions also play an essential role.

Consequently, in the implementation of the finite element method one must store geometric information about the boundary of the given domain Ω as well as information regarding the differential equation and the boundary conditions.

The finite element method is based on a decomposition of the given domain Ω into subdomains Ω_j, $j = 1, \ldots, M$. The details of this decomposition are needed in the execution of the finite element method, as are the precise shape functions and the types of degrees of freedom involved. This data is used to generate the discrete problem from the Ritz-Galerkin equations, and is also used later to extract information from the approximate solution via its representation in terms of a basis in the discrete space.

Typically, three steps are evident in the development of a finite element code:

1. Description of the given problem, generation of an initial mesh;
2. Generation of the discrete problem, solution of the discrete problem, a posteriori error estimation, mesh refinement;
3. Postprocessing of the results obtained, graphical presentation.

Step 1 is sometimes called *pre-processing* in terminology resembling the *postprocessing* of step 3. If at the end of step 2 the estimated error is too large, the current mesh is refined and step 2 is repeated. (This is the strategy followed in the h-version of the finite element method, which is the most widely used; in the p- and hp-versions a slightly different approach is used—see Section 4.9.4.)

In this Section we shall discuss three aspects of the implementation of the program just described: the description of the given problem, the generation of the discrete problem, and questions related to mesh generation and mesh refinement.

In a finite element code several very different subproblems have to be solved, which explains why every code contains a large number of special algorithms. Furthermore, for each of these subproblems, many different algorithms are available. For instance, every code uses an algorithm to generate an initial mesh, but if one searches for mesh generation algorithms using the internet, one gets a flood of information about codes dealing only with that subject. Thus it is unsurprising that there does not exist a universal finite element code. Nevertheless there are many fairly well-developed codes that solve relatively large classes of problems.

In the following discussion our attention is mainly focused on the code PLTMG [Ban98] (piecewise linear triangle multi grid). This code was one of the first to use adaptive grid generation (see Section 4.7) and multigrid to solve the discrete problems efficiently. Let us also mention ALBERTA [SS05] and the very compact description of a finite element code using MATLAB in [3].

4.3.2 Description of the Problem

The first step in the implementation of a finite element method is the description of the given problem by data structures in a computer. Here one must describe the domain, the differential equation and the boundary conditions.

Because the finite element method is based on a decomposition of the given domain Ω into subdomains Ω_j, $j = 1, \ldots, M$, one also needs information about these subdomains. Let us assume, for simplicity, that the subdomains are convex polyedra and the degrees of freedom of the shape functions used are associated with the vertices of these polyedra. (Points with which the degrees of freedom are associated are called knots.) Then one needs at least two lists to store the necessary information: a list of the geometrical locations of the polyedra, i.e., the coordinates of their vertices, and a second list that associates certain vertices in the first list with each subdomain.

We give some details for our example from Section 4.1. Number the subdomains and the $\overline{N} = 21$ mesh points as in Figure 4.10.

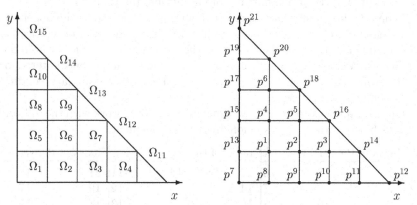

Figure 4.10 Enumeration of subdomains and knots

Then we obtain the following list for the coordinates of the vertices or knots p^i, $i = 1, \ldots, 21$:

i	x_i	y_i		i	x_i	y_i		i	x_i	y_i
1	0.2	0.2		8	0.2	0.0		15	0.0	0.4
2	0.4	0.2		9	0.4	0.0		16	0.6	0.4
3	0.6	0.2		10	0.6	0.0		17	0.0	0.6
4	0.2	0.4		11	0.8	0.0		18	0.4	0.6
5	0.4	0.4		12	1.0	0.0		19	0.0	0.8
6	0.2	0.6		13	0.0	0.2		20	0.2	0.8
7	0.0	0.0		14	0.8	0.2		21	0.0	1.0

Next, we store the indices of the vertices appearing in each subdomain. Here each group of vertices is taken in a counter-clockwise order.

j	Index of the vertices				j	Index of the vertices				j	Index of the vertices		
1	7	8	1	13	6	1	2	5	4	11	11	12	14
2	8	9	2	1	7	2	3	16	5	12	3	14	16
3	9	10	3	2	8	15	4	6	17	13	5	16	18
4	10	11	14	3	9	4	5	18	6	14	6	18	20
5	13	1	4	15	10	17	6	20	19	15	19	20	21

As the given domain is polygonal and the decomposition has the property $\overline{\Omega} = \bigcup_{j=1}^{M} \overline{\Omega}_j$, no additional description of the domain is necessary.

If one uses degrees of freedom that are not located at the vertices of the subdomains or if, for instance, so-called isoparametric finite elements are used (see Section 4.5), then one has to store additional information.

When the boundary conditions are of Dirichlet type and are homogeneous, if one enumerates the vertices in such a way that the interior vertices are numbered before the boundary vertices, then it is easy to handle the boundary conditions. When more complicated boundary conditions are encountered, one has to store segments of the boundary and the type of boundary condition corresponding to each segment.

Our example has 15 boundary segments. One could work with the following list which shows the vertices belonging to each segment:

k	Index of the vertices		k	Index of the vertices		k	Index of the vertices	
1	7	8	6	12	14	11	21	19
2	8	9	7	14	16	12	19	17
3	9	10	8	16	18	13	17	15
4	10	11	9	18	20	14	15	13
5	11	12	10	20	21	15	13	7

Given different boundary conditions on various parts of the boundary, we could complete this list by an indicator of the type of boundary condition, say 1 for Dirichlet, 2 for Neumann and 3 for Robin boundary conditions.

Sometimes it is appropriate to store additional information, such as the numbering of the vertices of adjacent elements. This can be used effectively if for instance in an adaptive algorithm one has to implement local mesh refinement.

If a large amount of information is stored then it is often possible to use that information algorithmically in a specific way. On the other hand, it may be more effective to compute desired data several times instead of permanently storing all available information. Thus one has to decide on a compromise between storage restrictions and a comprehensive structured modular finite element code.

It is clear that a complete description of the given problem contains programs to compute the coefficients of the given differential equation, its right-hand side and the functions determining the boundary conditions.

If the given problem has a complicated structure—for instance if one has a system of partial differential equations, or different differential equations in different subdomains coupled by some conditions on the interface between the subdomains—then, of course, additional tools are needed to describe this.

4.3.3 Generation of the Discrete Problem

Let us describe the generation of the discrete problem for the abstract variational equation whose theory we have already studied:

Find $u \in V$ such that

$$a(u, v) = f(v) \qquad \text{for all } v \in V.$$

For the discretization we use a conforming version of the Ritz-Galerkin method that is based on a subspace $V_h \subset V$. Then the discrete problem is

Find $u_h \in V_h$ such that

$$a(u_h, v_h) = f(v_h) \qquad \text{for all } v_h \in V_h.$$

Let $\{\varphi_i\}_{i=1}^{\hat{N}}$ be a basis for V_h. The discrete problem is equivalent to the linear system

$$A_h \, u_h = f_h$$

with $A_h = (a(\varphi_k, \varphi_i))_{i,k=1}^{\hat{N}}$ and $f_h = (f(\varphi_i))_{i=1}^{\hat{N}}$. The solution of this linear system produces the coefficients $(u_i)_{i=1}^{\hat{N}}$ in the representation

$$u_h(x) = \sum_{i=1}^{\hat{N}} u_i \, \varphi_i(x)$$

of the approximate solution.

Unlike spectral methods based on the Ritz-Galerkin method, the finite element method uses ansatz functions with small local support. Consequently the stiffness matrix A_h is sparse, i.e., relatively few entries of A_h are non-zero.

How is the stiffness matrix generated?

Typically, one computes the contributions of *each element* to the stiffness matrix, then the full matrix is got by summing these contributions. The same procedure is used to compute the vector on the right-hand side. The contribution of a single element to the stiffness matrix is called the *element stiffness matrix*. There are two ways of computing the entries of the element stiffness matrix: one can compute the desired integrals directly over the element or transform it to a *reference element* where the integrals are computed. The process of generating the stiffness matrix and right-hand-side vector is known as *assembling*.

We sketch the assembly of the stiffness matrix for the example of Section 4.1. There the bilinear form $a(\cdot, \cdot)$ and right hand side $f(\cdot)$ were given by

$$a(u, v) = \int_\Omega \nabla u \nabla v \, dx \qquad \text{and} \qquad f(v) = \int_\Omega fv \, dx.$$

Then the element stiffness matrix associated with each element Ω_j has the form

$$A_h^j = (a_{ik}^j)_{i,k \in I_j}$$

where

$$a_{ik}^j := \int_{\Omega_j} \nabla \varphi_i \nabla \varphi_k \, dx \qquad \text{and} \qquad I_j := \{ i : \, supp\, \varphi_i \cap \Omega_j \neq \emptyset \}.$$

Analogously,

$$f^j = (f_i^j)_{i \in I_j}, \qquad \text{with} \qquad f_i^j := \int_{\Omega_j} f \varphi_i \, dx,$$

gives the contribution of Ω_j to the right-hand side. The additivity of the integrals clearly leads to

$$A_h = (a_{ik})_{i,k=1}^{\hat{N}} \qquad \text{with} \qquad a_{ik} = \sum_{j=1,\, i \in I_j,\, k \in I_j}^M a_{ik}^j$$

and

$$f_h = (f_i)_{i=1}^{\hat{N}} \qquad \text{with} \qquad f_i = \sum_{j=1,\, i \in I_j}^M f_i^j.$$

Next we explicitly compute these element stiffness matrices. In our special case of linear and bilinear elements, the number $\hat{N}$ equals the number of mesh points $\overline{N}$. Moreover, our earlier list on page 204, which shows the vertices appearing in each subdomain, provides us with the index set I_j of vertices of each element $\overline{\Omega}_j$.

Let us first compute the element stiffness matrix related to the bilinear form above for a conforming linear triangular element and a given triangle

$K = \overline{\Omega}_j$. For simplicity assume that the indices of the vertices of this triangle are $1, 2, 3$. As previously mentioned, the required integrals are often computed after transforming onto a reference element. This transformation is also used to estimate projection errors; see the next section.

The transformation

$$\begin{pmatrix} x \\ y \end{pmatrix} = F_j \begin{pmatrix} \xi \\ \eta \end{pmatrix} := \begin{pmatrix} x_1 \\ y_1 \end{pmatrix} + \xi \begin{pmatrix} x_2 - x_1 \\ y_2 - y_1 \end{pmatrix}$$

$$+ \eta \begin{pmatrix} x_3 - x_1 \\ y_3 - y_1 \end{pmatrix}, \quad \begin{pmatrix} \xi \\ \eta \end{pmatrix} \in K', \tag{3.1}$$

maps the reference triangle

$$K' = \left\{ \begin{pmatrix} \xi \\ \eta \end{pmatrix} : \xi \geq 0, \eta \geq 0, \xi + \eta \leq 1 \right\}$$

onto K. The functional determinant of this transformation is given by

$$D_j = \begin{vmatrix} x_2 - x_1 & x_3 - x_1 \\ y_2 - y_1 & y_3 - y_1 \end{vmatrix}. \tag{3.2}$$

Note the connection with the area T of triangle K given by $2T = |D_j|$.
Now derivatives with respect to x and y can be replaced by derivatives with respect to the new variables ξ and η:

$$\begin{pmatrix} \frac{\partial}{\partial x} \\ \frac{\partial}{\partial y} \end{pmatrix} = \begin{pmatrix} \frac{\partial \xi}{\partial x} & \frac{\partial \eta}{\partial x} \\ \frac{\partial \xi}{\partial y} & \frac{\partial \eta}{\partial y} \end{pmatrix} \begin{pmatrix} \frac{\partial}{\partial \xi} \\ \frac{\partial}{\partial \eta} \end{pmatrix}. \tag{3.3}$$

By differentiating (3.1) with respect to x and y, we get

$$\begin{pmatrix} 1 \\ 0 \end{pmatrix} = \begin{pmatrix} x_2 - x_1 & x_3 - x_1 \\ y_2 - y_1 & y_3 - y_1 \end{pmatrix} \begin{pmatrix} \frac{\partial \xi}{\partial x} \\ \frac{\partial \eta}{\partial x} \end{pmatrix}, \quad \begin{pmatrix} 0 \\ 1 \end{pmatrix} = \begin{pmatrix} x_2 - x_1 & x_3 - x_1 \\ y_2 - y_1 & y_3 - y_1 \end{pmatrix} \begin{pmatrix} \frac{\partial \xi}{\partial y} \\ \frac{\partial \eta}{\partial y} \end{pmatrix}.$$

Hence

$$\frac{\partial \xi}{\partial x} = \frac{(y_3 - y_1)}{D_j}, \quad \frac{\partial \eta}{\partial x} = \frac{(y_1 - y_2)}{D_j},$$

$$\frac{\partial \xi}{\partial y} = \frac{(x_1 - x_3)}{D_j}, \quad \frac{\partial \eta}{\partial y} = \frac{(x_2 - x_1)}{D_j}.$$

Thus (3.3) becomes

$$\begin{pmatrix} \frac{\partial}{\partial x} \\ \frac{\partial}{\partial y} \end{pmatrix} = \frac{1}{D_j} \begin{pmatrix} y_3 - y_1 & y_1 - y_2 \\ x_1 - x_3 & x_2 - x_1 \end{pmatrix} \begin{pmatrix} \frac{\partial}{\partial \xi} \\ \frac{\partial}{\partial \eta} \end{pmatrix}.$$

The three nodal basis functions on the reference triangle are

$$\tilde{\varphi}_1(\xi, \eta) = 1 - \xi - \eta, \qquad \tilde{\varphi}_2(\xi, \eta) = \xi, \qquad \tilde{\varphi}_3(\xi, \eta) = \eta.$$

Transforming the integrals $\int_K \nabla \varphi_i \nabla \varphi_k \, dx$ to the reference triangle K' yields eventually the element stiffness matrix

$$A_h^j = \frac{1}{2|D_j|} \begin{pmatrix} y_2 - y_3 & x_3 - x_2 \\ y_3 - y_1 & x_1 - x_3 \\ y_1 - y_2 & x_2 - x_1 \end{pmatrix} \begin{pmatrix} y_2 - y_3 & y_3 - y_1 & y_1 - y_2 \\ x_3 - x_2 & x_1 - x_3 & x_2 - x_1 \end{pmatrix}.$$

For a general triangle Ω_j with vertices p^i, p^k and p^l, i.e., $I_j = \{i, k, l\}$, the entries of the element stiffness matrix for linear elements are given by

$$a_{ik}^j = \frac{1}{2|D_j|}[(x_i - x_l)(x_l - x_k) + (y_i - y_l)(y_l - y_k)] \quad \text{if } i \neq k$$

and

$$a_{ii}^j = \frac{1}{2|D_j|}[(x_k - x_l)^2 + (y_k - y_l)^2].$$

If the vertices $\{i, k, l\}$ are oriented in a positive way then $D_j > 0$.

Remark 4.18. (linear FEM, maximum principle and FVM)
An elementary manipulation of the above results shows the following:
If a triangle K has vertices p^i, p^j and p^k, then

$$\int_K \nabla \varphi_i \nabla \varphi_k = -\frac{1}{2} \cot \gamma_{ik}, \qquad (3.4)$$

where γ_{ik} denotes the angle opposite the edge $p^i p^k$. Moreover, the formula

$$\cot \alpha + \cot \beta = \frac{\sin(\alpha + \beta)}{\sin \alpha \sin \beta}$$

implies that if a triangulation has the property that the sum of the two angles opposite each interior edge is always less than π and the angles opposite the boundary edges are not obtuse, and the triangulation is not degenerate due to occurring narrow strips as part of the original domain, then the discretization of the Laplacian generated by linear finite elements satisfies a maximum principle.
We recall that in Chapter 2 this property was shown for the discretization of the Laplacian using a finite volume method based on Voronoi boxes. Exploiting (3.4) to compare the discretizations, one finds in fact that linear finite elements and finite volumes based on Voronoi boxes generate the same stiffness matrix for the discrete problem.   □

The element stiffness matrix for rectangular elements with bilinear shape functions can be computed analogously.
In our example we have rectangles with edges parallel to the coordinate axes. Set

$$\Omega_j = \text{conv } \{p^1, p^2, p^3, p^4\} = [x_1, x_2] \times [y_2, y_3],$$
$$\Delta x_j := x_2 - x_1 \quad \text{and} \quad \Delta y_j := y_3 - y_2.$$

Then nodal basis functions lead to

$$A_h^j = \frac{1}{6\Delta x_j \Delta y_j} \begin{pmatrix} 2\Delta x_j^2 + 2\Delta y_j^2 & \Delta x_j^2 - 2\Delta y_j^2 & -\Delta x_j^2 - \Delta y_j^2 & -2\Delta x_j^2 + \Delta y_j^2 \\ \Delta x_j^2 - 2\Delta y_j^2 & 2\Delta x_j^2 + 2\Delta y_j^2 & -2\Delta x_j^2 + \Delta y_j^2 & -\Delta x_j^2 - \Delta y_j^2 \\ -\Delta x_j^2 - \Delta y_j^2 & -2\Delta x_j^2 + \Delta y_j^2 & 2\Delta x_j^2 + 2\Delta y_j^2 & \Delta x_j^2 - 2\Delta y_j^2 \\ -2\Delta x_j^2 + \Delta y_j^2 & -\Delta x_j^2 - \Delta y_j^2 & \Delta x_j^2 - 2\Delta y_j^2 & 2\Delta x_j^2 + 2\Delta y_j^2 \end{pmatrix}.$$

For the computation of element stiffness matrices for other types of elements, see [Sch88] and [GRT93].

Boundary conditions of Neumann or Robin type directly influence the structure of the weak formulation. In simple cases Dirichlet boundary conditions can however be immediately taken into consideration: assuming that function values are the degrees of freedom of the shape functions used, one assigns those degrees of freedom associated with boundary knots the corresponding values from the Dirichlet data. The remaining unknowns in the discrete equations correspond to the degrees of freedom at the interior mesh points.

The computation of the right-hand side vector f_h requires, in general, the application of numerical quadrature. The same is, of course, usually true for the element stiffness matrix if the given differential operator has non-constant coefficients or if one has inhomogeneous Neumann or Robin boundary conditions. We shall discuss numerical quadrature in Section 4.5.

If the mesh has a regular structure one can, alternatively to our description above, assemble the stiffness matrix in a mesh-point oriented way. This is possible because for such meshes information on adjacent elements and adjacent mesh points is easily made available. For unstructured meshes, however, one has to store such information during the mesh generation process.

The stiffness matrix generated by finite elements is sparse. Its structure depends on the geometry of the decomposition of the domain and on the shape functions used. If this decomposition is very regular, then for differential operators with constant coefficients the discrete problems generated by finite elements using several standard shape functions turn out to coincide with well-known finite difference methods discussed in Chapter 2.

When any finite-dimensional space V_h is combined with the finite element method for the discretization, then the choice of basis in V_h will influence the structure and conditioning of the discrete problem generated. One often wishes to use the multigrid method (see Chapter 8) to solve this discrete problem. Now multigrid uses a family of meshes. Therefore in this situation it makes sense to define the shape functions first on the coarsest mesh and then to add local shape functions related to the new mesh points that arise in the mesh refinement process. As we mentioned previously, we call this a hierarchical basis. Hierarchical basis functions play an important role in both adaptive finite element methods and the preconditioning process for the discrete problem.

4.3.4 Mesh Generation and Manipulation

The implementation of the finite element method requires a suitable decomposition of the given domain $\Omega \subset \mathbb{R}^n$ into subdomains Ω_j, for $j = 1, \ldots, M$, that have a geometrically simple shape: triangles and quadrilaterals in 2D, tetrahedra and parallelepipeds in 3D.

A significant advantage of the finite element method over the finite difference method is that on every mesh the discrete problem can be generated in a relatively simple and systematic way. Consequently problems with fairly complicated geometry can be readily handled by finite elements. Information about the solution—such as knowledge of corner singularities or boundary layers—can be used to create a suitable locally adapted mesh. For finite element methods, moreover, the theory of a posteriori error estimation is well developed; this allows judicious local mesh refinement in regions where the error in the computed solution is still too large.

In a typical finite element code the following three factors determine the construction of the final mesh used to compute the approximate solution:

1. Generation of the initial mesh;
2. Local evaluation of the current mesh using the current computed solution;
3. Mesh refinement.

We shall discuss item 2—a posteriori error estimation—in detail in Section 4.7.

Here we first examine initial mesh generation and subsequent mesh refinement. Note that many adaptive codes use mesh coarsening as well as mesh refinement. Furthermore, we shall not discuss the so-called *r-method* where the knots are moved to optimal positions while preserving the number of mesh points.

Generation of an Initial Mesh

Initial decompositions of a given domain are constructed using a mesh generator, which is a natural component of any finite element package. Each mesh generator is a compromise between universality, simple applicability, transparency of data structures and robustness of code. Mesh generators often incorporate heuristic features as well.

In what follows we sketch some basic principles of mesh generation for the two-dimensional case, using triangles for the decomposition. The following three strategies are often used to generate initial meshes:

Strategy 1: The given domain Ω is first covered by a uniform quadrilateral mesh and then triangulated. Then it is locally adapted to the boundary of Ω by shifting mesh points near that boundary. Figure 4.11 illustrates the technique; in this example mesh points are moved only in the y-direction.

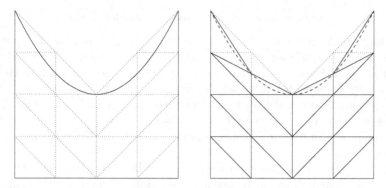

Figure 4.11 Shifting of mesh points near the boundary

Strategy 2: The given domain Ω or parts of the given domain are transformed onto a reference triangle or rectangle. The reference element is uniformly decomposed into triangles. Then the inverse transformation yields a decomposition of Ω whose subdomains have in general curved boundaries. In the final step the curved elements are replaced by triangles.

Figure 4.12 demonstrates the technique. The first figure shows the given domain, the second one the triangulated reference element. Back transformation yields the third figure, and the final figure shows the triangulation obtained.

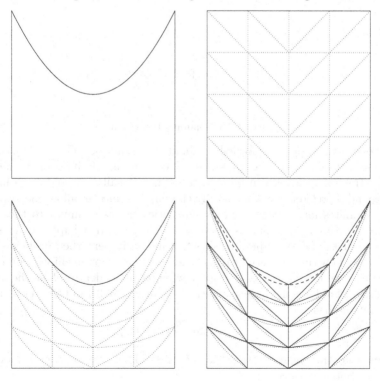

Figure 4.12 Mesh generation by transformation

Strategy 3: We begin by generating a set of points in Ω that are the vertices of the initial triangulation (this set could be generated using strategies 1 or 2). Then an algorithm to construct a *Delaunay triangulation* is launched. Here it is necessary to fix sufficiently many points on the boundary of Ω.

A Delaunay triangulation triangulates the convex hull of a given set of points in such a way that every triangle has the following property: no other mesh point lies in the circumcircle passing through the three vertices of the triangle.

In 2D a Delaunay triangulation has the following additional attribute: among all triangulations of the given set of points, it maximizes the smallest angle appearing in any triangle. This is a valuable property in the convergence theory for finite element methods; see Section 4.4.

Figure 4.13 shows a Delaunay triangulation of the points in the final diagram of Figure 4.12.

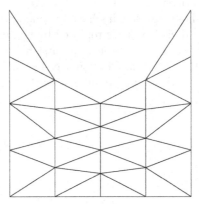

Figure 4.13 Delaunay triangulation

If one has a priori knowledge about the solution of the differential equation—for example the presence of corner singularities or boundary layers—this information can be used to define a locally adapted mesh in the corresponding regions. When carefully chosen, such meshes allow one to prove error estimates for problems with singularities that are similar to those for problems without singularities on standard meshes. In Chapter 6 we shall discuss in detail layer-adapted meshes for singularly perturbed problems.

Here we now turn to the case of a corner singularity for an elliptic boundary value problem. Let us assume that for a given second-order problem, the origin is a corner of a domain whose geometry is as in Figure 3.1 of Chapter 3. Then typically one has

$$u - \lambda\, r^{\pi/\omega} \sin\frac{\pi\theta}{\omega} \in H^2(\Omega).$$

Assume that the angle ω satisfies $\omega \in (\pi, 2\pi)$. Then the solution u does not lie in the Sobolev space $H^2(\Omega)$.

Now the convergence theory for linear finite elements (which will be presented in Section 4.4) tells us that we cannot expect the method to be first-order convergent when measured in the $H^1(\Omega)$ norm. But if $u \in H^2(\Omega)$ were true, then one would have first-order convergence on standard meshes. Therefore we ask: can one construct a special mesh that preserves first-order convergence in $H^1(\Omega)$?

From the theoretical point of view, problems with corner singularities are more difficult to handle because one uses the weighted Soboloev space $H^{2,\alpha}(\Omega)$, which is the set of all functions $w \in H^1(\Omega)$ that satisfy

$$r^\alpha D^\beta w \in L_2(\Omega) \quad \text{for} \quad |\beta| = 2.$$

It can be shown for our corner singularity problem that

$$u \in H^{2,\alpha}(\Omega) \quad \text{for} \quad \alpha > 1 - \pi/\omega.$$

Based on this analytical behaviour near the corner we construct a special mesh. Triangles that have the origin as a vertex are decomposed using subdivision points chosen according to the following recipe:

$$\left(\frac{i}{n}\right)^\gamma, \quad \text{with} \quad \gamma = 1/(1-\alpha), \quad \text{for } i = 0, 1, \ldots, n.$$

(Observe that $\gamma = 1$ defines a standard mesh.) Figure 4.14 shows such a local mesh with $\gamma = 2$ (i.e., $\alpha = 1/2$) and $n = 4$. This is a sensible choice if $\omega < 2\pi$.

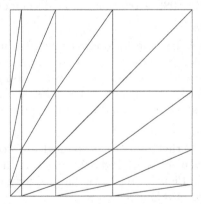

Figure 4.14 geometrically graded mesh

A mesh like this that is successively refined as one goes deeper into the region of difficulty is called a *graded mesh*.

Interpolation error estimates in weighted Sobolev spaces can be used to prove error estimates on geometrically graded meshes along the lines of the standard results in Section 4.4. For our example, linear finite elements on the above geometrically graded mesh will indeed yield first-order convergence in the $H^1(\Omega)$ norm; see [Gri85].

Mesh Refinement

(I) Global uniform refinement

The given domain Ω is covered with a coarse admissible initial mesh and then uniformly refined. Each triangle is decomposed into 4 congruent subtriangles and each quadrilateral into 4 congruent subquadrilaterals. Near the boundary some local variation of this rule may be needed. This type of refinement automatically guarantees the admissibility of the finer meshes generated. Figure 4.15 shows two refinement levels starting from a coarse mesh.

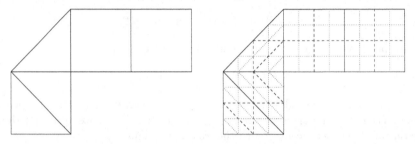

Figure 4.15 global uniform refinement

(II) Local refinement

Suppose that we want to refine locally a given triangular mesh.

First we present the technique that is used in the mesh generation process of the code PLTMG [Ban98]. A triangle can be decomposed into subtriangles in the following two ways:

1. Decomposition of a triangle into four congruent subtriangles using the midpoints of the edges. These triangles are called red triangles.
2. Decomposition of a triangle into two triangles using a midpoint of one edge. These triangles are called green triangles, and they appear in pairs. In the following figures this second type of decomposition is indicated by a dotted line.

As a rule, triangles that need to be refined are decomposed using red refinement; green refinement is invoked only to ensure the admissibility of the resulting triangulation. Figure 4.16 shows the need for green refinement if one triangle is decomposed into four red ones.

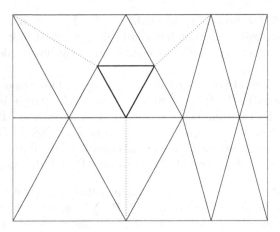

Figure 4.16 red-green refinement

To avoid degeneration of the triangles (i.e., large maximum angles), a further direct refinement of a green pair is not allowed. If it becomes necessary later to decompose one of the green pair in order to guarantee admissibility, then the green pair are first reunited to form one triangle after which that triangle is decomposed into 4 congruent red triangles. Figure 4.17 shows a typical example of red-green refinement in a sequence of diagrams.

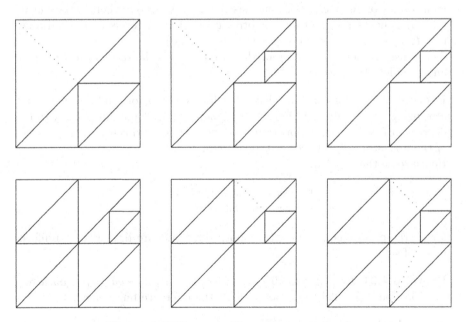

Figure 4.17 Decomposition of a green triangle

The above strategy ensures that on the sequence of locally refined meshes the minimal angle of all triangles is uniformly bounded below away from zero. On the other hand, the possibility of birth and death of green triangles means that successive meshes are not automatically nested. If nested meshes are algorithmically necessary, then more complicated refinement rules will achieve this; see [Ban98].

There are alternatives to *red-green refinement*. One important strategy is *bisection*. Here, in the two-dimensional case, each triangle is decomposed into two triangles by using the midpoint of its longest edge. This technique is implemented in, for example, the code ALBERTA [SS05]. Of course, one must still consider strategies to guarantee mesh admissibility and to avoid geometrical degeneration; see [10] for both the 2D and 3D cases.

Exercise 4.19. Let A be a linear, formally selfadjoint, positive semidefinite second-order differential operator and $(T, P_T, \sum_T)$ a finite element.
a) Verify that the element stiffness matrix E_T generated by A is symmetric and positive semidefinite.
b) Prove that E_T is invariant with respect to translation and rotation of T if the operator A is isotropic, i.e., invariant with respect to a congruence transformation.
c) Let the operator A be isotropic and homogeneous of degree q. Let the element T' be geometrically similar to the element T. What is the relationship between the corresponding element matrices?
Remark: The operator A is homogeneous of degree q if under the similarity transformation $\lambda t = x$, $\lambda \in \mathbb{R}$, one has $Au(x) = \lambda^q A_i v(t)$, where $v(t) := u(\lambda t)$ and A_i denotes the operator A formally written in terms of the variables $t = (t_1, t_2, \cdots, t_N)$ instead of $x = (x_1, x_2, \cdots, x_N)$.
d) Discuss problems b) and c) for the case that A is the sum of the Laplacian and the identity operator.

Exercise 4.20. Compute the element stiffness matrix for the Laplacian operator using the so-called Hermite T10-triangle: cubic shape functions with ten degrees of freedom—function values at the vertices and centroid, and values of the first-order derivatives at the vertices.
Hence solve the problem

$$\begin{cases} -\triangle u = 1 & \text{in} \quad \Omega = (0,1)^2, \\ \quad u = 0 & \text{on} \quad \partial \Omega, \end{cases}$$

for a criss-cross triangulation (i.e., both diagonals are drawn) of a uniform mesh of squares.

Exercise 4.21. Consider solving the problem $\quad -\triangle u + cu = f \quad$ on a rectangular mesh parallel to the axes, using the reference finite element

$$\hat{T} = [-1,1]^2, \quad P = Q_1(T), \quad \sum = \{\sigma_i : \sigma_i(u) = u(\pm 1, \pm 1), \quad i = 1, \cdots, 4\}.$$

What difference method is generated?

Exercise 4.22. Compute a basis for the Bogner-Fox-Schmit element on the reference element $\hat{T} = [0, 1]^2$. Deduce the element stiffness matrix for the Laplacian.

4.4 Convergence Theory of Conforming Finite Elements

4.4.1 Interpolation and Projection Error in Sobolev Spaces

For a conforming finite element method with $V_h \subset V$, the lemmata of Lax-Milgram and Cea have given us already the following information: if the given variational equation satisfies the assumptions of the Lax-Milgram lemma, then the continuous and discrete problems each have a unique solution (u, u_h respectively) and the error (measured in the norm of the underlying space V) satisfies the estimate

$$\|u - u_h\| \leq c \inf_{v_h \in V_h} \|u - v_h\|$$

for some positive constant c.

But for a given space V_h, it is often troublesome to estimate the error of the best approximation on the right-hand side here. If instead we choose some arbitrary projection $\Pi^* u \in V_h$ of the exact solution, then

$$\|u - u_h\| \leq c \inf_{v_h \in V_h} \|u - v_h\| \leq c \|u - \Pi^* u\|.$$

This bound enables us to concentrate on estimating the projection error, for it turns out that in general the results obtained are asymptotically optimal, i.e., the error of the best approximation and the error of some suitably chosen projection differ only by a multiplicative constant. As a projection one often takes for convenience some interpolant of the exact solution. To pursue this analysis, we have to address the following central questions:

Given u in some Sobolev space, how is its interpolant or projection in V_h defined? What estimates can then be proved for the interpolation or projection error?

In this Section we shall study conforming methods, which are characterized by the assumptions that $V_h \subset V$, the boundary of the given domain is polygonal, and all integrals arising in the method are computed exactly. If any of these three assumptions is not satisfied then we can no longer appeal to the conforming theory, and in that case the nonconforming feature will contribute an additional error that will be discussed in Section 4.5.

Assume that we have a Lagrange element with knots $p^1, ..., p^M$ and nodal basis $\varphi^1, ..., \varphi^M$. Then the mapping $\Pi : V \mapsto V_h$

$$\Pi u := \sum_{i=1}^{M} u(p^j) \varphi_j \in V_h$$

is an interpolation operator that is well defined provided that the function values $u(p^j)$ are well defined for a given u. For example, if u lies in $H^2(\Omega)$ where Ω is a two- or three-dimensional domain, then Sobolev embedding theory implies that one can immediately define an *interpolant* of u in V_h, given a nodal Lagrange basis in V_h. If u is less smooth but we need a projection of u into V_h, then quasi-interpolants can be used; see Section 4.7.

Now let us assume that V_h is a finite element space of Lagrange elements of type P_k on a triangulation of a given domain. Such elements are an important example of an *affine family* of finite elements: the shape functions and degrees of freedom on an arbitrary element can be defined by an affine-linear mapping of the shape functions and degrees of freedom on a reference element. Let $\Pi_Z u$ be a linear continuous projector into V_h, which is first defined on the reference element and then transformed to each element by an affine-linear mapping. We want to estimate the projection error

$$\|u - \Pi_Z u\|.$$

This is done in three steps:

- transformation onto the reference element;
- estimation on the reference element, often using the Bramble-Hilbert lemma;
- inverse transformation to the finite element.

Before studying the general case, we carry out these three steps for a simple example. Let K be a triangle whose vertices are

$$(x_0, y_0), \quad (x_0 + h, y_0), \quad (x_0, y_0 + h).$$

Given a function $u \in H^2$, we want to estimate the interpolation error in the H^1-seminorm if its interpolant u^I is linear on K and coincides with u at the vertices.

The transformation

$$\xi = \frac{x - x_0}{h}, \quad \eta = \frac{y - x_0}{h}$$

maps K onto the reference triangle E whose vertices are $(0,0)$, $(1,0)$ and $(0,1)$. Transforming the integrals and using the chain rule yields

$$|u - u^I|_{1,K}^2 = 2|K|\frac{1}{h^2}|u - u^I|_{1,E}^2, \tag{4.1}$$

where $|K|$ denotes the measure of K.

We estimate the interpolation error on the reference triangle E:

$$(u - u^I)_x(x, y) = u_x(x, y) - (u(1, 0) - u(0, 0))$$

$$= \int_0^1 [u_x(x, y) - u_x(\xi, y) + u_x(\xi, y) - u_x(\xi, 0)] \, d\xi$$

$$= \int_0^1 \left(\int_\xi^x u_{xx}(\mu, y) \, d\mu + \int_0^y u_{xy}(\xi, \nu) \, d\nu \right) d\xi.$$

Hence

$$\|(u - u^I)_x\|_{0,E}^2 \leq C \, |u_x|_{1,E}^2 \tag{4.2}$$

and consequently

$$|u - u^I|_{1,E}^2 \leq C|u|_{2,E}^2.$$

Transforming this back to K and recalling (4.1) yields the desired result

$$|u - u^I|_{1,K}^2 \leq C \, h^2 \, |u|_{2,K}^2.$$

In this simple example a direct estimate on K would also have been possible, but in general it is often simpler to transform to a reference element.

Now we start to investigate the general case in 2D. Let the reference element be the triangle

$$K' = \left\{ \begin{pmatrix} \xi \\ \eta \end{pmatrix} \; : \; \xi \geq 0, \, \eta \geq 0, \, \xi + \eta \leq 1 \right\}. \tag{4.3}$$

It is clear that each triangle $K = \overline{\Omega}_j$ is the image of a one-to-one affine mapping of the reference triangle, $F_j : K' \longrightarrow K$, where

$$x = F_j(p) = B_j p + b^j, \quad p = \begin{pmatrix} \xi \\ \eta \end{pmatrix}. \tag{4.4}$$

To simplify the notation we drop the index j from F_j, the matrix B_j and the vector b^j.

If the given triangle has vertices $\begin{pmatrix} x_1 \\ y_1 \end{pmatrix}$, $\begin{pmatrix} x_2 \\ y_2 \end{pmatrix}$ and $\begin{pmatrix} x_3 \\ y_3 \end{pmatrix}$, then

$$\begin{pmatrix} x \\ y \end{pmatrix} = F(p) = \begin{pmatrix} x_2 - x_1 & x_3 - x_1 \\ y_2 - y_1 & y_3 - y_1 \end{pmatrix} \begin{pmatrix} \xi \\ \eta \end{pmatrix} + \begin{pmatrix} x_1 \\ y_1 \end{pmatrix}. \tag{4.5}$$

Figure 4.18 illustrates the mappings F and F^{-1}.

Setting

$$v(p) = u(F(p)), \quad p \in K', \tag{4.6}$$

we see that each function $u(x)$, for $x \in K$, is mapped to a function $v(p)$ defined on the reference element. If F is differentiable, then the chain rule yields

$$\nabla_p v(p) = F'(p) \nabla_x u(F(p)). \tag{4.7}$$

Let h be the maximum length of the edges of the triangles in the triangulation of the given domain Ω. Then there exists a constant $c > 0$ such that

$$\|F'(p)\| \leq ch \qquad \text{for all } p \in K'. \tag{4.8}$$

To transform Sobolev norms on K to Sobolev norms on K' we need the functional determinant

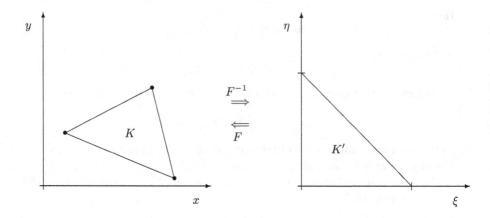

Figure 4.18 Transformation on the reference element

$$s(p) := \det F'(p).$$

We assume that

$$s(p) > 0 \qquad \text{for all } p \in K'.$$

The change-of-variable rules for transforming integrals yield

$$\int_K u^2(x)\, dx = \int_{K'} v^2(p) s(p)\, dp. \tag{4.9}$$

Then (4.9) implies the following estimate for $u \in L_2(K)$:

$$\left\{ \inf_{p \in K'} s(p) \right\}^{1/2} \|v\|_{0,K'} \leq \|u\|_{0,K} \leq \left\{ \sup_{p \in K'} s(p) \right\}^{1/2} \|v\|_{0,K'}.$$

When transforming norms that involve derivatives, the situation becomes more complicated because of (4.7)). We first restrict ourselves to the linear-affine mapping (4.4) and write $\|B\|$ for the norm on B induced by the maximum norm $\|\cdot\|$ on $\mathbb{R}^2$.

Lemma 4.23. *Let*

$$x = F(p) = Bp + b, \qquad p \in K', \tag{4.10}$$

be a one-to-one affine mapping from the reference element K' onto K. The following relations hold between functions defined on K and the corresponding functions (4.6) defined on K': for $l = 0, 1, \dots,$

i) $u \in H^l(K) \iff v \in H^l(K')$,

ii) $|v|_{l,K'} \leq c \, \|B\|^l \, |\det B|^{-1/2} \, |u|_{l,K},$

$\quad |u|_{l,K} \leq c \, \|B^{-1}\|^l \, |\det B|^{1/2} \, |v|_{l,K'}.$

Proof: First we verify the validity of the estimate ii) for smooth functions u. The function v is then smooth as well.

From (4.6) we get

$$\frac{\partial v}{\partial p_j} = \sum_{i=1}^{2} \frac{\partial u}{\partial x_i} \frac{\partial x_i}{\partial p_j}$$

and then (4.10) yields

$$\left| \frac{\partial v}{\partial p_j} \right| \leq \|B\| \max_i \left| \frac{\partial u}{\partial x_i} \right|.$$

For any multi-index α, this estimate used recursively implies

$$|[D^\alpha v](p)| \leq \|B\|^{|\alpha|} \max_{\beta, |\beta|=|\alpha|} |[D^\beta u](x(p))|, \qquad p \in K'.$$

Taking into account the equivalence of the maximum and L_2 norms in the finite-dimensional space $\mathbb{R}^{l+1}$ (there are $l+1$ various derivatives of order l for functions of 2 variables), it follows that there exists a constant c such that

$$\sum_{|\alpha|=l} |[D^\alpha v](p)|^2 \leq c \|B\|^{2l} \sum_{|\beta|=l} |[D^\beta u](x(p))|^2, \qquad p \in K'. \qquad (4.11)$$

The value of the constant c depends on l and on the dimension n of the space in which K lies (we are dealing here with $n = 2$). Using (4.11), the standard rules for transforming integrals yield

$$|v|_{l,K'}^2 = \int_{K'} \sum_{|\alpha|=l} |[D^\alpha v](p)|^2 \, dp \leq c\|B\|^{2l} \int_{K'} \sum_{|\beta|=l} |[D^\beta u](x(p))|^2 \, dp$$

$$\leq c\|B\|^{2l} |\det B|^{-1} \int_K \sum_{|\beta|=l} |[D^\beta u](x)|^2 \, dx$$

$$= c\|B\|^{2l} |\det B|^{-1} |u|_{l,K}^2.$$

This proves the first inequality in ii); the second is handled similarly starting from

$$p = B^{-1}x - B^{-1}b.$$

Thus ii) is proved for sufficiently smooth functions u and v. The density of the space $C^l(\overline{K})$ in $H^l(K)$ then implies ii) for functions in $H^l(K)$.

Finally, the equivalence i) follows immediately from ii). ∎

Next we estimate $\|B\|$ and $\|B^{-1}\|$.

Lemma 4.24. *Let the assumptions of Lemma 4.23 be satisfied with K' our fixed reference element. Let ρ denote the radius of the largest ball one can inscribe in K, and R the radius of the smallest ball that contains K. Then there exists a constant c such that*

$$\|B\| \le cR, \qquad \|B^{-1}\| \le c\rho^{-1}.$$

Proof: Because K' is fixed, there are fixed positive numbers ρ' and R' that represent the corresponding radii of balls for K'. Then there exists a $p_0 \in K'$ such that

$$p_0 + p \in K'$$

for arbitrary p with $\|p\| = \rho'$. For the corresponding points

$$x_0 = Bp_0 + b \qquad \text{and} \qquad x = B(p_0 + p) + b$$

assigned by (4.10), one has $x_0, x \in K$. Our hypotheses imply that

$$\|x - x_0\| \le 2R.$$

Hence

$$\|B\| = \frac{1}{\rho'} \sup_{\|p\|=\rho'} \|Bp\| \le \frac{1}{\rho'}\|x - x_0\| \le 2\frac{R}{\rho'}.$$

If we interchange the roles of K and K', the second estimate follows similarly. ∎

When Lemma 4.24 is invoked in finite element analyses, we obtain estimates that contain the diameter h_K of the element K as well as the radius of its inscribed ball ρ_K. The result of the lemma can be simplified in such a way that h_K appears instead of ρ_K in the final estimate if the class of triangulations satisfies

$$\frac{h_K}{\rho_K} \le \sigma \tag{4.12}$$

for some fixed constant σ. A family of triangulations $\mathcal{Z}$ that satisfies (4.12) for all elements K of the family with a single fixed constant σ is called *quasi-uniform*. (We remark that in the literature a variety of terminology is used for (4.12): the terms *regular* triangulation and *shape-regular* triangulation are common.)

For a triangular mesh in 2D, the condition (4.12) is equivalent to the *minimal angle condition* of Zlamal [128]:

$$\min \alpha \ge \underline{\alpha} > 0, \tag{4.13}$$

where α is an arbitrary angle of a triangle from the family of triangulations under consideration and $\underline{\alpha} > 0$ is a fixed lower bound for the angles in that class.

It should be noted that for a single triangulation with a finite number of elements (4.12) and (4.13) are trivially satisfied; the point is that these conditions must hold for a family of triangulations with $h := \max h_K \to 0$ because we want to prove convergence results as $h \to 0$, and on such a family the conditions are not trivial.

So far we have studied an affine family of finite elements and used affine transformations between the elements and a reference element. But for quadrilateral elements one needs more general mappings, as otherwise one could study only meshes consisting of parallelograms. Moreover, for curved domains more general elements require more general classes of transformations: see Section 4.5.5. Here we simply mention that for some classes of nonlinear transformations one can prove results similar to those for affine mappings.

Let

$$x = F(p), \qquad p \in K', \tag{4.14}$$

be a nonlinear transformation between a reference element K' and an element K of a triangulation $\mathcal{Z}$ of the given domain. We say that F is regular if there exist constants c_0, c_1, c_2 such that for all $u \in H^r(K)$ one has

$$|v|_{r,K'} \leq c_1 \left\{ \inf_{p \in K'} s(p) \right\}^{-1/2} h^r \|u\|_{r,K}, \tag{4.15}$$

$$|v|_{r,K'} \geq c_2 \left\{ \sup_{p \in K'} s(p) \right\}^{-1/2} h^r |u|_{r,K}, \tag{4.16}$$

$$0 < \frac{1}{c_0} \leq \frac{\sup\limits_{p \in K'} s(p)}{\inf\limits_{p \in K'} s(p)} \leq c_0; \tag{4.17}$$

here $s(p) = det\, F'(p)$. In the case of an affine transformation (4.5) the functional determinant $s(p)$ is constant, so affine transformations on quasi-uniform meshes are always regular, and in fact (4.15) can then be sharpened to

$$|v|_{r,K'} \leq c_1 \left\{ \inf_{p \in K'} s(p) \right\}^{-1/2} h^r |u|_{r,K}. \tag{4.18}$$

A detailed discussion of regular transformations can be found in [MW77].

On the reference element the projection error is often estimated using the following lemma. We say that a functional q on the space V is sublinear and bounded if

$$|q(u_1 + u_2)| \leq |q(u_1)| + |q(u_2)| \text{ and } |q(u)| \leq C \|u\|$$

for all $u_1, u_2, u \in V$ and some constant C.

Lemma 4.25 (Bramble/Hilbert). *Let $B \subset R^n$ be a domain with Lipschitz boundary and let q be a bounded sublinear functional on $H^{k+1}(B)$. Assume that*

$$q(w) = 0 \qquad \text{for all } w \in P_k. \tag{4.19}$$

Then there exists a constant $c = c(B) > 0$, which depends on B, such that

$$|q(v)| \le c|v|_{k+1,B} \qquad \text{for all } v \in H^{k+1}(B). \tag{4.20}$$

Proof: Let $v \in H^{k+1}(B)$ be an arbitrary function. We claim that one can construct a polynomial $w \in P_k$ such that

$$\int_B D^\alpha(v + w)\, dx = 0 \qquad \text{for all } |\alpha| \le k. \tag{4.21}$$

To justify this assertion, observe that any polynomial of degree k can be written as

$$w(x) = \sum_{|\beta| \le k} c_\beta x^\beta$$

with coefficients $c_\beta \in \mathbb{R}$. Here $\beta = (\beta_1, ..., \beta_n)$ denotes a multi-index and $x^\beta := \prod_{i=1}^n x_i^{\beta_i}$.

The linearity of the generalized derivative D^α and (4.21) mean that we require

$$\sum_{|\beta| \le k} c_\beta \int_B D^\alpha x^\beta\, dx = -\int_B D^\alpha v\, dx, \qquad |\alpha| \le k. \tag{4.22}$$

This is a linear set of equations for the unknown coefficients c_β, $|\beta| \le k$, of our polynomial w. As $D^\alpha x^\beta = 0$ for all multi-indices α, β with $\alpha_i > \beta_i$ for at least one $i \in \{1, ..., n\}$, the linear system (4.22) is triangular. It can be solved recursively, starting with those indices β for which $\beta_j = k$ for some $j \in \{1, ..., n\}$: for these β one gets

$$c_\beta = -\frac{1}{k!\, meas(B)} \int_B D^\beta v\, dx.$$

The other coefficients then follow recursively from (4.22). We have now verified that for each v there is a polynomial $w \in P_k$ satisfying (4.21).

By Poincaré's inequality,

$$\|u\|_{k+1,B}^2 \le c\left\{ |u|_{k+1,B}^2 + \sum_{|\alpha| \le k} \left| \int_B D^\alpha u\, dx \right|^2 \right\} \qquad \text{for all } u \in H^{k+1}(B).$$

Setting $u = v + k$ and invoking (4.20) yields

$$\|v + w\|_{k+1,B}^2 \le c|v + w|_{k+1,B}^2 = c|v|_{k+1,B}^2$$

because $w \in P_k$. The sublinearity of q implies

$$|q(v)| \leq |q(v+w)| + |q(w)| = |q(v+w)|$$

by (4.19). As q is bounded we get finally

$$|q(v)| \leq c\,\|v+w\|_{k+1,B} \leq c\,|v|_{k+1,B}$$

as desired. ∎

Remark 4.26. The Bramble-Hilbert lemma is usually proved for bounded linear functionals. It is occasionally useful to know a variant of it for continuous bilinear forms $S : H^{k+1}(B) \times H^{r+1}(B) \longrightarrow \mathbb{R}$ that satisfy

$$i)\ S(u,v) = 0 \ \text{ for all } u \in H^{k+1}(B),\ v \in P_r\,,$$
$$ii)\ S(u,v) = 0 \ \text{ for all } u \in P_k,\ v \in H^{r+1}(B).$$

For such S one can prove that

$$|S(u,v)| \leq c\,|u|_{k+1,B}|v|_{r+1,B} \quad \text{ for all } u \in H^{k+1}(B),\ v \in H^{r+1}(B). \quad (4.23)$$

Next we apply the Bramble-Hilbert lemma to estimate the difference between a given function and its projection onto polynomials of degree k.

Lemma 4.27. *Let $k \geq r$ and $\Pi : H^{k+1}(B) \longrightarrow P_k \subset H^r(B)$ a continuous linear projector. Then there exists a constant $c > 0$ such that*

$$\|v - \Pi v\|_{r,B} \leq c\,|v|_{k+1,B} \quad \text{ for all } v \in H^{k+1}(B).$$

This result follows immediately from the Bramble-Hilbert lemma because $\|\cdot\|_r$ is a bounded sublinear functional.

We are now ready to estimate the projection error for an affine family of finite elements. For $0 \leq r \leq k+1$ set

$$\|v\|_{r,h,\Omega}^2 = \sum_{j=1}^{m} \|v\|_{r,\Omega_j}^2\,.$$

Theorem 4.28. *Consider an affine family of finite elements of type P_k on a quasi-uniform triangulation $\mathcal{Z}$. Let $\Pi_{\mathcal{Z}} : H^{k+1}(\Omega) \longrightarrow P_{\mathcal{Z},k}(\Omega)$, which is defined piecewise on the triangulation $\mathcal{Z}$ of Ω, be a projector onto piecewise polynomials of degree at most k. Then there exists a constant $c > 0$ such that*

$$\|u - \Pi_{\mathcal{Z}} u\|_{r,h,\Omega} \leq c\,h^{k+1-r}|u|_{k+1,\Omega} \quad \text{for } 0 \leq r \leq k+1. \quad (4.24)$$

Proof: Let Ω_j, $j = 1, \ldots, m$, be the elements of the triangulation $\mathcal{Z}$. Clearly

$$\|u - \Pi_{\mathcal{Z}} u\|_{r,h,\Omega}^2 = \sum_{j=1}^{m} \|u - \Pi_{\Omega_j} u\|_{r,\Omega_j}^2\,. \quad (4.25)$$

We estimate the error on the subdomain $\Omega_j = K$. Transforming to the reference element K', we compute as follows:

Step 1: Lemmas 4.23 and 4.24 yield

$$\|u - \Pi u\|_{r,K} \leq c \,|\det B|^{1/2} \rho_K^{-r} \|u - \Pi u\|_{r,K'}.$$

Step 2: An application of Lemma 4.27 on the reference element gives

$$\|u - \Pi u\|_{r,K'} \leq c \,|u|_{k+1,K'}.$$

Step 3: Transforming back to K, one has

$$|u|_{k+1,K'} \leq c \,|\det B|^{-1/2} h_K^{k+1} |u|_{k+1,K}.$$

Combining these inequalities we get (4.24) since the triangulation is quasi-uniform. ∎

If $\Pi_{\mathcal{Z}} u \in H^r(\Omega)$, one can replace the piecewise-defined norm of (4.25) by the standard norm. Suppose, for instance, that $u \in H^2(\Omega)$ and we consider its linear interpolant Πu on a quasi-uniform triangular mesh. Then Theorem 4.28 tells us that

$$\|u - \Pi u\|_1 \leq ch^1 |u|_2, \quad \|u - \Pi u\|_0 \leq ch^2 |u|_2.$$

A result similar to Theorem 4.28 can be proved for regular transformations, but in practice it may be demanding to verify that a transformation is regular.

Finally we mention that, as well as the isotropic result of Theorem 4.28, anisotropic results are known. What is the difference between *isotropic* and *anisotropic elements*? If we consider a triangle or a rectangle in 2D and characterize its geometry only by its diameter then we have an isotropic view, because we ignore the possibility that the length of the element in the perpendicular direction can be much smaller. Of course triangles that satisfy the minimal angle condition cannot degenerate in this sense. It turns out that for problems with edge singularities or boundary layers the application of anisotropic elements can be advantageous; see Chapter 6.

For simplicity, let us study bilinear elements on a rectangular mesh of lines parallel to the axes. We do not assume that on our family of meshes the lengths h_1 and h_2 of the edges of a rectangle have the property that h_1/h_2 is bounded. Thus long thin rectangles—anisotropic ones—are allowed. For isotropic bilinear elements one can prove interpolation error estimates similar to the estimates above for linear elements if $u \in H^2$. What happens in the anisotropic case?

The transformation technique gives

$$\|(u - \Pi u)_y\|_0^2 \leq C \left(h_1^2 + h_2^2 + \left(\frac{h_1^2}{h_2}\right)^2 \right) |u|_2^2.$$

It seems anisotropic elements are bad if $h_1 \gg h_2$. But this is not the best result available. A better result is obtained if on the reference element we discard the estimate

$$\|(u - \Pi u)_y\|_0^2 \leq C|u|_2^2$$

in favour of the sharper estimate (see [6] or [KN96]; for linear elements compare (4.2))

$$\|(u - \Pi u)_y\|_0^2 \leq C|u_y|_1^2.$$

Then the transformation technique yields

$$\|u - \Pi u\|_1^2 \leq C(h_1^2|u_x|_1^2 + h_2^2|u_y|_1^2).$$

This estimate shows that anisotropic elements can be useful if the derivatives of the given function in different directions—here the x- and y-directions—have very dissimilar magnitudes.

A similar result for linear elements on anisotropic triangles in 2D shows that one can replace the minimal angle condition by the *maximal angle condition*:

$$\alpha \leq \bar{\alpha} \qquad \text{with a constant } \bar{\alpha} < \pi.$$

For the general theory of anisotropic elements in 2D and 3D, see [6] and [Ape99].

4.4.2 Hilbert Space Error Estimates

Cea's lemma and our bounds on the Sobolev space interpolation error can now be combined to give error estimates for conforming finite element methods.

Let $\Omega \subset \mathbb{R}^n$ be a given polygonal domain. For some subspace V of $H^m(\Omega)$ consider the following weak formulation of an elliptic boundary value problem: Find $u \in V$ such that

$$a(u, v) = f(v) \qquad \text{for all } v \in V. \tag{4.26}$$

Here $f : V \to \mathbb{R}$ is a continuous linear functional and $a : V \times V \to \mathbb{R}$ is a V-elliptic continuous bilinear form.

Let the finite element space V_h be a space of polynomial shape functions on a quasi-uniform triangulation $\mathcal{Z}$ that is defined by affine mappings from a reference element K'. Let h denote the maximum diameter of the elements of the triangulation. We assume that $P_{\mathcal{Z},k}(\Omega) \subset V_h \subset V$ for some $k \geq 0$.

Theorem 4.29. *Assume that the solution u of the given problem (4.26) satisfies the regularity condition $u \in V \cap H^{k+1}(\Omega)$ with $k \geq m$. Then the discrete problem*

$$a(u_h, v_h) = f(v_h) \qquad \text{for all } v_h \in V_h \tag{4.27}$$

has a unique solution $u_h \in V_h$. The error in the solution of the finite element method satisfies the following bound:

$$\|u - u_h\|_{m,\Omega} \leq c\, h^{k+1-m} |u|_{k+1,\Omega}, \tag{4.28}$$

where c is a positive constant.

Proof: Because $V_h \subset V$, the Lax-Milgram lemma implies existence and uniqueness of the discrete solution $u_h \in V_h$ of (4.27).
Then Cea's lemma yields

$$\|u - u_h\|_{m,\Omega} \leq c \inf_{v_h \in V_h} \|u - v_h\|_{m,\Omega} \leq c\|u - \Pi_{\mathcal{Z}} u\|_{m,h,\Omega}.$$

Apply Theorem 4.28 to get the final estimate (4.28). ∎

We are mainly interested in elliptic boundary value problems that are second order, i.e., $m = 1$ and $V \subset H^1(\Omega)$, so we restate Theorem 4.29 for the special case of homogeneous Dirichlet boundary conditions.

Corollary 4.30. *Let $V = H_0^1(\Omega)$ and let the solution u of (4.26) satisfy the regularity condition $u \in V \cap H^{k+1}(\Omega)$ with $k \geq 1$.*
Moreover, let the assumptions of Theorem 4.28 be valid.
Then the discrete problem

$$a(u_h, v_h) = f(v_h) \qquad \text{for all } v_h \in V_h$$

has a unique solution $u_h \in V_h$ and the error of this solution satisfies the bound

$$\|u - u_h\|_{1,\Omega} \leq c\, h^k |u|_{k+1,\Omega}. \tag{4.29}$$

Let us go further and consider the discretization of Poisson's equation with homogeneous Dirichlet boundary conditions using linear finite elements:

Corollary 4.31. *Let $\Omega \subset \mathbb{R}^n$ be polygonal. Let $u \in V \cap H^2(\Omega)$, where $V = H_0^1(\Omega)$, be the weak solution of*

$$-\Delta u = f \quad \text{in } \Omega, \qquad u|_\Gamma = 0,$$

i.e., the solution of the variational equation

$$\int_\Omega \nabla u \nabla v \, dx = \int_\Omega f v \, dx \qquad \text{for all } v \in V.$$

Let this problem be discretized by a conforming finite element method, namely linear elements on a triangulation that satisfies the maximal angle condition. Then the error in the finite element solution satisfies

$$\|u - u_h\|_{1,\Omega} \leq c\, h |u|_{2,\Omega}. \tag{4.30}$$

Remark 4.32. It is possible to compute the values of the constants in the error estimates of Corollaries 4.30 and 4.31, but the effort required for this computation increases with k. For linear finite elements, see [82].

Even if these constants are known, the a priori error estimates of the corollaries only indicate the asymptotic behaviour of the error as $h \to 0$ and do not provide an explicit bound because in most cases it is also difficult to compute a bound for $|u|_{k+1,\Omega}$.   □

Remark 4.33. In Chapter 3 we discussed in detail the principal techniques used to incorporate the given boundary conditions: natural boundary conditions influence the weak formulation of the problem while in principle one should satisfy the essential boundary conditions by the chosen ansatz.

If it is difficult to satisfy the essential boundary conditions in this way, then alternative approaches are available. Some examples of such techniques are

- weakly imposed boundary conditions and nonconforming methods;
- mixed finite elements;
- penalty methods.

We shall discuss the basic principles of these methods in Sections 4.5, 4.6 and 4.8.   □

Theorem 4.29 provides a discretization error estimate only in the norm of the given space V. But the projection error estimates in Theorem 4.28 tell us that for fixed k the asymptotic behaviour of the projection error depends on r, i.e., on the maximal order of the derivatives appearing in the Sobolev space norm in which the error is measured. Does the discretization error have the same property?

Let us take as an example the situation of Corollary 4.30. Is it possible to prove not only the error estimate (4.29) in the H^1 norm, but also the bound

$$\|u - u_h\|_{0,\Omega} \leq c\, h^{k+1} |u|_{k+1,\Omega}? \tag{4.31}$$

(We conjecture that this bound is valid because it is the bound for the interpolation error in L_2.) Note that one cannot derive this estimate directly from Theorem 4.29 by setting $V = L_2(\Omega)$, because then the bilinear form $a(\cdot,\cdot)$ is unbounded.

Aubin and Nitsche devised a technique for proving error estimates in L_2 that is based on a duality argument (it is sometimes called *Nitsche's trick*). The key idea is to introduce an auxiliary function $w \in V$ that is the solution of

$$a(u - u_h, w) = (u - u_h, u - u_h). \tag{4.32}$$

Here, as usual, $(\cdot,\cdot)$ denotes the scalar product in $L_2(\Omega)$. The L_2 norm that we want to estimate appears on the right-hand side of (4.32).

More generally, we consider the following variational equation: Find $w \in V$ such that for a given $g \in L_2(\Omega)$ one has

$$a(v, w) = (g, v) \qquad \text{for all } v \in V. \tag{4.33}$$

This variational equation is the *adjoint* or *dual* problem of the original problem (4.26).

One can easily verify that the bilinearity, continuity and V-ellipticity properties of the bilinear form $a(\cdot, \cdot)$ will also be valid for the bilinear form $a^*(\cdot, \cdot)$ defined by interchanging the arguments:

$$a^*(u, v) := a(v, u) \qquad \text{for all } u, \ v \in V.$$

Furthermore,

$$|(g, v)| \le \|g\|_{0,\Omega}\|v\|_{0,\Omega} \le c\|g\|_{0,\Omega}\|v\| \qquad \text{for all } v \in V$$

implies that $(g, \cdot) \in V^*$ for $g \in L_2(\Omega)$. Thus we can apply the Lax-Milgram lemma to deduce that the adjoint problem (4.33) has a unique solution.

In addition we assume that $w \in H^2(\Omega)$ with

$$|w|_{2,\Omega} \le c\|g\|_{0,\Omega} \tag{4.34}$$

with some positive constant c. The validity of this regularity assumption and a priori estimate depends on the properties of the differential operator, the geometry of the given domain Ω and the given boundary conditions.

Theorem 4.34. *Let $V \subset H^1(\Omega)$. Assume that the solution u of the given problem (4.26) satisfies $u \in V \cap H^{k+1}(\Omega)$, where $k \ge 1$. Moreover, let the assumptions of Theorem 4.28 be satisfied.*
Given $g \in L_2(\Omega)$, assume that the solution w of the adjoint problem (4.33) lies in $H^2(\Omega)$ and satisfies the a priori estimate (4.34).
Then the L_2-error of the finite element method can be estimated by

$$\|u - u_h\|_{0,\Omega} \le c\,h^{k+1}\,|u|_{k+1,\Omega}. \tag{4.35}$$

Proof: As $V_h \subset V$, by Galerkin orthogonality we have

$$a(u - u_h, v_h) = 0 \qquad \text{for all } v_h \in V_h. \tag{4.36}$$

Now $V_h \subset V \hookrightarrow L_2(\Omega)$, so $u - u_h \in L_2(\Omega)$. The adjoint problem

$$a(v, w) = (u - u_h, v) \qquad \text{for all } v \in V \tag{4.37}$$

has a unique solution $w \in V$. By our hypotheses

$$|w|_{2,\Omega} \le c\|u - u_h\|_{0,\Omega}. \tag{4.38}$$

Choose $v = u - u_h$ in (4.37). Recalling (4.36) also, it follows that

$$\|u - u_h\|_{0,\Omega}^2 = (u - u_h, u - u_h) = a(u - u_h, w) = a(u - u_h, w - v_h)$$

$$\le M\|u - u_h\|_{1,\Omega}\|w - v_h\|_{1,\Omega} \qquad \text{for all } v_h \in V_h. \tag{4.39}$$

Corollary 4.30 tells us that for the H^1 error one has

$$\|u - u_h\|_{1,\Omega} \le c\,h^k \,|u|_{k+1,\Omega}. \tag{4.40}$$

But Theorem 4.28 and (4.38) imply that the projection error of w satisfies

$$\|w - \Pi_h w\|_{1,\Omega} \le c\,h\,|w|_{2,\Omega} \le c\,h\,\|u - u_h\|_{0,\Omega}.$$

Setting $v_h = \Pi_h w$ in (4.39) and taking (4.40) into account, we obtain finally

$$\|u - u_h\|_{0,\Omega} \le c\,h^{k+1}\,|u|_{k+1,\Omega}. \quad \blacksquare$$

The convergence results presented for finite elements based on piecewise polynomials of degree k yield optimal orders of convergence only if the solution of the given problem is relatively smooth, i.e., $u \in H^{k+1}(\Omega)$. As we saw in Chapter 3, very smooth solutions are not to be expected in domains with corners and for problems with mixed boundary conditions. In cases like this, graded meshes as discussed in Section 4.3.4 are one possible means of preventing a degradation of the convergence order.

If one knows only that $u \in H^1(\Omega)$, then convergence of the finite element method can often still be proved, as the next result shows:

Theorem 4.35. *Let $\Omega \subset \mathbb{R}^2$ be a polygonal domain, with $V \subset H^1(\Omega)$, and let $H^2(\Omega)$ be dense in V with respect to the norm $\|\cdot\|_{1,\Omega}$.*
Consider piecewise linear finite elements on a quasi-uniform triangulation. Let u_h denote the discrete solution. Then

$$\lim_{h \to 0} \|u - u_h\|_{1,\Omega} = 0.$$

Proof: Let $\varepsilon > 0$ be arbitrary. As $H^2(\Omega)$ is dense in V, there exists $w \in H^2(\Omega)$ such that

$$\|u - w\|_{1,\Omega} \le \varepsilon.$$

By Theorem 4.28 the projection error satisfies

$$\|w - \Pi_h w\|_{1,\Omega} \le c\,h\,|w|_{2,\Omega}.$$

Choosing $h > 0$ sufficiently small, it follows that

$$\|w - \Pi_h w\|_{1,\Omega} \le \varepsilon.$$

The triangle inequality now gives

$$\|u - \Pi_h w\|_{1,\Omega} \le \|u - w\|_{1,\Omega} + \|w - \Pi_h w\|_{1,\Omega} \le 2\varepsilon.$$

Finally, applying Cea's lemma we get

$$\|u - u_h\|_{1,\Omega} \le c\varepsilon.$$

This proves the theorem because ε is an arbitrarily small positive number. $\blacksquare$

Remark 4.36. (Optimal approximation and finite element methods)
Let A be a subset of a normed space X. In approximation theory the N-width
of A in X is defined by

$$d_N(A, X) = \inf_{E_N} \sup_{f \in A} \inf_{g \in E_N} \|f - g\|_X,$$

where the first infimum is over all N-dimensional subspaces E_N of X.
The quantity $d_N(A, X)$ measures how well A can be approximated by N-dimensional subspaces of X.
In [73] it is proved that if $X = H_0^1(\Omega)$ and

$$A := \big\{ u \in X : \quad -\triangle u + cu = f, \, f \in H^s, \, \|f\|_s = 1 \big\},$$

then for d-dimensional Ω, the N-width $d_N(A, X)$ satisfies the bound

$$d_N \geq C N^{-(s-1)/d}.$$

This tells us for example that when $s = 0$, linear finite elements yield asymptotically optimal rates in the H^1 norm; thus from this point of view there
is no reason to consider more complicated finite-dimensional approximation
spaces.   □

Remark 4.37. (Interpolants or polynomial approximation)
For Lagrange elements and $u \in H^2$ one often uses standard interpolation to
define projections into the given finite element space. Other approaches are
also possible: see, for instance, Chapter 4 in [BS94]. We shall discuss other
projectors when we present a posteriori error estimators in Section 4.7.

   In analyzing *meshless discretizations*, Melenk [88] describes those properties of general systems of ansatz functions that are needed to prove optimal
approximation results. This characterization is also based on polynomial approximation in Sobolev spaces.   □

4.4.3 Inverse Inequalities and Pointwise Error Estimates

If instead of H^1 and L_2 one is interested in the error of a finite element
approximation at a specific point or measured in the L_∞ norm, then the
interpolation or projection error hints at what best one can expect.
For instance, for linear elements on a quasi-uniform triangulation one can
derive, analogously to Theorem 4.28, the interpolation error bounds

$$\|u - \varPi u\|_\infty \leq \begin{cases} Ch & \text{for} \quad u \in H^2, \\ Ch^2 & \text{for} \quad u \in W_\infty^2. \end{cases}$$

The question now is: can one prove a similar result for the finite element error
$\|u - u_h\|_\infty$?
   The proof of optimal L_∞ error estimates is much more difficult than the
arguments we have seen for error estimates in the H^1 or L_2 norms. Therefore

we first present a simple approach based on inverse inequalities, which gives first-order convergence under restrictive assumptions, and later sketch a more complicated approach that leads to optimal convergence rates.

Inverse inequalities relate different norms or seminorms in finite element spaces; in these inequalities, surprisingly, stronger norms are bounded by multiples of weaker norms. This can be done because every finite element space is finite-dimensional and in a finite-dimensional space all norms are equivalent. But the constants in such inequalities will depend on the dimension, and because the dimension of a finite element space V_h is related to h the constants in inverse inequalities also depend on h.

As an example of an inverse inequality we prove:

Lemma 4.38. *Let V_h be the space of piecewise linear finite elements on a quasi-uniform triangulation of a two-dimensional polygonal domain. Assume that the family of triangulations satisfies the inverse assumption*

$$\frac{h}{h_K} \le \nu \quad \text{for all elements } K,$$

where ν is some positive constant. Then there exists a constant C such that

$$||v_h||_\infty \le \frac{C}{h}||v_h||_0 \quad \text{for all} \quad v_h \in V_h.$$

Proof: Denote by $v_{i,K}$ the values of the linear function $v_h|_K$ at the knots, by $v_{ij,K}$ the values at the midpoints of the edges, and by $v_{111,K}$ the value at the centroid. Then a calculation yields

$$||v_h||_{0,K}^2 = \frac{1}{60}\left(3\sum_i v_{i,K}^2 + 8\sum_{i<j} v_{ij,K}^2 + 27v_{111,K}^2\right)(meas\ K)$$
$$\ge ch_K^2(\max_i |v_{i,K}|)^2,$$

where the quasi-uniformity of the triangulation was used. But

$$||v_h||_{\infty,K} = \max_i |v_{i,K}|,$$

so we have proved the local inverse inequality

$$||v_h||_{\infty,K} \le \frac{C}{h_K}||v_h||_{0,K} \quad \text{for all} \quad v_h \in V_h.$$

The inverse assumption now implies the global inverse inequality.   ∎

We follow the phrasing of some publications that call triangulations that satisfy the inverse assumption *uniform*. Of course, this does not mean that all mesh elements are identical. It should be noticed that uniform meshes do not allow arbitrarily fine local refinements.

Local inverse inequalities are often proved similarly to our derivation of estimates for the projection error in finite element spaces: first one maps the element onto a reference element, then the equivalence of norms in a finite-dimensional space on that element is used, and finally one maps back to the given element.

The following inverse inequalities are frequently invoked:

$$||v_h||_\infty \leq \frac{C}{h^{n/2}}||v_h||_0 \quad \text{for all} \quad v_h \in V_h \quad \text{(where the domain lies in } \mathbb{R}^n\text{)}$$

and

$$|v_h|_1 \leq \frac{C}{h}||v_h||_0 \quad \text{for all} \quad v_h \in V_h.$$

See Theorem 17.2 in [CL91] for a much more general result.

In general, one tries to avoid using global inverse inequalities because the inverse assumption seriously restricts the class of possible triangulations: in particular, local grid refinement is excluded. There are many special results concerning the validity of inverse inequalities under weaker assumptions; see for instance [41].

Now we set out to prove pointwise (i.e., L_∞) error estimates for linear elements. Writing u for the true solution and u_h for the computed solution, Lemma 4.38 implies the following inequalities:

$$||u - u_h||_\infty \leq ||u - \Pi u||_\infty + ||u_h - \Pi u||_\infty$$
$$\leq ||u - \Pi u||_\infty + \frac{C}{h}||u_h - \Pi u||_0$$
$$\leq ||u - \Pi u||_\infty + \frac{C}{h}\left[||u - \Pi u||_0 + ||u - u_h||_0\right].$$

Our previous results for the interpolation error immediately imply

Theorem 4.39. *If $u \in H^2(\Omega)$, $\Omega \subset \mathbb{R}^n$, $n \leq 3$ and $||u - u_h||_0 \leq Ch^2||u||_2$, then for linear finite elements on a uniform triangulation one has*

$$||u - u_h||_\infty \leq Ch||u||_2.$$

Remark 4.40. If the solution u lies only in $H^2(\Omega)$, then this first-order L_∞ error bound is optimal, as is demonstrated by an example in [Ran04]. □

A possible way of avoiding the inverse assumption is to instead invoke the *discrete Sobolev inequality*. In the two-dimensional case with linear finite elements on an arbitrary triangulation, this inequality states that

$$||v_h||_\infty \leq C|\ln h_{min}|^{1/2}||v_h||_1 \quad \text{for all} \quad v_h \in V_h.$$

See [79] for the proof of this result; in [Xu89] one finds the corresponding result in n space dimensions assuming quasi-uniformity of the triangulation.

If one wants to prove pointwise convergence of order greater than one for linear elements while assuming more regularity, say $u \in W^{2,\infty}$, then the analysis becomes deeper. One can for example use a Green's function.

Fix $x_0 \in \Omega$. Then the Green's function $G \in W^{1,1}(\Omega)$ with respect to x_0 satisfies the variational equation

$$a(v, G) = v(x_0) \quad \text{for all} \quad v \in W^{1,\infty}(\Omega).$$

Consequently the error at the point x_0 has the representation

$$(u - u_h)(x_0) = a(u - u_h, G).$$

Next we introduce the finite element approximation $G_h \in V_h$ of G; this is defined by

$$a(v_h, G_h) = v_h(x_0) \quad \text{for all} \quad v_h \in V_h.$$

Then by definition of u^h we have

$$(u - u_h)(x_0) = a(u - u_h, G - G_h) = a(u - v_h, G - G_h)$$

for arbitrary $v_h \in V_h$, where the definitions of G and G_h were used in the second equality. It follows that

$$\|u - u_h\|_\infty \leq \|G - G_h\|_{W_1^1} \inf_{v_h \in V_h} \|u - v_h\|_{W_\infty^1}.$$

The estimation of

$$\|G - G_h\|_{W_1^1}$$

is technically difficult.

Remark 4.41. For triangular elements of type k, Scott [108] proved that

$$\|G - G_h\|_{W_1^1} \leq C \begin{cases} h |\ln h| & \text{for} \quad k = 1, \\ h & \text{for} \quad k \geq 2. \end{cases}$$

Together with the interpolation error estimates

$$\|u - \Pi u\|_{W_\infty^1} \leq C h^k |u|_{W_\infty^{k+1}} \quad (k \geq 1),$$

this yields

$$\|u - u_h\|_\infty \leq C \begin{cases} h^2 |\ln h| & \text{for} \quad u \in W_\infty^2 \text{ and } k = 1, \\ h^{k+1} & \text{for} \quad u \in W_\infty^{k+1} \text{ and } k \geq 2. \end{cases}$$

Note that there is no logarithmic factor for $k \geq 2$. □

We now sketch a modification of Scott's technique for the case $k = 1$ that slightly reduces the technical difficulties and is due to Frehse and Rannacher

[50]. The basic idea is a regularization of the Green's function. We define a regularised Green's function g as follows: Find $g \in V = H_0^1(\Omega)$ such that

$$a(v, g) = (v, \delta^h) \quad \text{for all} \quad v \in V.$$

Here δ^h is some approximation of the δ-distribution and it can be chosen in various ways. We follow the approach of [75]. If $x_0 \in K$, where K is an element of the triangulation, define $\delta^h \in P_1(K)$ by

$$q(x_0) = \int_K \delta^h q \quad \text{for all} \quad q \in P_1(K). \tag{4.41}$$

It follows that

$$\int_K \delta^h = 1 \quad \text{and} \quad \max_K |\delta^h| \leq \frac{C}{\text{meas } K}.$$

By (4.41) we have

$$(u - u_h)(x_0) = (u - u^I)(x_0) + (u^I - u_h)(x_0) = (u - u^I)(x_0) + \int_K (u^I - u_h)\delta_h.$$

Hence

$$\|u - u_h\|_\infty \leq C\|u - u^I\|_\infty + |(u - u_h, \delta^h)|.$$

From the definition of g,

$$(u - u_h, \delta^h) = a(u - u_h, g).$$

Introducing the finite element approximation g_h of g, we obtain

$$|(u - u_h, \delta^h)| \leq \|g - g_h\|_{W_1^1} \inf_{v_h \in V_h} \|u - v_h\|_{W_\infty^1}.$$

At first sight the final estimate now seems to be simple because $g \in H^2$. But it is not possible to deduce immediately that $\|g - g_h\|_1 = O(h)$ (and consequently $\|g - g_h\|_{W_1^1} = O(h)$) because g depends on h and one must study this dependence carefully!

After a precise analysis one gets the following results:

$$|g|_\infty \leq C(|\ln h| + 1),$$
$$\|\nabla g\|_0 \leq C|\ln h|^{1/2},$$
$$\|g\|_{W_1^2} \leq C(|\ln h| + 1),$$
$$\|g\|_2 \leq Ch^{-1}.$$

The last estimate shows that a deeper study is necessary to estimate $\|g - g_h\|_{W_1^1}$. To do this one introduces a carefully chosen weight function σ and uses

$$\|g - g_h\|_{W_1^1} \le \|\sigma^{-1}\|_0 \cdot \|\sigma \nabla(g - g_h)\|_0 .$$

An optimal choice of σ finally leads to

$$\|g - g_h\|_{W_1^1} \le Ch|\ln h|;$$

see [Ran04] for details.

Remark 4.42. A precise analysis of an example [65] shows that the estimate

$$\|u - u_h\|_\infty \le Ch^2|\ln h|$$

for linear elements cannot be improved: the factor $\ln h$ is indeed necessary.
□

Remark 4.43. On certain equidistant triangular meshes the finite element method with linear elements generates a discrete problem that is equivalent to the familiar five-point difference scheme. The error estimate for the finite element method in the L_∞ norm is consequently an error estimate for the pointwise error of the five-point difference scheme. But in contrast to the standard analysis of finite difference methods, the analysis of finite element methods requires less smoothness of the exact solution.   □

Exercise 4.44. Let u be a sufficiently smooth function on $[0, 1]$ and u_I the piecewise linear continuous interpolant to u on a given mesh.
a) Prove the following representation of the interpolation error:

$$u(x) - u_I(x) = \frac{1}{x_i - x_{i-1}} \int_{x_{i-1}}^{x} \int_{x_{i-1}}^{x_i} \int_{\eta}^{\zeta} u''(\xi)\, d\xi\, d\eta\, d\zeta$$

for all $x \in (x_{i-1}, x_i)$.
b) Estimate the interpolation error in terms of the properties of u'', assuming for instance that $u \in H^2(0, 1)$.

Exercise 4.45. Fix $\xi \in (0, 1)$. Consider the boundary value problem

$$-au'' = 1 \qquad \text{in } (0, \xi) \cup (\xi, 1),$$
$$u(0) = u(1) = 0,$$
$$u'(\xi - 0) = 2u'(\xi + 0),$$

with

$$a = \begin{cases} 1 & \text{in } (0, \xi), \\ 2 & \text{in } (\xi, 1). \end{cases}$$

Using, e.g., the weak formulation, it is not difficult to compute the exact solution.
Choose $\xi = (1 + h)/2$ and discretize the problem with linear elements on an equidistant uniform mesh with mesh width h. What does one obtain for the discretization error? Compare with the general theory.

Exercise 4.46. Consider the function $f(t) = \sin t$ on the interval $[0, \pi/2]$. Compute the best approximation of f in the space $L_2(0, \pi/2)$ by piecewise linear continuous functions on an equidistant mesh.

4.5 Nonconforming Finite Element Methods

4.5.1 Introduction

Consider the continuous problem: Find $u \in V$ such that

$$a(u, v) = f(v) \qquad \text{for all } v \in V. \tag{5.1}$$

A conforming discretization of this variational problem using finite elements is characterized by two properties. First, the finite element space V_h is a subset of V; second, the variational equation for the discrete solution uses the same bilinear form $a(\cdot, \cdot)$ and the same linear form $f(\cdot)$ as the continuous problem. Consequently, the discrete solution satisfies

$$a(u_h, v_h) = f(v_h) \qquad \text{for all } v_h \in V_h. \tag{5.2}$$

In certain circumstances it turns out that these requirements are too rigorous, e.g., in each of the following situations:

- it is difficult to construct a finite dimensional space with $V_h \subset V$ (for example for differential operators of higher order);
- it is not possible to compute the necessary integrals based on $a(\cdot, \cdot)$ or $f(\cdot)$ exactly, so one must use quadrature rules;
- it is difficult to incorporate the essential inhomogeneous boundary conditions or it is vital to describe precisely the non-polygonal boundary of the domain.

Any finite element method that is not directly based on the discretization of (5.1) by (5.2) with $V_h \subset V$ is called a *nonconforming finite element method*. This terminology is used sometimes in the literature only for the case $V_h \not\subset V$, but we want to consider all nonconforming aspects that we have described. Thus to solve approximately the given problem (5.1) we consider the discrete problem

$$a_h(u_h, v_h) = f_h(v_h) \qquad \text{for all } v_h \in V_h. \tag{5.3}$$

Here V_h is a finite-dimensional Hilbert space with norm $\|\cdot\|_h$, while $a_h(\cdot, \cdot) : V_h \times V_h \longrightarrow \mathbb{R}$ is a continuous bilinear form that is uniformly V_h-elliptic, i.e., there exists a constant $\tilde{\gamma} > 0$ independent of h such that

$$\tilde{\gamma} \|v_h\|_h^2 \leq a_h(v_h, v_h) \qquad \text{for all } v_h \in V_h, \tag{5.4}$$

and $f_h : V_h \longrightarrow R$ is a continuous linear functional.

For a nonconforming method with $V_h \not\subset V$ it is not clear if $a_h(\cdot, \cdot)$ is defined on $V \times V$ or f_h is defined on V. Since $V_h \not\subset V$ is possible, in addition to problem (5.1) in the space V and the discrete problem (5.3) in V_h we consider two larger spaces Z, Z_h with

$$V \hookrightarrow Z \qquad \text{and} \qquad V_h \hookrightarrow Z_h \hookrightarrow Z \,.$$

We assume that $a_h(\cdot, \cdot)$ and f_h are defined on $Z_h \times Z_h$ and Z_h, respectively. These imbeddings of spaces allows one to estimate the distance between the discrete solution $u_h \in V_h$ and the exact solution $u \in V$ in the norm $||| \cdot |||$ of the space Z.

Our assumptions mean that the Lax-Milgram lemma can be applied to the discrete problem (5.3). Hence (5.3) has a unique solution. Analogously to Cea's lemma we now derive an abstract error estimate. Choose an arbitrary $z_h \in V_h$. Then the properties of the bilinear form $a_h(\cdot, \cdot)$ and the variational equation (5.3) yield

$$a_h(u_h - z_h, v_h) = f_h(v_h) - a_h(z_h, v_h) \qquad \text{for all } v_h \in V_h. \qquad (5.5)$$

As $f_h(\cdot) - a_h(z_h, \cdot) \in V_h^*$, by taking $v_h = u_h - z_h$ it follows that

$$\tilde{\gamma} \|u_h - z_h\|_h^2 \leq \|f_h(\cdot) - a_h(z_h, \cdot)\|_{*,h} \|u_h - z_h\|_h.$$

That is,

$$\|u_h - z_h\|_h \leq \frac{1}{\tilde{\gamma}} \|f_h(\cdot) - a_h(z_h, \cdot)\|_{*,h} \qquad \text{for all } z_h \in V_h.$$

In this calculation $\| \cdot \|_{*,h}$ denotes the usual norm in the dual space V_h^*, viz.,

$$\|w\|_{*,h} := \sup_{v_h \in V_h} \frac{|w(v_h)|}{\|v_h\|_h} \qquad \text{for } w \in V_h^*. \qquad (5.6)$$

The assumption $V_h \hookrightarrow Z$ and the triangle inequality yield

$$|||u - u_h||| \leq |||u - z_h||| + \frac{1}{\tilde{\gamma}} \|f_h(\cdot) - a_h(z_h, \cdot)\|_{*,h} \qquad \text{for all } z_h \in V_h. \quad (5.7)$$

Thus we have proved the following abstract estimate for nonconforming methods:

Lemma 4.47. *For the abstract nonconforming method described above, the discretization error $u - u_h$ is bounded by*

$$|||u - u_h||| \leq \inf_{z_h \in V_h} \left\{ |||u - z_h||| + \frac{1}{\tilde{\gamma}} \|f_h(\cdot) - a_h(z_h, \cdot)\|_{*,h} \right\}.$$

In the following subsections we shall consider specific cases and derive particular results from this general theory.

4.5.2 Ansatz Spaces with Low Smoothness

To obtain $v_h \in H^1(\Omega)$ in a conforming method one uses globally continuous finite elements; to have $v_h \in H^2(\Omega)$ one needs finite elements that are continuously differentiable on $\Omega \subset \mathbb{R}^n$ when $n = 2$. Nonconforming methods have the advantage that it is possible to use ansatz functions with less global regularity.

Let us set $Z_h = V + V_h := \{ z = v + v_h : v \in V, \ v_h \in V_h \}$. Let a_h be a symmetric, non-negative bilinear form on $Z_h \times Z_h$ that satisfies

$$|a_h(z_h, w_h)| \le M_0 |||z_h|||_h \, |||w_h|||_h \qquad \text{for all } z_h, \, w_h \in Z_h$$

with a constant M_0 that is independent of h. Here $|||\cdot|||_h$ denotes the seminorm defined by

$$|||z|||_h := a_h(z, z)^{1/2} \qquad \text{for all } z \in Z_h.$$

Define

$$\|v_h\|_h = |||v_h|||_h \qquad \text{for all } v_h \in V_h.$$

These assumptions imply the following well-known result:

Lemma 4.48 (Second Lemma of Strang). *There exists a constant $c > 0$ such that*

$$|||u - u_h|||_h \le c \left\{ \inf_{z_h \in V_h} |||u - z_h|||_h + \|f_h(\cdot) - a_h(u, \cdot)\|_{*,h} \right\}.$$

Proof: From (5.5) and the bilinearity of $a_h(\cdot, \cdot)$ we obtain

$$a_h(u_h - z_h, v_h) = a_h(u - z_h, v_h) + f_h(v_h) - a_h(u, v_h) \qquad \text{for all } z_h, \, v_h \in V_h.$$

Choosing $v_h = u_h - z_h$, it follows that

$$|||u_h - z_h|||_h \le M_0 |||u - z_h|||_h + \|f_h(\cdot) - a_h(u, \cdot)\|_{*,h}.$$

Now the triangle inequality yields

$$|||u - u_h|||_h \le (1 + M_0)|||u - z_h|||_h + \|f_h(\cdot) - a_h(u, \cdot)\|_{*,h} \qquad \text{for all } \ z_h \in V_h$$

and the result of the lemma follows. ∎

The first term in this estimate is, as in Cea's lemma, the approximation error; the second term in Lemma 4.48 is called the *consistency error*. This error arises because the identity

$$a_h(u, v_h) = f_h(v_h) \quad \text{for all} \quad v_h \in V_h \tag{5.8}$$

is not necessarily true for nonconforming methods. If, however, (5.8) does hold, we say that the finite element method is *consistent*. Consistency implies the Galerkin orthogonality property

$$a_h(u - u_h, v_h) = 0 \quad \text{for all} \quad v_h \in V_h.$$

Remark 4.49. In many concrete situations $|||\cdot|||_h$ is even a norm on Z_h. □

Next we study a special nonconforming element, apply Lemma 4.48 and investigate the consistency error.

Example 4.50. (Crouzeix-Raviart element) Let $\Omega \subset \mathbb{R}^2$ be a polygonal domain. We consider the boundary value problem

$$-\Delta u = f \quad \text{in } \Omega, \qquad u|_\Gamma = 0, \tag{5.9}$$

which we assume has a unique solution $u \in H^2(\Omega)$.

We decompose Ω using a family of triangles $\mathcal{Z}_h = \{\Omega_i\}_{i=1}^M$ that we assume is admissible and quasi-uniform. Let the finite element space V_h be the set of all piecewise linear functions that are continuous at every midpoint of every interior edge of the triangulation and zero at the midpoints of the edges on the boundary Γ. In general $V_h \not\subset H^1(\Omega)$ and the method is nonconforming.

The Crouzeix-Raviart element has important applications in fluid dynamics: see [BF91, GR89]. Using it to solve (5.9) is somewhat artificial but allows us to examine the basic features of a nonconforming method in a simple setting.

Denote the midpoints of the interior edges by $p^j \in \Omega$, $j = 1, \ldots, N$, and the midpoints of the boundary edges by $p^j \in \Gamma$, $j = N+1, \ldots, \overline{N}$. The finite element space is then

$$V_h = \left\{ v_h \in L_2(\Omega) : \begin{array}{l} v_h|_{\Omega_i} \in P_1(\Omega_i) \,\forall i, \; v_h \text{ continuous at } p^j, \, j = 1, \ldots, N, \\ v_h(p^j) = 0, \, j = N+1, \ldots, \overline{N} \end{array} \right\}.$$

It is clear that functions $v_h \in V_h$ fail, in general, to be continuous on Ω. Moreover it is not difficult to check that

$$V_h \not\subset H^1(\Omega).$$

Figure 4.19 shows the typical behaviour of a function $v_h \in V_h$ on two adjacent triangles.

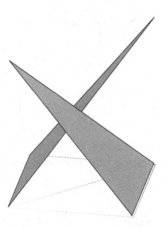

Figure 4.19 the nonconforming P_1-element

Furthermore, any function in V_h does not necessarily satisfy the homogeneous Dirichlet boundary condition of (5.9).

The bilinear form associated with (5.9) is

$$a(u,v) = \int_\Omega \nabla u \nabla v \, dx \qquad \text{for all } u, v \in V := H_0^1(\Omega).$$

This can be extended to $V_h \times V_h$ by a piecewise definition of the integral, i.e., set

$$a_h(u_h, v_h) = \sum_{i=1}^{M} \int_{\Omega_i} \nabla u_h \nabla v_h \, dx \qquad \text{for all } u_h, v_h \in V_h.$$

The bilinear form a_h is well defined on $V \times V$, and in fact

$$a_h(u,v) = a(u,v) \qquad \text{for all } u, v \in V.$$

There is no reason to change the definition of the linear form associated with f:

$$f_h(v) := f(v) := \int_\Omega f v \, dx \qquad \text{for all } v \in V_h + V.$$

We analyse the nonconforming method using the piecewise H^1 seminorm induced by our symmetric bilinear form a_h. Strang's second lemma shows that the main problem is an investigation of the consistency error $\|f_h(\cdot) - a_h(u,\cdot)\|_{*,h}$, where $u \in H^2(\Omega)$ is the solution of (5.9). The differential equation (5.9) yields

$$f_h(v_h) - a_h(u, v_h) = \sum_{i=1}^{M} \left\{ \int_{\Omega_i} f v_h \, dx - \int_{\Omega_i} \nabla u \nabla v_h \, dx \right\}$$

$$= -\sum_{i=1}^{M} \left\{ \int_{\Omega_i} \Delta u \, v_h \, dx + \int_{\Omega_i} \nabla u \nabla v_h \, dx \right\}. \tag{5.10}$$

Let us introduce the notation $v_i(\cdot) = v_h|_{\Omega_i}$, $i = 1, \dots, M$. Green's formula yields

$$\int_{\Omega_i} \nabla u \nabla v_h \, dx = \int_{\Gamma_i} \frac{\partial u}{\partial n_i} v_i \, ds - \int_{\Omega_i} \Delta u \, v_h \, dx, \tag{5.11}$$

where v_i is defined on the boundary $\Gamma_i = \partial \Omega_i$ by continuous extension and n_i denotes the outward-pointing unit normal vector on Γ_i. From (5.10) and (5.11) one gets

$$f_h(v_h) - a_h(u, v_h) = -\sum_{i=1}^{M} \int_{\Gamma_i} \frac{\partial u}{\partial n_i} v_i(s) \, ds. \tag{5.12}$$

In the following we bring the mean values along edges into the game. For each edge $e \subset \Gamma_i$ define $\bar{v}_i|_e \in \mathbb{R}$ by

$$\int_e (v_i - \bar{v}_i)ds = 0.$$

The linearity of v_i on each e means that $\bar{v}_i$ is the value of v_i at the midpoint of e. The midpoint continuity imposed on each Crouzeix-Raviart element and (5.12) then imply that

$$f_h(v_h) - a_h(u, v_h) = -\sum_{i=1}^{M} \int_{\Gamma_i} \frac{\partial u}{\partial n_i} (v_i - \bar{v}_i) \, ds,$$

since the midpoint contributions to each edge e from the two triangles sharing e cancel owing to the opposing directions of their outward-pointing normals. Now define the continuous piecewise linear function u^I on each triangle by linear interpolation at the vertices. Then $\partial u^I / \partial n_i$ is constant on each edge of Γ_i so

$$f_h(v_h) - a_h(u, v_h) = -\sum_{i=1}^{M} \int_{\Gamma_i} \frac{\partial(u - u^I)}{\partial n_i} (v_i - \bar{v}_i) \, ds. \tag{5.13}$$

Next we apply the Cauchy-Schwarz inequality to (5.13) and need information on both the integrals that emerge. A trace theorem and the Bramble-Hilbert lemma yield

$$\int_{\Gamma_i} |\nabla(u - u^I)|^2 ds \le c\, h \, |u|_{2,\Omega_i}^2. \tag{5.14}$$

Similarly

$$\int_{\Gamma_i} |v_i - \bar{v}_i|^2 ds \le c\, h \, |v_h|_{1,\Omega_i}^2. \tag{5.15}$$

Invoking this pair of inequalities, we get the consistency error estimate

$$|f_h(v_h) - a_h(u, v_h)| \le c\, h \, |u|_2 \, \|v_h\|_h$$

because $\|v_h\|_h^2 = \sum_i |v_h|_{1,\Omega_i}^2$.

The approximation error causes no trouble. The space V_h contains the piecewise linear globally continuous C^0 elements, and $||| \cdot |||_h^2 = \sum_i | \cdot |_{1,\Omega_i}^2$, so one gets easily

$$\inf_{z_h \in V_h} |||u - z_h|||_h \le c\, h \, \|u\|_{H^2(\Omega)}.$$

By Lemma 4.48 the final error estimate for the Crouzeix-Raviart element now follows:

$$|||u - u_h|||_h \le c\, h \, \|u\|_{H^2(\Omega)}.$$

We recognize the same order of convergence as that attained by the globally continuous P_1 element. $\square$

This convergence analysis is a typical example of the analysis of non-conforming elements and can extended to many other concrete examples: see [Cia78, GR89]. In [Sch97] Schieweck discusses a class of nonconforming elements on quadrilateral domains that includes the well-known Rannacher-Turek element [98]. Moreover in [Sch97] and [GR89] the application of nonconforming elements to the numerical solution of the Navier-Stokes equations is discussed in detail.

Remark 4.51. During the early stages of the development of the analysis of nonconforming methods with $V_h \not\subset V$, several authors introduced as sufficient conditions—on a more or less heuristic basis—local *Patch test* conditions on neighbouring elements. But later it turned out that these tests were insufficient to guarantee convergence of nonconforming methods. In [113, 110] this problem was systematically investigated and generalized patch tests were developed that guarantee convergence.   □

4.5.3 Numerical Integration

In this subsection we consider the setting where the finite element space satisfies $V_h \subset V$ but the Galerkin equations are formulated using a modified bilinear form $a_h(\cdot, \cdot)$ and a linear form f_h that are not necessarily defined on $V \times V$ and V respectively. This situation often arises when integrals are approximated by quadrature rules. In our abstract theory we then choose $Z = V$ and $||| \cdot ||| = \| \cdot \|$, the norm in V. Lemma 4.47 yields the estimate

$$\|u - u_h\| \le c \inf_{z_h \in V_h} \left\{ \|u - z_h\| + \|f_h(\cdot) - a_h(u, \cdot)\|_{*,h} \right\}.$$

Alternatively, one can prove

Lemma 4.52 (First Lemma of Strang). *Let $V_h \subset V$ and let the bilinear form $a_h(\cdot, \cdot)$ be uniformly V_h-elliptic. Then there exists a constant $c > 0$ such that*

$$\|u - u_h\| \le c \left[\inf_{z_h \in V_h} \left\{ \|u - z_h\| + \|a(z_h, \cdot) - a_h(z_h, \cdot)\|_{*,h} \right\} + \|f - f_h\|_{*,h} \right].$$

Proof: Combining (5.5) and (5.1) and using $V_h \subset V$, for all v_h and $z_h \in V_h$ we get

$$a_h(u_h - z_h, v_h) = a(u, v_h) - a_h(z_h, v_h) + f_h(v_h) - f(v_h)$$
$$= a(u - z_h, v_h) + a(z_h, v_h) - a_h(z_h, v_h) + f_h(v_h) - f(v_h).$$

Setting $v_h = u_h - z_h$ and invoking V_h-ellipticity yields

$$\tilde{\gamma} \|u_h - z_h\|^2 \le M \|u - z_h\| \|u_h - z_h\| + \|a(z_h, \cdot) - a_h(z_h, \cdot)\|_{*,h} \|u_h - z_h\|$$

$$+ \|f_h - f\|_{*,h} \|u_h - z_h\|.$$

The Lemma now follows from a triangle inequality. ∎

As an example of the application of quadrature rules we consider the elliptic boundary value problem

$$-\operatorname{div}(A(x)\operatorname{grad} u) = f \quad \text{in } \Omega, \qquad u|_\Gamma = 0.$$

We assume that $A : \bar{\Omega} \to \mathbb{R}$ is sufficiently smooth and

$$A(x) \ge \alpha > 0 \quad \text{for all } x \in \Omega,$$

where $\alpha > 0$ is some constant. The corresponding variational equation is

$$a(u, v) = f(v) \qquad \text{for all } v \in H_0^1(\Omega)$$

with

$$a(u, v) = \int_\Omega A(x)\nabla u(x)\nabla v(x)\, dx, \tag{5.16}$$

$$f(v) = \int_\Omega f(x)v(x)\, dx. \tag{5.17}$$

Without numerical integration the discrete solution u_h is a solution of

$$a(u_h, v_h) = f(v_h) \qquad \text{for all } v_h \in V_h,$$

but, in general, it is impossible to compute exactly the stiffness matrix and the right-hand side of this linear system.

The finite element method yields only an approximation of the exact solution; thus it makes no sense to demand quadrature rules that approximate the integrals with extremely high precision. The quadrature rule used should be suitable for the finite element method, in the sense that both quadrature and finite element errors are of the same order. (We shall not discuss the alternative approach of applying techniques of symbolic computation; see [Bel90].)

We now study concrete quadrature rules for approximating integrals of the form

$$\int_\Omega z(x)\, dx = \sum_{i=1}^{M} \int_{\Omega_i} z(x)\, dx$$

with globally continuous functions z that are smooth on each subdomain Ω_i. Recalling that the generation of the discrete problem uses transformations to a reference element, we follow this basic principle and first describe the quadrature rule on the reference element. Then the transformation mapping will induce a quadrature formula on each element.

Suppose that the element $K = \overline{\Omega}_i$ is mapped onto the reference element K'. Then in the two-dimensional case we have to study quadrature rules for a reference triangle and a reference quadrilateral, i.e., for

$$K' = \left\{ \begin{pmatrix} \xi \\ \eta \end{pmatrix} : \xi \geq 0,\, \eta \geq 0,\, \xi + \eta \leq 1 \right\}$$

and

$$K' = \left\{ \begin{pmatrix} \xi \\ \eta \end{pmatrix} : \xi \in [0,1],\, \eta \in [0,1] \right\}.$$

On a rectangle, the construction of quadrature rules uses a tensor product structure. Because Gauss-Legendre quadrature is both popular and accurate, one simply takes the Gauss-Legendre points in each coordinate direction.

The construction of quadrature rules on triangles is more complicated. One often applies the strategy of fixing a set of functions (usually polynomials) for which the quadrature rule is required to be exact, then deriving the quadrature formula by integrating an interpolating polynomial from this set. Let us demonstrate this technique for a quadratic polynomial on a triangle. We use the same notation as in the definition of the quadratic triangular finite element of subsection 4.2.2.

Lemma 4.53. *Let z_h denote the quadratic function uniquely defined by its values z_α at the points p^α where $|\alpha| = 2$, i.e.,*

$$z_h(p^\alpha) = z_\alpha, \qquad |\alpha| = 2,$$

on the triangle $K = conv\{p^1, p^2, p^3\}$. Then

$$\int_K z_h(x)\, dx = \frac{1}{3}\, (meas\, K)\, (z_{110} + z_{011} + z_{101}). \tag{5.18}$$

Proof: In barycentric coordinates on K' one can write

$$z_h(\lambda) = z_{200}\lambda_1(2\lambda_1 - 1) + z_{020}\lambda_2(2\lambda_2 - 1) + z_{002}\lambda_3(2\lambda_3 - 1)$$
$$+ z_{110}\, 4\lambda_1\lambda_2 + z_{011}\, 4\lambda_2\lambda_3 + z_{101}\, 4\lambda_1\lambda_3.$$

Taking the symmetries here into account, we have only to compute the integrals

$$I_{200} := \int_0^1 \int_0^{1-\xi} \xi(2\xi - 1)\, d\xi d\eta \quad \text{and} \quad I_{110} := 4 \int_0^1 \int_0^{1-\xi} \xi\eta\, d\xi d\eta.$$

One gets $I_{200} = I_{020} = I_{002} = 0$ and $I_{110} = I_{011} = I_{101} = 1/6$. Observe that $meas\, K' = 1/2$; the transformation back to K produces the formula (5.18). ∎

Lemma 4.53 yields the following quadrature rule for approximating the integral (5.17). Let us denote by q^{ij}, $i = 1, 2, 3$, the midpoints of the edges of the triangles Ω_j, $j = 1, \ldots, M$, in the decomposition of the domain Ω. Then (5.18) leads to the definition

$$f_h(v_h) := \frac{1}{3} \sum_{j=1}^{M} \left\{ (meas\ \Omega_j) \sum_{i=1}^{3} f(q^{ij}) v_h(q^{ij}) \right\}. \tag{5.19}$$

One can define $a_h(\cdot, \cdot)$ analogously.

To deduce from Lemma 4.52 an error estimate for the finite element method with quadrature, one must prove the V_h-ellipticity of the bilinear form $a_h(\cdot, \cdot)$ and bound the quadrature errors. In the present case of quadratic elements, V_h-ellipticity is easy: first use $a(x) \geq \alpha$, then the remaining integral is integrated exactly by the quadrature rule because the product of linear functions is quadratic. There still remains the quadrature error and as an example we estimate this when (5.17) is approximated by (5.19).

Let $E_h : C(\overline{\Omega}) \longrightarrow \mathbb{R}$ denote the error functional associated with (5.19), i.e.,

$$E_h(z) := \int_{\Omega} z(x)\, dx - \frac{1}{3} \sum_{j=1}^{M} \left\{ (meas\ \Omega_j) \sum_{i=1}^{3} q^{ij} \right\} \qquad \text{for } z \in C(\overline{\Omega}). \tag{5.20}$$

Then

$$E_h(z) = 0 \qquad \text{for all } z \in P_{2,h}(\Omega),$$

where $P_{2,h}(\Omega)$ is the space of globally continuous, piecewise quadratic polynomials on the given triangulation. The Bramble-Hilbert lemma will be used to estimate the continuous linear functional E_h. Let $\Omega_j \in \mathcal{Z}_h$ be a fixed triangle of the triangulation. The standard mapping of $K := \overline{\Omega}_j$ onto the reference element K', combined with an application of the Bramble-Hilbert lemma on K', yields

$$\left| \int_{\Omega_j} z(x)\, dx - \frac{1}{3} \sum_{i=1}^{3} z(q^{ij}) \right| \leq c\, |z|_{3,\Omega_j} h^3.$$

Now a triangle inequality implies

$$|E_h(z)| \leq \sum_{j=1}^{M} \left| \int_{\Omega_j} z(x)\, dx - \frac{1}{3} (meas\ \Omega_j) \sum_{i=1}^{3} z(q^{ij}) \right|$$
$$\leq c\, h^3 \sum_{j=1}^{M} |z|_{3,\Omega_j}.$$

Assume that the finite element space consists of piecewise linears, i.e., $v_h \in V_h = P_{1,h}(\Omega)$. We replace z by the product $f v_h$. Now

$$|f\, v_h|_{l,\Omega_j} \le c \sum_{s=0}^{l} |f|_{s,\infty,\Omega_j} |v_h|_{l-s,\Omega_j}$$

and

$$|v_h|_{m,\Omega_j} = 0, \qquad m \ge 2,$$

so

$$|f\, v_h|_{3,\Omega_j} \le c \|f\|_{3,\infty,\Omega} \|v_h\|_{1,\Omega_j}.$$

Applying a Cauchy-Schwarz inequality, it follows that

$$|E_h(f v_h)| \le c\, h^3 \|f\|_{3,\infty,\Omega} \sqrt{M} \left(\sum_{j=1}^{M} \|v_h\|_{1,\Omega_j}^2 \right)^{1/2}.$$

Hence, assuming the triangulation is quasi-uniform, one has

$$|E_h(f v_h)| \le c\, h^2 \|f\|_{3,\infty,\Omega} \|v_h\|_{1,\Omega} \qquad \text{for all } v_h \in V_h. \tag{5.21}$$

Now (5.19)–(5.21) give immediately

$$|f(v_h) - f_h(v_h)| = |E_h(f v_h)| \le c\, h^2 \|f\|_{3,\infty,\Omega} \|v_h\|_{1,\Omega}.$$

Thus we get finally

$$\|f(\cdot) - f_h(\cdot)\|_{*,h} \le c\, h^2 \|f\|_{3,\infty,\Omega}. \tag{5.22}$$

A similar estimate for the application of the quadrature rule (5.18) to the bilinear form $a(\cdot, \cdot)$ and a standard interpolation error estimate lead via Lemma 4.52 to the following error estimate for linear elements:

$$\|u - u_h\|_1 \le c \left(|u|_{2,\Omega}\, h + \|f\|_{3,\infty,\Omega}\, h^2 \right).$$

It is evident from the mismatch in powers of h that the interplay between the quadrature rule used and the finite element error is not optimal in this estimate. A more sophisticated analysis shows that for piecewise linear elements any quadrature rule that integrates constants exactly will produce an error estimate that includes the term $|u|_{2,\Omega}\, h$, and that our quadrature rule (5.18) with quadratic finite elements yields in fact

$$\|u - u_h\|_1 \le c \left(|u|_{3,\Omega} + \|f\|_{2,\infty,\Omega} \right) h^2. \tag{5.23}$$

More generally, the following result can be found in [Cia78]:
Assume that triangular P_k-elements are used with a quadrature rule that has positive weights and integrates polynomials of degree $2k - 2$ exactly. Then

$$\|u - u_h\|_1 \le c \left(|u|_{k+1,\Omega} + \|f\|_{k,\infty,\Omega} \right) h^k.$$

In the following table we present some quadrature rules for triangles that are exact for polynomials up to degree 5:

position of the nodes	coordinates	weights	exact for polynomials of degree
	$(\frac{1}{3},\frac{1}{3})$	$\frac{1}{2}$	1
	$(0,0)$, $(1,0)$, $(0,1)$	$\frac{1}{6}$, $\frac{1}{6}$, $\frac{1}{6}$	1
	$(\frac{1}{2},0)$, $(\frac{1}{2},\frac{1}{2})$, $(0,\frac{1}{2})$	$\frac{1}{6}$, $\frac{1}{6}$, $\frac{1}{6}$	2
	$(\frac{1}{6},\frac{1}{6})$, $(\frac{4}{6},\frac{1}{6})$, $(\frac{1}{6},\frac{4}{6})$	$\frac{1}{6}$, $\frac{1}{6}$, $\frac{1}{6}$	2
	$(0,0)$, $(1,0)$, $(0,1)$ $(\frac{1}{2},0)$, $(\frac{1}{2},\frac{1}{2})$, $(0,\frac{1}{2})$ $(\frac{1}{3},\frac{1}{3})$	$\frac{3}{120}$, $\frac{3}{120}$, $\frac{3}{120}$ $\frac{8}{120}$, $\frac{8}{120}$, $\frac{8}{120}$ $\frac{27}{120}$	3
	$(0,0)$, $(1,0)$, $(0,1)$ $(0,\frac{3+\sqrt{3}}{6})$, $(0,\frac{3-\sqrt{3}}{6})$ $(\frac{3+\sqrt{3}}{6},0)$, $(\frac{3-\sqrt{3}}{6},0)$ $(\frac{3+\sqrt{3}}{6},\frac{3-\sqrt{3}}{6})$ $(\frac{3-\sqrt{3}}{6},\frac{3+\sqrt{3}}{6})$ $(\frac{1}{3},\frac{1}{3})$	$-\frac{1}{120}$, $-\frac{1}{120}$, $-\frac{1}{120}$ $\frac{1}{20}$, $\frac{1}{20}$ $\frac{1}{20}$, $\frac{1}{20}$ $\frac{1}{20}$ $\frac{1}{20}$ $\frac{9}{40}$	4

position of the nodes	coordinates	weights	exact for polynomials of degree
	$(\frac{6-\sqrt{15}}{21}, \frac{6-\sqrt{15}}{21})$	$\frac{155-\sqrt{15}}{2400}$	5
	$(\frac{9+2\sqrt{15}}{21}, \frac{6-\sqrt{15}}{21})$	$\frac{155-\sqrt{15}}{2400}$	
	$(\frac{6-\sqrt{15}}{21}, \frac{9+2\sqrt{15}}{21})$	$\frac{155-\sqrt{15}}{2400}$	
	$(\frac{6+\sqrt{15}}{21}, \frac{6+\sqrt{15}}{21})$	$\frac{155+\sqrt{15}}{2400}$	
	$(\frac{9-2\sqrt{15}}{21}, \frac{6+\sqrt{15}}{21})$	$\frac{155+\sqrt{15}}{2400}$	
	$(\frac{6+\sqrt{15}}{21}, \frac{9-2\sqrt{15}}{21})$	$\frac{155+\sqrt{15}}{2400}$	
	$(\frac{1}{3}, \frac{1}{3})$	$\frac{9}{80}$	

For rectangular Q_k-elements the situation is slightly different. One gets (5.23) if *the quadrature weights are positive, the set of quadrature nodes contains a subset on which $Q_k \cap P_{2k-1}$ is unisolvent, and the quadrature rule is exact on Q_{2k-1}.*

On rectangles one prefers to use Gauss-Legendre or Gauss-Lobatto quadrature rules. These formulas are usually described on the interval $[-1, 1]$. In the following table we present their weights, nodes and degree of exactness for the one-dimensional integral

$$\int_{-1}^{1} \zeta(x)\, dx \approx \sum_{j=1}^{q} c_j\, \zeta(\xi_j)\, ;$$

on rectangles a simple tensor product of these formulas is used.

Gauss-Legendre formulas

q	ξ_j	c_j	exactness degree
1	0	2	1
2	$\pm\frac{1}{3}\sqrt{3}$	1	3
3	$\pm\frac{1}{5}\sqrt{5}$ 0	5/9 8/9	5
4	$\pm\sqrt{(15+2\sqrt{30})/35}$ $\pm\sqrt{(15-2\sqrt{30})/35}$	$1/2-\sqrt{30}/36$ $1/2+\sqrt{30}/36$	7

Gauss-Lobatto formulas

q	ξ_j	c_j	exactness degree
2	± 1	1	1
3	± 1 0	1/3 4/3	3
4	± 1 $\pm\frac{1}{5}\sqrt{5}$	1/6 5/6	5
5	± 1 $\pm\frac{1}{7}\sqrt{21}$ 0	1/10 49/90 32/45	7

The analysis of finite element methods combined with numerical integration is given in full in [Cia78], and the book of Engels [Eng80] discusses in detail the construction of quadrature formulas.

4.5.4 The Finite Volume Method Analysed from a Finite Element Viewpoint

Consider the two-dimensional boundary value problem

$$-\Delta u + cu = f \quad \text{in } \Omega \subset R^2, \qquad u|_\Gamma = 0, \tag{5.24}$$

where the domain Ω is polygonal.

As we saw in Chapter 2, Section 5, the finite volume discretization of (5.24) based on Voronoi boxes is given by

$$-\sum_{j\in N_i}\frac{m_{ij}}{d_{ij}}(u_j - u_i) + \left(\int_{\Omega_i}c\right)u_i = \int_{\Omega_i}f. \tag{5.25}$$

This subsection addresses the following question: *can (5.25) be interpreted as a finite element method and consequently be analysed using finite element techniques?*

Assume that a weakly acute triangulation of Ω is given. This is automatically a Delaunay triangulation, so one can construct Voronoi boxes using the vertices of the triangulation and the perpendiculars at the midpoints of the edges. Thus the given triangulation $\{\Omega_j\}$ yields a dual decomposition of the domain Ω that comprises Voronoi boxes D_i. Figure 4.20 shows a typical configuration.

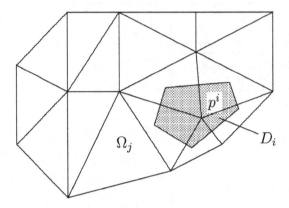

Figure 4.20 Dual decomposition into Voronoi boxes

Remark 4.54. (General dual boxes)
In [59] the author introduces a general class of boxes—*Donald boxes*—that we shall not discuss in detail. Nevertheless it is important to note that these are a viable alternative to Voronoi boxes. In contrast to Voronoi boxes, which are of circumcentric type, Donald boxes are of barycentric type: the line segments connecting the barycenter of a triangle with the midpoints of its edges are combined to form the boundary of each box. Thus each line segment on the boundary of a Voronoi box is in general replaced by two line segments on the boundary of a Donald box.

Donald boxes have the following property:
For each triangle T with vertex p_i surrounded by the Donald box Ω_i, one has the

$$\text{equilibrium condition}: \quad meas(\Omega_i \cap T) = \frac{1}{3} meas(T). \tag{5.26}$$

When dual boxes satisfy the equilibrium condition it turns out that the corresponding finite volume method has certain improved convergence properties. □

Let $V_h \subset H_0^1(\Omega)$ be the space of piecewise linear finite elements on the given Delaunay triangulation. We aim to formulate a finite element method that is equivalent to the finite volume method on the dual mesh of Voronoi boxes. First, for each $w_h \in V_h$ we have the property $w_h(p_i) = w_i$. With the notation

$$\underline{g}_i := \frac{1}{meas\,\Omega_i} \int_{\Omega_i} g, \quad \text{and} \quad m_i = meas\,\Omega_i,$$

one can write (5.25) in the form

$$\sum_i v_i \left[\left(-\sum_{j \in N_i} \frac{m_{ij}}{d_{ij}} (u_j - u_i) \right) + \underline{c}_i u_i m_i \right] = \sum_i \underline{f}_i v_i m_i.$$

Now define

$$a_h(u_h, v_h) := \sum_i v_i \left[\left(-\sum_{j \in N_i} \frac{m_{ij}}{d_{ij}} (u_j - u_i) \right) + \underline{c}_i u_i m_i \right], \qquad (5.27)$$

$$f_h(v_h) := \sum_i \underline{f}_i v_i m_i. \qquad (5.28)$$

Then the FVM can be written as a finite element method:
Find $u_h \in V_h$ such that

$$a_h(u_h, v_h) = f_h(v_h) \quad \text{for all } v_h \in V_h.$$

For problems that contain convection it is also possible to write down a corresponding finite element discretization: see Chapter 6.

Remark 4.55. We observed in Remark 4.18 that in our formulation the discretizations of the Laplace operator by the finite volume method and by linear finite elements are identical. The discretizations of the reaction and source terms are different, however.

The finite element discretization of the reaction term cu has the disadvantage that the mass matrix generated has, in general, positive off-diagonal elements. On the other hand for $c \geq 0$ the finite volume method preserves the M-matrix property on any mesh. If in the finite element method with linear elements one replaces the FEM discretization of the reaction term by its FVM discretization, this modification is called *mass lumping*. See also Chapter 5.
□

Now we can analyse the error of the finite volume method in the H^1 norm, using Lemma 4.52. The formula

$$(\nabla u_h, \nabla v_h) = \sum_i v_i \left(-\sum_{j \in N_i} \frac{m_{ij}}{d_{ij}} (u_j - u_i) \right)$$

implies that the bilinear form $a_h(\cdot, \cdot)$ is V_h-elliptic on $V_h \times V_h$. Thus we need only estimate the consistency errors arising in the reaction and source terms, viz.,

$$\left| (f, v_h) - \sum_i \underline{f}_i v_i m_i \right| \quad \text{and} \quad \left| (cu_h, v_h) - \sum_i c_i u_i v_i m_i \right|.$$

Consider the first term. Let us define $\bar{w}_h$ by

$$\bar{w}_h|_{\Omega_i} = w_i \quad \text{for } w_h \in V_h.$$

Then one has

Lemma 4.56.
$$\|v_h - \bar{v}_h\|_0 \le Ch|v_h|_1 \quad for \ v_h \in V_h.$$

Proof: The result follows immediately from

$$(v_h - \bar{v}_h)(x) = \nabla v_h (x - x_i) \quad for \ x \in \Omega_i \cap T$$

and the Cauchy-Schwarz inequality. ∎

Next,

$$|(f, v_h) - f_h(v_h)| = |(f, v_h) - (\underline{f}, \bar{v}_h)| \le |(f, v_h - \bar{v}_h)| + |(f - \underline{f}, \bar{v}_h)|.$$

Lemma 4.56 then gives

$$|(f, v_h) - f_h(v_h)| \le Ch\|f\|_1\|v_h\|_1.$$

A similar estimate for the reaction term cu yields

Theorem 4.57. *Consider the finite volume method (5.25) with dual Voronoi boxes based on a weakly acute triangulation of a convex polygonal domain Ω. If the triangulation satisfies the minimal angle condition, then the H^1-norm error of the finite volume method satisfies*

$$\|u - u_h\|_1 \le C\,h\,\|f\|_1. \tag{5.29}$$

For the finite element method one can prove second-order convergence of the error in the L_2 norm. To obtain the same result for a finite volume method, one uses the equilibrium condition—see [Bey98] and [CL00], which contain many detailed results for finite volume methods. An introduction to finite volume methods is also given in [KA03].

4.5.5 Remarks on Curved Boundaries

If the given domain Ω is two-dimensional but not polygonal then it is impossible to decompose it exactly into triangles and rectangles. What should one do?

The simplest approach is to use a polygonal domain to approximate Ω. This procedure is efficient for linear elements, in the sense that in the $H^1(\Omega)$ norm one gets first-order convergence, just like the case where Ω is polygonal. But for quadratic elements this approximation leads to convergence of order only $3/2$ in $H^1(\Omega)$ instead of the second-order convergence that is attained when the domain is polygonal.

For better convergence properties with higher-order elements one must improve the approximation of the boundary by introducing curved elements. To achieve this there are two main approaches: one can approximate the boundary (in this context most works consider curved isoparametric elements) or work with more general curved elements that fit the boundary exactly.

In what follows we explain the basic concepts of isoparametric approximation of the boundary for quadratic triangular elements. For the multi-index $\alpha = (\alpha_1, \alpha_2, \alpha_3)$ with $|\alpha| = 2$, fix interpolation points p^α on the boundary of the curved triangle Ω_j that forms part of the decomposition of the given domain.

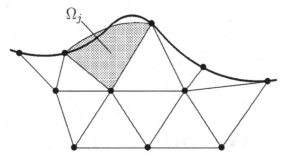

Figure 4.21 curved triangular element

Denote by $\Psi_\alpha(\underline{\lambda}) = \Psi_\alpha(\lambda_1, \lambda_2, \lambda_3)$ the local basis functions that satisfy $\Psi_\alpha(\beta/2) = \delta_{\alpha\beta}$, where as usual λ_1, λ_2 and λ_3 are the barycentric coordinates in Ω_j. Then $K = \overline{\Omega}_j$ can be approximately represented by

$$x = \sum_{|\alpha|=2} \Psi_\alpha(\underline{\lambda})p^\alpha, \qquad \sum_{i=1}^{3} \lambda_i = 1, \qquad \lambda_i \geq 0, \quad i = 1, \ldots, 3. \qquad (5.30)$$

Eliminating (for instance) λ_3 and setting $\lambda_1 = \xi$ and $\lambda_2 = \eta$ yields

$$x = \sum_{|\alpha|=2} \Psi_\alpha(\xi, \eta, 1 - \xi - \eta)p^\alpha, \qquad (5.31)$$

which defines a nonlinear mapping $F_j : K' \to \tilde{K}$ of the reference triangle

$$K' = \left\{ \begin{pmatrix} \xi \\ \eta \end{pmatrix} : \xi \geq 0, \ \eta \geq 0, \ \xi + \eta \leq 1 \right\}$$

onto some approximation $\tilde{K}$ of $K = \overline{\Omega}_j$.

For quadratic interpolation we know already the local basis functions Ψ_α:

$$\Psi_{200}(\underline{\lambda}) = \lambda_1(2\lambda_1 - 1), \qquad \Psi_{110}(\underline{\lambda}) = 4\lambda_1\lambda_2,$$

$$\Psi_{020}(\underline{\lambda}) = \lambda_2(2\lambda_2 - 1), \qquad \Psi_{011}(\underline{\lambda}) = 4\lambda_2\lambda_3,$$

$$\Psi_{002}(\underline{\lambda}) = \lambda_3(2\lambda_3 - 1), \qquad \Psi_{101}(\underline{\lambda}) = 4\lambda_1\lambda_3.$$

Thus the functions used for the transformation of the reference element belong to the same class as the functions used in the quadratic finite element; we therefore say that this approach is *isoparametric*.

The mapping (5.31) maps the reference triangle K' onto a "triangle" K that in general has three curved edges; see Figure 4.22. Note that for the approximation of the boundary it usually suffices to have one curved edge.

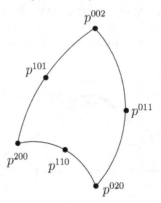

Figure 4.22 Isoparametric element K

Consider for example the curved edge with vertices p^{200} and p^{020}. This edge can be described by

$$\mathcal{C} = \left\{ x(\xi) = \xi(2\xi - 1)p^{200} + 4\xi(1 - \xi)p^{110} + (1 - \xi)(1 - 2\xi)p^{020} \; : \xi \in [0, 1] \right\}.$$

How does one choose p^{200}, p^{020} and p^{110}? The points p^{200} and p^{020} are the vertices of where Ω_j (say) intersects the boundary Γ of Ω. One way to select p^{110} is to take the point where the perpendicular bisector of the line segment joining the vertices p^{200} and p^{020} intersects Γ.

Let us assume that the mapping F_j of the reference triangle K' onto each element Ω_j is defined. Then, using the local basis functions v on K', the basis functions u on Ω_j are defined by

$$u(x) = v(F_j^{-1}(x)), \qquad x \in \Omega_j.$$

The basis functions on Ω_j do have a complicated structure (if, e.g., one has two quadratic equations in two unknowns, then solving for these unknowns does not yield a simple formula); furthermore, these expressions must apparently be inserted in the local basis functions on the reference element. But an explicit knowledge of these basis functions is not necessary: basis functions on the reference element and the mapping F_j allow us to compute the quantities that we need.

For the theory of isoparametric finite elements, see [Cia78] and the work of Lenoir [83]. Lenoir describes a practical procedure for the triangulation of n-dimensional domains by isoparametric simplicial elements that preserve the optimal order of accuracy. See also [14], where a general interpolation theory is presented that includes curved elements that fit the boundary exactly.

Exercise 4.58. Consider an admissible triangulation $\{T\}$ of a two-dimensional polygonal domain Ω with $h = \max_{T \in h}\{diam\ T\}$.
Let $Q_h(u)$ be a quadrature rule for $J(u) = \int_\Omega u dx$ that is elementwise exact for piecewise polynomials P_l of degree l. Assume that $u \in C^m(\bar{\Omega})$.
Estimate the quadrature error $|J(u) - Q_h(u)|$ as a function of l and m.

Exercise 4.59. Construct as simple as possible a quadrature rule that is P_2-exact on quadrilaterals
a) under the condition that the vertices of the quadrilateral are nodes of the rule
b) without condition a).

Exercise 4.60. Discretize the boundary value problem

$$-u'' + a(x)u' + b(x)u = 0 \quad \text{in } (0,1), \quad u(0) = \alpha, \quad u(1) = \beta,$$

using linear finite elements on an equidistant mesh. Discuss the structure, consistency and stability of the resulting difference scheme if the integrals are computed using each of the following quadrature formulas:

a) midpoint rule $\int_0^1 u(t)dt \approx u(0.5)$,

b) one-sided rectangular rule $\int_0^1 u(t)dt \approx u(0)$,

c) trapezoidal rule $\int_0^1 u(t)dt \approx [u(0) + u(1)]/2$,

d) Simpson's rule $\int_0^1 u(t)dt \approx [u(0) + 4u(0.5) + u(1)]/6$.

Exercise 4.61. Assume that $f \in W_q^{k+1}(\Omega) \subset C(\bar{\Omega})$. Let $\sum \hat{\omega}_\nu \hat{f}(\hat{b}_\nu)$ be a quadrature rule for $\int_{\hat{T}} \hat{f}(\hat{x})d\hat{x}$ that is exact on $P_k(\hat{T})$. Then by affine transformations we have also the quadrature rule $J_h(f) = \sum_T \sum_\nu \omega_{T\nu} f(b_{T\nu})$ on a given triangular mesh $\{T\}$. Use the Bramble-Hilbert lemma to prove that

$$\left| \int_\Omega f(x)dx - J_h(f) \right| \leq Ch^{k+1}|f|_{k+1,q}.$$

Exercise 4.62. On a given triangular mesh, begin with the quadrature rule that uses the vertices of a triangle then improve it by on subdividing each triangle into four congruent subtriangles, applying the quadrature rule on each, then performing one extrapolation step. What new quadrature rule is generated?

Exercise 4.63. Consider the L_2 error when taking account of numerical integration in our standard problem: find $u \in V \subset H$ such that $a(u,v) = f(v)$.
a) Prove: If φ_g is defined by $a(v, \varphi_g) = (g, v)_H$ with $H = L_2(\Omega)$, then

$$|u - u_h|_0 \leq \sup_{g \in H} \frac{1}{|g|_0} \inf_{\varphi_h \in V_h} \{M\|u - u_h\|\,\|\varphi_g - \varphi_h\|$$

$$+ |a(u_h, \varphi_h) - a_h(u_h, \varphi_h)| + |f(\varphi_h) - f_h(\varphi_h)|\},$$

where $u_h \in V_h$ is the discrete solution obtained using $a_h(\cdot, \cdot)$ and $f_h(\cdot)$.
b) Consider the boundary value problem

$$-\triangle u = f \quad \text{in} \quad \Omega, \qquad u|_\Gamma = 0,$$

in a convex polygonal domain. Suppose that we use a finite element discretization with numerical integration. What quadrature rules guarantee an optimal order of convergence in the L_2 norm?

Exercise 4.64. A domain Ω with a curved smooth boundary is approximated by a polygonal domain $\tilde{\Omega}_h$. Prove that this procedure is efficient for linear elements in the sense that in the H^1 norm one gets the same order of convergence as when the original domain is polygonal.

Exercise 4.65. Compute the local basis functions for the isoparametric quadratic triangular element whose knots are $\{(0,0), (0,1), (1,0), (\frac{1}{2}, \frac{1}{2}), (\frac{1}{4}, 0), (0, \frac{1}{4})\}$.

Exercise 4.66. Discretize $-\triangle u = 1$ in the unit circle with a homogeneous natural boundary condition, using only one isoparametric quadratic quadrilateral element.

4.6 Mixed Finite Elements

4.6.1 Mixed Variational Equations and Saddle Points

The finite element method for the numerical treatment of elliptic boundary value problems is based on the variational formulation of the given problem. This variational problem is discretized by the Ritz-Galerkin technique. In the standard approach, essential boundary conditions (and additional conditions such as the incompressibility requirement in the Navier-Stokes equations) are treated explicitly as restrictions. These restrictions influence the definition of the underlying space V and consequently have to be taken into account when constructing the discrete space V_h if the method is conforming.

The theory of optimization problems in functional analysis opens the way to a different technique: restrictions can be taken into account by introducing *Lagrange multipliers*. Convex analysis, and in particular duality theory, provide us with the necessary tools; see [ET76], and for a detailed discussion of mixed methods [BF91].

To explain the basic ideas, let us consider an abstract model problem. Let V and W be two real Hilbert spaces with respective scalar products $(\cdot, \cdot)_V$ and $(\cdot, \cdot)_W$ and induced norms $\| \cdot \|_V$ and $\| \cdot \|_W$. When there is no danger of misunderstanding we shall omit these subscripts.

Assume that $a(\cdot, \cdot) : V \times V \to \mathbb{R}$ and $b(\cdot, \cdot) : V \times W \to \mathbb{R}$ are given continuous bilinear forms such that

$$|a(u,v)| \le \alpha \|u\| \|v\| \qquad \text{for all } u, v \in V,$$
$$|b(v,w)| \le \beta \|v\| \|w\| \qquad \text{for all } v \in V, \ w \in W, \tag{6.1}$$

with some constants $\alpha > 0$ and $\beta > 0$. Let us set

$$Z = \{\, v \in V \ : \ b(v,w) = 0 \ \text{ for all } w \in W \,\}. \tag{6.2}$$

As $b(\cdot, \cdot)$ is a continuous bilinear form, it follows that $Z \subset V$ is a closed linear subspace of V. Consequently, equipped with the scalar product $(\cdot, \cdot)_V$, the set Z is itself a Hilbert space. Assume that the bilinear form $a(\cdot, \cdot)$ is Z-elliptic, i.e., that there exists a constant $\gamma > 0$ such that

$$\gamma \|z\|_V^2 \le a(z,z) \qquad \text{for all } z \in Z. \tag{6.3}$$

Let $f \in V^*$ and $g \in W^*$ be two linear functionals. Set

$$G = \{\, v \in V \ : \ b(v,w) = g(w) \ \text{ for all } w \in W \,\}. \tag{6.4}$$

Now we are able to formulate our *abstract model problem*: Find $u \in G$ such that

$$a(u,z) = f(z) \qquad \text{for all } z \in Z. \tag{6.5}$$

We say that this problem is a *variational equation with constraints*.

First we present an existence result:

Lemma 4.67. *Assume that G is nonempty. Then the model problem (6.5) has a unique solution $u \in G$, and this solution satisfies the bound*

$$\|u\| \le \frac{1}{\gamma} \|f\|_{V^*} + \left(\frac{\alpha}{\gamma} + 1 \right) \|v\|_V \quad \text{for all } v \in G. \tag{6.6}$$

Proof: Fix an arbitrary element $v \in G$. The bilinearity of $b(\cdot, \cdot)$, (6.2) and (6.4) imply that

$$v + z \in G \quad \text{for all } z \in Z.$$

On the other hand, any $u \in G$ can be written as

$$u = v + \tilde{z} \tag{6.7}$$

for some $\tilde{z} \in Z$. Hence problem (6.5) is equivalent to: find $\tilde{z} \in Z$ such that

$$a(\tilde{z}, z) = f(z) - a(v, z) \qquad \text{for all } z \in Z. \tag{6.8}$$

Our assumptions allow us to apply the Lax-Milgram lemma to (6.8). Consequently (6.8) has a unique solution $\tilde{z}$ and the same is true of problem (6.5).

Taking $z = \tilde{z}$ in (6.8), it follows from (6.1) and (6.3) that

$$\|\tilde{z}\| \le \frac{1}{\gamma} (\|f\|_{V^*} + \alpha \|v\|_V).$$

By (6.7) and the triangle inequality we then get (6.6). ∎

Example 4.68. Let us consider the elliptic boundary value problem

$$-\Delta u = f \quad \text{in } \Omega, \qquad u|_\Gamma = g, \tag{6.9}$$

where as usual Γ denotes the boundary of Ω. Setting $V = H^1(\Omega)$ and $W = L_2(\Gamma)$, define the following forms:

$$a(u, v) = \int_\Omega \nabla u \nabla v \, dx, \qquad f(v) = \int_\Omega fv \, dx,$$

$$b(v, w) = \int_\Gamma vw \, ds, \qquad g(w) = \int_\Gamma gw \, ds.$$

Then $Z = H_0^1(\Omega)$ and

$$G = \left\{ v \in H^1(\Omega) : \int_\Gamma vw \, ds = \int_\Gamma gw \, ds \quad \text{for all } w \in L_2(\Gamma) \right\}.$$

The Dirichlet conditions in problem (6.9) can also be written as

$$u \in V, \quad Tu = g,$$

where $T : V \to H^{1/2}(\Gamma) \subset L_2(\Gamma)$ is the trace operator of Section 3.2. The condition $G \neq \emptyset$ of Lemma 4.67 is equivalent to $g \in H^{1/2}(\Gamma)$. Thus it is more appropriate to choose $W = H^{-1/2}(\Gamma)$ instead of $W = L_2(\Gamma)$. $\square$

Next, we associate our constrained variational equation with the following extended variational equation *without constraints*:
Find a pair $(u, p) \in V \times W$ such that

$$\begin{aligned}
a(u, v) + b(v, p) &= f(v) \quad \text{for all } v \in V, \\
b(u, w) \qquad\quad &= g(w) \quad \text{for all } w \in W.
\end{aligned} \tag{6.10}$$

This system is also known as the *mixed variational equation*. To clarify the relationship between the constrained problem and the mixed formulation, we first have

Lemma 4.69. *If $(u, p) \in V \times W$ is a solution of the mixed variational problem (6.10), then u is a solution of the constrained variational equation (6.5).*

Proof: The second part of the variational equation (6.10) is equivalent to $u \in G$.

If we choose $v \in Z \subset V$, then by (6.2) one obtains $b(v, p) = 0$ for all $p \in W$. The first part of (6.10) therefore yields

$$a(u, v) = f(v) \qquad \text{for all } v \in Z.$$

In summary, u is a solution of (6.5). ∎

To gain more insight, we now investigate the following question: given a solution $u \in G$ of the constrained problem (6.5), must there exist $p \in W$ such that the pair $(u, p) \in V \times W$ is a solution of (6.10)?

The orthogonal complement $Z^{\perp}$ of Z with respect to the scalar product in V is

$$Z^{\perp} := \{ v \in V \; : \; (v, z) = 0 \quad \text{for all } z \in Z \}.$$

Now $Z^{\perp}$ is a closed linear subspace of V, and the Hilbert space V can be decomposed as a direct sum:

$$V = Z \oplus Z^{\perp}.$$

But $u \in G$ and (6.5) holds, so $(u, p) \in U \times W$ is a solution of the system (6.10) if and only if

$$b(v, p) = f(v) - a(u, v) \qquad \text{for all } v \in Z^{\perp}. \tag{6.11}$$

As $b(v, \cdot) \in W^*$ for each $v \in V$, we can define a linear continuous operator $B : V \to W^*$ by setting

$$\langle Bv, w \rangle = b(v, w) \qquad \text{for all } v \in V, \; w \in W. \tag{6.12}$$

In terms of B we have the following existence result for (6.11):

Lemma 4.70. *Suppose that for some constant $\delta > 0$ the operator B satisfies the bound*

$$\|Bv\|_{W^*} \geq \delta \, \|v\|_V \qquad \text{for all } v \in Z^{\perp}. \tag{6.13}$$

Then there exists a solution $p \in W$ of the variational problem (6.11).

Proof: Let $j : W^* \to W$ denote the Riesz representation operator. Then we define a continuous symmetric bilinear form $d : V \times V \to \mathbb{R}$ by

$$d(s, v) := \langle Bs, jBv \rangle = (jBs, jBv) \qquad \text{for all } s, \, v \in V. \tag{6.14}$$

The hypothesis (6.13) implies that

$$d(v, v) = \|jBv\|_W^2 = \|Bv\|_{W^*}^2 \geq \delta^2 \|v\|^2 \qquad \text{for all } v \in Z^{\perp},$$

i.e., $d(\cdot, \cdot)$ is $Z^{\perp}$-elliptic. By the Lax-Milgram lemma there exists $y \in Z^{\perp}$ such that

$$d(y, v) = f(v) - a(u, v) \qquad \text{for all } v \in Z^{\perp}.$$

Hence $p \in W$ defined by $p := jBy$ is a solution of (6.11), as can be seen by combining the relationships already established. ∎

Remark 4.71. The condition (6.13) requires the operator $B : V \to W^*$ to have closed range. The well-known *closed range theorem* for linear operators (see [Yos66]) is the abstract basis for the investigation above. It can be shown that condition (6.13) is equivalent to

$$\sup_{v \in V} \frac{b(v, w)}{\|v\|} \geq \delta \|w\| \qquad \text{for all } w \in Y^\perp \tag{6.15}$$

and some constant $\delta > 0$, where $Y^\perp$ is the orthogonal complement of

$$Y := \{ z \in W : b(v, z) = 0 \quad \text{for all } v \in V \}.$$

□

Next we present an important existence and stability result for the system of variational equations (6.10).

Theorem 4.72. *Let G be nonempty. Assume that the condition (6.15) is satisfied. Then the pair of variational equations (6.10) has at least one solution $(u, p) \in V \times W$. The first component $u \in V$ is uniquely determined, and the following estimates are valid:*

$$\|u\|_V \leq \frac{1}{\gamma} \|f\|_{V^*} + \frac{1}{\delta}\left(\frac{\alpha}{\gamma} + 1\right) \|g\|_{W^*} \tag{6.16}$$

and

$$\inf_{y \in Y} \|p + y\|_W \leq \frac{1}{\delta}\left(\frac{\alpha}{\gamma} + 1\right) \|f\|_{V^*} + \frac{\alpha}{\delta^2}\left(\frac{\alpha}{\gamma} + 1\right) \|g\|_{W^*}. \tag{6.17}$$

Proof: From Lemma 4.67 we know that the constrained variational equation (6.5) has a unique solution $u \in G$. As the conditions (6.13) and (6.15) are equivalent, Lemma 4.70 implies existence of a $p \in W$ such that the pair $(u, p) \in V \times W$ is a solution of the system (6.10).

By Lemma 4.69 the first component of any solution pair of (6.10) is also a solution of (6.5), so u is uniquely determined.

Now $u \in V$ can be written in a unique way as

$$u = \tilde{z} + \tilde{v}$$

with $\tilde{z} \in Z$ and $\tilde{v} \in Z^\perp$. From the bilinearity of $b(\cdot, \cdot)$ and (6.2), (6.12) and (6.13), one sees that

$$\|g\|_{W^*} = \|Bu\|_{W^*} = \|B\tilde{v}\|_{W_*} \geq \delta \|\tilde{v}\|_V.$$

That is, $\|\tilde{v}\| \leq \frac{1}{\delta} \|g\|_{W^*}$. But $u \in G$ implies $\tilde{v} \in G$, so Lemma 4.67 yields the estimate

$$\|u\| \leq \frac{1}{\gamma} \|f\|_{V^*} + \frac{1}{\delta}\left(\frac{\alpha}{\gamma} + 1\right) \|g\|_{W^*}.$$

Next, consider the second component $p \in W$, which satisfies the variational equation (6.11). The definition of $Z^\perp$ implies that

$$|b(v,p)| \leq |f(v)| + |a(u,v)| \leq (\|f\|_{V^*} + \alpha\|u\|)\|v\| \quad \text{for all } v \in Z^\perp.$$

Consequently, using the definition of Y and recalling that $V = Z \oplus Z^\perp$, for all $y \in Y$ we get

$$\sup_{v \in V} \frac{b(v, p+y)}{\|v\|} = \sup_{v \in V} \frac{b(v,p)}{\|v\|} \leq \|f\|_{V^*} + \alpha\|u\|.$$

Hence (6.15) gives

$$\inf_{y \in Y} \|p + y\|_W \leq \frac{1}{\delta}\|f\|_* + \frac{\alpha}{\delta}\|u\|.$$

The earlier estimate (6.16) now yields (6.17). ∎

Remark 4.73. When $Y = \{0\}$, the solution $(u, p) \in V \times W$ of (6.10) is unique. In this case both (6.16) and (6.17) are stability estimates that exhibit the influence of perturbations of f and g on the solution pair (u, p).   □

In the case of a Z-elliptic symmetric bilinear form $a(\cdot, \cdot)$ condition (6.5) is a necessary and sufficient condition to guarantee that $u \in G$ solves the variational problem

$$\min_{v \in G} J(v), \quad \text{where } J(v) := \frac{1}{2}a(v,v) - f(v). \tag{6.18}$$

Recalling (6.4) and taking into account the reflexivity property $W^{**} = W$ enjoyed by Hilbert spaces, the Lagrange functional associated with (6.18) is

$$L(v,w) = J(v) + b(v,w) - g(w) \quad \text{for all } v \in V, \ w \in W. \tag{6.19}$$

This functional allows us to state a sufficient condition for optimality in the form of a saddle point criterion. A pair $(u, p) \in V \times W$ is called a *saddle point* of the Lagrange functional $L(\cdot, \cdot)$ if

$$L(u,w) \leq L(u,p) \leq L(v,p) \quad \text{for all } v \in V, \ w \in W. \tag{6.20}$$

Saddle points are related to the solution of problem (6.18) in the following way.

Lemma 4.74. *Let $(u, p) \in V \times W$ be a saddle point of the Lagrange functional $L(\cdot, \cdot)$. Then $u \in V$ is a solution of (6.18).*

Proof: First we verify that $u \in G$. The first inequality in (6.20) implies that

$$b(u, w) - g(w) \leq b(u, p) - g(p) \qquad \text{for all } w \in W. \tag{6.21}$$

As W is a linear space it follows that

$$w := p + y \in W \qquad \text{for all } y \in W.$$

Now (6.21) and the linearity of $b(u, \cdot)$ and $g(\cdot)$ yield the inequality

$$b(u, y) - g(y) \leq 0 \qquad \text{for all } y \in W.$$

But $y \in W$ implies $-y \in W$, so again appealing to linearity we must have

$$b(u, y) - g(y) = 0 \qquad \text{for all } y \in W,$$

i.e., $u \in G$.

The second inequality in the saddle point definition (6.20) and the specification (6.4) of the set G of admissible solutions give us

$$\begin{aligned}
J(u) = J(u) + b(u, p) - g(u) &= L(u, p) \\
&\leq L(v, p) \\
&= J(v) + b(v, p) - g(p) = J(v) \qquad \text{for all } v \in G.
\end{aligned}$$

That is, u is a solution of (6.18). ∎

Using (6.20) and the inequality

$$\sup_{w \in W} \inf_{v \in V} L(v, w) \leq \inf_{v \in V} \sup_{w \in W} L(v, w) \tag{6.22}$$

one can associate a further variational problem with (6.18). Let us define

$$\underline{L}(w) := \inf_{v \in V} L(v, w) \qquad \text{for all } w \in W \tag{6.23}$$

and

$$\overline{L}(v) := \sup_{w \in W} L(v, w) \qquad \text{for all } v \in V. \tag{6.24}$$

The functionals $\underline{L} : W \to \overline{\mathbb{R}}$ and $\overline{L} : V \to \overline{\mathbb{R}}$ defined in this way are permitted to take the values "$-\infty$" and "$+\infty$". Thus it is necessary to use as image space the extended real numbers $\overline{\mathbb{R}} := \mathbb{R} \cup \{-\infty\} \cup \{+\infty\}$ instead of $\mathbb{R}$. It is not a problem to equip $\overline{\mathbb{R}}$ with an appropriate arithmetic; see [ET76, Zei90], where the interested reader will also find detailed results on the duality theory in general.

The definitions of $\overline{L}(\cdot)$, $L(\cdot, \cdot)$ and G imply that

$$\overline{L}(v) = \begin{cases} J(v) & \text{if } v \in G, \\ \infty & \text{otherwise.} \end{cases}$$

Consequently the formally unconstrained variational problem

$$\inf_{v \in V} \overline{L}(v) \tag{6.25}$$

is equivalent to (6.18). We associate with (6.25) the optimization problem

$$\sup_{w \in W} \underline{L}(w), \tag{6.26}$$

which we call the *dual variational problem* of (6.25)—and hence, also of (6.18).

When (6.25) and (6.26) are discretized with a conforming finite element method, inequality (6.22) immediately yields

Lemma 4.75. *Let $V_h \subset V$ and $W_h \subset W$. Then*

$$\sup_{w_h \in W_h} \underline{L}(w_h) \le \sup_{w \in W} \underline{L}(w) \le \inf_{v \in V} \overline{L}(v) \le \inf_{v_h \in V_h} \overline{L}(v_h).$$

In the particular case of the Poisson equation, Trefftz [114] proposed a conforming discretization method for the dual problem (6.26) in which the computation of the functional $\underline{L}(\cdot)$ was based on the Green's function of the given problem. The use here of Ritz's method and the bounds of Lemma 4.75 allow one to enclose the error to a desired degree of exactness.

An interesting application of the bounds provided by Lemma 4.75 for the optimal value of the variational problem (6.18) is their use in defining error estimators for the Ritz method in the case of symmetric bilinear forms (in Section 4.7 we shall discuss the general case of bilinear forms that are not necessarily symmetric). Chapter 11.4 of [NH81] contains a detailed discussion of this technique, which originated with Synge in 1957 and is sometimes called the "hypercircle".

4.6.2 Conforming Approximation of Mixed Variational Equations

We now discuss the Galerkin finite element method discretization of the mixed variational equation (6.10), which is restated here:

$$
\begin{aligned}
a(u,v) + b(v,p) &= f(v) \quad \text{for all } v \in V, \\
b(u,w) &= g(w) \quad \text{for all } w \in W.
\end{aligned}
$$

Choose finite element spaces $V_h \subset V$ and $W_h \subset W$. Then the discrete problem is:
Find $(u_h, p_h) \in V_h \times W_h$ such that

$$
\begin{aligned}
a(u_h, v_h) + b(v_h, p_h) &= f(v_h) && \text{for all } v_h \in V_h, \\
b(u_h, w_h) &= g(w_h) && \text{for all } w_h \in W_h.
\end{aligned}
\tag{6.27}
$$

This finite element method for (6.10), which is also a finite element method for (6.5), is called the *mixed finite element method*.

We study the solvability of (6.27) and the stability of its solutions, as was done already in the continuous case. Set

$$G_h = \{\, v_h \in V_h \ : \ b(v_h, w_h) = g(w_h) \quad \text{for all } w_h \in W_h \,\}$$

and

$$Z_h = \{\, v_h \in V_h \ : \ b(v_h, w_h) = 0 \quad \text{for all } w_h \in W_h \,\}.$$

Note that although one has $V_h \subset V$ and $W_h \subset W$, nevertheless in general

$$G_h \not\subset G \qquad \text{and} \qquad Z_h \not\subset Z.$$

Consequently the Z-ellipticity of $a(\cdot, \cdot)$ does not automatically imply its Z_h-ellipticity. We therefore assume that a constant $\gamma_h > 0$ exists such that

$$\gamma_h \, \|v_h\|^2 \le a(v_h, v_h) \qquad \text{for all } v_h \in Z_h, \tag{6.28}$$

and that a constant $\delta_h > 0$ exists with

$$\sup_{v_h \in V_h} \frac{b(v_h, w_h)}{\|v_h\|} \ge \delta_h \, \|w_h\| \qquad \text{for all } w_h \in Y_h^\perp \tag{6.29}$$

where Y_h is defined by

$$Y_h := \{\, z_h \in W_h \ : \ b(v_h, z_h) = 0 \quad \text{for all } v_h \in V_h \,\}.$$

Under these assumptions it is clear that the argument of Theorem 4.72 can be carried over to the discrete problem (6.27), yielding

Theorem 4.76. *Let G_h be nonempty and assume that conditions (6.28) and (6.29) are satisfied. Then the mixed finite element discretization (6.27) has at least one solution $(u_h, p_h) \in V_h \times W_h$. The first component $u_h \in V_h$ is uniquely determined, and the following estimates are valid:*

$$\|u_h\| \le \frac{1}{\gamma_h} \|f\|_* + \frac{1}{\delta_h}\left(\frac{\alpha}{\gamma_h} + 1\right) \|g\|_*, \tag{6.30}$$

$$\inf_{y_h \in Y_h} \|p_h + y_h\| \le \frac{1}{\delta_h}\left(\frac{\alpha}{\gamma_h} + 1\right) \|f\|_* + \frac{\alpha}{\delta_h^2}\left(\frac{\alpha}{\gamma_h} + 1\right) \|g\|_*. \tag{6.31}$$

To simplify the arguments we next study the convergence behaviour of the mixed finite element method in the case $Y_h = \{0\}$. Then one can apply the methodology of Section 3.4 to analyse the Ritz-Galerkin method, obtaining the following result:

Theorem 4.77. *Assume that the continuous problem (6.10) and the discrete problem (6.27) satisfy the assumptions of Theorem 4.72 and 4.76, respectively.*

Moreover, let the conditions (6.28) and (6.29) hold uniformly with respect to h, i.e., there exist constants $\tilde{\gamma} > 0$ and $\tilde{\delta} > 0$ such that

$$\tilde{\gamma}\|v_h\|^2 \leq a(v_h, v_h) \qquad \text{for all } v_h \in Z_h \tag{6.32}$$

and

$$\sup_{v_h \in V_h} \frac{b(v_h, w_h)}{\|v_h\|} \geq \tilde{\delta}\|w_h\| \qquad \text{for all } w_h \in W_h. \tag{6.33}$$

Then the mixed finite element method satisfies the following error estimate: there exists a constant $c > 0$ such that

$$\max\{\|u - u_h\|, \|p - p_h\|\} \leq c \left\{ \inf_{v_h \in V_h} \|u - v_h\| + \inf_{w_h \in W_h} \|p - w_h\| \right\}. \tag{6.34}$$

Proof: To begin, observe that the hypothesis (6.33) guarantees that $Y_h = \{0\}$. Analogously to the proof of Lemma 3.45, for arbitrary $\tilde{v}_h \in V_h$ and $\tilde{w}_h \in W_h$ one gets

$$a(u_h - \tilde{v}_h, v_h) + b(v_h, p_h - \tilde{w}_h) = a(u - \tilde{v}_h, v_h) + b(v_h, p - \tilde{w}_h) \quad \forall v_h \in V_h,$$
$$b(u_h - \tilde{v}_h, w_h) = b(u - \tilde{v}_h, w_h) \qquad\qquad \forall w_h \in W_h.$$

Theorem 4.76 then implies, since $Y_h = \{0\}$, that

$$\|u_h - \tilde{v}_h\| \leq \frac{\alpha}{\tilde{\delta}}\|u - \tilde{v}_h\| + \frac{\beta}{\tilde{\delta}}\left(\frac{\alpha}{\tilde{\gamma}} + 1\right)\|p - \tilde{w}_h\|$$

and

$$\|p_h - \tilde{w}_h\| \leq \frac{\alpha}{\tilde{\delta}}\left(\frac{\alpha}{\tilde{\gamma}} + 1\right)\|u - \tilde{v}_h\| + \frac{\alpha\beta}{\tilde{\delta}^2}\left(\frac{\alpha}{\tilde{\gamma}} + 1\right)\|p - \tilde{w}_h\|$$

for arbitrary $\tilde{v}_h \in V_h$ and $\tilde{w}_h \in W_h$. The triangle inequalities

$$\|u - u_h\| \leq \|u - \tilde{v}_h\| + \|u_h - \tilde{v}_h\|$$

and

$$\|p - p_h\| \leq \|p - \tilde{w}_h\| + \|p_h - \tilde{w}_h\|,$$

now give us the estimate (6.34). ■

Remark 4.78. The constant c in (6.34) can easily be determined explicitly in terms of the constants α, β, $\tilde{\gamma}$ and $\tilde{\delta} > 0$ by tracing this dependence through the the proof. □

Remark 4.79. The condition (6.33) is called the *Babuška-Brezzi condition* or *LBB condition* (where LBB stands for Ladyshenskaja-Babuška-Brezzi). It plays a fundamental role in the convergence analysis of mixed finite element methods. The Babuška-Brezzi condition requires the discrete spaces V_h and W_h to be compatible with respect to the bilinear form $b(\cdot,\cdot)$. Roughly speaking, (6.33) says that the space V_h (for the primal variables) has to be sufficiently rich while the space W_h (for the dual variables) should not be so large that it imposes too many constraints.

In particular (6.33) guarantees that $G_h \neq \emptyset$.   □

Remark 4.80. If instead of the Babuška-Brezzi condition (6.33) one has only

$$\sup_{v_h \in V_h} \frac{b(v_h, w_h)}{\|v_h\|} \geq \delta_h \|w_h\| \qquad \text{for all } w_h \in W_h$$

for some $\delta_h > 0$ with $\lim_{h \to 0} \delta_h = 0$, then the proof of Theorem 4.77 shows that the primal component u_h converges to u if

$$\lim_{h \to 0} \left[\frac{1}{\delta_h} \inf_{w_h \in W_h} \|p - w_h\| \right] = 0.$$

The stronger condition

$$\lim_{h \to 0} \frac{1}{\delta_h} \left[\inf_{v_h \in V_h} \|u - v_h\| + \frac{1}{\delta_h} \inf_{w_h \in W_h} \|p - w_h\| \right] = 0$$

implies the convergence of both components, viz.,

$$\lim_{h \to 0} \|u - u_h\| = 0 \qquad \text{and} \qquad \lim_{h \to 0} \|p - p_h\| = 0.$$

□

Next we apply our theory to the important example of the Stokes problem, examining its mixed finite element discretization and making a special choice of the spaces V_h and W_h in order to satisfy the Babuška-Brezzi condition.

Let $\Omega \subset R^2$ be a given polygonal domain. Set

$$V = H_0^1(\Omega) \times H_0^1(\Omega) \qquad \text{and} \qquad W = \left\{ w \in L_2(\Omega) : \int_\Omega w \, dx = 0 \right\},$$

with $\|\underline{v}\|_V = \|v_1\|_1 + \|v_2\|_1$ for $\underline{v} = (v_1, v_2) \in V$ and $\|\cdot\|_W = \|\cdot\|_0$. Then W is a closed subspace of $L_2(\Omega)$ and is itself a Hilbert space. We introduce

$$a(\underline{u}, \underline{v}) := \sum_{i=1}^{2} \int_\Omega \nabla u_i \nabla v_i \, dx \qquad \text{for all } \underline{u}, \underline{v} \in V, \tag{6.35}$$

where $\underline{u} = (u_1, u_2)$ and $\underline{v} = (v_1, v_2)$. Set

$$b(\underline{v}, w) := - \int_{\Omega} w \operatorname{div} \underline{v} \, dx \qquad \text{for all } \underline{v} \in V, \ w \in W.$$

Then

$$G := \left\{ \underline{v} \in V : \int_{\Omega} w \operatorname{div} \underline{v} \, dx = 0 \quad \text{for all } w \in W \right\}. \tag{6.36}$$

The homogeneity of the constraints implies that $Z = G$. Finally, given functions f_1 and $f_2 \in L_2(\Omega)$, define the functional $f : V \to R$ by

$$f(\underline{v}) = \sum_{i=1}^{2} \int_{\Omega} f_i v_i \, dx \qquad \text{for all } \underline{v} \in V.$$

The Stokes problem can now be formulated as a constrained variational equation:
Find $\underline{u} \in G$ such that

$$a(\underline{u}, \underline{v}) = f(\underline{v}) \qquad \text{for all } \underline{v} \in Z. \tag{6.37}$$

Written out in full, the mixed formulation of (6.37) is

$$\sum_{i=1}^{2} \int_{\Omega} \nabla u_i \nabla v_i \, dx - \int_{\Omega} p \operatorname{div} \underline{v} \, dx = \sum_{i=1}^{2} \int_{\Omega} f_i v_i \, dx \qquad \text{for all } \underline{v} \in V,$$

$$- \int_{\Omega} w \operatorname{div} \underline{u} \, dx = 0 \qquad \text{for all } w \in W.$$

If the solution is sufficiently regular so that we can integrate by parts, then it is not difficult to return to the classical formulation of the Stokes problem:

$$-\Delta \underline{u} + \nabla p = \underline{f} \ \text{ in } \Omega,$$

$$\operatorname{div} \underline{u} = 0 \ \text{ in } \Omega,$$

$$\underline{u}|_{\Gamma} = 0.$$

Here the primal variables $\underline{u} = (u_1, u_1)$ correspond to the velocity, while the dual variable p is simply the pressure. Thus the mixed formulation is not merely a mathematical construct but reflects a physically natural model; in particular the dual variable has a concrete physical interpretation.

Now assume, for simplicity, that the given domain is rectangular so we can use a uniform rectangular mesh $\mathcal{Z}_h = \{ \Omega_i : i = 1, \ldots, M \}$ with mesh width h. Choose piecewise biquadratic functions for the approximation of the velocity $\underline{u}$ and piecewise constant functions for the pressure p, i.e.,

$$V_h := \left\{ \underline{v}_h \in C(\overline{\Omega})^2 : \underline{v}_h|_{\Omega_i} \in [Q_2(\overline{\Omega}_i)]^2, \ \Omega_i \in \mathcal{Z}_h \right\}$$

and

$$W_h := \left\{ \underline{w}_h \in W \; : \; \underline{w}_h|_{\Omega_i} \in P_0(\overline{\Omega}_i), \; \Omega_i \in \mathcal{Z}_h \right\}.$$

For brevity, this element pair is referred to as the $Q_2 - P_0$ *element*.

Prior to verifying the validity of the discrete Babuška-Brezzi condition (6.33), we show that condition (6.15) is satisfied for the continuous problem (6.37).

Let $w \in W$ be arbitrary but fixed. Using regularity it can be shown [Tem79, GR89] that there exists $\tilde{\underline{v}} \in V$ such that

$$-\operatorname{div} \tilde{\underline{v}} = w \qquad \text{and} \qquad \|\tilde{\underline{v}}\|_V \leq c \, \|w\|_W \tag{6.38}$$

for some constant $c > 0$. It follows that

$$\sup_{\underline{v} \in V} \frac{b(\underline{v}, w)}{\|\underline{v}\|_V} \geq \frac{b(\tilde{\underline{v}}, w)}{\|\tilde{\underline{v}}\|_V} = \frac{\|w\|_W^2}{\|\tilde{\underline{v}}\|_V} \geq \frac{1}{c} \, \|w\|_W.$$

Thus (6.15) is valid. Next we study the discrete Babuška-Brezzi condition for our chosen pair of finite element spaces.

Let $w_h \in W_h$ be arbitrary but fixed. As $W_h \subset W$, there exists $\tilde{\underline{v}} \in V$ with

$$-\operatorname{div} \tilde{\underline{v}} = w_h \qquad \text{and} \qquad \|\tilde{\underline{v}}\| \leq c \, \|w_h\|. \tag{6.39}$$

The bilinear form $a : V \times V \to R$ is V-elliptic and $V_h \subset V$, so there exists a unique $\tilde{\underline{v}}_h \in V_h$ defined by

$$a(\tilde{\underline{v}}_h, \underline{v}_h) = a(\tilde{\underline{v}}, \underline{v}_h) \qquad \text{for all } \underline{v}_h \in V_h.$$

By Cea's lemma

$$\|\tilde{\underline{v}}_h\| \leq c \, \|\tilde{\underline{v}}\|,$$

where the positive constant c is independent of $\tilde{\underline{v}}_h$ and $\tilde{\underline{v}}$. It follows that

$$\|\tilde{\underline{v}}_h\| \leq c \, \|w_h\|. \tag{6.40}$$

Next we define a special "interpolant" $\hat{\underline{v}}_h \in V_h$ by

$$\hat{\underline{v}}_h(x^i) = \tilde{\underline{v}}_h(x^i), \; i = 1, \dots, \overline{N},$$

$$\int_{\Omega_j} \hat{\underline{v}}_h \, dx = \int_{\Omega_j} \tilde{\underline{v}} \, dx, \; j = 1, \dots, M,$$

$$\int_{\Gamma_{jk}} \hat{\underline{v}}_h \, ds = \int_{\Gamma_{jk}} \tilde{\underline{v}} \, ds, \; j, k = 1, \dots, M. \tag{6.41}$$

Here x^i, $i = 1, \dots \overline{N}$ denote all grid points and $\Gamma_{jk} = \overline{\Omega}_j \bigcap \overline{\Omega}_k$ denotes any interior edge of our decomposition $\mathcal{Z}_h$. Remark that it is not difficult to show that the used nine degrees of freedom over each of the rectangles define uniquely the interpolant.

Thus, we have

$$\|w_h\|^2 = \int_\Omega w_h \operatorname{div} \underline{\tilde{v}} \, dx = \sum_{j=1}^M \int_{\Omega_j} w_h \operatorname{div} \underline{\tilde{v}} \, dx. \qquad (6.42)$$

Now $w_h \in W_h$ is constant on each subdomain Ω_j, so integration by parts yields

$$\int_{\Omega_j} w_h \operatorname{div} \underline{\tilde{v}} \, dx = \int_{\Gamma_j} w_h \, \underline{\tilde{v}} \cdot \underline{n}_j \, ds, \qquad j = 1, \dots, M, \qquad (6.43)$$

where $\underline{n}_j$ is the outer unit normal vector on $\Gamma_j := \partial \Omega_j$. But the trace of w_h on the boundary Γ_j is also constant; thus (6.41) implies that

$$\int_{\Gamma_j} w_h \, \underline{\tilde{v}} \cdot \underline{n}_j \, ds = \int_{\Gamma_j} w_h \, \underline{\tilde{v}}_h \cdot \underline{n}_j \, ds, \qquad j = 1, \dots, M.$$

Recalling (6.43), integration by parts yields

$$\int_{\Omega_j} w_h \operatorname{div} \underline{\tilde{v}} \, dx = \int_{\Omega_j} w_h \operatorname{div} \underline{\tilde{v}}_h \, dx, \qquad j = 1, \dots, M.$$

Hence (6.42) gives us

$$\|w_h\|^2 = \sum_{j=1}^M \int_{\Omega_j} w_h \operatorname{div} \underline{\tilde{v}}_h \, dx.$$

Invoking (6.39) and (6.40), we finally conclude that

$$\sup_{\underline{v}_h \in V_h} \frac{b(\underline{v}_h, w_h)}{\|\underline{v}_h\|} \geq \frac{b(\underline{\tilde{v}}_h, w_h)}{\|\underline{\tilde{v}}_h\|} = \frac{\|w_h\|^2}{\|\underline{\tilde{v}}_h\|} \geq c \|w_h\| \qquad \text{for all } w_h \in W_h$$

with some constant $c > 0$. That is, for the $Q_2 - P_0$ pair of spaces the Babuška-Brezzi condition is satisfied.

One should be aware that on the other hand the $Q_1 - P_0$ *element* is unstable; see, e.g., [Bra01].

The above discussion of the $Q_2 - P_0$ element is fairly representative: first, it is not trivial to verify the Babuška-Brezzi condition for a given pair of spaces, and second, this verification is often carried out with the tools used here for the $Q_2 - P_0$ element. For more details and other pairs of finite element spaces, see [BF91, Bra01, Sch97].

4.6.3 Weaker Regularity for the Poisson and Biharmonic Equations

The formulation of certain problems as mixed variational equations has two significant advantages: it permits a weakening of the smoothness requirements in the underlying function spaces (this is essential for higher-order differential equations), and sometimes the new variables introduced in the mixed formulation are just as important as the original variables and can be approximated better in the mixed formulation because they are treated as independent variables.

In this subsection we discuss two typical examples, the Poisson and biharmonic equations.

We start with the following boundary value problem for the Poisson equation:

$$-\Delta z = f \quad \text{in } \Omega, \qquad z|_\Gamma = 0, \tag{6.44}$$

where $\Omega \subset \mathbb{R}^n$ is bounded and has boundary Γ. Let us assume that its unique weak solution lies in $H_0^1(\Omega) \cap H^2(\Omega)$. Then

$$-\int_\Omega w \Delta z \, dx = \int_\Omega f w \, dx \qquad \text{for all } w \in L_2(\Omega). \tag{6.45}$$

Setting $\underline{u} := \nabla z$ it follows that $\underline{u} \in H(div; \Omega)$.

For the mixed formulation we choose $V := H(div; \Omega)$ and $W := L_2(\Omega)$. The norm in V is defined by

$$\|\underline{v}\|_V^2 := \sum_{i=1}^n \|v_i\|_{0,\Omega}^2 + \|\operatorname{div} \underline{v}\|_{0,\Omega}^2, \tag{6.46}$$

where $\underline{v} = (v_1, \dots, v_n)$. Next, introduce the continuous bilinear mapping $b : V \times W \to R$ defined by

$$b(\underline{v}, w) := \int_\Omega w \operatorname{div} \underline{v} \, dx \qquad \text{for all } \underline{v} \in V, \ w \in W. \tag{6.47}$$

Let us define

$$G := \left\{ \underline{v} \in V : b(\underline{v}, w) = -\int_\Omega f w \, dx \quad \text{for all } w \in W \right\}$$

and, correspondingly,

$$Z := \{ \underline{v} \in V : b(\underline{v}, w) = 0 \quad \text{for all } w \in W \}.$$

From the assumptions of the problem we see that if z is the solution of problem (6.44) then $\nabla z \in G$. Finally, the continuous bilinear form $a : V \times V \to R$ is given by

$$a(\underline{u}, \underline{v}) := \sum_{i=1}^{n} \int_{\Omega} u_i v_i \, dx \qquad \text{for all } \underline{u}, \underline{v} \in V.$$

The definitions (6.46) and (6.48) clearly imply that the bilinear form $a(\cdot, \cdot)$ is Z-elliptic.

Now we consider the constrained variational equation:
Find $\underline{u} \in G$ such that

$$a(\underline{u}, \underline{v}) = 0 \qquad \text{for all } \underline{v} \in Z. \tag{6.48}$$

To clarify the relationship between problems (6.44) and (6.48) we prove

Lemma 4.81. *If the solution z of problem (6.44) satisfies $z \in H_0^1(\Omega) \cap H^2(\Omega)$ then $\underline{u} := \nabla z$ is a solution of the constrained variational equality (6.48). Moreover, $\underline{u}$ is the only solution of (6.48) in G.*

Proof: We saw already that $\nabla z \in G$. We show that $\underline{u} := \nabla z$ is a solution of the variational equation (6.48). For integration by parts yields

$$a(\underline{u}, \underline{v}) = \int_{\Omega} \underline{v} \cdot \nabla z \, dx = \int_{\Gamma} z \underline{v} \cdot \underline{n} \, ds - \int_{\Omega} z \operatorname{div} \underline{v} \, dx \qquad \text{for all } \underline{v} \in Z.$$

But $z \in H_0^1(\Omega)$ forces

$$\int_{\Gamma} z \underline{v} \cdot \underline{n} \, ds = 0$$

and the definition of the space Z also gives

$$\int_{\Omega} z \operatorname{div} \underline{v} \, dx = 0.$$

That is, $\underline{u} = \nabla z$ is a solution of (6.48). Uniqueness of this solution follows from Lemma 4.67. ∎

The space $V = H(div; \Omega)$ and the constrained variational equation (6.48) allow the use of solutions of (6.44) that are weaker than their classical counterparts. Written out explicitly, the mixed form (6.48) is

$$\sum_{i=1}^{n} \int_{\Omega} u_i v_i \, dx + \int_{\Omega} z \operatorname{div} \underline{v} \, dx = 0 \qquad \text{for all } \underline{v} \in H(div; \Omega),$$

$$\int_{\Omega} w \operatorname{div} \underline{u} \, dx \qquad\qquad = -\int_{\Omega} fw \, dx \quad \text{for all } w \in L_2(\Omega). \tag{6.49}$$

Remark 4.82. The essential boundary condition $z|_{\Gamma} = 0$ of the given problem (6.44) is not explicitly stated in the mixed formulation (6.49). In the mixed approach this boundary condition is a natural boundary condition. □

Next we discuss a finite element method for the discretization of (6.49), assuming that the given domain $\Omega \subset \mathbb{R}^2$ is polygonal. For the characterization of finite element spaces with $V_h \subset V = H(div; \Omega)$ we recall Lemma 4.3. Let $\mathcal{Z}_h = \{\Omega_j\}_{j=1}^M$ be an admissible decomposition of Ω into triangles. Setting $\underline{v}_j := \underline{v}|_{\Omega_j}$, condition (2.8) of Lemma 4.3, viz.,

$$(\underline{v}_j - \underline{v}_k) \cdot \underline{n}_{jk} = 0 \qquad \text{for all } j, k = 1, \ldots, M \text{ with } \Gamma_{jk} \neq \emptyset, \qquad (6.50)$$

is sufficient for $\underline{v} \in H(div; \Omega)$. For the components v_1 and v_2 of the ansatz functions $\underline{v}_h$ on the triangulation $\mathcal{Z}_h$ we choose piecewise polynomials of degree l. Then the projection $\underline{v}_h \cdot \underline{n}$ on any edge is a polynomial of degree l. The ansatz functions are determined by prescribing corresponding values at interior points of the edges. For instance, for piecewise linear functions one fixes two degrees of freedom on every edge. The corresponding Lagrange basis functions can be computed; see, e.g., [GRT93].

We now describe simple ansatz functions $\underline{\varphi} : \mathbb{R}^2 \to \mathbb{R}^2$ of the form

$$\underline{\varphi}(x) = (\varphi_1(x), \varphi_2(x)) = (a + bx_1, c + bx_2) \qquad (6.51)$$

with coefficients $a, b, c \in \mathbb{R}$. If

$$\mathcal{G} := \{ x \in \mathbb{R}^2 \ : \ \alpha x_1 + \beta x_2 = \gamma \}$$

denotes an arbitrary straight line, then its normal is $\underline{n} = \begin{pmatrix} \alpha \\ \beta \end{pmatrix}$, so

$$\underline{n} \cdot \underline{\varphi}(x) = \alpha(a + bx_1) + \beta(c + bx_2) = constant \qquad \text{for all } x \in \mathcal{G}.$$

That is, the projection of the special ansatz (6.51) on any edge is a constant.

Denote by y^i, $i = 1, \ldots, s$, the midpoints of the edges of our triangulation $\mathcal{Z}_h$. Let $\underline{n}^i$, $i = 1, \ldots, s$, be a unit vector normal to the edge containing y^i. Define the ansatz functions $\underline{\varphi}_k : \Omega \to \mathbb{R}^2$ by the following conditions:

i) $\underline{\varphi}_k|_{\Omega_j}$ has the form (6.51)

ii) $[\underline{\varphi}_k \cdot \underline{n}^i]$ is continuous at y^i and it holds $[\underline{\varphi}_k \cdot \underline{n}^i](y^i) = \delta_{ik}$ for $i, k = 1, \ldots, s$.

For example, consider the triangle $K = \{ x \in \mathbb{R}_+^2 \ : \ x_1 + x_2 \leq 1 \}$ with the midpoints of its edges denoted by y^1, y^2, y^3. Then the basis functions just defined are given by

$$\underline{\varphi}_1(x) = (x_1, -1 + x_2)$$
$$\underline{\varphi}_2(x) = (\sqrt{2}x_1, \sqrt{2}x_2) \qquad \text{for all } x \in int\,K.$$
$$\underline{\varphi}_3(x) = (-1 + x_1, x_2)$$

It is easy to verify that

$$\underline{\varphi}_k \in H(div; \Omega), \qquad k = 1, \ldots, s.$$

Consequently,

$$V_h = \operatorname{span}\{\underline{\varphi}_k\}_{k=1}^s \qquad (6.52)$$

is a reasonable finite element space for the discretization (6.49) as $V_h \subset V$. The degrees of freedom are $\underline{v}_h \cdot \underline{n}(y^i)$, $i = 1, \dots, s$. To approximate the space W we simply set

$$W_h := \{\, w_h \in L_2(\Omega) \; : \; w_h|_{\Omega_j} \in P_0(\Omega_j)\,\}, \qquad (6.53)$$

i.e., W is approximated by piecewise constants.

When the spaces V_h and W_h from (6.52) and (6.53) are used in the mixed finite element method for the problem (6.49), one can prove: if the solution is sufficiently smooth and the mesh is quasi-uniform, then the errors in the L_2 norm are bounded by

$$\|u - u_h\| \le c\,h \qquad \text{and} \qquad \|z - z_h\| \le c\,h.$$

If instead standard piecewise linear approximations are used for both components of the vector functions in V_h, one can then prove that

$$\|u - u_h\| \le c\,h^2 \qquad \text{and} \qquad \|z - z_h\| \le c\,h.$$

For the proof of these error estimates and further examples of related finite element spaces see [BF91].

Remark 4.83. The latter convergence result is remarkable because it shows that the direct discretization of $u = \nabla z$ by the mixed finite element method yields a better approximation for the derivative ∇z than for the function z itself. When a standard finite element method is used to solve (6.44), the approximation of the solution z is usually more accurate than the approximation of the derivatives of z.   □

Next we consider, as an example of a higher-order differential equation, the following boundary value problem for the biharmonic equation:

$$\Delta^2 z = f \quad \text{in } \Omega, \qquad z|_\Gamma = \frac{\partial z}{\partial \underline{n}}\Big|_\Gamma = 0. \qquad (6.54)$$

The standard weak formulation (see Chapter 3.3) of this problem is based on the space $H^2(\Omega)$; thus any standard conforming finite element method must use elements in $C^1(\Omega)$ which are complicated to construct. Consequently it is preferable to apply nonconforming or mixed methods when solving (6.54).

The substitution $u := \Delta z$ transforms this fourth-order problem into a system of two second-order problems:

$$\begin{aligned} \Delta z &= u \\ \Delta u &= f \end{aligned} \quad \text{in } \Omega, \qquad z|_\Gamma = \frac{\partial z}{\partial \underline{n}}\Big|_\Gamma = 0.$$

The usual integration by parts of the first equation yields

$$\int_\Gamma v \frac{\partial z}{\partial \underline{n}}\, ds - \int_\Omega \nabla v \nabla z\, dx = \int_\Omega uv\, dx \qquad \text{for all } v \in H^1(\Omega).$$

Now the homogenous Neumann boundary condition for z implies that

$$\int_\Omega uv\, dx + \int_\Omega \nabla v \nabla z\, dx = 0 \qquad \text{for all } v \in H^1(\Omega).$$

At the next step, the second equation is also written in a weak form. Then, denoting the dual variable by p as in the abstract problem (6.10), the weakend mixed formulation of (6.54) is

Find a pair $(u, p) \in H^1(\Omega) \times H_0^1(\Omega)$ such that

$$
\begin{aligned}
\int_\Omega uv\, dx + \int_\Omega \nabla v \nabla p\, dx &= 0 && \text{for all } v \in H^1(\Omega), \\
\int_\Omega \nabla u \nabla w\, dx &= - \int_\Omega fw\, dx && \text{for all } w \in H_0^1(\Omega).
\end{aligned}
\tag{6.55}
$$

If $V := H^1(\Omega)$, $W := H_0^1(\Omega)$ and the bilinear forms $a : V \times V \to \mathbb{R}$ and $b : V \times W \to \mathbb{R}$ are defined by

$$a(u, v) := \int_\Omega uv\, dx \qquad \text{for all } u,\, v \in V$$

and

$$b(v, w) := \int_\Omega \nabla v \nabla w\, dx \qquad \text{for all } v \in V,\, w \in W,$$

then (6.55) is a special case of the abstract mixed variational equation (6.10).

Using the above transformation from (6.54) to (6.55) it is easy to prove

Lemma 4.84. *Let $z \in H_0^2(\Omega) \cap H^3(\Omega)$. Then $(u, p) := (\Delta z, z) \in V \times W$. If in addition z is a solution of the variational equation*

$$\int_\Omega \Delta z\, \Delta y\, dx = \int_\Omega fy\, dx \qquad \text{for all } y \in H_0^2(\Omega),$$

which is related to (6.54), then (u, p) is a solution of the mixed variational equation (6.55).

Unfortunately the bilinear form $a(\cdot, \cdot)$ fails to be Z-elliptic on the subspace

$$Z := \Big\{ v \in V : \int_\Omega \nabla v \nabla w\, dx = 0 \quad \text{for all } w \in W \Big\}.$$

Consequently our general theory, which assumes uniform Z_h-ellipticity, is inapplicable. Nevertheless, the invocation of inverse inequalities for specific finite element spaces allows one to prove

$$\gamma_h \|v_h\|^2 \leq a(v_h, v_h) \qquad \text{for all } v_h \in Z_h$$

and some $\gamma_h > 0$, but $\lim_{h \to 0} \gamma_h = 0$. This leads to suboptimal error estimates, which can be improved using special techniques: see, e.g., [105]. The property $\lim_{h \to 0} \gamma_h = 0$ means that the discrete problem is ill-posed and will require special solution strategies.

It is interesting to observe that the verification of the Babuška-Brezzi conditions for (6.55) is trivial: for $W \hookrightarrow V$ and $b(w, w)$ induces an equivalent norm on W, which yields

$$\sup_{v \in V} \frac{b(v, w)}{\|v\|} \geq \frac{b(w, w)}{\|w\|} \geq c \frac{\|w\|^2}{\|w\|} = c \|w\| \qquad \text{for all } w \in W,$$

and the related discrete problem can be analysed analogously.

Plate problem models based on Kirchhoff's hypothesis lead also to boundary value problems for fourth-order differential equations. In [Bra01] the author discusses in detail both Kirchhoff and Mindlin-Reissner plates and, for example, the relationship between mixed finite element methods and the popular nonconforming DKT-elements for Kirchhoff plates.

4.6.4 Penalty Methods and Modified Lagrange Functions

The agreeable properties of elliptic problems often come from their close relationship to convex variational problems. For instance, in the case of a symmetric bilinear form $a(\cdot, \cdot)$ the associated variational equation is equivalent to a convex minimization problem. On the other hand, mixed variational equations usually lead to a saddle-point problem (see (6.20)) with convex-concave behaviour.

The resulting difficulties that arise in solving mixed variational equations can be handled by applying special elimination techniques that are closely related to traditional approaches—such as penalty methods—for handling classical optimization problems with constraints. We now present some basic results for penalty methods.

Let $a : V \times V \to \mathbb{R}$ and $b : V \times W \to \mathbb{R}$ be given continuous bilinear forms where $a(\cdot, \cdot)$ is symmetric and $|a(u, v)| \leq \alpha \|u\| \|v\|$ for all $u, v \in V$. Given $f \in V^*$ and $g \in W^*$, consider the variational problem (6.18), i.e.,

$$\text{find } \min_{v \in G} J(v), \quad \text{where } J(v) := \frac{1}{2} a(v, v) - f(v) \tag{6.56}$$

and

$$G := \{ v \in V : b(v, w) = g(w) \quad \text{for all } w \in W \}.$$

We assume that the bilinear form $a(\cdot, \cdot)$ is Z-elliptic on

$$Z := \{ v \in V : b(v, w) = 0 \quad \text{for all } w \in W \}$$

with ellipticity constant $\gamma > 0$. The operator $B : V \to W^*$ defined by (6.12) allows us to rewrite the constraint of problem (6.56) in the equivalent form $Bv = g$.

A *penalty method* modifies the functional J that is to be minimized by adding to it an extra term that becomes large when the constraints are violated. Here the modified unconstrained problem can be written as

$$\min_{v \in V} J_\rho(v), \quad \text{where } J_\rho(v) := J(v) + \frac{\rho}{2} \|Bv - g\|_*^2. \tag{6.57}$$

Here $\rho > 0$ is the penalty parameter.

To simplify the notation we identify W^* with W. Then

$$\|Bv - g\|_*^2 = (Bv - g, Bv - g) = (Bv, Bv) - 2(Bv, g) + (g, g).$$

Consequently, the objective functional $J_\rho(\cdot)$ of (6.57) has the form

$$J_\rho(v) = \frac{1}{2} a(v, v) + \frac{\rho}{2} (Bv, Bv) - f(v) - \rho(Bv, g) + \frac{\rho}{2} (g, g). \tag{6.58}$$

Now consider the associated bilinear form $a_\rho : V \times V \to \mathbb{R}$ defined by

$$a_\rho(u, v) := a(u, v) + \rho(Bu, Bv) \quad \text{for all } u, v \in V. \tag{6.59}$$

To analyse this we shall decompose the space V into a direct sum of subspaces. First, $Z \neq \emptyset$ because $0 \in Z$. For each $v \in V$, Lemma 4.67 (with G replaced by Z) implies the existence of a unique $\tilde{v} \in Z$ such that

$$a(\tilde{v}, z) = a(v, z) \quad \text{for all } z \in Z. \tag{6.60}$$

Hence one can define a projector $P : V \to Z$ by $Pv := \tilde{v}$. Now every element $v \in V$ can be decomposed as

$$v = Pv + (I - P)v.$$

In other words, the space V can be written as the direct sum $V = Z \oplus \tilde{Z}$ where

$$\tilde{Z} := \{ y \in V : y = (I - P)v \quad \text{for some } v \in V \}.$$

Choosing $z = Pv$ in (6.60), one then gets immediately

$$a((I - P)v, Pv) = 0 \quad \text{for all } v \in V. \tag{6.61}$$

Lemma 4.85. *Assume that there exists $\sigma > 0$ such that*

$$\|Bv\| \geq \sigma \|v\| \quad \text{for all } v \in \tilde{Z}.$$

Then the bilinear form $a_\rho(\cdot, \cdot)$ defined in (6.59) is uniformly V-elliptic for all ρ with $\rho \geq \bar{\rho}$, where $\bar{\rho} := (\alpha + \gamma)/\sigma > 0$.

Proof: From (6.59), (6.61), and the definition of Z, one has

$$a_\rho(v, v) = a_\rho(Pv + (I - P)v, Pv + (I - P)v)$$
$$= a(Pv, Pv) + a((I - P)v, (I - P)v) + \rho(B(I - P)v, B(I - P)v)$$
$$\geq \gamma \|Pv\|^2 + (\rho\sigma - \alpha) \|(I - P)v\|^2.$$

Here we recall that $\gamma > 0$ is the ellipticity constant of $a(\cdot, \cdot)$ on Z while α is an upper bound for the norm of this bilinear form. Assume that $\rho \geq \overline{\rho}$. Then

$$a_\rho(v, v) \geq \gamma \left[\|Pv\|^2 + \|(I - P)v\|^2\right] \qquad \text{for all } v \in V.$$

The equivalence of norms in $\mathbb{R}^2$ and the triangle inequality yield

$$a_\rho(v, v) \geq \tfrac{\gamma}{2} \left[\|Pv\| + \|(I - P)v\|\right]^2$$
$$\geq \tfrac{\gamma}{2}\|Pv + (I - P)v\|^2 = \tfrac{\gamma}{2}\|v\|^2 \qquad \text{for all } v \in V.$$

∎

Next we present sufficient conditions for the convergence of this penalty method.

Theorem 4.86. *Assume that the assumptions of Lemma 4.85 are fulfilled, that $G \neq \emptyset$ and choose $\overline{\rho} > 0$ as in Lemma 4.85. Then for each $\rho \geq \overline{\rho}$ the penalty problem (6.57) has a unique solution $u_\rho \in V$. As $\rho \to \infty$, the function u_ρ converges to the solution u of the constrained variational problem (6.56).*

Proof: By virtue of Lemma 4.67 and $G \neq \emptyset$, problem (6.56) has a unique solution u. The objective functional $J_\rho(\cdot)$ of (6.57) is convex for $\rho \geq \overline{\rho}$, so $u_\rho \in V$ is a solution of (6.57) if and only if

$$\langle J_\rho'(u_\rho), v \rangle = 0 \qquad \text{for all } v \in V,$$

which is equivalent to the variational equation

$$a_\rho(u_\rho, v) = f(v) + \rho(Bv, g) \qquad \text{for all } v \in V. \tag{6.62}$$

By construction the bilinear form $a_\rho(\cdot, \cdot)$ is continuous. When $\rho \geq \overline{\rho}$, Lemma 4.85 guarantees that the bilinear form $a_\rho(\cdot, \cdot)$ is V-elliptic. Therefore the Lax-Milgram lemma ensures existence and uniqueness of the solution u_ρ of (6.62).

The optimality of u_ρ for (6.57) and $Bu = g$ imply that

$$J(u) = J_\rho(u) \geq J_\rho(u_\rho).$$

That is,

$$\frac{1}{2}a(u, u) - f(u) \geq J(u_\rho) + \frac{\rho}{2}\|Bu_\rho - Bu\|_*^2$$

$$= \frac{1}{2}a(u_\rho, u_\rho) + \frac{\rho}{2}(B(u_\rho - u), B(u_\rho - u)) - f(u_\rho).$$

Rearranging, we get

$$f(u_\rho - u) - a(u, u_\rho - u) \geq \frac{1}{2} a_\rho(u_\rho - u, u_\rho - u) \geq \frac{\gamma}{2} \|u_\rho - u\|^2, \quad (6.63)$$

where the last inequality holds since $u_\rho - u \in Z$. The continuity of $a(\cdot, \cdot)$ now gives

$$\frac{\gamma}{2} \|u_\rho - u\|^2 \leq (\|f\|_* + \alpha \|u\|) \|u_\rho - u\|.$$

Thus the sequence $\{u_\rho\}_{\rho \geq \overline{\rho}}$ is bounded. The reflexivity of the Hilbert space V then implies that $\{u_\rho\}_{\rho \geq \overline{\rho}}$ is weakly compact.

The functional $J_{\overline{\rho}}(\cdot)$ is convex and continuous. It is hence weakly lower semicontinuous; see [Zei90]. Therefore the weak compactness of $\{u_\rho\}_{\rho \geq \overline{\rho}}$ implies the existence of $\mu \in \mathbb{R}$ such that

$$J_{\overline{\rho}}(u_\rho) \geq \mu \qquad \text{for all } \rho \geq \overline{\rho}.$$

Recalling the optimality of u_ρ for problem (6.57) and the structure of $J_\rho(\cdot)$, we get

$$\begin{aligned} J(u) = J_\rho(u) &\geq J_\rho(u_\rho) = J_{\overline{\rho}}(u_\rho) + \frac{\rho - \overline{\rho}}{2} \|Bu_\rho - g\|^2 \\ &\geq \mu + \frac{\rho - \overline{\rho}}{2} \|Bu_\rho - g\|^2. \end{aligned} \qquad (6.64)$$

It follows that

$$\lim_{\rho \to \infty} \|Bu_\rho - g\| = 0,$$

i.e.,

$$\lim_{\rho \to \infty} [b(u_\rho, w) - g(w)] = 0 \qquad \text{for each } w \in W.$$

Associate with $b : V \times W \to \mathbb{R}$ the operator $B^* : W \to V^*$ defined by

$$\langle B^* w, v \rangle_V = b(v, w) \qquad \text{for all } v \in V, \ w \in W.$$

Then

$$\lim_{\rho \to \infty} [\langle B^* w, u_\rho \rangle - g(w)] = 0 \qquad \text{for each } w \in W.$$

Let $\overline{u}$ be an arbitrary weak accumulation point of $\{u_\rho\}_{\rho \geq \overline{\rho}}$. We obtain

$$\langle B^* w, \overline{u} \rangle - g(w) = 0 \qquad \text{for all } w \in W.$$

This is equivalent to

$$b(\overline{u}, w) = g(w) \qquad \text{for all } w \in W;$$

that is, $\overline{u} \in G$. Furthermore

$$\begin{aligned} J(u) = J_\rho(u) &\geq J_\rho(u_\rho) = J_{\overline{\rho}}(u_\rho) + \frac{\rho - \overline{\rho}}{2} \|Bu_\rho - g\|^2 \\ &\geq J_{\overline{\rho}}(u_\rho) \qquad \text{for all } \rho \geq \overline{\rho}. \end{aligned}$$

The lower semicontinuity of $J_{\overline{\rho}}(\cdot)$ then yields

$$J(u) \geq J_{\overline{\rho}}(\overline{u}) \geq J(\overline{u}).$$

We have now shown that $\overline{u}$ is a solution of the constrained problem (6.56). But this problem has a unique solution u. The weak compactness of $\{u_\rho\}_{\rho \geq \overline{\rho}}$ therefore gives us $u_\rho \rightharpoonup u$ for $\rho \to \infty$. Combining this weak convergence with (6.63), we see that in fact $\|u_\rho - u\| \to 0$ as $\rho \to \infty$. ∎

In practice one approximates the solution of the penalty method using a finite element space V_h. Then one needs to solve the finite-dimensional penalty problem

$$\min_{v_h \in V_h} J_\rho(v_h). \tag{6.65}$$

In the conforming case $V_h \subset V$, although Lemma 4.85 ensures the existence of a unique solution $u_{\rho h} \in V_h$ for each $\rho \geq \overline{\rho}$, it turns out that for large values of the parameter ρ the problem (6.65) is ill conditioned. Thus it becomes necessary to consider a good choice of the parameter ρ, which will also depend on the discretization parameter in the finite element method. We shall deal with this question in Chapter 7 when we examine variational inequalities; see also [57].

A way of avoiding large penalty parameters in the problems (6.57) and (6.65) is to apply an iterative technique based on modified Lagrange functionals. For the original problem (6.56) the standard Lagrange functional $L(\cdot, \cdot)$ was defined in (6.19). The *modified Lagrange functional* $L_\rho : V \times W \to \mathbb{R}$ is defined by

$$L_\rho(v, w) := L(v, w) + \frac{\rho}{2} \|Bv - g\|_*^2$$

$$= J(v) + (Bv - g, w) + \frac{\rho}{2} (Bv - g, Bv - g), \tag{6.66}$$

$$v \in V, \, w \in W.$$

The saddle points of $L(\cdot, \cdot)$ and $L_\rho(\cdot, \cdot)$ are closely related:

Lemma 4.87. *Every saddle point of the Lagrange functional $L(\cdot, \cdot)$ is also a saddle point of the modified Lagrange functional $L_\rho(\cdot, \cdot)$ for each positive value of the parameter ρ.*

Proof: Let $(u, p) \in V \times W$ be a saddle point of $L(\cdot, \cdot)$. By Lemma 4.74, u is a solution of the variational problem (6.56). In particular $Bu = g$. The saddle point inequality for $L(\cdot, \cdot)$ yields

$$L_\rho(u, w) = L(u, w) + \frac{\rho}{2} \|Bu - g\|_*^2 = L(u, w)$$

$$\leq L(u, p) = L_\rho(u, p) \leq L(v, p)$$

$$\leq L(v, p) + \frac{\rho}{2} \|Bv - g\|_*^2 = L_\rho(v, p) \qquad \text{for all } v \in V, \, w \in W.$$

That is, the pair $(u,p) \in V \times W$ is also a saddle point of $L_\rho(\cdot,\cdot)$. ∎

The *Uzawa algorithm* (see [BF91]), which was originally formulated for the Lagrange functional $L(\cdot,\cdot)$, alternately minimizes over V and computes a gradient step to maximize over W. This technique can also be applied to the modified Lagrange functional $L_\rho(\cdot,\cdot)$ and yields the *modified Lagrange method*:

Step 1:   Choose $p^0 \in W$ and $\rho > 0$. Set $k = 0$.

Step 2:   Find $u^k \in V$ such that

$$L_\rho(u^k, p^k) = \min_{v \in V} L_\rho(v, p^k). \qquad (6.67)$$

Step 3:   Set

$$p^{k+1} := p^k + \rho(Bu^k - g), \qquad (6.68)$$

increment k to $k+1$ and go to step 2.

Remark 4.88. If one chooses $\rho \geq \bar{\rho}$ with $\bar{\rho}$ as in Lemma 4.85, then the variational problem (6.67) has a unique solution u^k, and this solution satisfies the equation

$$a(u^k, v) + \rho(Bu^k, Bv) + (Bv, p^k) = f(v) + \rho(Bv, g) \quad \text{for all } v \in V. \ (6.69)$$

Equation (6.68) can be written in the alternative form

$$(p^{k+1}, w) = (p^k, w) + \rho[b(u^k, w) - g(w)] \qquad \text{for all } w \in W. \ \square \quad (6.70)$$

Both the penalty method and the modified Lagrange method can be interpreted as regularized mixed variational equations. Let us first consider the penalty method. Assume that $(u_\rho, p_\rho) \in V \times W$ is a solution of the mixed variational equation

$$\begin{aligned}
a(u_\rho, v) + b(v, p_\rho) &= f(v) && \text{for all } v \in V, \\
b(u_\rho, w) - \tfrac{1}{\rho}(p_\rho, w) &= g(w) && \text{for all } w \in W.
\end{aligned} \qquad (6.71)$$

Recalling the definition of B, and making the identification $W^* = W$, we can solve the second equation to get

$$p_\rho = \rho(Bu_\rho - g). \qquad (6.72)$$

Substituting this result into the first equation of (6.71) yields

$$a(u_\rho, v) + \rho((Bu_\rho - g), Bv) = f(v) \qquad \text{for all } v \in V.$$

This equation for $u_\rho \in V$ is equivalent to (6.62). Alternatively, starting from (6.62), one can define p_ρ by (6.72) and derive (6.71). Thus the penalty method

(6.57) is equivalent to the regularized mixed variational equation (6.71). We remark that this regularization is simply the standard Tychonov regularization for ill-posed problems.

The modified Lagrange method described above can also be derived from a regularized mixed variational equation. Unlike the penalty method, we now use a sequential Prox regularization (see [BF91], [Mar70]). Then the modified Lagrange method is equivalent to the following iteration:

$$a(u^k, v) + \quad b(v, p^{k+1}) \quad = f(v) \qquad \text{for all } v \in V,$$
$$b(u^k, w) - \tfrac{1}{\rho}(p^{k+1} - p^k, w) = g(w) \qquad \text{for all } w \in W.$$

Elimination of p^{k+1} from the second equation gives (6.68). Substituting this result into the first equation we get (6.69), which is a necessary and sufficient condition for achievement of the minimum in (6.67).

From Theorem 4.86 we know that the penalty method converges as $\rho \to \infty$. Now we investigate its precise convergence behaviour. This analysis uses the equivalence of the penalty method and the mixed variational formulation (6.71).

Lemma 4.89. *Assume that the hypotheses of Lemma 4.85 are satisfied, and that*

$$\sup_{v \in V} \frac{b(v, w)}{\|v\|} \geq \delta \|w\| \qquad \text{for all } w \in W$$

and some constant $\delta > 0$. Then the difference between the solution u_ρ of the penalty problem (6.57) and the solution u of the original problem (6.56) is bounded by

$$\|u - u_\rho\| \leq c\rho^{-1} \qquad \text{and} \qquad \|p - p_\rho\| \leq c\rho^{-1} \qquad \text{for all } \rho \geq \bar{\rho},$$

where p and p_ρ are defined in (6.10) and (6.71) respectively, while $\bar{\rho}$ is defined in Lemma 4.85.

Proof: By Theorem 4.72, the mixed variational equation (6.10) has at least one solution $(u, p) \in V \times W$ and the component u is uniquely determined. Lemma 4.69 shows that u is a solution of the original problem (6.56).

Assume that $\rho \geq \bar{\rho}$ where $\bar{\rho}$ is specified in Lemma 4.85. By Theorem 4.86 the regularized mixed problem (6.71) has a solution (u_ρ, p_ρ) where the penalized solution u_ρ is unique. The argument following (6.71) shows that u_ρ is a solution of (6.57).

Subtraction of (6.10) from (6.71) yields

$$a(u_\rho - u, v) + b(v, p_\rho - p) = 0 \qquad \text{for all } v \in V,$$
$$b(u_\rho - u, w) \qquad\qquad = \tfrac{1}{\rho}(p_\rho, w) \qquad \text{for all } w \in W.$$

Applying Theorem 4.72 to this system, we get

$$\|u - u_\rho\| \le c\rho^{-1}\|p_\rho\| \tag{6.73}$$

and

$$\|p - p_\rho\| \le c\rho^{-1}\|p_\rho\| \tag{6.74}$$

for some constant $c > 0$.

In the proof of Theorem 4.86 we demonstrated the existence of a constant $c^* > 0$ such that

$$\|u - u_\rho\| \le c^*.$$

Now (6.72) yields

$$\|p_\rho\| \le c.$$

The desired estimates are now immediate from (6.73) and (6.74). ∎

The previous lemma tells us the convergence behaviour of the penalty method for the continuous problem, but in practice we need information about the discrete version. Choose some finite element space V_h and discretize (6.57); this means that we have to study the finite-dimensional variational problem

$$\min_{v_h \in V_h} J_\rho(v_h), \quad \text{where } J_\rho(v_h) = J(v_h) + \frac{\rho}{2}\|Bv_h - g\|_*^2. \tag{6.75}$$

If $\rho \ge \bar{\rho}$ then this problem has a unique solution $u_{\rho h} \in V_h$.

Lemma 4.90. *Let $\rho \ge \bar{\rho}$ and $V_h \subset V$. Assume the hypotheses of Lemma 4.85. For the solutions of (6.57) and (6.75) one has*

$$\|u_\rho - u_{\rho h}\| \le c\rho^{1/2} \inf_{v_h \in V_h} \|u_\rho - v_h\|$$

for some constant $c > 0$.

Proof: By Lemma 4.85 the bilinear form $a_\rho(\cdot, \cdot)$ associated with $J_\rho(\cdot)$ is uniformly V-elliptic. Since $V_h \subset V$, it is also V_h-elliptic, with the same ellipticity constant $\gamma/2$. The definition (6.59) of $a_\rho(\cdot, \cdot)$ implies that

$$|a_\rho(u, v)| \le \alpha\|u\|\,\|v\| + \rho\beta^2\|u\|\,\|v\|$$

where the constants α and β are upper bounds on the norms of the bilinear forms $a(\cdot, \cdot)$ and $b(\cdot, \cdot)$. The improved version of Cea's lemma for symmetric problems now yields

$$\|u_\rho - u_{\rho h}\| \le \sqrt{\frac{2(\alpha + \rho\beta^2)}{\gamma}} \inf_{v_h \in V_h} \|u_\rho - v_h\| \qquad \text{for all } \rho \ge \bar{\rho}. \quad ∎$$

Combining Lemmas 4.89 and 4.90, we have a result that shows the dependence of the error in the discretized penalty method on both ρ and h.

Theorem 4.91. *Let the hypotheses of Lemma 4.89 and Lemma 4.90 be fulfilled. Then there exist positive constants c_1 and $c_2 > 0$ such that the error of the discretized penalty method (6.75) for the solution of (6.56) satisfies the estimate*

$$\|u - u_{\rho h}\| \leq c_1\,\rho^{-1/2} + c_2\,\rho^{1/2} \inf_{v_h \in V_h} \|u - v_h\| \qquad \text{for all } \rho \geq \overline{\rho}.$$

Proof: Two triangle inequalities and Lemmas 4.89 and 4.90 give

$$\|u - u_{\rho h}\| \leq \|u - u_\rho\| + \|u_\rho - u_{\rho h}\|$$

$$\leq c\rho^{-1} + c\rho^{1/2} \inf_{v_h \in V_h} \|u_\rho - u + u - v_h\|$$

$$\leq c\rho^{-1} + c\rho^{1/2}\rho^{-1} + c\rho^{1/2} \inf_{v_h \in V_h} \|u - v_h\| \qquad \text{for all } \rho \geq \overline{\rho}.$$

The estimate is proved. ∎

Remark 4.92. Theorem 4.91 indicates a reduced convergence order for the penalty method, compared with the Ritz finite element method applied without constraints. If, for instance,

$$\inf_{v_h \in V_h} \|u - v_h\| = O(h^p),$$

then the choice $\rho = \rho(h) = h^{-p}$ (in order to optimize the bound of Theorem 4.91) yields only the convergence rate

$$\|u - u_{\rho h}\| = O(h^{p/2}).$$

This is a well-known disadvantage of the penalty method which can be observed numerically. It is a result of the asymptotic ill-posedness of the penalty method as $\rho \to \infty$.

As we shall see shortly, one can avoid this order reduction by discretizing the mixed formulation (6.71) using a pair of spaces that satisfy the Babuška-Brezzi condition.   □

Choose $V_h \subset V$ and $W_h \subset W$. Denote by $(\overline{u}_{\rho h}, \overline{p}_{\rho h}) \in V_h \times W_h$ the solution of the regularized mixed variational equation

$$a(\overline{u}_{\rho h}, v_h) + b(v_h, \overline{p}_{\rho h}) = f(v_h) \qquad \text{for all } v_h \in V_h,$$
$$b(\overline{u}_{\rho h}, w_h) - \tfrac{1}{\rho}(\overline{p}_{\rho h}, w_h)_h = g(w_h) \qquad \text{for all } w_h \in W_h. \tag{6.76}$$

Here let $(\cdot, \cdot)_h : W_h \times W_h \to R$ be a continuous bilinear form that satisfies the condition

$$\underline{\sigma}\,\|w_h\|^2 \leq (w_h, w_h)_h \leq \overline{\sigma}\,\|w_h\|^2 \qquad \text{for all } w_h \in W_h \tag{6.77}$$

for some constants $\overline{\sigma} \geq \underline{\sigma} > 0$ that are independent of h.

Theorem 4.93. *Let the pair of spaces $V_h \subset V$ and $W_h \subset W$ satisfy the Babuška-Brezzi condition. Assume that (6.77) holds.*
Then the error between the solution $(\overline{u}_{\rho h}, \overline{p}_{\rho h}) \in V_h \times W_h$ of (6.76) and the solution $(u, p) \in V \times W$ of the mixed variational equation (6.10) is bounded as follows:

$$\max\{\, \|u - \overline{u}_{\rho h}\|, \|p - \overline{p}_{\rho h}\| \,\} \leq c\{\, \rho^{-1} + \inf_{v_h \in V_h} \|u - v_h\| + \inf_{w_h \in W_h} \|p - w_h\| \,\}$$

for all $\rho \geq \hat{\rho}$, where $\hat{\rho} \geq \overline{\rho}$.

Proof: Let us denote by $(u_h, p_h) \in V_h \times W_h$ the solution of the discrete mixed variational equation (6.27). Theorem 4.77 yields

$$\max\{\, \|u - p_h\|, \|p - p_h\| \,\} \leq c\{\, \inf_{v_h \in V_h} \|u - v_h\| + \inf_{w_h \in W_h} \|p - w_h\| \,\}. \quad (6.78)$$

Subtracting (6.27) and (6.76), one gets

$$a(u_h - \overline{u}_{\rho h}, v_h) + b(v_h, p_h - \overline{p}_{\rho h}) = 0 \qquad \text{for all } v_h \in V_h,$$

$$b(u_h - \overline{u}_{\rho h}, w_h) \qquad\qquad = -\frac{1}{\rho}(\overline{p}_{\rho h}, w_h)_h \qquad \text{for all } w_h \in W_h.$$

Then the Babuška-Brezzi condition, Theorem 4.76 and (6.77) deliver the estimate

$$\max\{\, \|u_h - \overline{u}_{\rho h}\|, \|p_h - \overline{p}_{\rho h}\| \,\} \leq c\rho^{-1}\|\overline{p}_{\rho h}\| \qquad (6.79)$$

for some $c > 0$.

It remains only to verify that $\|\overline{p}_{\rho h}\|$ is bounded. Choose $\hat{\rho} \geq \overline{\rho}$ such that $c\hat{\rho}^{-1} < 1$ for the constant $c > 0$ from (6.79). Then (6.79) implies

$$\|\overline{p}_{\rho h}\| - \|p_h\| \leq c\hat{\rho}^{-1}\|\overline{p}_{\rho h}\| \qquad \text{for all } \rho \geq \hat{\rho}.$$

Consequently

$$\|\overline{p}_{\rho h}\| \leq c\|p_h\| \qquad \text{for all } \rho \geq \hat{\rho}$$

and some constant $c > 0$. Invoking (6.78), we see that for some constant c^* one has

$$\|\overline{p}_{\rho h}\| \leq c^* \qquad \text{for all } h > 0, \ \rho \geq \hat{\rho}.$$

The theorem now follows from (6.78), (6.79) and a triangle inequality. ∎

Remark 4.94. Unlike Theorem 4.91 for the discretized penalty method (6.75), the terms in the error estimate of Theorem 4.93 for the method (6.76) separate the effects of the discretization error and the penalty parameter. If we choose

$$\rho^{-1} = O\Big(\inf_{v_h \in V_h} \|u - v_h\| + \inf_{w_h \in W_h} \|p - w_h\| \Big),$$

then the penalty method (6.76) has the same order of convergence as the discretization of the mixed variational equation (6.10) by (6.27); no reduction in order occurs.   □

Remark 4.95. Condition (6.77) ensures that the second equation of (6.76) can be solved for $\overline{p}_{\rho h} \in W_h$, and this solution is unique. Consequently (6.76) can be interpreted as the penalty method

$$\min_{v_h \in V_h} J_{\rho h}(v_h), \quad \text{where } J_{\rho h}(v_h) := J(v_h) + \frac{\rho}{2}\|Bv_h - g\|_h^2,$$

and $\|\cdot\|_h$ can be viewed as an approximation of $\|\cdot\|_*$. $\square$

4.7 Error Estimators and Adaptive FEM

The error estimates of Section 4.4, which are of the form

$$\|u - u_h\| \leq Ch^p \||u\||,$$

have several disadvantages:

- their proofs often assume that $h \leq h_0$, but h_0 is not explicitly known so it is not clear whether the estimate is valid for any specific mesh used in practice;
- the constant C is in general unknown and can be computed explicitly without great effort only for simple elements;
- the norm $\||u\||$ of the exact solution is unknown;
- discretization with piecewise polynomials of degree k assumes that for optimal convergence rates in, e.g., the H^1 norm, one has $u \in H^{k+1}(\Omega)$, but this degree of regularity of u is often unrealistic.

Beginning with the pioneering work of Babuška and Rheinboldt at the end of the 1970s, much work has gone into constructing estimates of the error $\|u - u_h\|$ of the finite element approximation u_h that are bounded by a *local* quantity η that is *computable* from u_h:

$$\|u - u_h\| \leq D\eta. \tag{7.1}$$

While the error estimates from Section 4.4 are known as a priori bounds, inequalities like (7.1) are called *a posteriori* error estimates. The quantity η is called an *error estimator*. If η satisfies

$$D_1 \eta \leq \|u - u_h\| \leq D_2 \eta \tag{7.2}$$

for some constants D_1 and D_2, then the error estimator is called *efficient* and *reliable*—these terms refer to the first and second inequalities respectively. This basic idea was later modified by replacing $\|u - u_h\|$ by an objective functional $J(u - u_h)$ which has to be controlled. Here J represents the main quantity of interest in the practical problem that is being solved by the finite element method; see Section 4.7.2.

In an *adaptive* FEM an error estimator is used to control the computational process. For the h-version FEM based on grid refinement (see Section 4.9.4 for the p-and hp-versions of the FEM) the basic steps of an adaptive algorithm are:

(1) Solve the problem on the current mesh;
(2) Estimate the contribution to the error on each element K using the local error estimator η_K;
(3) Modify the mesh using the information of Step 2 and return to Step 1.

Grid refinement is the standard tool in Step 3, but coarsening procedures and movement of nodes can also be used. If one desires to refine those elements having a relatively large error contribution, there are several strategies for the selection of these elements. A widely-used strategy is the following bulk criterion:

Choose a parameter θ with $0 < \theta < 1$, then determine a subset $\mathcal{T}^*$ of the current triangulation $\mathcal{T}$ for which

$$\left(\sum_{K \in \mathcal{T}^*} \eta_K^2 \right)^{1/2} \geq \Theta \eta, \quad \text{where} \quad \eta = \left(\sum_{K \in \mathcal{T}} \eta_K^2 \right)^{1/2}. \tag{7.3}$$

We shall not describe the full technical details of several adaptive finite element methods. Instead we examine in depth the most important component of every adaptive algorithm: the error estimator.

Many different error estimators are currently in use. We begin in Section 4.7.1 with the classical residual estimator. In Section 4.7.2 the popular averaging technique and the basic concepts of goal-oriented error estimation are discussed. Further estimators and detailed statements concerning the relationships between them can be found in [Ver96, AO00, BR03].

For simplicity, we restrict ourselves to the model problem

$$-\Delta u = f \quad \text{in } \Omega, \quad u = 0 \quad \text{on } \partial\Omega, \tag{7.4}$$

in a two-dimensional polygonal domain Ω, and consider its discretization using piecewise linear finite elements.

4.7.1 Residual Estimators

If $\tilde{x}$ is a known approximate solution of the linear system $Ax = b$, then to control the error $x - \tilde{x}$ it is reasonable to start from the equation

$$A(x - \tilde{x}) = b - A\tilde{x},$$

where one can compute the residual $b - A\tilde{x}$. To proceed further, we need some information about $\|A^{-1}\|$. Turning to the finite element method, it is our aim to develop an analogous approach for its a posteriori analysis.

Let $u_h \in V_h$ be the finite element approximation of u in (7.4). Then for arbitrary $v \in V := H_0^1(\Omega)$ one has

$$(\nabla(u - u_h), \nabla v) = (f, v) - (\nabla u_h, \nabla v) =: \langle R(u_h), v \rangle. \tag{7.5}$$

The residual $R(u_h)$ is unfortunately an element of the dual space $V^* = H^{-1}(\Omega)$ of V. Clearly (7.5) implies

$$|u - u_h|_1^2 \le \|R(u_h)\|_{-1} \|u - u_h\|_1,$$

but the computation of the H^{-1} norm of the residual is difficult. One therefore tries to prove an estimate

$$|\langle R(u_h), v \rangle| \le C \eta \|v\|_1, \tag{7.6}$$

because Friedrich's inequality then yields

$$\|u - u_h\|_1 \le C \eta.$$

The first step towards proving (7.6) is a transformation of the equation defining the residual by means of integration by parts:

$$\langle R(u_h), v \rangle = \sum_K \int_K (f + \Delta u_h) v - \sum_K \int_{\partial K} (n \cdot \nabla u_h) v,$$

where Ω is partitioned into elements K and n is the outward-pointing unit normal to ∂K. (For linear elements one has, of course, $\Delta u_h = 0$; we include this term nevertheless because it plays a role when higher-order elements are used.) Next, we introduce the element-oriented and edge-oriented residuals defined by

$$r_K(u_h) := (f + \Delta u_h)|_K \quad \text{and} \quad r_E(u_h) := [n_E \cdot \nabla u_h]_E,$$

where E is a generic edge of the triangulation, and $[\cdot]$ denotes the jump of the (discontinuous) normal derivative of u_h across the edge E. In this new notation we have

$$\langle R(u_h), v \rangle = \sum_K \int_K r_K v - \sum_E \int_E r_E v. \tag{7.7}$$

The error orthogonality of the Galerkin method allows us to replace v on the left-hand side of (7.5) by $v - v_h$, for arbitrary $v_h \in V_h$, and then we have

$$\langle R(u_h), v \rangle = \sum_K \int_K r_K (v - v_h) - \sum_E \int_E r_E (v - v_h).$$

Hence

$$|\langle R(u_h), v \rangle| \le \sum_K \|r_K\|_{0,K} \|v - v_h\|_{0,K} + \sum_E \|r_E\|_{0,E} \|v - v_h\|_{0,E}. \tag{7.8}$$

To continue, we would like to choose v_h in such a way that we can bound $v - v_h$ for any $v \in V$, but we cannot use a standard interpolant because it is not in general defined for $v \in V$. This is a technical difficulty that can be handled by the introduction of generalized interpolants or *quasi-interpolants*; see [EG04] for a detailed discussion of quasi-interpolants.

To define such an interpolant and to formulate approximation error estimates, let us define the following notation:

- ω_K: the set of all elements that share at least one vertex with the element K
- ω_E: the set of all elements that share at least one vertex with the edge E.

We assume that the triangulation is quasi-uniform, which implies that the number of elements in ω_K or ω_E is bounded by a fixed constant. Let h_K denote the diameter of K and h_E the length of the edge E.

Lemma 4.96. *Let the triangulation be quasi-uniform. Then to each $v \in V$ corresponds a quasi-interpolant $I_h v \in V_h$ such that*

$$\|v - I_h v\|_{0,K} \le C h_K |v|_{1,\omega_K} \quad \text{and} \quad \|v - I_h v\|_{0,E} \le C h_E^{1/2} |v|_{1,\omega_E}$$

Proof: Let $v \in V$. The construction of $I_h\, v$ is achieved in two steps:

(1) Given a vertex x_j, let ω_{x_j} denote the set of all elements for which x_j is a vertex. Let P_j denote the L_2 projector from V to the constant functions on ω_{x_j}.

(2) Set $I_h\, v := \sum_j (P_j\, v)(x_j)\varphi_j$, where $\{\varphi_j\}$ is the usual nodal basis of V_h.

Now the desired estimates are easy consequences of the Bramble-Hilbert lemma. ■

Returning to (7.8), choose v_h to be our quasi-interpolant and invoke the estimates of Lemma 4.96; this gives

$$|\langle R(u_h), v \rangle| \le C \left\{ \sum_K h_K^2 \|r_K\|_{0,K}^2 + \sum_E h_E \|r_E\|_{0,E}^2 \right\}^{1/2} \|v\|_1.$$

This bound is of the desired form (7.6). Thus, define the *residual error estimator*

$$\eta := \left[\sum_K \eta_K^2 \right]^{1/2}, \quad \text{where} \quad \eta_K^2 := h_K^2 \|r_K\|_{0,K}^2 + \frac{1}{2} \sum_{E \subset K} h_E \|r_E\|_{0,E}^2.$$

The error estimator η is computable, local and—by its construction—reliable. It remains to investigate its efficiency.

The proof of efficiency of η is based on a technique of Verfürth [Ver96]. Assume for simplicity (as otherwise additional terms arise) that f is elementwise constant. Then both residuals are bounded above by the error; we give the details of the this analysis only for the element residual.

Consider the bubble function $b_K = 27\lambda_1\lambda_2\lambda_3$ (where the λ_i are, as usual, barycentric coordinates on K). Then b_K vanishes on the boundary of K. In (7.5) and (7.7) choose v by setting

$$v|_K := v_K = r_K b_K = f_K r_K;$$

this yields

$$\int_K \nabla(u - u_h)\nabla v_K = \int_K r_K\, v_K.$$

But $(r_K, v_K) = c\|r_K\|_{0,K}^2$ so it follows that

$$c\|r_K\|_{0,K}^2 \le |u - u_h|_{1,K}|r_K\, b_K|_{1,K}.$$

In the second factor on the right-hand side, replace the H^1 seminorm by $\|r_K b_K\|_0$ via a local inverse inequality. Then we obtain the desired estimate

$$h_K\|r_K\|_{0,K} \le C\,|u - u_h|_{1,K}. \tag{7.9}$$

An edge bubble function can be used in a similar way to prove an analogous estimate for the edge residual:

$$h_E^{1/2}\|r_E\|_{0,E} \le C\,|u - u_h|_{1,\omega_E}.$$

In summary, the efficiency of the residual error estimator is proved.

Remark 4.97. The values of the constants in Lemma 4.96 are stated in [33]. In [34] it is proved that for lower-order elements the edge residuals are the dominant terms in the error estimator η.   □

Remark 4.98. (Convergence of adaptive FEM)
For a long time a proof of convergence of an adaptive algorithm remained an open problem. The first proof of such a result is due to Dörfler [46], who used a residual error estimator and the bulk criterion (7.3). His analysis assumes that the initial mesh is already sufficiently fine to control data oscillations, which are defined by

$$osc(f, \mathcal{T}_h) := \left\{ \sum_{T \in \mathcal{T}_h} \|h(f - f_T)\|_{0,T}^2 \right\}^{1/2},$$

where f_T is the mean value of f on an element T.

Later it became clear that error reduction requires conditions on the refinement; moreover, an extended bulk criterion also takes data oscillations into account; see [93]. Examples in [93] show the significance for the convergence behaviour of generating new interior mesh points during the refinement process.

Binev, Dahmen and DeVore [16] proved optimal convergence rates for an adaptive algorithm with optimal complexity. In their algorithm both mesh-coarsening and mesh-refinement steps are taken into account. Stevenson [112] simplified the algorithm by combining [16] and [93] in such a way that—at least for our linear model problem—coarsening steps are unnecessary.   □

4.7.2 Averaging and Goal-Oriented Estimators

In an extensive numerical comparison [9] of several error estimators, the so-called *ZZ-error estimator* of Zienkiewicz and Zhu worked surprisingly well. Its basic idea is to replace the gradient of the exact solution u in the error term $\nabla u - \nabla u_h$ by a computable reconstruction or recovery Ru_h. Then the local error estimator can be defined by

$$\eta_K := \|R_h u_h - \nabla u_h\|_{0,K}. \tag{7.10}$$

If $Ru_h \in V_h$ and, analogously to the proof of Lemma 4.96, we set

$$Ru_h := \sum_j (Pu_h)(x_j)\,\varphi_j$$

for some projector P, then there are various choices for P. Let us, for example, take the L_2 projection of ∇u_h on ω_{x_j}. Then

$$(Pu_h)(x_j) = \frac{1}{meas(\omega_{x_j})} \sum_{K \subset \omega_{x_j}} \nabla u_h|_K \, meas(K)$$

and we obtain the ZZ-error estimator.

It turns out that this particular Ru_h has the property of yielding a super-convergent approximation of ∇u on certain meshes (see also Section 4.8.3). Chapter 4 of [AO00] contains a detailed discussion of those properties a recovery operator should have and shows how one can analyse the resulting error estimators if additional superconvergence results are available.

In [31, 32] it emerged that it is not necessary to have superconvergence properties to prove, for instance, reliability of the ZZ-error estimator. In fact, *every averaging* yields reliable a posteriori error control. To explain the underlying idea, let us define an error estimator by best approximation of the discrete gradient:

$$\eta := \min_{q_h \in V_h} \|\nabla u_h - q_h\|_0.$$

In practice it is important to note that the estimate (7.11) of the next Lemma trivially remains true if the best approximation is replaced by some averaging process.

Lemma 4.99. *Assume that the L_2 projector into V_h is H^1 stable on the given triangulation. Then*

$$\|\nabla(u - u_h)\|_0 \le c\eta + HOT, \tag{7.11}$$

where η is the above discrete gradient best approximation estimator and HOT represents higher-order terms.

Proof: Let P denote the L_2 projector into V_h. Set $e := u - u_h$. The error orthogonality of the Galerkin method implies the identity

$$\|\nabla e\|_0^2 = (\nabla u - q_h, \nabla(e - Pe)) + (q_h - \nabla u_h, \nabla(e - Pe)) \tag{7.12}$$

for all $q_h \in V_h$. Now choose q_h such that $\eta = \|\nabla u_h - q_h\|$. The second term on the right-hand side of (7.12) is estimated using the assumed H^1-stability and Cauchy-Schwarz:

$$|(q_h - \nabla u_h, \nabla(e - Pe))| \le c\eta \|\nabla e\|_0.$$

To estimate the first term we integrate ∇u by parts and introduce Δu_h, which equals zero locally:

$$(\nabla u - q_h, \nabla(e - Pe)) = (f, e - Pe) + \sum_K \int_K \nabla \cdot (q_h - \nabla u_h)(e - Pe)$$

$$= (f - Pf, e - Pe) + \sum_K \int_K \nabla \cdot (q_h - \nabla u_h)(e - Pe).$$

Now the application of an local inverse inequality and a standard approximation error estimate $\|e - Pe\|_{0,K} \le c h_K |e|_{1,K}$ together yield

$$|(\nabla u - q_h, \nabla(e - Pe))| \le \|f - Pf\|_0 \|e - Pe\|_0 + c\|q_h - \nabla u_h\|_0 |e|_1.$$

Combining our estimates, we are done. ∎

Conditions for the H^1 stability of the L_2-projection are found in, e.g., [29]. Roughly speaking, standard meshes have this property. A detailed discussion of specific averaging operators appears in [31, 32, 30].

Up to this point we have discussed error estimators for the H^1 seminorm of the error in the context of elliptic second-order boundary value problems. Error estimators also exist for other norms, especially L_2 and L_∞: see [Ver96, AO00, BR03].

The goal of a numerical simulation is often the effective and accurate computation of a target quantity that can be described as a functional of the solution of a given boundary value problem. For instance, consider a viscous incompressible flow around a body, modelled by the Navier-Stokes equations. If the goal of the computation is an accurate approximation of the drag (or lift) coefficient, then

$$J(v, p) = c_{drag} = a \int_S n^T (2\nu\tau - pI) b \, ds$$

supplies a target functional (which is a surface integral over the boundary S of the given body) and it does not make much sense to ignore this and control the error in a global norm of the velocity or pressure. Instead, the error control should guarantee that the quantity J of interest is computed accurately. Such error estimators are called *goal-oriented*.

Consider thus the variational problem

$$a(u, v) = f(v) \quad \text{for all} \quad v \in V \tag{7.13}$$

and its finite element discretization under standard assumptions. Given a target functional $J(\cdot)$, we seek a reliable a posteriori estimator for $|J(u) - J(u_h)|$. Let us introduce the auxiliary problem:
Find $w \in V$ such that

$$a(v, w) = J(v) \quad \text{for all} \quad v \in V.$$

This is a dual or adjoint problem that is auxiliary to (7.13)—compare the Nitsche technique for L_2 error estimates or the approach sketched in Section 4.4.3 for deriving pointwise error estimates. Now

$$J(u - u_h) = a(u - u_h, w),$$

and for arbitrary $w_h \in V_h$, it follows that

$$J(u - u_h) = a(u - u_h, w - w_h). \tag{7.14}$$

For our example (7.1), analogously to Section 4.7.1, we can bound the right-hand side by

$$|J(u - u_h)| \leq \sum_K \|r_K\|_{0,K} \|w - w_h\|_{0,K} + \sum_E \|r_E\|_{0,E} \|w - w_h\|_{0,E}.$$

Now, depending on the precise problem under consideration, there are several possible ways of computing (approximately) or estimating $\|w - w_h\|$; see [BR03] for some examples. But in some cases the numerical approximation of the solution of this dual problem requires meshes different from those used in the original problem, and one has to balance carefully all terms that contribute to the error.

The method sketched here is the dual weighted residual method (DWR). The first convergence results for an adaptive scheme based on this method are in [42].

4.8 The Discontinuous Galerkin Method

In 1973 Reed and Hill [99] introduced the discontinuous Galerkin method for first-order hyperbolic problems. In the following years many researchers used this idea not only for first-order hyperbolic equations but also for the time discretization of unsteady problems. Independently of this work, in the 1970s the first versions of the discontinuous Galerkin method for elliptic problems appeared, using discontinuous ansatz functions and inadmissible meshes. In recent years more advanced versions of the discontinuous Galerkin finite element method (dGFEM) have become popular.

The main advantages of the dGFEM over the standard FEM are its flexibility with respect to the mesh and the local ansatz functions used, especially in combination with hp-methods (higher-order ansatz functions can be used in subdomains where the solution of the given problem is smooth). The possibilities of incorporating boundary conditions and the treatment of convective terms also favour the dGFEM. Furthermore, the local nature of the dGFEM simplifies the parallelization of dGFEM codes. A disadvantage is the large number of degrees of freedom present compared with a conventional FEM.

We shall give an introduction to the method that includes a typical convergence analysis, restricting ourselves to simple ansatz functions (piecewise linears and bilinears) on an admissible mesh. More general results appear, for instance, in [7, 11, 38, 71].

It is our intention to explain also the advantages of the dGFEM for problems with dominant convection, so we introduce a small parameter $\varepsilon > 0$ that multiplies the Laplacian in the given boundary value problem. In this chapter we shall investigate only how the constants arising in the error estimates vary with ε, while ignoring the dependence of the solution u on ε; the consequences of this solution dependence will be revealed in Chapter 6.

4.8.1 The Primal Formulation for a Reaction-Diffusion Problem

Let us consider the linear reaction-diffusion problem

$$-\varepsilon \triangle u + cu = f \quad \text{in} \quad \Omega, \tag{8.1a}$$

$$u = 0 \quad \text{on} \quad \Gamma, \tag{8.1b}$$

while assuming that $c \geq c^* > 0$ and Ω is a two-dimensional polygonal domain.

Let $\mathcal{T}$ be an admissible decomposition of Ω into triangles or parallelograms κ with

$$\bar{\Omega} = \bigcup_{\kappa \in \mathcal{T}} \kappa.$$

(In general, it is not necessary to assume admissibility of the decomposition; in [71], for instance, one can have one hanging node per edge.)

To each element $\kappa \in \mathcal{T}$ we assign a nonnegative integer s_κ and define the composite Sobolev space of the order $\mathbf{s} = \{s_\kappa : \kappa \in \mathcal{T}\}$ by

$$H^{\mathbf{s}}(\Omega, \mathcal{T}) = \left\{ v \in L^2(\Omega) : v|_\kappa \in H^{s_\kappa}(\kappa) \, \forall \kappa \in \mathcal{T} \right\}.$$

The associated norm and semi-norm are

$$\|v\|_{\mathbf{s}, \mathcal{T}} = \left(\sum_{\kappa \in \mathcal{T}} \|v\|^2_{H^{s_\kappa}(\kappa)} \right)^{\frac{1}{2}}, \qquad |v|_{\mathbf{s}, \mathcal{T}} = \left(\sum_{\kappa \in \mathcal{T}} |v|^2_{H^{s_\kappa}(\kappa)} \right)^{\frac{1}{2}}.$$

If $s_\kappa = s$ for all $\kappa \in \mathcal{T}$, we write instead $H^s(\Omega, \mathcal{T})$, $\|v\|_{s, \mathcal{T}}$ and $|v|_{s, \mathcal{T}}$. If $v \in H^1(\Omega, \mathcal{T})$ then the composite gradient $\nabla_{\mathcal{T}} v$ of a function v is defined by $(\nabla_{\mathcal{T}} v)|_\kappa = \nabla(v|_\kappa)$, $\kappa \in \mathcal{T}$.

We assume that each element $\kappa \in \mathcal{T}$ is the affine image of a reference element $\hat{\kappa}$, viz., $\kappa = F_\kappa(\hat{\kappa})$. One could use two reference elements but for simplicity we just study the case of a reference triangle.

The finite element space is defined by

$$S(\Omega, \mathcal{T}, \mathbf{F}) = \left\{ v \in L^2(\Omega) : v|_\kappa \circ F_\kappa \in P_1(\hat{\kappa}) \right\} ; \tag{8.2}$$

here $\mathbf{F} = \{ F_\kappa : \kappa \in \mathcal{T} \}$ and $P_1(\hat{\kappa})$ is as usual the space of linear polynomials on $\hat{\kappa}$. Note that the functions in $S(\Omega, \mathcal{T}, \mathbf{F})$ may be discontinuous across element edges.

Let $\mathcal{E}$ be the set of all edges of the given triangulation $\mathcal{T}$, with $\mathcal{E}_{int} \subset \mathcal{E}$ the set of all interior edges $e \in \mathcal{E}$ in Ω. Set $\Gamma_{int} = \{ x \in \Omega : x \in e$ for some $e \in \mathcal{E}_{int} \}$. Let the elements of $\mathcal{T}$ be numbered sequentially: $\kappa_1, \kappa_2, \dots$. Then for each $e \in \mathcal{E}_{int}$ there exist indices i and j such that $i > j$ and $e = \bar{\kappa}_i \cap \bar{\kappa}_2$. Set $\kappa := \kappa_i$ and $\kappa' := \kappa_j$. Define the jump (which depends on the enumeration of the triangulation) and average of each function $v \in H^1(\Omega, \mathcal{T})$ on $e \in \mathcal{E}_{int}$ by

$$[v]_e = v|_{\partial \kappa \cap e} - v|_{\partial \kappa' \cap e} , \qquad \langle v \rangle_e = \frac{1}{2} \left(v|_{\partial \kappa \cap e} + v|_{\partial \kappa' \cap e} \right) .$$

Furthermore, to each edge $e \in \mathcal{E}_{int}$ we assign a unit normal vector ν directed from κ to κ'; if instead $e \subset \Gamma$ then we take the outward-pointing unit normal vector μ on Γ. When there is no danger of misinterpretation we omit the indices in $[v]_e$ and $\langle v \rangle_e$.

We shall assume that *the solution u of (8.1) lies in $H^2(\Omega) \subset H^2(\Omega, \mathcal{T})$.* (For more general problems it is standard to assume that $u \in H^2(\Omega, \mathcal{T})$ and that both u and $\nabla u \cdot \nu$ are continuous across all interior edges, where ν is a normal to the edge.) In particular we have

$$[u]_e = 0 , \qquad \langle u \rangle_e = u , \qquad e \in \mathcal{E}_{int} , \qquad \kappa \in \mathcal{T} .$$

Multiply the differential equation (8.1) by a (possibly discontinuous) test function $v \in H^1(\Omega, \mathcal{T})$ and integrate over Ω:

$$\int_\Omega (-\varepsilon \Delta u + cu) \, v \, dx = \int_\Omega f v \, dx . \tag{8.3}$$

First we consider the contribution of $-\varepsilon \Delta u$ to (8.3). Let μ_κ denote the outward-pointing unit normal to $\partial \kappa$ for each $\kappa \in \mathcal{T}$. Integration by parts and elementary transformations give us

$$\int_\Omega (-\varepsilon \Delta u) \, v \, dx = \sum_{\kappa \in \mathcal{T}} \varepsilon \int_\kappa \nabla u \cdot \nabla v \, dx - \sum_{\kappa \in \mathcal{T}} \varepsilon \int_{\partial \kappa} (\nabla u \cdot \mu_\kappa) \, v \, ds$$

$$= \sum_{\kappa \in \mathcal{T}} \varepsilon \int_\kappa \nabla u \cdot \nabla v \, dx - \sum_{e \in \mathcal{E} \cap \Gamma} \varepsilon \int_e (\nabla u \cdot \mu) \, v \, ds$$

$$- \sum_{e \in \mathcal{E}_{int}} \varepsilon \int_e \left(((\nabla u \cdot \mu_\kappa) v)|_{\partial \kappa \cap e} + ((\nabla u \cdot \mu_{\kappa'}) v)|_{\partial \kappa' \cap e} \right) ds .$$

The sum of the integrals over $\mathcal{E}_{int}$ can be written as

$$\sum_{e\in\mathcal{E}_{int}} \varepsilon \int_e \Big(((\nabla u \cdot \mu_\kappa)v)|_{\partial\kappa\cap e} + ((\nabla u \cdot \mu_{\kappa'})v)|_{\partial\kappa'\cap e}\Big)\,ds$$

$$= \sum_{e\in\mathcal{E}_{int}} \varepsilon \int_e \Big(((\nabla u \cdot \nu)v)|_{\partial\kappa\cap e} - ((\nabla u \cdot \nu)v)|_{\partial\kappa'\cap e}\Big)\,ds$$

$$= \sum_{e\in\mathcal{E}_{int}} \varepsilon \int_e \Big(\langle\nabla u \cdot \nu\rangle_e[v]_e + [\nabla u \cdot \nu]_e\langle v\rangle_e\Big)\,ds$$

$$= \sum_{e\in\mathcal{E}_{int}} \varepsilon \int_e \langle\nabla u \cdot \nu\rangle_e[v]_e\,ds.$$

With the abbreviations

$$\sum_{e\in\mathcal{E}_{int}} \varepsilon \int_e \langle\nabla u \cdot \nu\rangle_e[v]_e\,ds = \varepsilon \int_{\Gamma_{int}} \langle\nabla u \cdot \nu\rangle[v]\,ds$$

and

$$\sum_{e\in\mathcal{E}\cap\Gamma} \varepsilon \int_e (\nabla u \cdot \mu)\,v\,ds = \varepsilon \int_\Gamma (\nabla u \cdot \mu)\,v\,ds\,,$$

we get

$$\int_\Omega (-\varepsilon\Delta u)\,v\,dx = \sum_{\kappa\in\mathcal{T}} \varepsilon \int_\kappa \nabla u \cdot \nabla v\,dx$$

$$- \varepsilon \int_\Gamma (\nabla u \cdot \mu)\,v\,ds - \varepsilon \int_{\Gamma_{int}} \langle\nabla u \cdot \nu\rangle[v]\,ds. \qquad (8.4)$$

Finally, add or subtract to the right-hand side the terms

$$\varepsilon \int_\Gamma u(\nabla v \cdot \mu)\,ds \qquad \text{and} \qquad \varepsilon \int_{\Gamma_{int}} [u]\langle\nabla v \cdot \nu\rangle\,ds$$

and the penalty terms

$$\int_\Gamma \sigma uv\,ds \qquad \text{and} \qquad \int_{\Gamma_{int}} \sigma[u][v]\,ds\,.$$

All these terms vanish for the exact solution u. The penalty parameter σ is piecewise constant:

$$\sigma|_e = \sigma_e, \qquad e \in \mathcal{E}\,.$$

We now have

$$\int_{\Omega} (-\varepsilon \Delta u) v \, dx = \sum_{\kappa \in \mathcal{T}} \varepsilon \int_{\kappa} \nabla u \cdot \nabla v \, dx$$

$$+ \varepsilon \int_{\Gamma} (\pm u(\nabla v \cdot \mu) - (\nabla u \cdot \mu)v) \, ds + \int_{\Gamma} \sigma u v \, ds$$

$$+ \varepsilon \int_{\Gamma_{int}} (\pm [u]\langle \nabla v \cdot \nu \rangle - \langle \nabla u \cdot \nu \rangle [v]) \, ds + \int_{\Gamma_{int}} \sigma [u][v] \, ds .$$

This ends the rearrangement of (8.3). We can now give the primal formulation of the discontinuous Galerkin methods with interior penalties:

$$\begin{cases} \text{Find} \quad u_h \in S(\Omega, \mathcal{T}, \mathbf{F}) \text{ such that} \\ B_{\pm}(u_h, v_h) = L(v_h) \text{ for all } v_h \in S(\Omega, \mathcal{T}, \mathbf{F}), \end{cases} \tag{8.5}$$

where

$$L(w) = \sum_{\kappa \in \mathcal{T}} \int_{\kappa} f w \, dx$$

and the underlying bilinear forms are

$$B_{\pm}(v, w) = \sum_{\kappa \in \mathcal{T}} \left(\varepsilon \int_{\kappa} \nabla v \cdot \nabla w \, dx + \int_{\kappa} c v w \, dx \right) \tag{8.6}$$

$$+ \varepsilon \int_{\Gamma} (\pm v(\nabla w \cdot \mu) - (\nabla v \cdot \mu)w) \, ds + \int_{\Gamma} \sigma v w \, ds$$

$$+ \varepsilon \int_{\Gamma_{int}} (\pm [v]\langle \nabla w \cdot \nu \rangle - \langle \nabla v \cdot \nu \rangle [w]) \, ds + \int_{\Gamma_{int}} \sigma [v][w] \, ds ,$$

If one chooses B_- here, then the bilinear form is symmetric and one has the SIP (symmetric with interior penalties) method. With B_+ one obtains a nonsymmetric bilinear form—the NIP method. Later we discuss the pros and cons of these methods.

Example 4.100. Let us consider the simple problem

$$-u'' + cu = f, \qquad u(0) = u(1) = 0,$$

with constant c. Discretize this using NIP with piecewise linear elements on an equidistant mesh of diameter h, taking $\sigma = 1/h$.

Our approximation is discontinuous; we denote by u_i^- and u_i^+ the discrete values at the mesh point x_i. After some computation the following difference stencil like representation emerges (where the vertical lines separate the + and - components):

$$-\frac{1}{h^2} \quad -\frac{2}{h^2} \left| \frac{4}{h^2} + \frac{2c}{3} \quad -\frac{2}{h^2} + \frac{c}{3} \right| + \frac{1}{h^2}$$

$$-\frac{1}{h^2} \left| -\frac{2}{h^2} + \frac{c}{3} \quad \frac{4}{h^2} + \frac{2c}{3} \right| -\frac{2}{h^2} \quad -\frac{1}{h^2} .$$

To explain this notation, the full version of the first line is the expression

$$-\frac{1}{h^2}u_{i-1}^+ - \frac{2}{h^2}u_i^- + \left(\frac{4}{h^2} + \frac{2c}{3}\right)u_i^+ + \left(-\frac{2}{h^2} + \frac{c}{3}\right)u_{i+1}^- + \frac{1}{h^2}u_{i+1}^+.$$

If the approximation were continuous then the stencil would simplify to the standard central difference scheme.

For the dGFEM and linear elements, a vector (u_i^-, u_i^+) is assigned to every mesh point, so the discontinuous Galerkin method generates a vector-valued difference scheme.   □

Remark 4.101. (The flux formulation of dGFEM)
As well as the primal formulation of the dGFEM, one can use the so-called flux formulation. Like mixed finite element methods, the starting point (for our problem) is

$$\theta = \nabla u, \qquad -\varepsilon\nabla\cdot\theta + cu = f.$$

The element-based variational form of this is

$$\int_\kappa \theta\cdot\tau = -\int_\kappa u\,\nabla\cdot\tau + \int_{\partial\kappa} u\,\mu_\kappa\cdot\tau,$$

$$-\varepsilon\int_\kappa \theta\cdot\nabla v + \int_\kappa cuv = \int_\kappa fv + \int_{\partial\kappa}\theta\cdot\mu_\kappa v.$$

Introducing discrete discontinuous finite element spaces, we get the flux formulation of the dGFEM: Find u_h, θ_h such that

$$\int_\kappa \theta_h\cdot\tau_h = -\int_\kappa u_h\,\nabla\cdot\tau_h + \int_{\partial\kappa}\hat{u}_\kappa\mu_\kappa\cdot\tau_h,$$

$$-\varepsilon\int_\kappa \theta_h\cdot\nabla v_h + \int_\kappa cu_hv_h = \int_\kappa fv_h + \int_{\partial\kappa}\hat{\theta}_\kappa\cdot\mu_\kappa v_h.$$

This formulation depends on the *numerical fluxes* $\hat{\theta}_\kappa$ and $\hat{u}_\kappa$ that approximate $\theta = \nabla u$ and u on $\partial\kappa$. Their definitions are of critical importance. In [7], nine variants of the dGFEM are discussed based on various choices of $\hat{\theta}_\kappa$ and $\hat{u}_\kappa$; for each variant the corresponding primal formulation is derived and the properties of the different methods are discussed.   □

Only the primal form of the dGFEM will be examined here. Next we turn to the treatment of convective terms.

4.8.2 First-Order Hyperbolic Problems

Let us consider the pure convection problem

$$b\cdot\nabla u + cu = f \quad \text{in } \Omega, \tag{8.7a}$$

$$u = g \quad \text{on } \Gamma_-, \tag{8.7b}$$

under the assumption that $c - (\operatorname{div} b)/2 \geq \omega > 0$. Here Γ_- denotes the inflow part of the boundary Γ of Ω, which is characterized by $b \cdot \mu < 0$ where μ is the outward-pointing unit normal to Γ.

For more than 20 years it has been recognised that standard finite elements for the discretization of (8.7) are unsatisfactory [74]: their stability properties are poor and the convergence rates obtained are, in general, suboptimal. For instance, with piecewise linear elements the L_2 error is typically only $O(h)$.

Thus to solve (8.7) with a finite element method, one needs to modify the standard method. Two effective mechanisms for this are the streamline diffusion finite element method (SDFEM) that will be discussed in Chapter 6 and the dGFEM.

First we establish some notation. Denote the inflow and the outflow parts of the boundary of an element $\partial \kappa$ by

$$\partial_- \kappa = \{x \in \partial \kappa : b(x) \cdot \mu_\kappa(x) < 0\}, \qquad \partial_+ \kappa = \{x \in \partial \kappa : b(x) \cdot \mu_\kappa(x) \geq 0\}.$$

Here $\mu_\kappa(x)$ is the outward-pointing unit vector normal to $\partial \kappa$ at the point $x \in \partial \kappa$.

For each element $\kappa \in \mathcal{T}$ and $v \in H^1(\kappa)$, denote by v_κ^+ the interior trace of $v|_\kappa$ on $\partial \kappa$. If $\partial_- \kappa \setminus \Gamma \neq \emptyset$ for some $\kappa \in \mathcal{T}$ then for each $x \in \partial_- \kappa \setminus \Gamma$ there exists a unique $\kappa' \in \mathcal{T}$ such that $x \in \partial_+ \kappa'$. In this way $\partial_- \kappa \setminus \Gamma$ is partitioned into segments, each of which is the intersection of $\bar{\kappa}$ with the outflow boundary of a unique κ'. Given a function $v \in H^1(\Omega, \mathcal{T})$ and a $\kappa \in \mathcal{T}$ with the property $\partial_- \kappa \setminus \Gamma \neq \emptyset$, define the outer trace v_κ^- of v with respect to κ on each segment of $\partial_- \kappa \setminus \Gamma$ to be the interior trace of $v_{\kappa'}^+$ with respect to that κ' for which $\partial_+ \kappa'$ intersects that segment of $\partial_- \kappa$. The jump of v across $\partial_- \kappa \setminus \Gamma$ is defined by

$$\lfloor v \rfloor_\kappa = v_\kappa^+ - v_\kappa^- .$$

Note that the jump $\lfloor \cdot \rfloor$ depends on the vector b; this is not the case for the jump $[\cdot]$.

Integration by parts yields the identity

$$\int_\kappa (b \cdot \nabla u) \, v \, dx$$

$$= \int_{\partial \kappa} (b \cdot \mu_\kappa) u \, v \, ds - \int_\kappa u \, \nabla \cdot (b \, v) \, dx$$

$$= \int_{\partial \kappa_-} (b \cdot \mu_\kappa) u v \, ds + \int_{\partial \kappa_+} (b \cdot \mu_\kappa) u v \, ds - \int_\kappa u \, \nabla \cdot (b \, v) \, dx.$$

For a continuous function it is irrelevant whether we use u, u^+ or u^-, but for a discontinuous function these are significantly different. *On the inflow boundary u is replaced by u^-* then the last term is again integrated by parts. The $\partial \kappa_+$ terms cancel and we get

$$\int_\kappa (b \cdot \nabla u)\, v\, dx$$

$$= \int_{\partial\kappa_-} (b \cdot \mu_\kappa) u^- v^+\, ds + \int_{\partial\kappa_+} (b \cdot \mu_\kappa) u\, v^+\, ds - \int_\kappa u\, \nabla \cdot (b\, v)\, dx$$

$$= \int_\kappa (b \cdot \nabla u)\, v\, dx - \int_{\partial_-\kappa \cap \Gamma^-} (b \cdot \mu_\kappa) u^+ v^+\, ds - \int_{\partial_-\kappa \backslash \Gamma} (b \cdot \mu_\kappa) \lfloor u \rfloor v^+\, ds.$$

This is the motivation for the following weak formulation of (8.7):

$$B_0(u,v) := \sum_{\kappa \in \mathcal{T}} \left(\int_\kappa (b \cdot \nabla u + cu)\, v\, dx \right.$$

$$- \int_{\partial_-\kappa \cap \Gamma} (b \cdot \mu_\kappa) u^+ v^+\, ds - \int_{\partial_-\kappa \backslash \Gamma} (b \cdot \mu_\kappa) \lfloor u \rfloor v^+\, ds \bigg)$$

$$= \sum_{\kappa \in \mathcal{T}} \left(\int_\kappa fv\, dx - \int_{\partial_-\kappa \cap \Gamma^-} (b \cdot \mu_\kappa) gv^+\, ds \right). \tag{8.8}$$

As usual the corresponding finite element method—here the dGFEM—is obtained by a restriction of (8.8) to the discrete spaces.

Example 4.102. Let us consider the simple test problem

$$u_x + u = f, \quad u(0) = A \quad \text{in } (0,1),$$

and discretize using dGFEM on a mesh $\{x_i\}$ with a piecewise constant approximation u_i on every subinterval. This generates the discrete problem

$$\frac{u_i - u_{i-1}}{h_i} + u_i = \frac{1}{h_i} \int_{x_{i-1}}^{x_i} f\, dx.$$

where $h_i = x_i - x_{i-1}$. We have here a clear improvement on the standard Galerkin method, which for piecewise linears generates a central differencing scheme which is unstable for this problem: dGFEM generates a stable approximation that is one-sided (the choice of side depends on the sign of b in (8.7)). □

Why is $B_0(\cdot, \cdot)$ superior to the bilinear form of the standard Galerkin FEM? It is not difficult to verify (with $c_0^2 := c - (\nabla \cdot b)/2$) that

$$B_0(v,v) = \sum_{\kappa \in \mathcal{T}} \left(\|c_0 v\|_{L^2(\kappa)}^2 + \tfrac{1}{2} \left(\|v^+\|_{\partial_-\kappa \cap \Gamma}^2 \right. \right.$$

$$+ \|v^+ - v^-\|_{\partial_-\kappa \backslash \Gamma}^2 + \|v^+\|_{\partial_+\kappa \cap \Gamma}^2 \bigg) \bigg).$$

Here we used the notation

$$\|v\|_\tau^2 = (v,v)_\tau \quad \text{where } (v,w)_\tau = \int_\tau |b \cdot \mu_\kappa| vw\, ds, \quad \tau \subset \partial\kappa. \tag{8.9}$$

Thus, compared with the standard Galerkin FEM, the dGFEM has improved stability properties: we control not only the L_2 norm—the additional terms in $B_0(\cdot,\cdot)$ deliver more stability. As a consequence, one also obtains error estimates in a stronger norm.

The Galerkin orthogonality property

$$B_0(u - u_h, v_h) = 0 \quad \forall v_h \in V_h$$

of the dGFEM means that we can carry out error estimates by following the familiar pattern of finite element analyses. In the next subsection this analysis will be sketched for convection-diffusion problems.

Remark 4.103. (jump-stabilization)
The discretization above can also be derived from the standard weak formulation by the addition of certain jump-penalty terms [38]. Consequently, the dGFEM for problem (8.7) is a kind of stabilized FEM. In [27] the advantages of this approach are discussed.   □

4.8.3 Error Estimates for a Convection-Diffusion Problem

The ideas of the previous two subsections will now be combined in a discussion of the dGFEM for the convection-diffusion problem

$$-\varepsilon\Delta u + b \cdot \nabla u + cu = f \quad \text{in} \quad \Omega, \tag{8.10a}$$

$$u = 0 \quad \text{on} \quad \Gamma. \tag{8.10b}$$

We assume the following hypotheses, which we call assumption (A):

1. $c - (\nabla \cdot b)/2 \geq c_0 > 0$ and Ω is polygonal
2. $u \in H^2(\Omega)$
3. the triangulation of Ω is admissible and quasi-uniform.

The bilinear form associated with problem (8.10) is

$$
\begin{aligned}
B_\pm(v, w) = \sum_{\kappa \in T} \Bigg(& \varepsilon \int_\kappa \nabla v \cdot \nabla w \, dx + \int_\kappa (b \cdot \nabla v + cv)w \, dx \\
& - \int_{\partial_- \kappa \cap \Gamma} (b \cdot \mu)v^+ w^+ \, ds - \int_{\partial_- \kappa \setminus \Gamma} (b \cdot \mu_\kappa)\lfloor v \rfloor w^+ \, ds \Bigg) \\
& + \varepsilon \int_\Gamma (\pm v(\nabla w \cdot \mu) - (\nabla v \cdot \mu)w) \, ds + \int_\Gamma \sigma vw \, ds \\
& + \varepsilon \int_{\Gamma_{int}} (\pm[v]\langle \nabla w \cdot \nu \rangle - \langle \nabla v \cdot \nu \rangle[w]) \, ds + \int_{\Gamma_{int}} \sigma[v][w] \, ds,
\end{aligned}
$$

for $v, w \in V := H^1(\Omega, T)$. Then the dGFEM is:

$$\begin{cases} \text{Find } u_h \in S(\Omega, \mathcal{T}, \mathbf{F}) \text{ such that} \\ B_\pm(u_h, v_h) = L(v_h) \text{ for all } v_h \in S(\Omega, \mathcal{T}, \mathbf{F}), \end{cases} \qquad (8.11)$$

where

$$L(w) = \sum_{\kappa \in \mathcal{T}} \int_\kappa fw \, dx \,.$$

For the error analysis it is important to observe that the dGFEM satisfies the Galerkin orthogonality condition

$$B_\pm(u - u_h, v) = 0, \qquad \text{for all} \quad v \in S(\Omega, \mathcal{T}, \mathbf{F}). \qquad (8.12)$$

Define the dG norm by

$$\|v\|_{dG}^2 = \sum_{\kappa \in \mathcal{T}} \left(\varepsilon \|\nabla v\|_{L^2(\kappa)}^2 + \|c_0 v\|_{L^2(\kappa)}^2 \right) + \int_\Gamma \sigma v^2 \, ds + \int_{\Gamma_{int}} \sigma [v]^2 \, ds$$

$$+ \frac{1}{2} \sum_{\kappa \in \mathcal{T}} \left(\|v^+\|_{\partial_-\kappa \cap \Gamma}^2 + \|v^+ - v^-\|_{\partial_-\kappa \setminus \Gamma}^2 + \|v^+\|_{\partial_+\kappa \cap \Gamma}^2 \right), \quad (8.13)$$

where we recall the notation defined in (8.9). A crucial question is whether $B_\pm(\cdot, \cdot)$ is elliptic on the discrete finite element space. One can see easily that

$$B_+(v, v) \ge c\|v\|_{dG}^2 \quad \forall v \in S(\Omega, \mathcal{T}, \mathbf{F})$$

for the nonsymmetric version on every triangulation and all $\sigma > 0$ (quasi-uniformity is not needed). For the symmetric version, however, to obtain ellipticity over $S(\Omega, \mathcal{T}, \mathbf{F})$ on a quasi-uniform triangulation one should choose

$$\sigma = \frac{\varepsilon}{h} \sigma_0 \qquad (8.14)$$

with some sufficiently large constant σ_0. Once ellipticity is established, the derivation of error estimates is more or less identical for the NIP and SIP methods. We shall present error estimates for NIP because with this method anisotropic meshes can be used without difficulties; this is important for problems whose solutions exhibit layers. With SIP on anisotropic meshes it is more demanding to obtain accurate numerical results: see [52].

Remark 4.104. If one wishes to prove optimal L_2 error estimates or to apply the DWR method to control the error, it is important that the method have the so-called adjoint consistency property. The SIP form is adjoint consistent, but NIP does not have this property; see [60] for details.   □

Now we analyse the error in the nonsymmetric dGFEM, using piecewise linear (discontinuous) elements. The error is

$$u - u_h = (u - \Pi u) + (\Pi u - u_h) \equiv \eta + \xi;$$

here Π is for the moment an arbitrary projector into the finite element space. As usual, Galerkin orthogonality implies

$$\|\xi\|_{dG}^2 = B_+(\xi, \xi) = -B_+(\eta, \xi).$$

We shall estimate $|B_+(\eta, \xi)|$ in a way that shows how the error depends only on the projection error with respect to different norms.

Choose the projection Π to be the L_2 projection onto the finite element space. Then by standard arguments for linear elements on a quasi-uniform triangulation we have

$$\|\eta\|_{L_2} \le C\, h^2\, \|u\|_{H^2(\Omega)} \quad \text{and} \quad \|\eta\|_{H^1(\Omega, T)} \le C\, h\, \|u\|_{H^2(\Omega)}.$$

Applying the Cauchy-Schwarz inequality to estimate the contribution $|B_0(\eta, \xi)|$ from the convective part, one can see that it remains only to bound

$$\sum_{\kappa \in T} \|\eta\|_{L_2(\kappa)} + \|\eta^-\|_{\partial_-\kappa \backslash \Gamma} + \|\eta^+\|_{\partial_+\kappa \cap \Gamma}$$

and

$$\sum_{\kappa \in T} \int_\kappa \eta (b \cdot \nabla \xi)\, dx. \tag{8.15}$$

To estimate the terms $\|\eta^-\|_{\partial_-\kappa \backslash \Gamma}$ and $\|\eta^+\|_{\partial_+\kappa \cap \Gamma}$ one can use the multiplicative trace inequality [44]

$$\|v\|_{L_2(\partial\kappa)}^2 \le C\left(\|v\|_{L_2(\kappa)}|v|_{H^1(\kappa)} + \frac{1}{h_\kappa}\|v\|_{L_2(\kappa)}^2\right), \quad v \in H^1(\kappa). \tag{8.16}$$

The choice of Π implies that the term (8.15) will vanish if

$$b \cdot \nabla_T v \in S(\Omega, T, \mathbf{F}) \quad \forall v \in S(\Omega, T, \mathbf{F}). \tag{8.17}$$

This condition is satisfied if b is piecewise linear. Otherwise, we approximate b by piecewise linears, then apply a triangle inequality and a local inverse inequality to obtain the estimate (for $b \in W^{1,\infty}$)

$$\left|\sum_{\kappa \in T} \int_\kappa \eta (b \cdot \nabla \xi)\, dx\right| \le C\, h^2 \|u\|_{H^2} \|\xi\|_{L_2}.$$

Let us summarize the error analysis up to this point:

Lemma 4.105. *Consider the pure convection problem (8.7) under the assumption that $c - (\nabla \cdot b)/2 \ge c_0 > 0$. Suppose that this problem is discretized on a quasi-uniform triangulation using the discontinuous Galerkin method with piecewise linear (discontinuous) elements. Then the error in the dG-norm related to $B_0(\cdot, \cdot)$ can be estimated by*

$$\|u - u_h\|_{dG_0} \le C\, h^{3/2} \|u\|_{H^2(\Omega)}.$$

The $O(h^{3/2})$ error in L_2 is an improvement over the $O(h)$ attained by the standard Galerkin FEM.

To continue the analysis and get an error estimate for the convection-diffusion problem (8.10), we have to estimate the remaining terms of $B(\eta, \xi)$. For the first term and the penalty terms one simply applies the Cauchy-Schwarz inequality. The remaining integrals on Γ and Γ_{int} are handled using the same technique, which will now be demonstrated for the integrals on Γ.

To estimate the expression

$$Z = \int_\Gamma \varepsilon \left(\eta(\nabla \xi \cdot \nu) - (\nabla \eta \cdot \nu)\xi \right) ds,$$

introduce an auxiliary positive parameter γ and then by the Cauchy-Schwarz inequality one gets

$$|Z| \le \left(\sum_{\kappa \in \mathcal{T}} \frac{\varepsilon}{\gamma} \|\eta\|_{L_2(\partial\kappa \cap \Gamma)}^2 \right)^{1/2} \left(\sum_{\kappa \in \mathcal{T}} \varepsilon \gamma \|\nabla \xi\|_{L_2(\partial\kappa \cap \Gamma)}^2 \right)^{1/2}$$
$$+ \left(\sum_{\kappa \in \mathcal{T}} \frac{\varepsilon^2}{\sigma} \|\nabla \eta\|_{L_2(\partial\kappa \cap \Gamma)}^2 \right)^{1/2} \left(\sum_{\kappa \in \mathcal{T}} \sigma \|\xi\|_{L_2(\partial\kappa \cap \Gamma)}^2 \right)^{1/2}.$$

The second term here can be directly estimated by $\|\xi\|_{dG}$. In the first term we replace the integrals on $\partial\kappa$ by integrals on κ via a local inverse inequality. To compensate for the powers of h that arise we choose $\gamma = O(h)$. Then

$$|Z| \le \left(\left(\sum_{\kappa \in \mathcal{T}} \frac{\varepsilon}{h_\kappa} \|\eta\|_{L_2(\partial\kappa \cap \Gamma)}^2 \right)^{1/2} + \left(\sum_{\kappa \in \mathcal{T}} \frac{\varepsilon^2}{\sigma} \|\nabla \eta\|_{L_2(\partial\kappa \cap \Gamma)}^2 \right)^{1/2} \right) \|\xi\|_{dG}.$$

These terms contribute $O(\varepsilon^{1/2} h)$ and $O(\varepsilon h^{1/2}/(\sigma^{1/2}))$ respectively to the error estimate. On the other hand, the estimates of the penalty terms are $O(\sigma^{1/2} h^{3/2})$. Balancing the influence of the various terms on σ, we conclude that the choice $\sigma = \sigma_0 \varepsilon/h$ for some positive constant σ_0 leads to the best rate of convergence.

Theorem 4.106. *Consider the convection-diffusion problem (8.10) under assumption (A). Suppose that this problem is discretized using the nonsymmetric discontinuous Galerkin method NIP and (discontinuous) piecewise linear elements. If one chooses the penalty parameter to be*

$$\sigma = \sigma_0 \frac{\varepsilon}{h} \tag{8.18}$$

for some constant $\sigma_0 > 0$, then the following error bound holds:

$$\|u - u_h\|_{DG}^2 \le C \left(\varepsilon h^2 + h^3 \right) \|u\|_{H^2(\Omega)}^2. \tag{8.19}$$

The same error estimate can be proved for the symmetric version SIP but one has to choose σ_0 sufficiently large so that the bilinear form is V-elliptic. More general error estimates for the hp-version of the method appear in [71] and [60]. The results of several numerical tests are described in [36].

Remark 4.107. (SIP and SDFEM)
In Chapter 6 we will meet an error estimate for the SDFEM that looks superficially like the above error estimate for dGFEM, but on closer inspection one sees that different norms are used to estimate the error.
It is therefore natural to ask whether one can, for instance, prove an error estimate for the dGFEM in a norm that is typical of the SDFEM. Indeed, in [53] the authors prove a sharpened version of (8.19): for the symmetric version SIP that is based on the bilinear form $B_-(v, w)$, they introduce the norm

$$|||v|||^2 = B_-(v, v) + \sum_{\kappa \in \mathcal{T}} diam(\kappa) \, ||b \cdot \nabla v||^2_{L^2(\kappa)}$$

then prove that

$$|||u - u_h|||^2 \leq C \max\{\varepsilon h^2 + h^3\} ||u||^2_{H^2(\Omega)} .$$

This result is remarkable because $||| \cdot |||$ contains a term that is typical of the SDFEM.   □

Remark 4.108. (L_∞-norm error estimates)
Error estimates in the L_∞ norm are proved in [75] for the symmetric version (SIP) of the dGFEM, but only the Poisson equation is studied.   □

4.9 Further Aspects of the Finite Element Method

4.9.1 Conditioning of the Stiffness Matrix

Consider the simplest stiffness matrix that arises in the finite element method: for the one-dimensional differential expression $-\frac{d^2}{dx^2}$ with homogeneous Dirichlet conditions, using piecewise linear finite elements on an equidistant mesh, the stiffness matrix is

$$A = \frac{1}{h} \begin{bmatrix} 2 & -1 & 0 & \cdots & 0 & 0 \\ -1 & 2 & -1 & \cdots & 0 & 0 \\ 0 & -1 & 2 & -1 & \cdots & 0 \\ \cdots & & \cdot & \cdot & \cdots & \cdot \\ 0 & 0 & \cdots & 0 & -1 & 2 \end{bmatrix} .$$

Let its dimension be $(N - 1) \times (N - 1)$, where $Nh = 1$. The eigenvalues of A are

$$\lambda_k = \frac{4}{h} \sin^2 \frac{k\pi h}{2} \quad \text{for } k = 1, \cdots, N - 1.$$

Consequently the smallest eigenvalue λ_1 satisfies $\lambda_1 \approx \pi^2 h$ and the largest eigenvalue is $\lambda_{N-1} \approx 4/h$. Thus A has (spectral) condition number $\kappa(A) = O(1/h^2)$.

We shall show that $\kappa(A)$ typically has this magnitude when one discretizes symmetric H_1-elliptic bilinear forms with piecewise linear finite elements in the two-dimensional case. For simplicity we assume the triangulation to be both quasi-uniform and uniform.

Let us write an arbitrary $v_h \in V_h$ (the piecewise linear space) in terms of the nodal basis $\{\varphi_i\}$ of V_h:

$$v_h = \sum_i \eta_i \varphi_i.$$

Then substituting v_h into the bilinear form generates a quadratic form:

$$a(v_h, v_h) = \eta^T A \eta.$$

We estimate the eigenvalues of A using a Rayleigh quotient. First, setting $\|\eta\|^2 = \sum_i \eta_i^2$, the application of a global inverse inequality yields

$$\frac{\eta^T A \eta}{\|\eta\|^2} = \frac{a(v_h, v_h)}{\|\eta\|^2} \leq C h^{-2} \frac{\|v_h\|_0^2}{\|\eta\|^2} \leq C.$$

The last inequality holds because on V_h the continuous L_2 norm $\|v_h\|_0^2$ is equivalent to the discrete norm $h^2 \|\eta\|^2$.

Second, by V-ellipticity we have

$$\frac{\eta^T A \eta}{\|\eta\|^2} = \frac{a(v_h, v_h)}{\|\eta\|^2} \geq \alpha \frac{\|v_h\|_1^2}{\|\eta\|^2} \geq \alpha \frac{\|v_h\|_0^2}{\|\eta\|^2} \geq C h^2.$$

That is, $\lambda_{max} \leq C$ and $\lambda_{min} \geq C h^2$; hence $\kappa(A)$ is $O(1/h^2)$.

In general: *The discretization of boundary value problems of order $2m$ by finite elements leads under the above assumptions on the triangulation to a stiffness matrix whose condition number $\kappa(A)$ is $O(1/h^{2m})$.*

Nevertheless note that the choice of basis in the finite element space V_h significantly influences the conditioning. For instance, hierarchical or wavelet bases are better behaved than nodal bases; see [Osw94, Chapter 4.2].

4.9.2 Eigenvalue Problems

Consider the following eigenvalue problem:
Find $u \in H_0^1(\Omega) = V$ and $\lambda \in \mathbb{R}$ such that

$$a(u, v) = \lambda (u, v) \quad \text{for all} \quad v \in V. \tag{9.1}$$

Here $a(\cdot, \cdot)$ is a given symmetric V-elliptic bilinear form.

As $a(\cdot, \cdot)$ is symmetric it is well known that its eigenvalues are real and, owing to the V-ellipticity, positive. Moreover they can be enumerated and have no finite accumulation point. Ordering the eigenvalues by

$$0 < \lambda_1 \leq \lambda_2 \leq \cdots \leq \lambda_k \leq \lambda_{k+1} \leq \cdots,$$

the Rayleigh quotient R allows the characterization

$$\lambda_k = \min_{v \perp E_{k-1}} R(v) = \min_{v \perp E_{k-1}} \frac{a(v, v)}{\|v\|_0^2},$$

where E_k is the space spanned by the first k eigenfunctions.

The eigenvalues can also be characterized by the min-max principle

$$\lambda_k = \min_{M_k \subset V} \left\{ \max_{v \in M_k} \frac{a(v, v)}{\|v\|_0^2}, \ \dim M_k = k. \right\}$$

Let us now discretize the eigenvalue problem (9.1) using continuous piecewise linear finite elements: Find $u_h \in V_h$ and $\lambda^h \in \mathbb{R}$ such that

$$a(u_h, v_h) = \lambda^h (u_h, v_h) \quad \text{for all} \quad v_h \in V_h.$$

This is equivalent to a standard matrix eigenvalue problem. Setting $N = \dim V_h$, we enumerate the discrete eigenvalues:

$$0 < \lambda_1^h \leq \lambda_2^h \leq \cdots \leq \lambda_N^h.$$

Then one has

Theorem 4.109. *If $u \in H^2(\Omega)$ and the triangulation used is quasi-uniform, then the discrete eigenvalues λ_k^h satisfy the estimate*

$$\lambda_k \leq \lambda_k^h \leq \lambda_k + C(\lambda_k h)^2.$$

For the proof observe first that the inequality $\lambda_k \leq \lambda_k^h$ follows immediately from the min-max principle above for $\dim V_h \geq k$ (i.e., for h sufficiently small).

It is much more complicated to verify the second inequality. We start from

$$\lambda_k^h \leq \max_{v \in M_k} \frac{a(v, v)}{\|v\|_0^2}$$

for any subspace M_k of V_h with $\dim M_k = k$. Let E_k^0 be the k-dimensional subspace spanned by the first k eigenfunctions of $a(\cdot, \cdot)$. Choose M_k to be the Ritz projection of E_k^0 into V_h. Later we shall show that $\dim M_k = k$ also. Denoting the Ritz projector by P, it follows that

$$\lambda_k^h \leq \max_{w \in E_k^0} \frac{a(Pw, Pw)}{\|Pw\|_0^2}.$$

One can write the numerator as

$$a(Pw, Pw) = a(w, w) - 2a(w - Pw, Pw) - a(w - Pw, w - Pw).$$

The central term on the right-hand side vanishes by Galerkin orthogonality, so

$$a(Pw, Pw) \leq a(w, w) \leq \lambda_k \quad \text{since } w \in E_k^0.$$

For the denominator we have

$$\|Pw\|_0^2 = \|w\|_0^2 - 2(w, w - Pw) + \|w - Pw\|_0^2. \tag{9.2}$$

We shall prove below that

$$\sigma_k^h := \max_{w \in E_k^0} \left| 2(w, w - Pw) - \|w - Pw\|_0^2 \right| \tag{9.3}$$

satisfies

$$0 \leq \sigma_k^h \leq C\lambda_k h^2 < \frac{1}{2} \tag{9.4}$$

for sufficiently small h. Incorporating this bound into (9.2) yields

$$\|Pw\|_0^2 \geq 1 - \sigma_k^h.$$

The desired estimate

$$\lambda_k^h \leq \lambda_k(1 + 2C\lambda_k h^2)$$

now follows from the inequality $1/(1 - x) \leq 1 + 2x$ for $x \in [0, 1/2]$ because $\sigma_k^h < 1/2$. The key ingredient of our proof is the inequality (9.4), which we now verify. Denote the normalised eigenfunctions of $a(\cdot, \cdot)$ by $\{w_i\}$. Let

$$w = \sum_{i=1}^{k} \alpha_i w_i$$

be an arbitrary member of E_k^0. Then

$$|(w, w - Pw)| = \left| \sum_i (\alpha_i w_i, w - Pw) \right|$$

$$= \left| \sum_i \alpha_i \lambda_i^{-1} a(w_i, w - Pw) \right|$$

$$= \left| \sum_i \alpha_i \lambda_i^{-1} a(w_i - Pw_i, w - Pw) \right|$$

$$\leq Ch^2 \left\| \sum_i \alpha_i \lambda_i^{-1} w_i \right\|_2 \|w\|_2.$$

The assumed H^2 regularity of u implies $\|w\|_2 \leq C\lambda_k$ and

$$\left\| \sum_i \alpha_i \lambda_i^{-1} w_i \right\|_2 \le C$$

follows from V-coercivity and

$$a\left(\sum_i \alpha_i \lambda_i^{-1} w_i, v \right) = \left(\sum_i \alpha_i w_i, v \right) \quad \text{for all } v \in V.$$

Consequently, we have

$$|(w, w - Pw)| \le C\lambda_k h^2 \quad \text{for all } w \in E_k^0.$$

On the other hand, the error of the Ritz projection in the L_2 norm yields

$$\|w - Pw\|_0^2 \le Ch^4 \|w\|_2^2 \le C\lambda_k^2 h^4.$$

Combining these bounds with (9.3), we have proved (9.4).

It remains only to show that the dimension of M_k is indeed k. For suppose that there exists $w^* \in E_k^0$ with $Pw^* = 0$. Without loss of generality $\|w^*\| = 1$ and it then follows that

$$1 = \|w^*\|_0^2 = \left| 2(w^*, w^* - Pw^*) - \|w^* - Pw^*\|_0^2 \right| \le \sigma_k^h < \frac{1}{2}.$$

This contradiction proves the bijectivity of the mapping from E_k^0 to M_k. ■

The finite element method with polynomials of local degree m gives the bound

$$\lambda_k^h \le \lambda_k + C\lambda_k^{m+1} h^{2m};$$

see [Hac03a]. This reference also estimates the accuracy with which the eigenfunctions are approximated:

$$\|w_k - w_k^h\|_1 \le C\lambda_k^{(m+1)/2} h^m$$

and

$$\|w_k - w_k^h\|_0 \le C\lambda_k^{(m+1)/2} h^{m+1}.$$

4.9.3 Superconvergence

The term *superconvergence* is used in the literature to describe various phenomena that are characterized by improved convergence rates. For instance, we say we have superconvergence when

(i) at special points x_i one has better convergence rates for $(u - u_h)(x_i)$ than for $\|u - u_h\|_\infty$. Similar effects can be detected for derivatives.

(ii) for some projector P into the finite element space, the convergence rates for the difference $\|u_h - Pu\|$ are better than for the finite element error $\|u_h - u\|$;

(iii) there exists a recovery operator R—e.g., for the gradient—such that the approximation of ∇u by Ru_h is of higher order than its approximation by ∇u_h.

Further examples of superconvergence phenomena are given in [Wah95] and [KNS98]. In what follows we prove some special results related to the classes (i) and (ii) above. Recovery is discussed in detail in [AO00, Chapter 4]; see also the recent polynomial recovery technique of [127]. As we mentioned in Section 4.7, the construction of a posteriori error estimators can be based on recovery techniques.

We begin with an examination of superconvergence points for the derivative of the solution in the one-dimensional case. Consider the boundary value problem

$$-(u' + bu)' = f, \quad u(0) = u(1) = 0,$$

and its finite element discretization by the space V_h of piecewise polynomials of degree at most $k \geq 1$ on an equidistant mesh $\{x_i\}$ of diameter h. We know that for sufficiently smooth solutions one has the error estimate

$$\|(u_h - u)'\|_\infty \leq C\, h^k$$

and our interest is the identification of points with an improved convergence rate.

First we prove :

Lemma 4.110. *Let $\tilde{u}_h \in V_h$ be the projection of u' into V_h induced by the scalar product associated with the H^1 seminorm. Then*

$$\|(u_h - \tilde{u}_h)'\|_\infty \leq C\, h^{k+1}.$$

Proof: Set $\theta = \tilde{u}_h - u_h$. Then

$$(\theta', v_h') = ((\tilde{u}_h - u)', v_h') - ((u_h - u)', v_h') = (b(u - u_h), v_h') \quad \text{for all } v_h \in V_h.$$

Hence

$$|(\theta', v_h')| \leq C\|u - u_h\|_\infty \|v_h\|_{W^{1,1}} \leq C\, h^{k+1} \|v_h\|_{W^{1,1}} \quad \text{for all } v_h \in V_h. \quad (9.5)$$

This estimate will be used to control $\|\theta'\|_\infty$ via the well-known L_∞-norm characterization

$$\|\theta'\|_\infty = \sup_{\|\psi\|_{L_1}=1} (\theta', \psi). \quad (9.6)$$

Let P denote the L_2 projection onto the space of (not necessarily globally continuous) piecewise polynomials of degree at most $k - 1$. Let $\psi \in L_1[0,1]$ be arbitrary. Then

$$(\theta', \psi) = (\theta', P\psi).$$

Define the auxiliary function $\varphi \in V_h$ by

$$\varphi(x) := \int_0^x P\psi - x \int_0^1 P\psi \quad \text{for } 0 \le x \le 1.$$

This function has the property that

$$(\theta', \psi) = (\theta', P\psi) = \left(\theta', \varphi' + \int_0^1 P\psi\right) = (\theta', \varphi').$$

But

$$\|\varphi\|_{W^{1,1}} \le C\|P\psi\|_{L_1} \le C\|\psi\|_{L_1}. \tag{9.7}$$

By combining (9.5), (9.6) and (9.7) the proof is complete. ∎

In fact for piecewise linear elements $\tilde{u}_h$ is the nodal interpolant of u.

Lemma 4.110 enables us to identify superconvergence points for the derivative of the discrete solution.

Theorem 4.111. *Let the points η_j be the zeros of the kth Legendre polynomial L_k on the element (x_i, x_{i+1}). Then*

$$|(u - u_h)'(\eta_j)| \le C h^{k+1}.$$

Proof: The definition of $\tilde{u}_h$ in Lemma 4.110 implies that if ψ is some arbitrary polynomial of degree $k - 1$ on (x_i, x_{i+1}) then

$$\int_{x_i}^{x_{i+1}} (\tilde{u}_h - u)'\psi = 0. \tag{9.8}$$

Now expand $(\tilde{u}_h - u)'$ in a Taylor series about the point $(x_i + x_{i+1})/2$. This yields a polynomial of degree k, which we express as a linear combination of Legendre polynomials, and a remainder:

$$(\tilde{u}_h - u)'(x) = \sum_{l=0}^{k} c_l L_l(x) + O(h^{k+1}).$$

Evaluating the coefficients c_l for $l \le k - 1$ by means of the orthogonality property of Legendre polynomials, the identity (9.8) tells us that each c_l is $O(h^{k+1})$. Since by definition $L_k(\eta_i) = 0$, one therefore obtains

$$(\tilde{u}_h - u)'(\eta_i) = \sum_{l=0}^{k-1} c_l L_l(\eta_i) + O(h^{k+1}) = O(h^{k+1}). \tag{9.9}$$

This estimate and Lemma 4.110 together give the desired result. ∎

One can show similarly that the zeros of L_k' are superconvergence points for $u - u_h$.

For more details, including a generalization of these results to the higher-dimensional tensor product case, see [Wah95, Chapter 6]. On simplices (e.g., triangles in 2D) the situation is more complicated.

Next we move on to Lin's technique of integral identities for the verification of superconvergence results of type (ii). This material appears in several books from the 1980s and 1990s but unfortunately these books are available only in Chinese with one recent exception [LL06].

Let V_h be the space of piecewise bilinear finite elements on a two-dimensional rectangular mesh of diameter h. Denote by $u^I \in V_h$ the bilinear nodal interpolant of u.

Theorem 4.112. *Let $u \in H_0^1(\Omega) \cap H^3(\Omega)$. Then*

$$|((u - u^I)_x, v_x)| + |((u - u^I)_y, v_y)| \le C h^2 \|u\|_3 \|v\|_1 \quad \text{for all } v \in V_h. \quad (9.10)$$

Proof: Let us first remark that a direct application of the Cauchy-Schwarz inequality and an invocation of standard interpolation error estimates (which assume only $u \in H^2$) yield a multiplicative factor h but not h^2.

We prove (9.10) for $((u - u^I)_x, v_x)$. Consider a mesh rectangle R with midpoint (x_r, y_r) and edge lengths $2h_r$ and $2k_r$. Define the auxiliary function

$$F(y) := \frac{1}{2} \left[(y - y_r)^2 - k_r^2 \right].$$

Then the following *Lin identity* is valid:

$$\int_R (u - u^I)_x \, v_x = \int_R \left[F u_{xyy} \, v_x - \frac{1}{3} (F^2)' u_{xyy} \, v_{xy} \right]. \quad (9.11)$$

Now, observing that $F = O(k_r^2)$ and $(F^2)' = O(k_r)^3$, the inequality (9.10) follows immediately from the Lin identity by applying a local inverse inequality to v_{xy} (this can be done since $v \in V_h$).

How does one verify (9.11)? We use the abbreviation $e := u - u^I$. As v is bilinear one gets

$$\int_R (e_x v_x)(x, y) \, dx \, dy = \int_R e_x(x, y) \left[v_x(x_r, y_r) + (y - y_r) v_{xy} \right] dx \, dy$$

$$= \left(\int_R e_x \right) v_x(x_r, y_r) + \left(\int_R (y - y_r) e_x(x, y) \, dx \, dy \right) v_{xy}.$$

Repeated integrations by parts give

$$\int_R F e_{xyy} = -\int_R F' e_{xy} = \int_R F'' e_x = \int_R e_x$$

because F vanishes identically and $\int e_x = 0$ along the top and bottom of R. Similarly

$$\int_R (y - y_r) e_x(x, y) = \frac{1}{6} \int_R (F^2)' e_{xyy}.$$

Consequently

$$\int_R e_x v_x = \int_R F(y) e_{xyy}(x, y) [v_x(x, y) - (y - y_r) v_{xy}] \, dx \, dy + \frac{1}{6} \int_R (F^2)' e_{xyy} v_{xy}.$$

But $(F^2)'(y) = 2(y - y_r) F(y)$ so (9.11) follows. ∎

Why is Theorem 4.112 valuable?
Consider, for instance, the Poisson equation with homogeneous Dirichlet conditions in a rectangular domain and discretized with bilinear finite elements. Since $a(v, w) = (\nabla v, \nabla w)$, Theorem 4.112 yields

$$\begin{aligned}
\alpha \|u_h - u^I\|_1^2 &\le (\nabla(u_h - u^I), \nabla(u_h - u^I)) \\
&= (\nabla(u - u^I), \nabla(u_h - u^I)) \le C h^2 \|u\|_3 \|u_h - u^I\|_1.
\end{aligned}$$

Now, assuming that $u \in H_0^1(\Omega) \cap H^3(\Omega)$, we have the interesting superconvergence estimate (compare with $\|u - u^I\|_1 = O(h)$ and $\|u - u_h\|_1 = O(h)$)

$$\|u_h - u^I\|_1 \le C h^2.$$

Similar Lin identities can be proved for certain other types of integrals, and some results can be extended to higher-order rectangular elements provided special interpolants are used; see, e.g., [84]. On triangular meshes the situation is more complicated but some results are known [KNS98].

4.9.4 p- and hp-Versions

The accuracy of a finite element approximation can be improved in several ways. If the mesh is refined while the structure of the local basis functions is unchanged, one has the *h-version* of the finite element method. Alternatively, one can work on a fixed mesh and increase the polynomial degree of the local basis functions used; this is the *p-version*. The *hp-version* is a combination of these strategies.

Even in the one-dimensional case the difference between the asymptotic convergence rates of the h-version and the p-version is evident. Let $\Omega = (a, b)$ and subdivide Ω into equidistant subintervals of diameter h. On this mesh let $V_h^p \subset H^1$ be the finite element space of piecewise polynomials of degree at most p. Then the following approximation error estimates hold true [106]: For each $u \in H^{k+1}(\Omega)$ there exists a $v_h^p \in V_h^p$ such that

$$|u - v_h^p|_1 \le C \frac{h^{\min(p,k)}}{p^k} |u|_{k+1},$$

$$\|u - v_h^p\|_0 \le C \frac{h^{\min(p,k)+1}}{p^{k+1}} |u|_{k+1}.$$

We deduce that *if k is large—i.e., the solution u is very smooth—then it is more advantageous to increase the polynomial degree (p → ∞) than to refine the mesh (h → 0)!*

Using standard analytical techniques these approximation rates can be converted into convergence rates and the convergence behaviour is called *spectral convergence* as $p \to \infty$.

If the solution u is analytic, then there exists $r > 0$ with

$$|u - v_h^p|_1^2 \leq C(r)\, p\, r^{-2p},$$
$$\|u - v_h^p\|_0^2 \leq C(r)\, p^{-1}\, r^{-2p}.$$

That is, one witnesses *exponential convergence*.

Often one deals with solutions u that are at least piecewise analytic. Moreover for singular solutions of the type $|x - x_0|^\alpha$, it can be shown that geometric mesh refinement near the singularity at $x = x_0$ combined with a judicious increase of the polynomial degree leads to good convergence rates; see [106].

Up to now not much is known about the stiffness matrices generated by the p-version of the FEM. Most of the existing results concentrate on the tensor product case. It seems to be typical that the condition number behaves like p^4 (in 2D) and p^6 (in 3D); see [89] and its references. In the two-dimensional case it is also proved that static condensation is a good preconditioner.

5

Finite Element Methods for Unsteady Problems

In this chapter we deal with the discretization of second-order initial-boundary value problems of parabolic and hyperbolic types. Unlike Chapter 2, the space discretization is handled by finite element methods. We shall begin with some basic properties that are used later to derive appropriate weak formulations of the problems considered. Then, moving on to numerical methods, various temporal discretizations are discussed for parabolic problems: one-step methods (Runge-Kutta), linear multi-step methods (BDF) and the discontinuous Galerkin method. An examination of similar topics for second-order hyperbolic problems follows. The initial value problems that are generated by semi-discretization methods are very stiff, which demands a careful choice of the time integration codes for such problems. Finally, at the end of the chapter some results on error control are given.

For first-order hyperbolic equations, see the variants of the finite element method that are presented in Chapters 4 (the discontinuous Galerkin method) and 6 (the streamline diffusion FEM).

Throughout this chapter differential operators such as ∇ and Δ involve only spatial derivatives.

5.1 Parabolic Problems

5.1.1 On the Weak Formulation

As a model problem we consider the initial-boundary value problem

$$
\begin{aligned}
\frac{\partial u}{\partial t} - \triangle u &= f && \text{in} && \Omega \times (0, T), \\
u(\cdot, 0) &= u_0 && \text{on} && \overline{\Omega}, \\
u &= 0 && \text{on} && \partial \Omega \times (0, T),
\end{aligned}
\tag{1.1}
$$

where Ω is a bounded domain in $\mathbb{R}^n$ for some $n \geq 1$, with piecewise smooth boundary $\partial \Omega$, and T is a positive constant. Multiplication of the differential equation by an arbitrary $v \in C_0^\infty(\Omega)$ then integration by parts yields

$$\frac{d}{dt} \int_\Omega u(x,t)v(x)\,dx + \int_\Omega \nabla u \cdot \nabla v\,dx = \int_\Omega fv\,dx. \qquad (1.2)$$

Set $V = H_0^1(\Omega)$ and $H = L_2(\Omega)$. For each fixed t the mapping $x \mapsto u(x,t)$ is considered as an element of the space V, and is denoted by $u(t) \in V$. Then for variable t the mapping $t \mapsto u(t) \in V$ is defined. Now (1.2) can be written in the following form:
Find $u(t) \in V$ that satisfies $u(0) = u_0$ and

$$\frac{d}{dt}(u(t),v) + a(u(t),v) = (f(t),v) \qquad \text{for all } v \in V. \qquad (1.3)$$

Here $(\cdot,\cdot)$ is the $L_2(\Omega)$ inner product and $a(u(t),v) := (\nabla u(t), \nabla v)$.

To give a precise statement of an existence theorem for problems of type (1.3), some analysis on spaces of vector-valued functions (like our $u(t) \in V$) is required. Here we present some results from the literature and for further reading refer to, e.g., [Zei90].

Let us consider a simple but important example. Let X be an arbitrary Banach space with norm $\|\cdot\|_X$. Then $L_2(0,T;X)$ denotes the space of all functions $u : (0,T) \to X$ for which the norm

$$\|u\|_{L_2(0,T;X)} := \left(\int_0^T \|u(t)\|_X^2\,dt \right)^{1/2}$$

is finite. Equipped with the norm $\|\cdot\|_{L_2(0,T;X)}$, the space $L_2(0,T;X)$ is in fact a Banach space. Set $Q = \Omega \times (0,T)$. If $f \in L_2(Q)$, then $f \in L_2(0,T; L_2(\Omega))$. Indeed, Fubini's theorem applied to such f implies that

$$\|f\|_{L_2(Q)}^2 = \int_0^T \left(\int_\Omega f^2(x,t)\,dx \right) dt.$$

Hence, for each $t \in [0,1]$, the function $x \mapsto f(x,t)$ is an element of $L_2(\Omega)$. Equivalently, $f(\cdot,t) \in L_2(\Omega)$. It is in this sense that we understand the connection between the right-hand side of the original problem (1.1) and the right-hand side of its weak formulation (1.3).

If X is reflexive and separable then $L_2(0,T;X^*)$ is the dual space of $L_2(0,T;X)$, where X^* denotes the dual space of X.

An *evolution triple* (or *Gelfand triple*) is a pair of imbeddings

$$V \subset H \subset V^*$$

such that

- V is a separable and reflexive Banach space;
- H is a separable Hilbert space;

- V is dense in H and continuously embedded in H, i.e.,

$$\|v\|_H \leq \text{constant } \|v\|_V \qquad \text{for all } v \in V .$$

In the context of unsteady problems, the typical example of such a triple is

$$V = H_0^1(\Omega), \quad H = L_2(\Omega), \quad V^* = H^{-1}(\Omega).$$

Assuming in the sequel that (V, H, V^*) is a Gelfand triple, a mapping $u \in L_2(0, T; V)$ has a generalized derivative $w \in L_2(0, T; V^*)$ if

$$\int_0^T \varphi'(t) u(t) dt = - \int_0^T \varphi(t) w(t) dt \qquad \text{for all } \varphi \in C_0^\infty(0, T) .$$

These are Bochner integrals for function-valued mappings; see [Zei90].

Armed with this definition of generalized derivatives, we define (similarly to the standard Sobolev spaces of functions) the Sobolev space $W_2^1(0, T; V, H)$ of function-valued mappings to be the space of $u \in L_2(0, T; V)$ that possess generalized derivatives $u' \in L_2(0, T; V^*)$. The norm on this space is

$$\|u\|_{W_2^1} := \|u\|_{L_2(0,T;V)} + \|u'\|_{L_2(0,T;V^*)}.$$

An important property is that the mapping $u : [0, T] \to H$ is continuous (with possible exceptions only on sets of measure zero). Consequently the initial condition $u(0) \in H$ that appears in parabolic problems is well defined.

Furthermore, the following *integration by parts formula* holds:

$$(u(t), v(t)) - (u(s), v(s)) = \int_s^t [\langle u'(\tau), v(\tau) \rangle + \langle v'(\tau), u(\tau) \rangle] d\tau,$$

where $\langle \cdot, \cdot \rangle : V^* \times V \to \mathbb{R}$ denotes the duality pairing.

Now we are able to state more precisely the generalized or weak formulation of a linear parabolic initial-boundary value problem. Let V be a Sobolev space, with $H_0^1(\Omega) \subset V \subset H^1(\Omega)$, that encompasses the boundary conditions of the given problem. Set $H = L_2(\Omega)$. Assume that we are given a bounded V-elliptic bilinear form $a(\cdot, \cdot) : V \times V \to \mathbb{R}$ and $f \in L_2(0, T; V^*)$. The weak formulation of the related parabolic problem is:
Find $u \in W_2^1(0, T; V, H)$, with $u(0) = u_0 \in H$, such that

$$\frac{d}{dt}(u(t), v) + a(u(t), v) = \langle f(t), v \rangle \text{ for all } v \in V . \tag{1.4}$$

This problem has a unique solution; see, e.g., [Zei90].

In particular for example (1.1) we have:
Given any $f \in L_2(Q)$ and $u_0 \in L_2(\Omega)$, there exists a unique solution $u \in L_2(0, T; H_0^1(\Omega))$ with $u' \in L_2(0, T; H^{-1}(\Omega))$.

Choosing $v = u(t)$ in (1.4), we obtain

$$\frac{1}{2}\frac{d}{dt}\|u(t)\|^2 + a(u(t), u(t)) = (f(t), u(t)),$$

where $\|\cdot\| = \|\cdot\|_{L_2(\Omega)}$. Hence the V-ellipticity (with coercivity constant α in the H^1-norm) of the bilinear form $a(\cdot, \cdot)$ and the Cauchy-Schwarz inequality yield

$$\frac{d}{dt}\|u(t)\| + \alpha\|u(t)\| \le \|f(t)\|.$$

Integrating in time leads to the a priori estimate

$$\|u(t)\| \le \|u_0\|e^{-\alpha t} + \int_0^t e^{-\alpha(t-s)}\|f(s)\|ds \quad \text{for } 0 \le t \le T. \tag{1.5}$$

As we already saw in the case of elliptic boundary value problems, an optimal rate of convergence of the solution obtained from a discretization method can be achieved only under the hypothesis of higher-order regularity of the solution u. Parabolic initial-boundary value problems have the *smoothing property* that the solution becomes more regular with increasing t.

Example 5.1. Consider the initial-boundary value problem

$$u_t - u_{xx} = 0, \quad u(0, t) = u(\pi, t) = 0,$$

$$u(x, 0) = u_0(x).$$

The separation of variables technique yields the following representation of the solution:

$$u(x, t) = \sum_{j=1}^{\infty} u_j e^{-j^2 t} \sin(jx) \quad \text{with} \quad u_j = \sqrt{\frac{2}{\pi}} \int_0^{\pi} u_0(x) \sin(jx) dx.$$

By Parseval's identity the behaviour of $\|u_t\|^2$ (higher-order derivatives can be studied in a similar manner) is directly related to the convergence behaviour of the series

$$\sum_{j=1}^{\infty} u_j^2 j^4 e^{-j^2 t}.$$

For $t \ge \sigma$, where σ is any fixed positive constant, the terms in the series are damped strongly by the exponential function. This implies the uniform boundedness of $\|u_t\|$ (and also higher-order temporal derivatives) as a function of $t \in [\sigma, T]$. But in the neighbourhood of $t = 0$ no such uniform boundedness property can be expected.
For example, if $u_0(x) = \pi - x$ then $u_j = c/j$. In this case we have

$$\|u_t\| \sim c t^{-\frac{3}{4}}, \quad \text{since} \sum_{j=1}^{\infty} u_j^2 j^4 e^{-j^2 t} = c^2 \left(\sum_{j=1}^{\infty} \frac{1}{j} \left(j t^{\frac{1}{2}} \right)^3 e^{-j^2 t} \right) \cdot t^{-\frac{3}{2}}.$$

Only if $u_0(\cdot)$ is sufficiently smooth with $u_0(0) = u_0(\pi) = 0$ will the Fourier coefficients u_j decrease more rapidly, and in such cases the solution is better behaved near $t = 0$. $\quad\square$

As a sample regularity result we state [Tho97, Chapter 3, Lemma 2]. Consider the initial-boundary value problem

$$\begin{aligned} \tfrac{\partial u}{\partial t} - \triangle u &= 0 \quad \text{in } \Omega \times (0,T), \\ u &= 0 \quad \text{on } \partial\Omega \times (0,T), \\ u &= u_0 \text{ for } t = 0 \, . \end{aligned} \tag{1.6}$$

Lemma 5.2. *If $u_0 \in L_2(\Omega)$, then for $t \geq \delta > 0$ and any natural k the $H^k(\Omega)$ norm of $u(t)$ is bounded; more precisely,*

$$\|u(t)\|_{H^k(\Omega)} \leq C\, t^{-\frac{1}{2}k} \|u_0\| \qquad \text{for } t > 0 \, . \tag{1.7}$$

Further results on the regularity of solutions are given in [Tho97].

5.1.2 Semi-Discretization by Finite Elements

The discretization of parabolic initial-boundary value problems by the method of finite elements could be done directly for both the spatial and temporal variables, just like the finite difference approach discussed in Chapter 2. Nevertheless there are certain advantages if, to begin with, only the spatial variables or only time is discretized. That is, we prefer to use *semi-discretization* and this technique will be analysed below. In the present section we first discretize the spatial variables; this type of semi-discretization is called the *(vertical) method of lines* (MOL). The alternative approach of first discretizing the temporal variable—Rothe's method—will be discussed later.

We consider the initial-boundary value problem

$$\begin{aligned} \tfrac{\partial u}{\partial t} + Lu &= f \quad \text{in} \quad \Omega \times (0,T), \\ u &= 0 \quad \text{on} \quad \partial\Omega \times (0,T), \\ u &= u_0(x) \text{ for } t = 0 \text{ and } x \in \Omega, \end{aligned} \tag{1.8}$$

where L is some uniformly elliptic differential operator in the spatial variables. Here $\Omega \subset \mathbb{R}^n$ is a bounded domain with smooth boundary $\partial\Omega$. Other types of boundary conditions can be treated similarly.

When the spatial variables in problem (1.8) are discretized by any discretization method (finite differences, finite volumes, finite elements, ...), this semi-discretization yields a system of ordinary differential equations.

Example 5.3. Let us consider the initial-boundary value problem

$$\begin{aligned} u_t - u_{xx} &= f(x,t) \qquad \text{in } (0,1) \times (0,T), \\ u|_{x=0} = u|_{x=1} &= 0, \\ u|_{t=0} &= u_0(x). \end{aligned}$$

We discretize in x on an equidistant grid $\{x_i : i = 0, \ldots, N\}$ of diameter h by applying finite differences as described in Chapter 2. For each i let $u_i(t)$ denote an approximation of $u(x_i, t)$ and set $f_i(t) = f(x_i, t)$. Then semi-discretization yields the ODE system

$$\frac{du_i}{dt} = \frac{u_{i-1} - 2u_i + u_{i+1}}{h^2} + f_i(t), \quad i = 1, \dots, N-1, \quad u_0 = u_N = 0,$$

with the initial condition $u_i(0) = u_0(x_i)$ for all i. $\qquad \square$

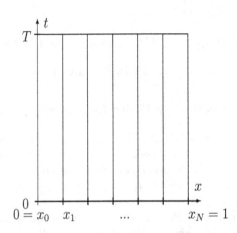

Figure 5.1 Vertical method of lines

Next consider a semi-discretization where a finite element method is used to discretize the spatial variables. To do this we use a weak formulation of (1.8). As in subsection 5.1.1, choose $V = H_0^1(\Omega)$, $H = L_2(\Omega)$ and $u_0 \in H$. Then the weak formulation of the problem is: find $u \in W_2^1(0, T; V, H)$ with $u(0) = u_0$ such that

$$\frac{d}{dt}(u(t), v) + a(u(t), v) = \langle f(t), v \rangle \qquad \text{for all } v \in V. \qquad (1.9)$$

Here $f \in L_2(0, T; V^*)$ and $a(\cdot, \cdot)$ is a continuous V-elliptic bilinear form on $V \times V$. Let $V_h \subset V$ be a conforming finite element space. (Semi-discretizations by non-conforming methods are not discussed here.) The semi-discrete analogue of (1.9) is:
Find $u_h(t) \in V_h$ with

$$\frac{d}{dt}(u_h(t), v_h) + a(u_h(t), v_h) = \langle f(t), v_h \rangle \text{ for all } v_h \in V_h \qquad (1.10)$$

and $u_h(0) = u_h^0 \in V_h$. Here $u_h^0 \in V_h$ is some approximation of the initial condition u_0 in V_h, e.g., an interpolant to u_0 from V_h provided this is well defined. An alternative choice of u_h^0 is the L_2 projection into V_h defined by

$$(u_h^0, v_h) = (u_0, v_h) \qquad \text{for all} \quad v_h \in V_h.$$

Let $\{\varphi_1, \cdots, \varphi_M\}$ be a basis for V_h. Then the numerical solution is

$$u_h(x, t) = \sum_{i=1}^{M} u_i(t)\varphi_i(x)$$

for some functions $\{u_i(t)\}$. When this is substituted into (1.10) one obtains the following system of ordinary differential equations:

$$\sum_{i=1}^{M} u_i'(t)(\varphi_i, \varphi_j) + \sum_{i=1}^{M} u_i(t)a(\varphi_i, \varphi_j) = \langle f(t), \varphi_j \rangle, \quad j = 1, \ldots, M.$$

For notational convenience we introduce the matrices and vectors

$$D = (d_{ij}), \ d_{ij} = (\varphi_j, \varphi_i),$$
$$A = (a_{ij}), \ a_{ij} = a(\varphi_j, \varphi_i),$$
$$\hat{f}(t) = (f_j), \ f_j = \langle f, \varphi_j \rangle, \ \hat{u}(t) = (u_j).$$

Then the above system of ordinary differential equations can be expressed as

$$D(\hat{u}(t))' + A\hat{u}(t) = \hat{f}(t). \tag{1.11}$$

The initial condition for $\hat{u}(0)$ must also be taken into account. If its L_2 projection into V_h is used, then in the finite element method this initial condition becomes

$$D\hat{u}(0) = u_0^* \quad \text{with} \quad u_0^* = (u_{0,j}), \ u_{0,j} = (u_0, \varphi_j).$$

Example 5.4. Consider again the initial-boundary value problem of Example 5.3. For the spatial discretization, suppose that we use piecewise linear finite elements on an equidistant mesh of diameter h with the usual hat function basis. Then the matrices D and A are

$$D = \frac{h}{6}\begin{bmatrix} 4 & 1 & \cdots & 0 \\ 1 & 4 & \cdots & \vdots \\ \vdots & \cdots & & \\ & \cdots & 4 & \\ 0 & \cdots & 1 & 4 \end{bmatrix}, \quad A = \frac{1}{h}\begin{bmatrix} 2 & -1 & \cdots & 0 \\ -1 & 2 & \cdots & \vdots \\ \vdots & \cdots & 2 & \\ 0 & \cdots & -1 & 2 \end{bmatrix}.$$

After this is normalized by dividing by h, the finite element discretization of the term u_{xx} is identical with its finite difference discretization. But the matrix D is no longer the unit matrix, unlike the semi-discretization by finite differences. In general semi-discretization by finite elements produces a linear system (1.11) where D is not diagonal. If instead for semi-discretization one uses the finite volume method, then D is a diagonal matrix. □

Remark 5.5. When more than one spatial variable is present, the advantages of finite element semi-discretizations over other methods negate the failing that D is not in general a diagonal matrix. The matrix D is diagonal only if the basis for the finite element space is orthogonal with respect to the $L_2(\Omega)$ scalar product. This is the case, e.g., for the non-conforming Crouzeix-Raviart element. *Wavelet*-based methods that have been intensively studied recently also have this property (see [Urb02]). For piecewise linear finite elements, *mass lumping* provides a means of diagonalizing the mass matrix D; this technique will be discussed later.   □

We now derive an error estimate for our finite element semi-discretization. Assume that the chosen finite element space V_h possesses the following approximation property for some $r > 1$:

$$\|v - \Pi_h v\| + h\|\nabla(v - \Pi_h v)\| \le Ch^s \|v\|_s, \ 1 \le s \le r, \ v \in H^s(\Omega), \quad (1.12)$$

where $\Pi_h : V \to V_h$ is an interpolation operator, h is the mesh diameter, and $\|\cdot\| = \|\cdot\|_{L_2(\Omega)}$. In the case of conforming linear finite elements this estimate is valid with $r = 2$.

Let $R_h : u \to V_h$ denote the *Ritz projection* operator, i.e., $R_h u \in V_h$ is defined by

$$a(R_h u, v_h) = a(u, v_h) \quad \text{for all } v_h \in V_h. \quad (1.13)$$

A common device in error estimates is to split the error $u - u_h$ into two components as

$$u - u_h = (u - R_h u) + (R_h u - u_h)$$

then to estimate each part separately.

Lemma 5.6. *For the Ritz projection we have*

$$\|\nabla(v - R_h v)\| \le Ch^{s-1} \|v\|_s \quad \text{for} \quad 1 \le s \le r, \ v \in H^s(\Omega).$$

If the solution w of the variational equation

$$a(w, v) = (f, v) \quad \text{for all} \quad v \in V$$

is in $H^2(\Omega)$ for all $f \in L_2(\Omega)$, then

$$\|v - R_h v\| \le Ch^s \|v\|_s \quad \text{for } 1 \le s \le r, \ v \in H^s(\Omega).$$

Proof: The first inequality follows immediately from

$$a(v - R_h v, \ v - R_h v) = a(v - R_h v, v - v_h) \quad \text{for all } v_h \in V_h,$$

on invoking coercivity, continuity and the usual $H^1(\Omega)$ interpolation error estimates (see Chapter 4).

The second part is simply the standard L_2 error estimate for elliptic problems. ■

Set $\rho = R_h u - u_h$. Then

$$(\rho_t, v_h) + a(\rho, v_h) = ((R_h u)_t, v_h) + a(R_h u, v_h) - ((u_h)_t, v_h) - a(u_h, v_h)$$
$$= ((R_h u)_t, v_h) + a(u, v_h) - \langle f(t), v_h \rangle$$

by (1.10) and (1.13). Hence

$$(\rho_t, v_h) + a(\rho, v_h) = ((R_h u - u)_t, v_h) \quad \text{for all } v_h \in V_h. \tag{1.14}$$

But $\rho \in V_h$, so we can take $v_h = \rho$ here. By virtue of $a(\rho, \rho) \geq \alpha \|\rho\|^2$ and the Cauchy-Schwarz inequality, this yields

$$\frac{1}{2} \frac{d}{dt} \|\rho\|^2 + \alpha \|\rho\|^2 \leq \|(u - R_h u)_t\| \cdot \|\rho\|$$

whence

$$\frac{d}{dt} \|\rho\| + \alpha \|\rho\| \leq \|(u - R_h u)_t\|.$$

Integration in time now leads to

$$\|\rho(t)\| \leq e^{-\alpha t} \|\rho(0)\| + \int_0^t e^{-\alpha(t-s)} \|(u - R_h u)_t\| ds.$$

To clarify this we invoke

$$\|\rho(0)\| \leq \|u_0 - u_h^0\| + Ch^r \|u_0\|_r$$

and, by Lemma 5.6, for each t one has

$$\|u(t) - R_h u(t)\| \leq Ch^r \|u(t)\|_r .$$

Then we get

$$\|\rho(t)\| \leq e^{-\alpha t} \|u_0 - u_h^0\| + Ch^r \left\{ e^{-\alpha t} \|u_0\|_r + \int_0^t e^{-\alpha(t-s)} \|u_t(s)\|_r \, ds \right\} .$$

Finally, the triangle inequality and Lemma 5.6 yield the following theorem.

Theorem 5.7. *If the solution u is sufficiently regular, then the $L_2(\Omega)$ error for the semi-discrete solution obtained using finite elements that satisfy (1.12) can be estimated for $0 \leq t \leq T$ by*

$$\|u(t) - u_h(t)\| \leq Ce^{-\alpha t} h^r \|u_0\|_r$$
$$+ Ch^r \left\{ \|u(t)\|_r + \int_0^t e^{-\alpha(t-s)} \|u_t(s)\|_r \, ds \right\} . \tag{1.15}$$

This result produces the expected order of convergence (e.g., $O(h^2)$ in the case of conforming linear finite elements) under suitable regularity hypotheses and also shows that the error in the initial approximation (u_0 may not be smooth) will be exponentially damped in time.

Naturally error estimates in the $H^1(\Omega)$ norm and (under appropriate assumptions) in the $L_\infty(\Omega)$ norm can also be proved, as we shall now demonstrate.

Let us consider the special case $a(v_1, v_2) = \int_\Omega \nabla v_1 \nabla v_2 \, d\Omega$. Then taking $v_h = \rho_t$ in (1.14) yields

$$\|\rho_t\|^2 + \frac{1}{2}\frac{d}{dt}\|\nabla \rho\|^2 \le \frac{1}{2}\|(R_h u - u)_t\|^2 + \frac{1}{2}\|\rho_t\|^2,$$

and consequently

$$\frac{d}{dt}\|\nabla \rho\|^2 \le \|(R_h u - u)_t\|^2. \tag{1.16}$$

Integration of this estimate gives immediately a $H^1(\Omega)$ error bound. Choose $u_h(0) = R_h u_0$; then $\nabla \rho(0) = 0$, and for piecewise linear finite elements it follows that

$$\|\nabla \rho\| \le Ch^2 \left(\int_0^t \|u_t(s)\|_2^2 \, ds \right)^{1/2}. \tag{1.17}$$

This is a superconvergence result. It will now be used to derive an $L_\infty(\Omega)$ error estimate via the discrete Sobolev inequality [79]. In the case $\Omega \subset \mathbb{R}^2$ we obtain

$$\|v_h\|_\infty := \|v_h\|_{L_\infty(\Omega)} \le C|\ln h|^{1/2}\|\nabla v_h\| \quad \text{for all } v_h \in V_h. \tag{1.18}$$

Since $\rho = R_h u - u_h \in V_h$, (1.17) and (1.18) yield

$$\|\rho\|_\infty \le Ch^2|\ln h|^{1/2} \left(\int_0^t \|u_t(s)\|_2^2 \, ds \right)^{1/2}.$$

As shown in Chapter 4, the Ritz projection satisfies

$$\|u - R_h u\|_\infty \le Ch^2|\ln h|\, \|u\|_{W_\infty^2}.$$

Now a triangle inequality gives

$$\|u - u_h\|_\infty \le Ch^2|\ln h|\, \|u\|_{W_\infty^2} + Ch^2|\ln h|^{1/2} \left(\int_0^t \|u_t(s)\|_2^2 \, ds \right)^{1/2}. \tag{1.19}$$

Other approaches to $L_\infty(\Omega)$ error estimates can be found in [Ike83, Dob78].

Remark 5.8. The above error estimates do not explain satisfactorily how any reduced regularity of the exact solution (in particular near the initial time $t = 0$) can affect the convergence of the method. In fact a lack of smoothness near $t = 0$ may lead to difficulties there, but a more precise analysis reveals that this does not affect the discretization error for $t \ge \delta > 0$; see [Tho97].
□

To finish this section we deal with the technique of mass lumping, which was briefly mentioned in Remark 5.5. We start from the semi-discrete problem, written as a system of ordinary differential equations in the form

$$D(\hat{u}(t))' + A\hat{u}(t) = \hat{f}(t),$$

where the matrix D is not a diagonal matrix (see Example 5.4). In the case of conforming linear finite elements *mass lumping* is a standard way of diagonalizing D, but for other finite elements almost no analytical results are known when the original mass matrix is replaced by a diagonal one.

In mass lumping, the matrix D is replaced by a diagonal matrix $\bar{D} = (\bar{d}_{ij})$ where each non-zero entry $\bar{d}_{ii}$ is obtained by summing the elements of the ith row of D, viz.,

$$\bar{d}_{ii} = \sum_k d_{ik}. \tag{1.20}$$

We analyse this theoretically for the case $n = 2$ where each finite element is a triangle. Now $D = (d_{ij})$ with $d_{ij} = (\varphi_j, \varphi_i)$ for all i and j, where the $\{\varphi_i\}$ are the usual nodal basis for the finite element space. We shall show first that the replacement of D by $\bar{D}$ is equivalent to the approximate evaluation of each integral (φ_j, φ_i) by applying on each triangle K of the triangulation T_h the quadrature formula

$$\int_K f dx \approx \frac{1}{3} (meas\,K) \sum_{k=1}^{3} f(P_{K_k}).$$

where the points P_{K_k} are the vertices of K. Define

$$(v, w)_h = \sum_{K \in T_h} \frac{1}{3} (meas\,K) \sum_{k=1}^{3} (vw)(P_{K_k}). \tag{1.21}$$

Clearly $(\varphi_j, \varphi_i)_h = 0$ for $i \neq j$ since the product $\varphi_j\varphi_i$ then vanishes at each vertex. On the other hand, suppose that $\varphi_i = 1$ at vertex P_i; then

$$(\varphi_i, \varphi_i)_h = \frac{1}{3} \, meas\,D_i,$$

where D_i is the union of all triangles that contain P_i as a vertex. Meanwhile,

$$\int_K \varphi_j\varphi_i = \frac{1}{12} \, meas\,K \text{ if } j \neq i \quad \text{and} \quad \int_K \varphi_i^2 = \frac{1}{6} \, meas\,K$$

when φ_j and φ_i do not vanish on triangle K. Hence $(\varphi_i, \varphi_i)_h = \sum_j (\varphi_j, \varphi_i)$. That is, the application of the quadrature formula to the entries (φ_j, φ_i) of D is equivalent to the diagonalization of D by mass lumping.

Thus mass lumping is equivalent to a modification of (1.10) to

$$\frac{d}{dt}(u_h(t), v_h)_h + a(u_h(t), v_h) = \langle f(t), v_h \rangle \quad \text{forall } v_h \in V_h. \tag{1.22}$$

To derive an error estimate for this modification, the next lemma is useful.

Lemma 5.9. *The norms defined by* $v \mapsto (v,v)^{\frac{1}{2}}$ *and* $v \mapsto (v,v)_h^{\frac{1}{2}}$ *are equivalent on the discrete space* V_h. *Furthermore,*

$$|(v,w)_h - (v,w)| \le Ch^2 \|\nabla v\| \cdot \|\nabla w\| \qquad \text{for all } v,w \in V_h. \qquad (1.23)$$

Proof: First we show the equivalence of the two norms on V_h. Let $v_h \in V_h$ be arbitrary. Let K be an arbitrary mesh triangle. On this triangle one can decompose v_h as

$$v_h = v_{1,K}\varphi_{1,K} + v_{2,K}\varphi_{2,K} + v_{3,K}\varphi_{3,K},$$

where the basis function $\varphi_{i,K}$ is associated with vertex i of K. Then a calculation gives (to simplify the notation we omit the index K)

$$\|v_h\|_K^2 = \frac{1}{6}(meas\,K)(v_1^2 + v_2^2 + v_3^2 + v_1v_2 + v_2v_3 + v_3v_1),$$

$$\|v_h\|_{h,K}^2 = \frac{1}{3}(meas\,K)(v_1^2 + v_2^2 + v_3^2),$$

where the subscript K on the left-hand sides means that the norms are computed only over K. By summing over K one can explicitly evaluate the norms $\|\cdot\|$ and $\|\cdot\|_h$ in the space V_h. The norm equivalence follows immediately; it is even uniform in h, the discretization parameter.

To prove the bound (1.23) we apply the usual techniques for estimates of quadrature errors. Our quadrature rule is exact when restricted to a triangle K and applied to linear polynomials on K, so the Bramble-Hilbert lemma applied to $f \in H^2(K)$ yields the bound

$$\left| \frac{1}{3}(meas\,K)\sum_{j=1}^{3} f(P_{K_j}) - \int_K f\,dx \right| \le Ch^2|f|_2.$$

Now choose $f = vw$ for arbitrary $v, w \in V_h$, so $v|_K$ and $w|_K$ are linear. We get

$$|(v,w)_{h,K} - (v,w)_K| \le Ch^2|vw|_{2,K} \le Ch^2\|\nabla v\|_K\|\nabla w\|_K.$$

Applying the Cauchy-Schwarz inequality for sums, inequality (1.23) follows. ∎

The semi-discrete problem with mass lumping satisfies a convergence result similar to Theorem 5.7 for the non-lumped method. We prove such a bound here with α in Theorem 5.7 replaced by zero, and of course $r = 2$.

Theorem 5.10. *If the solution* u *is sufficiently regular then the* L_2 *error for the semi-discrete solution with conforming linear finite elements and mass lumping can be estimated for* $0 \le t \le T$ *by*

$$\|u(t) - u_h(t)\| \le Ch^2\|u_0\|_2 + Ch^2\left\{ \|u(t)\|_2 + \left(\int_0^t \|u_t(s)\|_2^2\,ds\right)^{1/2} \right\}. \qquad (1.24)$$

Proof: The argument imitates the proof of Theorem 5.7. Set $\rho = R_h u - u_h$. Then

$$
\begin{aligned}
(\rho_t, v_h)_h + a(\rho, v_h) &= ((R_h u)_t, v_h)_h + a(R_h u, v_h) \\
&\quad - ((u_h)_t, v_h)_h - a(u_h, v_h) \\
&= ((R_h u)_t, v_h)_h + a(u, v_h) - (f, v_h) \\
&= ((R_h u)_t, v_h)_h - (u_t, v_h),
\end{aligned}
$$

i.e., $(\rho_t, v_h)_h + a(\rho, v_h) = ((R_h u)_t, v_h)_h - ((R_h u)_t, v_h) + ((R_h u - u)_t, v_h)$.

Choose $\rho = v_h$. Then (1.23) and a Cauchy-Schwarz inequality give

$$
\frac{1}{2} \frac{d}{dt} \|\rho\|_h^2 + \alpha \|\nabla \rho\|^2 \leq Ch^2 \|\nabla (R_h u)_t\| \cdot \|\nabla \rho\| + \|(R_h u - u)_t\| \cdot \|\rho\| .
$$

Next, we make use of

$$
\|u - R_h u\| \leq Ch^2 \|u\|_2
$$

and

$$
\|\nabla u - \nabla (R_h u)\| \leq Ch \|u\|_2 .
$$

Invoking $ab \leq \varepsilon a^2/2 + b^2/(2\varepsilon)$ and a Poincaré inequality, these inequalities yield

$$
\frac{1}{2} \frac{d}{dt} \|\rho\|_h^2 + \alpha \|\nabla \rho\|^2 \leq Ch^4 \|u_t\|_2^2 + \alpha \|\nabla \rho\|^2 .
$$

Hence

$$
\frac{1}{2} \frac{d}{dt} \|\rho\|_h^2 \leq Ch^4 \|u_t\|_2^2,
$$

For $0 \leq t \leq T$ this implies

$$
\|\rho(t)\|_h^2 \leq \|\rho(0)\|_h^2 + Ch^4 \int_0^t \|u_t(s)\|_2^2 \, ds.
$$

Now

$$
\begin{aligned}
\|\rho(0)\| = \|u_h(0) - R_h u(0)\| &\leq \|u_h(0) - u_0\| + \|u_0 - R_h u(0)\| \\
&\leq \|u_h^0 - u_0\| + Ch^2 \|u_0\|_2 .
\end{aligned}
$$

Recalling the equivalence of the norms proved in Lemma 5.9, we are done. ∎

It is also possible to derive error estimates in other norms.

An alternative approach to the analysis of errors of solutions obtained by mass-lumping is via the finite volume method. In one version of the finite volume method, starting from a given triangulation of the domain we constructed a dual grid comprising Donald boxes Ω_i that satisfy the equilibrium condition

$$
meas(\Omega_i \cap T) = \frac{1}{3} (meas \, T) \quad \text{for all boxes } \Omega_i \text{ and triangles } T.
$$

Define a *lumping* operator $\sim : C(\overline{\Omega}) \to L_\infty(\Omega)$ by

$$w \mapsto \tilde{w} := \sum_i w(P_i)\, \tilde{\phi}_{i,h}\,,$$

where $\tilde{\phi}_{i,h}$ is the characteristic function of the Donald box Ω_i. Then

$$(\tilde{\varphi}_j, \tilde{\varphi}_k) = 0 \qquad \text{for} \qquad j \neq k$$

and

$$(\tilde{\varphi}_j, \tilde{\varphi}_j) = \frac{1}{3}\, meas\ \Omega_j.$$

It follows that the semi-discrete lumped problem (1.22) is equivalent to

$$\frac{d}{dt}\big(\, \tilde{u}_h\,(t), \tilde{v}_h\,\big) + a(u_h(t), v_h) = \langle f(t), v_h \rangle \quad \text{for all } v_h \in V_h\,. \tag{1.25}$$

An error analysis can also be derived in this finite volume framework.

5.1.3 Temporal Discretization by Standard Methods

Semi-discretization of linear parabolic initial-boundary value problems produces an initial value problem for a system of ordinary differential equations (o.d.e.s), in the form

$$\frac{du_h}{dt} = B_h u_h + \hat{f}_h(t), \qquad u_h(0) = u_0. \tag{1.26}$$

Similarly, the semi-discretization of non-linear parabolic differential operators generates systems of the form

$$\frac{du_h}{dt} = F_h(t, u_h), \qquad u_h(0) = u_0. \tag{1.27}$$

To discretize (1.26) and (1.27), a naive approach is to apply some standard numerical method (e.g., as described in [HNW87]) for solving initial value problems. But the o.d.e. systems generated by semi-discretization are extremely *stiff systems*. This happens because—to take as an example the symmetric case—the matrix B_h has negative real eigenvalues, some of moderate size but others of very large absolute value (typically $O(1/h^2)$ in the case where the spatial elliptic component of the parabolic problem is second-order). Stiff o.d.e.s are discussed in detail in [HW91]. To avoid the enforced use of extremely small time step sizes, one must discretize stiff systems through numerical methods that have particular stability properties. Furthermore, formally higher-order discretization methods often suffer order reductions if applied to (1.26) or (1.27).

We begin our study of numerical methods for solving (1.26) or (1.27) by discussing *A-stability* and related basic properties of discretizations of initial value problems

$$\frac{du}{dt} = f(t, u(t)), \qquad u(0) = u_0 \tag{1.28}$$

on a mesh $\{t_n\}$ where $t_n = n\tau$ for $n = 0, 1, \ldots$ and some step size $\tau > 0$. First, consider *one-step methods* (OSM) for the temporal discretization. These methods have the form

$$u_{n+1} = u_n + \tau\phi(\tau, u_n, u_{n+1}). \tag{1.29}$$

Here u_n denotes the computed numerical approximation for $u(t_n)$ and the function ϕ defines the specific method.

A discretization method is *numerically contractive* if one always has

$$\|\tilde{u}^{n+1} - u^{n+1}\| \le \kappa\|\tilde{u}^n - u^n\| \tag{1.30}$$

for some constant $\kappa \in [0, 1]$ when $\tilde{u}^n$ and u^n are generated by the same method from different initial values.

A discretization method is *A-stable* if it is contractive when applied to all test problems

$$u' = \lambda u \quad \text{where } \lambda \in \mathbb{C} \text{ with } \operatorname{Re}\lambda \le 0.$$

If a one-step method is used to discretize $u' = \lambda u$ then

$$u_{n+1} = R(\tau\lambda)u_n$$

for some *stability function* $R(\cdot)$. Clearly A-stability is equivalent to the condition

$$|R(z)| \le 1 \qquad \text{for all } z \in \mathbb{C} \text{ with } \operatorname{Re} z \le 0. \tag{1.31}$$

Example 5.11. Consider

- the explicit Euler method $u_{n+1} = u_n + \tau f(t_n, u_n)$,
- the implicit Euler method $u_{n+1} = u_n + \tau f(t_{n+1}, u_{n+1})$,
- the midpoint rule $u_{n+1} = u_n + \tau f\left(\frac{t_n + t_{n+1}}{2}, \frac{u_n + u_{n+1}}{2}\right)$.

When applied to the test differential equation $u' = \lambda u$, these methods yield respectively

$$u_{n+1} = (1 + \tau\lambda)u_n \qquad\qquad \text{with } R(z) = 1 + z,$$

$$u_{n+1} = [1/(1 - \tau\lambda)]u_n \qquad \text{with } R(z) = 1/(1 - z),$$

$$u_{n+1} = [(2 + \tau\lambda)/(2 - \tau\lambda)]u_n \text{ with } R(z) = (2 + z)/(2 - z).$$

On considering the complex z-plane it is not difficult to see that the implicit Euler and midpoint methods are A-stable, but the explicit Euler method is not.   □

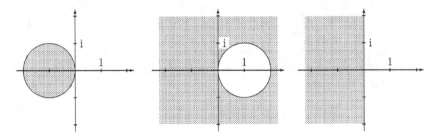

Figure 5.2 Stability regions

The stability region of any method is the set of all $z \in \mathbb{C}$ for which (1.31) holds. Stability regions for the three one-step methods considered above are drawn in Figure 5.2.

For certain applications it is not necessary that the stability region coincides with $\{z \in \mathbb{C} : \mathrm{Re}\, z \leq 0\}$, the left half of the complex plane. This leads to other forms of stability. For example, a method is called A_0-*stable* if

$$|R(z)| \leq 1 \quad \text{for all real } z < 0.$$

Remark 5.12. Some further stability notions are as follows. A method is

- *L-stable* if it is A-stable and in addition $\lim_{z \to \infty} R(z) = 0$;
- *strongly A_0-stable* if
  $|R(z)| < 1$ for $z < 0$ and $R(\infty) < 1$ both hold;
- *strongly A_δ-stable* for some $\delta \in (0, \pi/2)$ if
  $|R(z)| < 1$ on the set $\{z \in \mathbb{C} : |\arg z - \pi| \leq \delta\}$ and $|R(\infty)| < 1$ both hold,
- *L_δ-stable* if it is A_δ-stable and $R(\infty) = 0$. □

For the discretization of stiff linear systems of differential equation, it is preferable to use methods whose stability properties lie between A_0-stability and A-stability. Explicit one-step methods do not have this stability behaviour.

Within the class of implicit *Runge-Kutta methods* one finds A-stable methods of arbitrarily high order. The Gauss methods, Radau I A-method and Radau II A-method, and the Lobatto III C-methods have such properties. The s-stage versions of these methods have orders $2s$, $2s - 1$ and $2s - 2$, respectively. For further reading see [HNW87] and [HW91]; here we shall sketch only some basic ideas of Runge-Kutta methods.

An s-stage *Runge-Kutta method* is characterized by the Butcher array $\dfrac{c\,|\,A}{|\,b}$. This array of parameter data corresponds to the scheme

$$u_{n+1} = u_n + \tau \sum_{j=1}^{s} b_i k_i \quad \text{with} \quad k_i = f\left(t_n + c_i \tau, u_n + \tau \sum_{j=1}^{s} a_{ij} k_j\right). \quad (1.32)$$

In the methods named above, the parameters c_i are the roots of polynomials that are obtained in a certain way from Legendre polynomials. For example, in the case of the s-stage Gauss method,

$$P_s^*(c_i) = 0$$

where $P_s^*(x) := P_s(2x - 1)$ and P_s is the Legendre polynomial of degree s, while setting

$$C = \operatorname{diag}(c_i),$$
$$S = \operatorname{diag}(1, 1/2, \cdots, 1/s),$$

$$V = \begin{bmatrix} 1 & c_1 & \cdots & c_1^{s-1} \\ 1 & & & \\ & \cdot & & \\ & \cdot & & \\ & \cdot & & \\ 1 & c_s & \cdots & c_s^{s-1} \end{bmatrix},$$

one also has

$$b = (V^T)^{-1} S (1, \cdots, 1)^T \quad \text{and} \quad A = CVSV^{-1}.$$

The s-stage Gauss method has order $2s$, as already mentioned. This method is described for the values $s = 1$ and $s = 2$ by the respective Butcher arrays

$$\begin{array}{c|c} \frac{1}{2} & \frac{1}{2} \\ \hline & 1 \end{array} \qquad \text{and} \qquad \begin{array}{c|cc} \frac{1}{2} - \frac{1}{6}\sqrt{3} & \frac{1}{4} & \frac{1}{4} - \frac{1}{6}\sqrt{3} \\ \frac{1}{2} + \frac{1}{6}\sqrt{3} & \frac{1}{4} + \frac{1}{6}\sqrt{3} & \frac{1}{4} \\ \hline & \frac{1}{2} & \frac{1}{2} \end{array}.$$

In the stability analysis of Runge-Kutta methods one can make use of the fact that the stability function $R(\cdot)$ is now given explicitly by

$$R(z) = 1 + b^T (z^{-1} I - A)^{-1} e, \quad \text{where } e = (1, \cdots, 1)^T.$$

In the case of non-linear initial value problems, any implicit Runge-Kutta scheme requires the solution of systems of non-linear discrete equations. To avoid this imposition, linearly implicit methods have been introduced. The *Rosenbrock methods* form an important class of this type; they have the structure

$$u_{n+1} = u_n + \tau \sum_{j=1}^{s} b_i k_i^*,$$

$$k_i^* = f\left(t_n + \alpha_i \tau, u_n + \tau \sum_{j=1}^{i-1} \alpha_{ij} k_j^*\right) + \tau f_u(t_n, u_n) \sum_{j=1}^{i} \gamma_{ij} k_j^* + \tau \gamma_i f_t(t_n, u_n),$$

with

$$\alpha_i = \sum_{j=1}^{i-1} \alpha_{ij}, \qquad \gamma_i = \sum_{j=1}^{i} \gamma_{ij}.$$

At each step only a linear systems of equations needs to be solved, but the partial derivatives f_u and f_t (in the case of systems, their Jacobians) must be evaluated. The class of Rosenbrock methods contains L_δ-stable schemes of order $p \le s$.

Multi-step methods (MSM) are another important class of o.d.e. solvers. They have the general form

$$\frac{1}{\tau} \sum_{j=0}^{k} \alpha_j u_{m+j} = \sum_{j=0}^{k} \beta_j f(t_{m+j}, u_{m+j}). \tag{1.33}$$

Dahlquist has proved (see [HW91]) that *no explicit MSM (i.e., with $\beta_k = 0$) is A-stable. Furthermore, the maximal order of any A-stable MSM is only 2.*
   The implicit Euler method

$$\frac{u_{m+1} - u_m}{\tau} = f(t_{m+1}, u_{m+1})$$

is *A-stable* as we have seen. The second-order *BDF-method*

$$\frac{3u_{m+2} - 4u_{m+1} + u_m}{2\tau} = f(t_{m+2}, u_{m+2}).$$

ia also *A-stable*. The abbreviation BDF stands for **b**ackward **d**ifferentiation **f**ormula. Higher-order k-step BDF-methods can be written in the form

$$\sum_{l=1}^{k} \frac{1}{l} \nabla^l u_{m+k} = \tau\, f(t_{m+k}, u_{m+k}).$$

They have order k and for $2 < k \le 6$ they are A_δ-stable, but not *A-stable*.
   If simple temporal discretization schemes are used, then the total discretization error can be directly estimated without any splitting into spatial and temporal discretization errors. This avoids some difficulties in proving error bounds that are caused by the stiffness of the semi-discrete system. We sketch this direct estimate for the example of the discretization of the problem (1.10) by finite elements in space and a simple OSM for the temporal discretization: the θ scheme.
   Let τ again denote the discretization step size in time. Let $U^k \in V_h$ be the approximation of $u(\cdot)$ at time level $t_k = k \cdot \tau$. The approximation $U^{k+1} \in V_h$ at time t_{k+1} is defined by

$$\left(\frac{U^{k+1} - U^k}{\tau}, v_h\right) + a(\theta U^{k+1} + (1 - \theta)U^k, v_h) = \langle \hat{f}^k, v_h \rangle \ \forall\, v_h \in V_h, \tag{1.34}$$

$$U^0 = u_h^0,$$

where $\hat{f}^k := \theta f^{k+1} + (1-\theta)f^k$. The parameter θ that characterizes the method satisfies $0 \le \theta \le 1$. If $\theta = 0$ we get the explicit Euler method, while $\theta = 1$ gives the implicit Euler method.

For each k, problem (1.34) is a discrete elliptic boundary value problem that by the Lax-Milgram lemma has a unique solution $U^{k+1} \in V_h$. The matrix of the linear system that must be solved at each time level has the form $D + \tau \theta A$ where D denotes the mass matrix of the FEM and A is the stiffness matrix.

Similarly to the splitting

$$u - u_h = (u - R_h u) + (R_h u - u_h)$$

that was used in the analysis of semi-discretization by finite elements, we write

$$u(t_k) - U^k = (u(t_k) - R_h u(t_k)) + (R_h u(t_k) - U^k)$$

and define $\rho^k := R_h u(t_k) - U^k$. For the projection error one has

$$\|u(t_k) - R_h u(t_k)\| \le Ch^r \|u(t_k)\|_r \le Ch^r \left[\|u_0\|_r + \int_0^{t_k} \|u_t(s)\|_r \, ds \right]. \quad (1.35)$$

Next, an appropriate equation for ρ^k will be derived. By the definition of the continuous and discrete problems, and by exploiting properties of the Ritz projection, after some elementary transformations we obtain

$$\left(\frac{\rho^{k+1} - \rho^k}{\tau}, v_h \right) + a(\theta \rho^{k+1} + (1-\theta)\rho^k, v_h) = (w^k, v_h), \quad (1.36)$$

where

$$w^k := \frac{R_h u(t_{k+1}) - R_h u(t_k)}{\tau} - [\theta\, u_t(t_{k+1}) + (1-\theta)u_t(t_k)].$$

Rewriting this in a more convenient form for analysis,

$$
\begin{aligned}
w^k = {}& \left(\frac{R_h u(t_{k+1}) - R_h u(t_k)}{\tau} - \frac{u(t_{k+1}) - u(t_k)}{\tau} \right) \\
&+ \left(\frac{u(t_{k+1}) - u(t_k)}{\tau} - [\theta\, u_t(t_{k+1}) + (1-\theta)u_t(t_k)] \right).
\end{aligned}
\quad (1.37)
$$

To bound the second term here use a Taylor expansion, while the first term can be treated by reformulating it as

$$\frac{1}{\tau} \int_{t_k}^{t_{k+1}} [(R_h - I)u(s)]'\, ds$$

since the Ritz projection and temporal derivative commute.

To estimate ρ^{k+1} in (1.36), choose $v_h = \theta \rho^{k+1} + (1-\theta)\rho^k$ then bound the term $a(\cdot, \cdot)$ from below by zero. Assume that $\theta \ge 1/2$. Then

$$(\rho^{k+1} - \rho^k, \theta\rho^{k+1} + (1-\theta)\rho^k) = \theta\|\rho^{k+1}\|^2 + (1-2\theta)(\rho^{k+1}, \rho^k)$$
$$- (1-\theta)\|\rho^k\|^2$$
$$\geq \theta\|\rho^{k+1}\|^2 + (1-2\theta)\|\rho^{k+1}\|\,\|\rho^k\|$$
$$- (1-\theta)\|\rho^k\|^2$$
$$= (\|\rho^{k+1}\| - \|\rho^k\|)(\theta\|\rho^{k+1}\| + (1-\theta)\|\rho^k\|).$$

A Cauchy-Schwarz inequality now yields

$$\|\rho^{k+1}\| - \|\rho^k\| \leq \tau\|w^k\|,$$

i.e.,

$$\|\rho^{k+1}\| \leq \|\rho^k\| + \tau\|w^k\| \quad \text{for each } k.$$

This implies that

$$\|\rho^{k+1}\| \leq \|\rho^0\| + \tau\sum_{l=1}^{k}\|w^l\|. \tag{1.38}$$

From these calculations we infer

Theorem 5.13. *Let the exact solution u be sufficiently regular. Consider the complete discretization by finite elements in space and the θ scheme with $\theta \geq 1/2$ in time, where the finite elements satisfy (1.12). Then the $L_2(\Omega)$ error is bounded for $k = 1, 2, \ldots$ by*

$$\|u(t_k) - U^k\| \leq \|u_h^0 - u_0\| + Ch^r\left(\|u_0\|_r + \int_0^{t_k}\|u_t(s)\|_r\,ds\right)$$
$$+ \tau\int_0^{t_k}\|u_{tt}(s)\|\,ds. \tag{1.39}$$

In the case of piecewise linear finite elements, $r = 2$. If $\theta = \frac{1}{2}$ (the Crank-Nicolson method) then τ can be replaced by τ^2 in Theorem 5.13 provided that the exact solution has sufficient regularity.

Remark 5.14. The estimate (1.38) reflects the L_2-stability of the method. Analyses of finite difference methods also show that the restriction $\theta \geq 1/2$ is a natural.

In [QV94] stability and convergence results are given for the case $0 < \theta < 1/2$, under the step size restriction $\tau \leq ch^2$.

For $\theta > 1/2$, but not $\theta = 1/2$, a stronger form of stability holds. Here the damping influence of the $a(\cdot, \cdot)$ term has to be used to reduce the influence of the initial condition; cf. Theorem 5.7.

An analysis of the combination of the implicit Euler method with linear finite elements in [Ran04] includes the fairly general case where variable time steps and changes of the spatial grid from each time step to the next are permitted.   □

What about higher-order methods for the temporal discretization? In [Tho97, Chapters 7,8,9] one finds sufficient conditions for the property that a one-step temporal discretization method of order p lead to a bound

$$\|u(t_k) - U^k\| \leq C(h^r + \tau^p)$$

for the total error. The stability functions for stiff problems must satisfy some typical additional conditions; in [Tho97, Chapter 10] this question is addressed for BDF methods. These conditions are restrictive; in general, order reductions in higher-order methods are to be expected—see [95].

The construction of schemes based on linear finite elements for which discrete maximum principles are valid is examined in detail in [Ike83]. For an elliptic differential operator $-\Delta$ (negative Laplace operator), if the triangulation is of weakly acute type then the stiffness matrix A generated is an M-matrix. This may however not be the case for the matrix $\tau\sigma A + D$ that appears in the discrete parabolic problem. In each triangle the perpendicular bisectors of edges meet at a point; if $\hat{k}$ denotes the minimal length of such bisectors in the triangulation, then for problems of the form

$$\frac{\partial u}{\partial t} - \varepsilon \Delta u = f,$$

an M-matrix analysis shows that one has L_∞-stability if

$$6\varepsilon(1 - \theta)\tau \leq \hat{k}^2.$$

For the θ method, only the fully implicit case $\theta = 1$ satisfies this condition automatically. If mass lumping is applied then the stability condition changes slightly to

$$3\varepsilon(1 - \theta)\tau \leq \hat{k}^2.$$

Further variations including different versions of the discrete maximum principle and the addition of convective terms (i.e., problems of the form $\partial u/\partial t - \varepsilon \Delta u + b\nabla u = f$) are considered in the analysis of [Ike83].

5.1.4 Temporal Discretization with Discontinuous Galerkin Methods

One-step and multi-step methods are not the only class of methods for the temporal discretization of the problem

$$\frac{d}{dt}(u_h(t), v_h) + a(u_h(t), v_h) = \langle f(t), v_h \rangle \qquad \text{for all } v_h \in V_h. \tag{1.40}$$

Alternatively, one could apply a Galerkin method in time to this semi-discrete problem.

Let

$$0 = t_0 < t_1 < \cdots < t_M = T$$

be some grid on the time interval $[0, T]$ with the associated time steps $\tau_n :=$ $t_n - t_{n-1}$ for $n = 1, \ldots, M$. Let $W_{h,t}$ denote the space of piecewise polynomials of degree q in t that are defined on this grid with values in V_h. Then (1.40) could be discretized by finding $U \in W_{h,t}$ such that

$$\int_{t_{m-1}}^{t_m} [(U', v^*) + a(U, v^*)] \, dt = \int_{t_{m-1}}^{t_m} \langle f, v^* \rangle \, dt \qquad (1.41)$$

for some set of suitable test functions $v^* \in W_{h,t}$ and $m = 1, \ldots, M$.

Now $q + 1$ independent conditions are needed to determine the polynomial U of degree q on $[t_{m-1}, t_m]$. For each m and suitable functions w set

$$w_+^m = \lim_{s \to 0^+} w(t_m + s), \quad w_-^m = \lim_{s \to 0^-} w(t_m + s),$$

If one imposes continuity between neighbouring time intervals, i.e., one requires that $U_+^m = U_-^m$ for $m = 1, \ldots, M - 1$, then one could in theory choose v^* to be an arbitrary polynomial of degree $q - 1$. But such continuous Galerkin methods—which we abbreviate as $cG(q)$—are unsuitable because they treat space and time similarly, and continuous space-time elements should be avoided as illustrated by the next remark.

Remark 5.15. For $q = 1$ one obtains

$$\left(\frac{U^m - U^{m-1}}{\tau_m}, v \right) + \frac{1}{2} a(U^m + U^{m-1}, v) = \frac{1}{\tau_m} \int_{t_{m-1}}^{t_m} \langle f, v \rangle dt \qquad (1.42)$$
$$\text{for all } v \in V_h.$$

Thus the $cG(1)$ method is closely related to the Crank-Nicolson scheme, i.e., the θ scheme with $\theta = 1/2$. Its solutions are known to often exhibit unpleasant oscillatory behaviour as a consequence of the absence of a discrete maximum principle when the restrictive condition $\tau = O(h^2)$ is violated. $\square$

In the *discontinuous Galerkin method*, to couple the computed solution U between adjacent time intervals, continuity in t is relaxed by imposing it only weakly. Let

$$[v^m] = v_+^m - v_-^m$$

denote the jump in v at time t_m for $m = 1, \ldots, M - 1$. Set

$$A(w, v) := \sum_{m=1}^{M} \int_{t_{m-1}}^{t_m} ((w', v) + a(w, v)) dt + \sum_{m=2}^{M} \left([w^{m-1}], v_+^{m-1} \right) + (w_+^0, v_+^0),$$

$$L(v) \quad := \int_0^T \langle f, v \rangle dt + (u_0, v_+^0).$$

Then the numerical solution $U \in W_{h,t}$ is defined by the variational equation

$$A(U, v) = L(v) \qquad \text{for all } v \in W_{h,t}. \qquad (1.43)$$

The restrictions of the functions $v \in W_{h,t}$ to the different temporal subintervals are independent so (1.43) can be equivalently expressed as:
On each interval (t_{m-1}, t_m), a polynomial U of degree q in t with values in V_h (i.e., $U \in W_{h,t}$) satisfies

$$\int_{t_{m-1}}^{t_m} [(U', v) + a(U, v)]\, dt + (U_+^{m-1}, v_+^{m-1}) = \int_{t_{m-1}}^{t_m} \langle f, v \rangle dt + (U_-^{m-1}, v_+^{m-1}) \quad (1.44)$$

for all polynomials $v \in W_{h,t}$. Also, $U_-^0 = u_0$. We denote this method by $dG(q)$.

The existence and uniqueness of a solution of (1.44) will now be demonstrated. The problem is linear and finite-dimensional. The corresponding homogeneous problem with $f = 0$ and $U_-^{m-1} = 0$ is

$$\int_{t_{m-1}}^{t_m} [(U', v) + a(U, v)]dt + (U_+^{m-1}, v_+^{m-1}) = 0 \quad \text{for all } v \in W_{h,t}.$$

Choose $v = U$. Since

$$\int_{t_{m-1}}^{t_m} (U', U)dt = \frac{1}{2} \left(\|U_-^m\|^2 - \|U_+^{m-1}\|^2 \right)$$

we obtain

$$\frac{1}{2} \left(\|U_-^m\|^2 + \|U_+^{m-1}\|^2 \right) + \int_{t_{m-1}}^{t_m} a(U, U)dt = 0.$$

This implies $U = 0$. That is, the homogeneous problem has only the trivial solution. It follows that the inhomogeneous problem (1.44) has exactly one solution.

We now consider in detail the special cases $q = 0$ and $q = 1$ in $dG(q)$. If $q = 0$ (piecewise constant polynomials in t) let U^m denote the restriction of U to the subinterval (t_{m-1}, t_m). Then (1.44) has the form

$$\tau_m a(U^m, v) + (U^m, v) = \int_{t_{m-1}}^{t_m} \langle f, v \rangle\, dt + (U^{m-1}, v),$$

i.e.,

$$\left(\frac{U^m - U^{m-1}}{\tau_m}, v \right) + a(U^m, v) = \frac{1}{\tau_m} \int_{t_{m-1}}^{t_m} \langle f, v \rangle\, dt \quad \text{for all } v \in V_h. \quad (1.45)$$

A comparison of (1.45) with the θ scheme reveals that the discontinuous Galerkin method $dG(0)$, i.e., the case of piecewise constant approximation, is a modified implicit Euler method applied to the semi-discrete problem. An error estimate similar to Theorem 5.13 can be proved.

Now consider $q = 1$. For $t \in (t_{m-1}, t_m)$, write $U(t) = U_0^m + (t - t_{m-1})U_1^m/\tau_m$; then $U_-^m = U_0^m + U_1^m$ and $U_+^{m-1} = U_0^m$. Substituting this into (1.44) yields

$$\int_{t_{m-1}}^{t_m} \left[\frac{1}{\tau_m} (U_1^m, v) \right] + a \left(U_0^m + \frac{1}{\tau_m} (t - t_{m-1}) U_1^m, v \right) \, dt + (U_0^m, v_+^{m-1})$$

$$= \int_{t_{m-1}}^{t_m} \langle f, v \rangle dt + (U_0^{m-1} + U_1^{m-1}, v_+^{m-1}).$$

This identity holds for all linear polynomials in t with values in V_h, so we obtain two sets of variational equations—the first by the choice of v as constant in time and the second by taking v as linear in time, i.e., $v(t) := (t - t_{m-1})v/\tau_m$ for an arbitrary $v \in V_h$. The pair of equations is, for arbitrary $v \in V_h$,

$$(U_1^m, v) + \tau_m a(U_0^m, v) + \tfrac{1}{2}\tau_m a(U_1^m, v) + (U_0^m, v) =$$

$$= \int_{t_{m-1}}^{t_m} \langle f, v \rangle \, dt + (U_0^{m-1} + U_1^{m-1}, v)$$

and

$$\frac{1}{2}(U_1^m, v) + \frac{1}{2}\tau_m a(U_0^m, v) + \frac{1}{3}\tau_m a(U_1^m, v) = \frac{1}{\tau_m} \int_{t_{m-1}}^{t_m} (\tau - t_{m-1})\langle f(\tau), v \rangle \, d\tau.$$

Discontinuous Galerkin methods provide a systematic means of generating high-order temporal discretization schemes through Galerkin techniques. In [Joh88] the relationship of discontinuous Galerkin methods to certain implicit Runge-Kutta methods is discussed; see also Exercise 5.22 below.

To end this subsection, we analyse the error in the discontinuous Galerkin method for the temporal discretization of (1.9), which we recall is the problem

$$\frac{d}{dt}(u(t), v) + a(u(t), v) = \langle f(t), v \rangle \qquad \text{for all } v \in V.$$

In each interval (t_{m-1}, t_m) one seeks a polynomial U of degree q in t with values in V such that

$$\int_{t_{m-1}}^{t_m} [(U', v) + a(U, v)] \, dt + (U_+^{m-1}, v_+^{m-1}) =$$

$$= \int_{t_{m-1}}^{t_m} \langle f, v \rangle dt + (U_-^{m-1}, v_+^{m-1}) \tag{1.46}$$

holds for all polynomials v of degree q in t with values in V; furthermore $U_-^0 = u_0$. Because no spatial discretization is applied here, the $dG(0)$ method—i.e., (1.46)—can be considered as a generalization of Rothe's method, which will be discussed in Section 5.1.5.

This time we decompose the error as

$$U - u = \rho + \eta \qquad \text{with} \quad \rho := U - \tilde{u} \text{ and } \eta := \tilde{u} - u,$$

where $\tilde{u}$ is some interpolant of u that will be defined later. Then

$$\int_{t_{m-1}}^{t_m} [(\rho', v) + a(\rho, v)] \, dt + ([\rho^{m-1}], v_+^{m-1}) =$$

$$= -\int_{t_{m-1}}^{t_m} [(\eta', v) + a(\eta, v)] \, dt - ([\eta^{m-1}], v_+^{m-1}).$$

If $\tilde{u}$ is chosen to be the polynomial of degree q in t specified by

$$\tilde{u}(t_{m-1}) = u(t_{m-1}) \quad \text{and} \quad \int_{t_{m-1}}^{t_m} t^l \tilde{u}(t) \, dt = \int_{t_{m-1}}^{t_m} t^l u(t) \, dt$$
$$\text{for } l = 0, 1, \ldots, q - 1,$$

then the previous equation simplifies to

$$\int_{t_{m-1}}^{t_m} [(\rho', v) + a(\rho, v)] \, dt + ([\rho^{m-1}], v_+^{m-1}) = -\int_{t_{m-1}}^{t_m} a(\eta, v) dt.$$

Take $v = \rho$ here. Invoking the easily-verified inequality

$$\int_{t_{m-1}}^{t_m} \frac{d}{dt} \|\rho\|^2 + 2([\rho^{m-1}], \rho_+^{m-1}) \geq \|\rho_-^m\|^2 - \|\rho_+^{m-1}\|^2,$$

one gets

$$\|\rho_-^m\|^2 + 2\int_{t_{m-1}}^{t_m} a(\rho, \rho) \, dt \leq \|\rho_+^{m-1}\|^2 + 2\int_{t_{m-1}}^{t_m} |a(\rho, \eta)| \, dt.$$

This implies

$$\|\rho_-^m\|^2 \leq \|\rho_+^{m-1}\|^2 + c\int_{t_{m-1}}^{t_m} \|\eta(s)\|_1^2 \, ds.$$

Since $\rho^0 = 0$ can be assumed, standard interpolation error bounds now yield

$$\|U^m - u(t_m)\| \leq C \tau^{q+1} \left(\int_0^{t_m} |u^{(q+1)}(s)|_1^2 \, ds \right)^{1/2}. \tag{1.47}$$

Error estimates for the case of complete discretization, i.e., discretization in both time and space, can be found in [Tho97].

Exercise 5.16. Consider the initial-boundary value problem

$$u_t - u_{xx} = \sin x \quad \text{in} \quad (0, \pi/2) \times (0, 3),$$
$$u(x, 0) = 0,$$
$$u|_{x=0} = 0, \quad u_x|_{x=\pi/2} = 0.$$

Discretize this problem using piecewise linear finite elements in the x-direction and the Crank-Nicolson method for the temporal derivative. Find the system of linear equations that must be solved at each time level. Determine the discrete solutions for the particular discretization parameters $h = \pi/(2m)$ for $m = 8, 16, 32$ and $\tau = 0.2, 0.1, 0.05$ and compare these with the exact solution.

Exercise 5.17. Consider the initial-boundary value problem

$$u_t - \alpha \triangle u = f \quad \text{in} \quad \Omega \times (0, T),$$
$$u(x, 0) = u_0(x),$$
$$u = 0 \quad \text{on} \quad \partial \Omega \times (0, T)$$

with some $\alpha > 0$. Assume that the domain Ω is polygonal. For some weakly acute triangulation of Ω, let V_h be the space of conforming piecewise linear finite elements. The discretization is defined by the variational equation

$$\left(\frac{U^{k+1} - U^k}{\tau}, v_h \right) + \alpha(\sigma \nabla U^{k+1} + (1-\sigma)\nabla U^k, \nabla v_h) = (\sigma f^{k+1} + (1-\sigma)f^k, v_h)$$

for all $v_h \in V_h$, where $0 \le \sigma \le 1$; this defines the numerical solution $U^{k+1} \in V_h$ at t_{k+1}. Prove that the scheme is stable in the L_∞ norm if the condition

$$6a(1 - \sigma)\tau \le \hat{k}^2$$

holds, where $\hat{k}$ denotes the minimal length of all perpendicular bisectors of the sides of edges of triangles in the triangulation.

Exercise 5.18. Assume that $0 < \tau << 1$. Find upper and lower bounds for the eigenvalues of the matrix

$$A = \frac{1}{\tau} \begin{bmatrix} 2 & -1 & 0 & \cdots & & 0 \\ -1 & 2 & -1 & 0 & \cdots & 0 \\ 0 & \ddots & \ddots & \ddots & & 0 \\ 0 & \cdots & -1 & 2 & -1 & 0 \\ 0 & \cdots & 0 & -1 & 2 & -1 \\ 0 & & \cdots & 0 & -1 & 2 \end{bmatrix}$$

of dimension $(N - 1) \times (N - 1)$, where $\tau N = 1$.

Exercise 5.19. Discuss the order of consistency of all
a) explicit
b) diagonally-implicit
two-stage Runge-Kutta methods for the initial value problem

$$u'(t) = f(t, u), \quad u(t_0) = u_0.$$

Exercise 5.20. Construct a two-stage implicit Runge-Kutta scheme by means of the collocation principle, using the collocation points $t_{k,i} = t_k + c_i \tau$ for $i = 1, 2$ where $c_1 = 0$ and c_2 is arbitrary. What value of c_2 produces the maximal order of convergence and what order can be achieved?

Exercise 5.21. Determine whether the following Runge-Kutta methods are A-stable:

a) the implicit mid point rule

$$\begin{array}{c|c} 1/2 & 1/2 \\ \hline & 1 \end{array}$$

b) the two-stage Gauss method

$$\begin{array}{c|cc} (3-\sqrt{3})/6 & 1/4 & (3-\sqrt{3})/12 \\ (3+\sqrt{3})/6 & (3+2\sqrt{3})/12 & 1/4 \\ \hline & 1/2 & 1/2 \end{array}$$

Exercise 5.22. Apply the discontinuous Galerkin method with $q = 1$ to discretize

$$u' = \lambda u.$$

Find the associated stability function.

5.1.5 Rothe's Method (Horizontal Method of Lines)

In *Rothe's method* one applies a temporal semi-discretization to approximate a given parabolic initial-boundary value problem by a finite sequence of elliptic boundary value problems.

Example 5.23. Consider the initial-boundary value problem

$$\frac{\partial u}{\partial t} - \frac{\partial^2 u}{\partial x^2} = \sin x \quad \text{in} \quad (0, \pi) \times (0, T),$$
$$u(x, 0) = 0, \quad u(0, t) = u(\pi, t) = 0.$$

The exact solution is $u(x, t) = (1 - e^{-t}) \sin x$. Let us divide the interval $[0, T]$ by an equidistant mesh of step size $\tau = T/M$. Let $z_j(x)$ denote the computed approximation of $u(x, t_j)$ at each time level $t_j := j\tau$, $j = 0, \ldots, M$. These approximations will be defined iteratively by

$$\frac{z_j(x) - z_{j-1}(x)}{\tau} - z_j''(x) = \sin x, \quad z_j(0) = z_j(\pi) = 0, \quad z_0(x) = 0.$$

In this simple example we can solve for $z_j(x)$ explicitly, obtaining

$$z_j(x) = \left[1 - \frac{1}{(1+\tau)^j}\right] \sin x.$$

This approximate solution can be extended from its values at the grid points t_j to all $t \in [0, T]$ by setting

$$u^\tau(x, t) = z_{j-1}(x) + \frac{t - t_{j-1}}{\tau}[z_j(x) - z_{j-1}(x)] \quad \text{on} \quad [t_{j-1}, t_j] \quad \text{for } j = 1, \ldots, M.$$

Since

$$\lim_{\tau \to 0} \frac{1}{(1+\tau)^{t_j/\tau}} = e^{-t_j} \quad \text{it follows that} \quad \lim_{\tau \to 0} u^\tau(x, t) = u(x, t),$$

i.e., the approximate solution converges pointwise to the known exact solution as the step size tends to zero. □

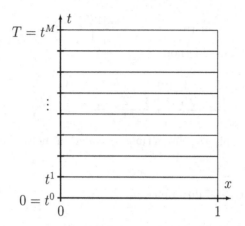

Figure 5.3 Horizontal Method of Lines

For the general description of Rothe's method, we consider the following weak formulation of a parabolic problem:

$$\frac{d}{dt}(u(t), v) + a(u(t), v) = \langle f(t), v \rangle \qquad \text{for all } v \in V \qquad (1.48)$$

with

$$u(0) = u_0 \in H \quad \text{and} \quad u \in W_2^1(0, T; V, H).$$

The time interval $[0, T]$ will be split into M subintervals $[t_{i-1}, t_i]$ with $t_i = \tau i$, where the step size is $\tau = T/M$. See Figure 5.3.

Define piecewise linear functions $\varphi_i(\cdot)$ by

$$\varphi_i(t) = \begin{cases} (t - t_{i-1})/\tau \text{ for } & t \in [t_{i-1}, t_i], \\ (t_{i+1} - t)/\tau \text{ for } & t \in [t_i, t_{i+1}], \\ 0 & \text{otherwise.} \end{cases}$$

An approximation of $u(x, t)$ is given by the *Rothe function*

$$u^\tau(x, t) := \sum_{i=0}^{M} z_i(x) \varphi_i(t),$$

where the approximations $z_i(x)$ of $u(x, t_i)$ are defined iteratively for $i = 1, \ldots, M$ by

$$\left(\frac{z_i - z_{i-1}}{\tau}, v \right) + a(z_i, v) = \langle f_i, v \rangle \quad \text{for all } v \in V. \qquad (1.49)$$

Remark 5.24. It's interesting to compare (1.49) with the temporal discretization by the implicit Euler scheme of the semi-discrete problems generated by the method of finite elements in space, which yields

$$\left(\frac{U^{k+1} - U^k}{\tau}, v_h\right) + a(U^{k+1}, v_h) = \langle f^{k+1}, v_h \rangle \quad \text{for all } v_h \in V_h .$$

Hence Rothe's method (1.49) can be considered as a continuous analogue of this other method where the spatial variables have not been discretized.   □

If the bilinear form $a(\cdot, \cdot) : V \times V \to \mathbb{R}$ is V-elliptic then the Lax-Milgram lemma implies immediately that (1.49) has a unique solution $z_{i+1} \in V$ for all $\tau > 0$.

In [Rek82] Rothe's method is used mainly to derive existence theorems for the original parabolic problem. As well as existence and uniqueness of the solution of (1.49), a priori estimates for z_{i+1} and for related functions are obtained to ensure certain limit behaviour as the time step size tends to zero. While these results are important in the theory of partial differential equations, they do not lie in our sphere of interest.

Following Rektorys [Rek82] we now sketch how bounds on the discretization error may be obtained. For simplicity let us assume that:

a) $V = H_0^1(\Omega)$

b) $a(\cdot, \cdot)$ is V-elliptic

c) $u_0 = 0$

d) $f \in V \cap H_2(\Omega)$ is *independent of time*

e) $|a(f, v)| \le C \|f\|_2 \|v\|$.

Fix $i \in \{1, \ldots, M\}$. A calculation shows that the error in the interval (t_{i-1}, t_i) satisfies

$$\frac{d}{dt}(u^\tau - u, v) + a(u^\tau - u, v) = \tau \left(\frac{z_i - 2z_{i-1} + z_{i-2}}{\tau^2} \cdot \frac{t_i - t}{\tau}, v \right) .$$

We shall estimate the L_2 norm of the error $u^\tau - u$ by invoking the a priori estimate (1.5) of Section 5.1. To do this some information about the magnitude of $(z_i - 2z_{i-1} + z_{i-2})/\tau^2$ is needed. This information will be derived step by step. Set

$$Z_i = \frac{z_i - z_{i-1}}{\tau} \quad \text{and} \quad s_i = \frac{Z_i - Z_{i-1}}{\tau} .$$

We begin with a priori estimates for Z_i and s_i. Now

$$a(z_1, v) + \frac{1}{\tau}(z_1, v) = (f, v) \quad \text{for all } v \in V,$$

so on taking $v = z_1$ one gets the bound

$$\|z_1\| \le \tau\gamma \quad \text{with} \quad \gamma = \|f\|.$$

Subtracting (1.49) for two consecutive indices yields

$$a(z_j - z_{j-1}, v) + \frac{1}{\tau}(z_j - z_{j-1}, v) = \frac{1}{\tau}(z_{j-1} - z_{j-2}, v) \quad \text{for all } v \in V.$$

Choose $v = z_j - z_{j-1}$; this gives

$$\|z_j - z_{j-1}\| \le \|z_{j-1} - z_{j-2}\|,$$

and by induction we get

$$\|z_j - z_{j-1}\| \le \tau\gamma \quad \text{and} \quad \|Z_j\| \le \gamma \quad \text{for } j = 1, \dots, M.$$

Next we analyse $Z_j - Z_{j-1}$ in a similar manner. From

$$a(Z_j - Z_{j-1}, v) + \frac{1}{\tau}(Z_j - Z_{j-1}, v) = \frac{1}{\tau}(Z_{j-1} - Z_{j-2}, v) \quad \text{for all } v \in V$$

we obtain

$$\|Z_j - Z_{j-1}\| \le \|Z_{j-1} - Z_{j-2}\| \quad \text{for } j = 3, \dots, M.$$

The identity

$$a(Z_2 - Z_1, v) + \frac{1}{\tau}(Z_2 - Z_1, v) = \frac{1}{\tau}(Z_1 - f, v)$$

leads to

$$\|Z_2 - Z_1\| \le \|Z_1 - f\|.$$

Now induction yields

$$\|Z_j - Z_{j-1}\| \le \|Z_1 - f\| \quad \text{for } j = 2, \dots, M.$$

Since

$$a(Z_1 - f, v) + \frac{1}{\tau}(Z_1 - f, v) = -a(f, v),$$

on taking $v = Z_1 - v$ our hypotheses imply that

$$\|Z_1 - f\| \le \tau\|f\|_2.$$

Consequently

$$\|s_j\| = \left\| \frac{Z_j - Z_{j-1}}{\tau} \right\| \le \|f\|_2 \quad \text{for } j = 2, \dots, M.$$

Combining these bounds with (1.5), we arrive at the following theorem.

Theorem 5.25. *Under our hypotheses the error of Rothe's method satisfies*

$$\|(u - u^\tau)(t)\| \le C\tau \quad \text{for all } t.$$

The assumptions made here are unnecessarily restrictive; they were imposed to simplify the analysis. Similar estimates can be derived under weaker hypotheses. For a detailed study of this task we refer to [Rek82, Kač85]. In these monographs error estimates in other norms are also derived.

To obtain a complete discretization the elliptic boundary value problems generated by (1.49) must be discretized, e.g., by means of some finite element method. Convergence results for this situation are derived in [Rek82] but no estimates for the rates of convergence are given. We have seen in Remark 5.24 that Rothe's method combined with a finite element discretization coincides with the application of the implicit Euler scheme to a finite element semi-discretization. Hence the error estimate of Theorem 5.13 is also valid for Rothe's method combined with a spatial finite element discretization.

As was mentioned already, Rothe's method provides an elegant tool for the derivation of existence results for time-dependent problems by their reduction to a sequence of stationary problems. Furthermore, other techniques applicable to stationary problems can in this way be carried over to unsteady problems. Nevertheless it should be noted that elliptic problems of the type

$$\tau a(z_i, v) + (z_i - z_{i-1}, v) = \tau(f_i, v) \quad \text{for all } v \in V$$

require a special analysis and appropriate numerical treatment because of the small parameter τ multiplying the highest-order term; such problems can be classified as singularly perturbed (see Chapter 6).

In generalizations of Rothe's method the usual implicit Euler scheme is replaced by other implicit schemes for the temporal discretization of (1.48). Special attention has been paid in the literature to the Rothe-Rosenbrock (see [Lan01]) and discontinuous Galerkin methods that we have discussed previously. Lang [Lan01] proved error estimates for Rothe-Rosenbrock methods and also for the case of a complete discretization by finite elements of the elliptic problems generated by Rosenbrock methods. In this context reductions of order of convergence may occur.

Exercise 5.26. Use Rothe's method to discretize the initial-boundary value problem

$$\frac{\partial u}{\partial t} - \frac{\partial^2 u}{\partial x^2} = \sin x \quad \text{in} \quad (0, \pi/2) \times (0, T)$$

with

$$u|_{t=0} = 0, \quad u|_{x=0} = 0, \quad u_x|_{x=\pi/2} = 0.$$

Compare the computed approximate solution with the exact solution.

5.1.6 Error Control

It is desirable that in the discretization of parabolic initial-boundary value problems the discretization error should be controlled efficiently, and preferably in an *adaptive* way. This would enable the automatic distribution of the discrete degrees of freedom for the purpose of minimizing the overall error. Various strategies for error control are used in the discretization of parabolic problems, and three of them will be described here. First, a variant of the residual-based estimator measured in a norm appropriate to the parabolic problem; second, a technique based on temporal discretization via a generalized Rothe's method; third, a goal-oriented estimator that uses information from an adjoint problem.

Let us consider the model problem

$$\frac{\partial u}{\partial t} - \Delta u = f \quad \text{in} \quad \Omega \times (0, T), \tag{1.50}$$

$$u = 0 \quad \text{on} \quad \partial\Omega \times (0, T),$$

$$u = u_0 \quad \text{for} \quad t = 0, x \in \Omega.$$

In the notation of Section 5.1.1, its weak formulation is:

Find $u \in W_2^1(0, T; V, H)$ such that

$$\frac{d}{dt}(u(t), v) + (\nabla u, \nabla v) = (f, v) \quad \text{for all } v \in V := H_0^1(\Omega)$$

with $u(0) = u_0 \in H := L_2(\Omega)$.

First we modify the a priori estimate (1.5). Assume that $f \in L_2(0, T; H^{-1})$. We shall derive a relationship between the norm of f in this space and the solution u. Taking $v = u$ as in the proof of (1.5) leads to

$$\frac{1}{2}\frac{d}{dt}\|u(t)\|_0^2 + |u(t)|_1^2 \leq \frac{1}{2}\|f(t)\|_{-1}^2 + \frac{1}{2}|u(t)|_1^2.$$

Integration then yields

$$\|u(t)\|_0^2 + \int_0^t |u(s)|_1^2 ds \leq \|u(0)\|_0^2 + \int_0^t \|f(s)\|_{-1}^2 ds \quad \text{for } 0 \leq t \leq T,$$

which implies the a priori bound

$$\||u\|| := \|u\|_{L_\infty(0,T;L_2)} + \|u\|_{L_2(0,T;H_0^1)}$$

$$\leq \sqrt{2}\left(\|u(0)\|_0^2 + \|f\|_{L_2(0,T;H^{-1})}\right)^{1/2}. \tag{1.51}$$

This inequality motivates us to search for a residual estimator for the norm $\|| \cdot \||$.

Consider the discretization of (1.50) by finite elements in space and by the implicit Euler scheme in time (the θ scheme with $\theta \geq 1/2$ could be studied in a similar manner):

$$\left(\frac{U_h^n - U_h^{n-1}}{\tau_n}, v_h\right) + (\nabla U_h^n, \nabla v_h) = (f^n, v_h) \quad \text{for all } v_h \in V_{h,n}, \tag{1.52}$$

$$U_h^0 = \pi u_0,$$

where πu_0 is the $L_2(\Omega)$ projection of u_0. As this notation indicates, we shall apply a variable temporal grid; furthermore, the finite element space $V_{h,n}$ used at the time level t_n may vary with n as the spatial grid may change.

Let $U_{h,\tau}$ be piecewise linear in t with $U_{h,\tau}(t_n) = U_h^n$, so $U_{h,\tau}$ is an extension of the discrete solution to all of $\bar{\Omega} \times [0, T]$. Now let us define the residual of $U_{h,\tau}$ in V^* by

$$\langle R(U_{h,\tau}), v \rangle := (f, v) - (\partial_t U_{h,\tau}, v) - (\nabla U_{h,\tau}, \nabla v). \tag{1.53}$$

We shall exploit (1.51) to show that the error up to time t_n can be estimated using $\|u_0 - \pi u_0\|_0$ and the norm of the residual in $L_2(0, t_n; H^{-1})$, yielding a residual estimator that estimates the norm of the error from above. A lower bound on this norm and details of the following argument can be found in [119].

Define the piecewise constant function $\tilde{f}$ by $\tilde{f} = f^n$ on each interval (t_{n-1}, t_n). Let $f_{h,\tau}$ denote the $L_2(\Omega)$ projection of $\tilde{f}$ on the finite element space $V_{h,n}$.

First we split the residual into spatial and temporal components. This will enable us to control adaptively but separately the grids in space and time. Set

$$\langle R_\tau(U_{h,\tau}), v \rangle = (\nabla(U_h^n - U_{h,\tau}), \nabla v) \quad \text{on } (t_{n-1}, t_n)$$

and

$$\langle R_h(U_{h,\tau}), v \rangle = (f_{h,\tau}, v) - \left(\frac{U_h^n - U_h^{n-1}}{\tau_n}, v \right) - (\nabla U_h^n, \nabla v) \quad \text{on } (t_{n-1}, t_n).$$

Then

$$R(U_{h,\tau}) = f - f_{h,\tau} + R_\tau(U_{h,\tau}) + R_h(U_{h,\tau}). \tag{1.54}$$

Clearly

$$\|R_\tau(U_{h,\tau})\|_{-1} = |U_h^n - U_{h,\tau}|_1$$

and it follows that the norm of $R_\tau(U_{h,\tau})$ in $L_2(0, t_n; H^{-1}(\Omega))$ is explicitly given by

$$\begin{aligned}
\|R_\tau(U_{h,\tau})\|_{L_2(t_{n-1}, t_n; H^{-1})}^2 &= \int_{t_{n-1}}^{t_n} \left(\frac{t_n - s}{\tau_n} \right)^2 |U_h^n - U_h^{n-1}|_1^2 \, ds \\
&= \frac{1}{3} \tau_n |U_h^n - U_h^{n-1}|_1^2.
\end{aligned} \tag{1.55}$$

The spatial residual satisfies the Galerkin orthogonality relation

$$\langle R_h(U_{h,\tau}), v_h \rangle = 0 \quad \text{for all } v_h \in V_{h,n}.$$

Consequently bounds for the $H^{-1}(\Omega)$ norm of $R_h(U_{h,\tau})$ can be obtained through the techniques of Chapter 4 and Section 7.1. For simplicity we omit the technicalities caused by the change of triangulation $\mathcal{T}_{h,n}$ from each time step to the next. Next, we briefly sketch results from [119] and refer for further details to this paper. Assume as usual that the triangulation $\mathcal{T}_{h,n}$ is quasi-uniform. We also suppose that the refinement $\tilde{\mathcal{T}}_{h,n}$ of the grid $\mathcal{T}_{h,n}$

$$\sup_n \sup_{K \in \tilde{\mathcal{T}}_{h,n}} \sup_{K' \in \mathcal{T}_{h,n}, K' \subset K} \frac{h_{K'}}{h_K} < \infty. \tag{1.56}$$

This condition excludes any abrupt change of grid. It can then be shown that

$$\|R_h(U_{h,\tau})\|_{-1} \le c\eta_h^n \tag{1.57}$$

with

$$(\eta_h^n)^2 = \sum_{K \in \tilde{\mathcal{T}}_{h,n}} h_K^2 \|R_K\|_{0,K}^2 + \sum_{E \in \tilde{\mathcal{T}}_{h,n}} h_E \|R_E\|_{0,E}^2,$$

where the residuals R_K, R_E are defined by

$$R_K := f_{h,\tau} - \frac{U_h^n - U_h^{n-1}}{\tau_n} + \Delta U_h^n, \qquad R_E := [n_E \cdot \nabla U_h^n]_E.$$

Now from (1.55) and (1.57) we get

Theorem 5.27. *Suppose that (1.56) is satisfied. Then in $(0, t_n)$ one has*

$$|||u - U_{h,\tau}||| \le c \left(\sum_1^n (\eta^m)^2 + \|f - f_{h,\tau}\|_{L_2(0,t_n;H^{-1})} + \|u_0 - \pi u_0\|_0^2 \right)^{1/2} \tag{1.58}$$

where

$$(\eta^m)^2 := \tau_m (\eta_h^m)^2 + \sum_{K \in \tilde{\mathcal{T}}_{h,m}} \tau_m |U_h^m - U_h^{m-1}|_{1,K}^2. \tag{1.59}$$

Our second error estimator for error control comes from from [18, 19, 20] and [Lan01]. The generalized Rothe method is its starting point. Assume that $f \in L_2(\Omega \times [0,T])$ and $u_0 \in L_2(\Omega)$. An abstract formulation of (1.50) is an ordinary differential equation posed in some Banach space. For the case considered here, the space $L_2(\Omega)$ can be taken as this Banach space (see [Tho97]) and the abstract differential equation has the form:
Find $u(t) \in L_2(\Omega)$ such that

$$u' - Au = f, \tag{1.60}$$
$$u(0) = u_0.$$

Rothe's method can be regarded as a discretization of the abstract ordinary differential equation (1.60) via the implicit Euler method; it determines the approximation u_{j+1} at each time level t_{j+1} from

$$\frac{u_{j+1} - u_j}{\tau} - Au_{j+1} = f_{j+1}. \tag{1.61}$$

Discretization methods for initial value problems have fully-developed strategies for error control; see [HNW87]. Here we shall concentrate on one-step methods. One tool for deriving error estimates (and hence step size control) is to apply simultaneously at each time step two separate one-step methods of different orders. Consider the step from t_j to $t_j + \tau$. Denote by

u^{k+1} the numerical approximation generated by the method of order $k + 1$, and by

u^k the numerical approximation generated by the method of order k

for $u(t_j + \tau)$. Then

$$\varepsilon_k := \|u^{k+1} - u^k\|$$

is an error estimator. For a given tolerance *tol* the step size is controlled by

$$\tau_{new} := \tau_{old} \left(\frac{tol}{\varepsilon_k}\right)^{1/(k+1)}$$

and the order of the method is chosen in such a way that the effort to perform the temporal integration is minimized. A rigorous analytical justification of this strategy for the case of the abstract initial value problem (1.50) appears in [18, 19, 20].

To apply the above strategy to the abstract ordinary differential equation (1.50), we need to find suitable one-step methods to generalize the standard Rothe method (1.61). Towards this end we appeal first to the stability function $R(\cdot)$ of one-step methods. Given a stability function R, define a related one-step method for the discretization of (1.60) with step size τ by

$$u_{j+1} := R(\tau A)u_j + (-I + R(\tau A))A^{-1}f. \tag{1.62}$$

We assume that R is selected in such a way that the method generated by (1.62) has appropriate stability properties.

Remark 5.28. It is easy to verify that the choice $R(z) = 1/(1 - z)$ in (1.62) yields (1.61). Similarly $R(z) = 1 + z$ yields the explicit Euler method

$$\frac{u_{j+1} - u_j}{\tau} - Au_j = f_j$$

and $R(z) = (2 + z)/(2 - z)$ produces the Crank-Nicolson method

$$\frac{u_{j+1} - u_j}{\tau} - A\frac{u_j + u_{j+1}}{2} = \frac{f_j + f_{j+1}}{2}. \quad \square$$

The one-step method (1.62) has formal order p if

$$e^z - R(z) = cz^{p+1} + O(z^{p+2}) \quad \text{as } z \to 0.$$

An appropriate choice of $R(\cdot)$ depends not only on the temporal discretization but also on the discretization error in space. If (as is widely preferred) an implicit one-step method is used for the temporal integration, then at each time level an elliptic boundary value problem must be solved. Finite element methods give a reasonable discretization of these elliptic problems. Many error estimators for the temporal discretization involve u^k and u^{k+1}, but these quantities cannot be computed exactly; thus in particular the quantity

$$\varepsilon_k = \|u^{k+1} - u^k\|$$

is unreliable and should not be used as an estimate for the expected error.

Bornemann [18, 19, 20] recommends the use of stability functions and associated one-step methods that allow a *direct* evaluation of some error estimator $\|\eta_k\|$ for the temporal discretization; then u^{k+1} is computed as

$$u^{k+1} = u^k + \eta_k.$$

In [19] it is shown that the iterative definition of a stability function R_k^L by

$$R_1^L(z) = \frac{1}{1-z},$$

$$\rho_1^L(z) = -\frac{z^2}{2(1-z)^2} R_1^L(z),$$

$$R_{i+1}^L(z) = R_i^L(z) + \rho_i^L(z),$$

$$\rho_{i+1}^L(z) = -\gamma_{i+1}^L \frac{z}{1-z} \rho_i^L(z)$$

gives an L_0-stable method of order k. The parameters γ_k^L here are defined by

$$\gamma_k^L := \frac{L_{k+1}(1)}{L_k(1)}$$

where $L_k(\cdot)$ is the Laguerre polynomial of degree k.

Consider a fixed time level t_j where the solution u_j has already been computed. If the approximations u_{j+1}^{k+1} and u_{j+1}^k at the $(j+1)$th time level are determined by methods of order $k+1$ and order k, respectively, that satisfy (1.62), then

$$u_{j+1}^{k+1} - u_{j+1}^k = (R_{k+1}^L - R_k^L)(\tau A)u_j + (R_{k+1}^L - R_k^L)(\tau A)A^{-1}f.$$

Hence the above iterative definition of R_{k+1}^L shows that

$$\eta_k = \rho_k^L(\tau A)\left(u_j + A^{-1}f\right).$$

Similarly

$$\eta_{k+1} = \rho_{k+1}^L(\tau A)\left(u_j + A^{-1}f\right).$$

But

$$\rho_{k+1}^L(z) = -\gamma_{k+1}^L \frac{z}{1-z} \rho_k^L(z)$$

so

$$\eta_{k+1} = -\gamma_{k+1}^L \frac{\tau A}{1 - \tau A} \eta_k . \tag{1.63}$$

This relation is the cornerstone of the error estimate. Namely, starting from u^1 the first error estimate η_1 is evaluated. If this is too large then the estimator η_2 is computed from (1.63) then added to u^1 to get u^2, and so on. The initial approximation u^1 and all the η_k are solutions of elliptic boundary value problems of the form

$$w - \tau A w = g. \tag{1.64}$$

The basic philosophy of the discretization in [Lan01] is similar to that discussed above, except that Lang [Lan01] uses embedded Rosenbrock methods instead of one-step methods.

For our model problem (1.64) is equivalent to:
Find $w \in V = H_0^1(\Omega)$ with

$$\tau(\nabla w, \nabla v) + (w, v) = (g, v) \quad \text{for all } v \in V, \tag{1.65}$$

where $(\cdot, \cdot)$ is the $L_2(\Omega)$ inner product. Error estimators for the spatial discretization error must take into account the dependence of the elliptic problem (1.65) upon the time step parameter τ. The use of a τ-weighted H^1 norm $\|\cdot\|_\tau$ defined by

$$\|v\|_\tau^2 := \tau |v|_1^2 + \|v\|_0^2$$

is important. Error estimators for the spatial discretization should be based upon this norm.

One integral component of the computer package KARDOS (Konrad-Zuse-Zentrum, Berlin) is a *hierarchical error estimator*. Variants of hierarchical estimators and their relationship to error estimators of different types are studied in detail in [Ver96]. The robustness of such error estimators for problems of the type (1.65) seems however to be not yet entirely clear. In [2] and [118] two other robust estimators for reaction-diffusion problems are analysed.

Here we sketch briefly the basic idea of hierarchical estimators. Let

$$V_h^2 = V_h^1 \oplus Z_h$$

be some hierarchical splitting of the finite element space V_h^2 that was generated from the original discrete space V_h^1 by the inclusion of extra basis functions. One could hope that the difference $w_h^1 - w_h^2$ of the finite element approximations w_h^i of the solution w of (1.65) that are generated by the spaces V_h^i for $i = 1, 2$ might provide some information about the accuracy of w_h^1. Indeed, one has

Lemma 5.29. *Assume that on the discretization grids one has*

$$\|w - w_h^2\|_\tau \le \beta \|w - w_h^1\|_\tau \quad \text{with} \quad \beta \in (0,1) \tag{1.66}$$

for some constant β that is independent of τ and h (in fact β depends on the approximation properties of the V_h^i). Then there exists a constant γ, which depends only on β, such that

$$\|w_h^1 - w_h^2\|_\tau \le \|w - w_h^1\|_\tau \le \gamma \|w_h^1 - w_h^2\|_\tau . \tag{1.67}$$

Proof: Set

$$a_\tau(v, u) := \tau(\nabla u, \nabla v) + (u, v).$$

Clearly

$$a_\tau(v, v) = \|v\|_\tau^2 .$$

Now

$$\|w - w_h^1\|_\tau^2 = a_\tau(w - w_h^1, w - w_h^1)$$
$$= a_\tau(w - w_h^2 + w_h^2 - w_h^1, w - w_h^2 + w_h^2 - w_h^1)$$
$$= \|w - w_h^2\|_\tau^2 + 2a_\tau(w - w_h^2, w_h^2 - w_h^1) + \|w_h^2 - w_h^1\|_\tau^2,$$

but $a_\tau(w - w_h^2, w_h^2 - w_h^1) = 0$. Thus

$$\|w - w_h^1\|_\tau^2 = \|w - w_h^2\|_\tau^2 + \|w_h^2 - w_h^1\|_\tau^2 . \tag{1.68}$$

The first inequality in (1.67) follows immediately. For the second, (1.66) and (1.68) give

$$\|w - w_h^1\|_\tau^2 \le \beta^2 \|w - w_h^1\|_\tau^2 + \|w_h^2 - w_h^1\|_\tau^2,$$

whence

$$\|w - w_h^1\|_\tau^2 \le \frac{1}{1 - \beta^2} \|w_h^2 - w_h^1\|_\tau^2 . \quad \blacksquare$$

Lemma 5.29 says that the computable quantity $\|w_h^1 - w_h^2\|_\tau$ is a reliable and robust error estimator, uniformly in τ, in the τ-weighted H^1 norm provided that the β-approximation inequality (1.66) holds. Unfortunately this inequality is not easily satisfied. Details of this approach can be found in [18, 19, 20] and [Lan01]. In the last of these references numerical tests are carried out and interesting practical applications are described.

The third method for error control that we examine is the use of goal-oriented indicators for the model problem (1.50). Here the starting point is the discretization by the discontinuous Galerkin method $dG(0)$.

If u is the solution of (1.50) then for any test function $v(\cdot, t) \in V = H_0^1(\Omega)$ one gets

$$A(u, v) = \int_0^T (f, v) dt + (u_0, v_+(0))$$

with

$$A(u,v) := \sum_{m=1}^{M} \int_{t_{m-1}}^{t_m} [(u_t,v) + (\nabla u, \nabla v)] + \sum_{m=2}^{M} ([u^{m-1}], v_+^{m-1}) + (u(0), v_+(0)).$$

The goal-oriented error estimator is based on a dual problem that is based on an integration by parts. In the case of piecewise differentiable v, integration by parts and the continuity of u yield

$$A(u,v) = \sum_{m=1}^{M} \int_{t_{m-1}}^{t_m} [-(u,v_t) + (\nabla u, \nabla v)] \tag{1.69}$$
$$- \sum_{m=2}^{M} (u_+^{m-1}, [v^{m-1}]) + (u(t_M), v_-(t_M)).$$

The discontinuous Galerkin method $dG(0)$ uses piecewise constant approximations in t. Let $U_h^n \in V_{h,n}$ be constant on (t_{n-1}, t_n). Then we obtain

$$(U_h^n - U_h^{n-1}, v_h) + \tau_n(\nabla U_h^n, \nabla v_h) = \int_{t_{n-1}}^{t_n} (f, v_h) \quad \text{for all } v_h \in V_{h,n}. \tag{1.70}$$

Suppose that the goal of the adaptive method is control of the $L_2(\Omega)$ error at the terminal time $t = t_M = T$. Set $e := u - u_{h,\tau}$, where $u_{h,\tau}$ denotes the piecewise constant function that coincides on (t_{m-1}, t_m) with U_h^m. Define the error functional J by

$$J(\varphi) := \frac{(\varphi_-^M, e_-^M)}{\|e_-^M\|_0}, \quad \text{so } J(e) = \|e_-^M\|_0.$$

The associated dual problem is

$$-\frac{\partial z}{\partial t} - \Delta z = 0 \quad \text{in} \quad \Omega \times (0,T), \tag{1.71}$$
$$z = 0 \quad \text{on} \quad \partial\Omega \times (0,T),$$
$$z = \frac{e_-^M}{\|e_-^M\|_0} \quad \text{for} \quad t = t_M.$$

From (1.69) one has the error functional representation

$$J(e) = A(e, z). \tag{1.72}$$

The Galerkin orthogonality of the method allows us to subtract any piecewise constant function v_h with values in $V_{h,m}$ on (t_{m-1}, t_m), i.e., we have also

$$J(e) = A(e, z - z_h) = \sum_{m=1}^{M} \int_{t_{m-1}}^{t_m} [(e_t, z - z_h) + (\nabla e, \nabla(z - z_h))]$$
$$+ \sum_{m=2}^{M} ([e^{m-1}], (z - z_h)_+^{m-1}) + (e(0), (z - z_h)_+(0)).$$

Now integrate by parts in the usual way to introduce the element residual $R(U_h) := f - (U_h)_t + \Delta U_h$ and the edge residuals. Then, setting $Q_{m,l} = K_{m,l} \times (t_{m-1}, t_m)$, by the Cauchy-Schwarz inequality we obtain

$$\|e_-^M\|_0 \leq \sum_{m=1}^{M} \sum_{K_{m,l} \in \mathcal{T}_{h,m}} \left\{ \|R(U_h^m)\|_{Q_{m,l}} \, \rho_{ml}^1 \right.$$

$$\left. + \frac{1}{2} \|[\partial_n R(U_h^m)]\|_{\partial Q_{m,l}} \, \rho_{ml}^2 + \|[U_h^{m-1}]\|_{K_{m,l}} \, \rho_{ml}^3 \right\}$$

with the weights

$$\rho_{ml}^1 = \|z - z_h\|_{Q_{m,l}}, \quad \rho_{ml}^2 = \|z - z_h\|_{\partial Q_{m,l}}, \quad \rho_{ml}^3 = \|(z - z_h)_+^{m-1}\|_{K_{m,l}};$$

all norms in these formulas are $L_2(\Omega)$ norms.

If z_h is chosen to be a suitable interpolant then we get a result like Theorem 5.27. Alternatively, the weights ρ_{ml}^k can be evaluated approximately by a numerical treatment of the dual problem (1.71). This is done in [BR03], which contains a wide range of results for different applications of goal-oriented error indicators to several model problems.

5.2 Second-Order Hyperbolic Problems

5.2.1 Weak Formulation of the Problem

In this section we examine the numerical treatment of second-order linear hyperbolic initial-boundary value problems by the finite element method. Our presentation largely follows the approach of [LT03], [15]. We shall concentrate our attention on the model problem

$$\begin{aligned} u_{tt} - \Delta u &= f \quad \text{in } \Omega \times (0, T], \\ u &= 0 \quad \text{in } \Gamma \times (0, T], \\ u(\cdot, 0) = u^0, \qquad u_t(\cdot, 0) &= v^0 \quad \text{on } \overline{\Omega}, \end{aligned} \tag{2.1}$$

and give only an outline of extensions to more general problems. In (2.1) the domain Ω lies in $\mathbb{R}^n$ and is assumed bounded with smooth boundary Γ, while $T > 0$ is a given constant. The functions $u^0, v^0 : \overline{\Omega} \to \mathbb{R}$ are given. The regularity of Ω guarantees that integration by parts can be applied in the spatial direction. We set $Q_T = \Omega \times (0, T]$ and $\Gamma_T = \Gamma \times (0, T]$. As for elliptic and for parabolic problems, a function $u \in C^{2,1}(Q_T) \cap C^{0,1}(\overline{Q}_T)$ is called a classical solution of (2.1) if all equations of (2.1) are satisfied. Derivatives at boundary points of the domain are understood as continuous extensions from inside Q_T. It should be noted that existence results for classical solutions require further assumptions on the domain Ω and on the functions u^0 and v^0.

The application of the finite element method is based on a weak form of
(2.1). To obtain this weak form, as in the parabolic case we set $V = H_0^1(\Omega)$
and $H = L_2(\Omega)$. Together with the space $V^* = H^{-1}(\Omega)$ that is dual to V
and making the identification $H^* = H$, these spaces form a Gelfand triple
(see page 318)

$$V \hookrightarrow H \hookrightarrow V^*.$$

Integrating by parts in V, the operator $-\Delta$ can be related to the bilinear form

$$a(u, w) := \int_\Omega \nabla u \cdot \nabla w \qquad \text{for all} \quad u, w \in V.$$

Hence (2.1) has the weak formulation

$$\left(\frac{d^2 u}{dt^2}, w \right) + a(u, w) = (f, w) \qquad \text{for all} \quad w \in V,$$
$$u(0) = u^0, \quad \frac{du}{dt}(0) = v^0, \tag{2.2}$$

where $(\cdot, \cdot)$ is the $L_2(\Omega)$ inner product. Here one wishes to find a $u \in$
$L_2(0, T; V)$, with $\frac{du}{dt} \in L_2(0, T; H)$ and $\frac{d^2 u}{dt^2} \in L_2(0, T; V^*)$, that satisfies
the variational equation and the initial conditions of (2.2).

Define a continuous linear operator $L : V \to V^*$ by

$$a(u, w) = \langle Lu, w \rangle \qquad \text{for all} \quad u, w \in V,$$

where $\langle \cdot, \cdot \rangle : V^* \times V \to \mathbb{R}$ denotes the duality pairing. Then the weak formulation (2.2) can be interpreted as the operator differential equation

$$\frac{d^2 u}{dt^2} + Lu = f, \qquad u(0) = u^0, \quad \frac{du}{dt}(0) = v^0 \tag{2.3}$$

in the space $L_2(0, T; V^*)$. For the general analysis of such operator equations
we refer to [GGZ74]. In the particular case considered here we have the following theorem (see, e.g., [Wlo87, Theorem 29.1]):

Theorem 5.30. *Let $f \in L_2(0, T; H)$, $u^0 \in V$ and $v^0 \in H$. Then the problem
(2.2) has a unique solution $u \in L_2(0, T; V)$ with $\frac{du}{dt} \in L_2(0, T; H)$. Furthermore, the mapping*

$$\{f, u^0, v^0\} \to \left\{ u, \frac{du}{dt} \right\}$$

from $L_2(0, T; H) \times V \times H$ to $L_2(0, T; V) \times L_2(0, T; V^*)$ is linear and continuous.

Introduce $v := u_t$ as an auxiliary function in $L_2(0, T; H)$. Then (2.2) can be
written as an equivalent first-order system, either in the weak form

$$(u_t, z) - (v, z) = 0 \qquad \text{for all} \quad z \in H,$$
$$(v_t, w) + a(u, w) = (f, w) \qquad \text{for all} \quad w \in V, \tag{2.4}$$

or in the classical form

$$\begin{pmatrix} u_t \\ v_t \end{pmatrix} = \begin{pmatrix} 0 & I \\ \Delta & 0 \end{pmatrix} \begin{pmatrix} u \\ v \end{pmatrix} + \begin{pmatrix} 0 \\ f \end{pmatrix} \quad \text{in } Q_T \quad \text{with} \quad u|_\Gamma = 0, \quad \begin{pmatrix} u \\ v \end{pmatrix}(\cdot, 0) = \begin{pmatrix} u^0(\cdot) \\ v^0(\cdot) \end{pmatrix}.$$

This problem can be treated numerically by an appropriate discretization method such as the implicit Euler scheme for a system of ordinary differential equations in function spaces; compare (2.59).

As in Section 5.1 for parabolic problems, a complete numerical treatment of hyperbolic initial-boundary value problems requires discretizations in space and time. In Section 2.7 we examined complete discretizations of (2.1) in both types of variable through difference schemes. Semi-discretization in space (the method of lines) or in time (Rothe's method) provides a better way of structuring the analysis. In the next subsection we focus on spatial discretization by finite element methods.

Before we analyse the discretization the following stability estimate is given.

Lemma 5.31. *Let (u, v) be the solution of the system*

$$(u_t, z) - (v, z) = (g, z) \qquad \text{for all} \quad z \in H,$$
$$(v_t, w) + a(u, w) = (f, w) \qquad \text{for all} \quad w \in V, \tag{2.5}$$

which is perturbed by some $g \in L_2(0, T; V)$. Then for each $t \in [0, T]$ one has the estimate

$$\Big(a(u(t), u(t)) + (v(t), v(t)) \Big)^{1/2} \leq \Big(a(u^0, u^0) + (v^0, v^0) \Big)^{1/2}$$
$$+ \int_0^t \big(\|f(s)\| + \|\nabla g(s)\| \big) \, ds. \tag{2.6}$$

In the particular case $f \equiv 0$ and $g \equiv 0$ one has the identity

$$a(u(t), u(t)) + (v(t), v(t)) = a(u^0, u^0) + (v^0, v^0). \tag{2.7}$$

Under additional smoothness assumptions one can prove this result by considering the derivative of the function $\Phi(t) := a(u(t), u(t)) + (v(t), v(t))$; compare the proof of Lemma 5.33. Equation (2.7) is a statement of the principle of conservation of energy for the solution of the homogeneous wave equation.

5.2.2 Semi-Discretization by Finite Elements

As for the parabolic problems considered in Section 5.1, a conforming semi-discretization in space replaces V by some finite-dimensional subspace $V_h :=$

span $\{\varphi_j\}_{j=1}^N$ with appropriately-chosen linearly independent basis functions $\varphi_j \in V$, $j = 1, \ldots, N$. The semi-discrete solution $u_h \in V_h$, written in terms of coefficient functions $u_j : [0, T] \to \mathbb{R}$, is

$$u_h(t) := \sum_{j=1}^N u_j(t)\, \varphi_j. \tag{2.8}$$

Here the unknown functions u_j are determined by the system of differential equations

$$\left(\frac{d^2}{dt^2}\, u_h(t), w_h\right) + a(u_h(t), w_h) = (f(\cdot, t), w_h) \quad \forall\, w_h \in V_h,\ t \in (0, T] \tag{2.9}$$

and the initial conditions

$$u_h(0) = u_h^0, \qquad \frac{d}{dt}\, u_h(0) = v_h^0, \tag{2.10}$$

where u_h^0 and $v_h^0 \in V_h$ are appropriate approximations of the initial data u^0 and v^0. Taking (2.8) and the structure of V_h into consideration, the semi-discrete problem (2.9) and (2.10) is—similarly to the parabolic case— equivalent to the initial value problem

$$D\,\hat{u}''(t) + A\,\hat{u}(t) = \hat{f}(t), \quad t \in (0, T] \quad \text{and} \quad \hat{u}(0) = u_h^0,\ \hat{u}'(0) = v_h^0 \tag{2.11}$$

for a system of ordinary differential equations. Here $\hat{u} = (u_j)$ denotes the unknown vector function with components u_j, $j = 1, \ldots, N$. The mass matrix D, the stiffness matrix A and the right-hand side $\hat{f}$ are defined by

$$D = (d_{ij}), \quad d_{ij} = (\varphi_j, \varphi_i),$$
$$A = (a_{ij}), \quad a_{ij} = a(\varphi_j, \varphi_i),$$
$$\hat{f}(t) = (f_i), \quad f_i = (f, \varphi_i).$$

The linear independence of the basis φ_j ensures the invertibility of the matrix D.

Example 5.32. As a simple example we consider the problem

$$u_{tt}(x, t) - \sigma^2 u_{xx}(x, t) = e^{-t} \text{ for } x \in (0, 1), \quad t > 0,$$
$$u(0, t) = u(1, t) = 0 \quad \text{for } t > 0, \tag{2.12}$$
$$u(x, 0) = u_t(x, 0) = 0 \quad \text{for } x \in [0, 1],$$

where $\sigma > 0$ is a constant. Let us choose the ansatz functions $\varphi_j(x) = \sin(j\,\pi\,x)$, $j = 1, \ldots, N$, which correspond to the classical discrete Fourier analysis in space. Now

$$a(u,y) = \int_0^1 u'(x)\, y'(x)\, dx$$

and the orthogonality of the φ_j yields

$$D = \frac{1}{2} I, \quad A = (a_{ij}) = \frac{1}{2}\operatorname{diag}(j^2\, \pi^2)$$

$$\text{and} \quad f_j(t) = e^{-t} \int_0^1 \sin(j\,\pi\, x)\, dx, \quad j = 1, \ldots, N.$$

Suppose that N is even, i.e., $N = 2\tilde{N}$ for some $\tilde{N} \in \mathbb{N}$. Then we obtain the uncoupled system of differential equations

$$\begin{aligned} u''_{2j-1}(t) + (2j-1)^2\pi^2\, u_{2j-1}(t) &= 4\, e^{-t}, \\ u''_{2j}(t) + (2j)^2\pi^2\, u_{2j}(t) &= 0 \end{aligned} \quad j = 1, 2, .., \tilde{N}. \qquad (2.13)$$

Solving these and applying the homogeneous initial conditions gives

$$u_l(t) = \begin{cases} \dfrac{4}{1+l^2\pi^2}\left(-\cos(l\,\pi\, t) + \dfrac{1}{l\pi}\sin(l\,\pi\, t) + e^{-t}\right)\sin(l\pi x) & \text{for } l = 2j-1, \\ 0 & \text{for } l = 2j, \end{cases}$$

with $j = 1, 2, \ldots, \tilde{N}$. $\square$

Unlike (1.11), problem (2.11) is a system of second-order differential equations. Its solution is significantly different from the solution of the system (1.11): in the case where $f \equiv 0$, the solution of the parabolic problem decays exponentially (see (1.5)), but the solution of (2.11) has the following semi-discrete analogue of the conservation of energy property (2.7):

Lemma 5.33. Let $f \equiv 0$. Then the function u_h that is the solution of (2.9) and (2.10) satisfies for each $t \in [0, T]$ the relation

$$a(u_h(t), u_h(t)) + (u'_h(t), u'_h(t)) = a(u_h^0, u_h^0) + (v_h^0, v_h^0). \qquad (2.14)$$

Proof: Set

$$\Phi(t) = a(u_h(t), u_h(t)) + (u'_h(t), u'_h(t)) \quad \text{for } 0 \le t \le T.$$

Then

$$\Phi'(t) = 2\, a(u_h(t), u'_h(t)) + 2\, (u'_h(t), u''_h(t)).$$

Choosing $w_h = u'_h(t)$ in (2.9), we obtain $\Phi'(t) = 0$ for all $t \in (0, T]$. Thus $\Phi(t)$ is constant, and the result follows from the initial data of the problem. ∎

The next result is a useful tool in estimating the growth of functions that satisfy an integral inequality.

Lemma 5.34 (Gronwall). *Let σ and ρ be continuous real functions with $\sigma \geq 0$. Let c be a non-negative constant. Assume that*

$$\sigma(t) \leq \rho(t) + c \int_0^t \sigma(s)\, ds \qquad \text{for all} \quad t \in [0, T].$$

Then

$$\sigma(t) \leq e^{ct}\, \rho(t) \qquad \text{for all} \quad t \in [0, T].$$

Proof: See, e.g., [GGZ74]. ∎

Let $R_h : V \to V_h$ again denote the Ritz projection defined by (1.13).

Theorem 5.35. *Assume that the solution $u \in L_2(0, T; V)$ of (2.2) has a second-order derivative $u'' \in L_2(0, T; H)$ and that $u(t) \in V \cap H^2(\Omega)$ for each $t \in [0, T]$. For the spatial discretization suppose that piecewise linear C^0 elements are used. Then there exists a constant $c > 0$ such that the solution u_h of the semi-discrete problem (2.9) and (2.10) satisfies the following estimate for all $t \in [0, T]$:*

$$\|u_h(t) - u(t)\|_0 + h\,|u_h(t) - u(t)|_1 + \|u_h'(t) - u'(t)\|_0$$
$$\leq c \left(|u_h^0 - R_h u^0|_1 + \|v_h^0 - R_h v^0\|_0 \right)$$
$$+ c\,h^2 \left(\|u(t)\|_0 + \|u'(t)\|_0 + \left(\int_0^t \|u''(s)\|_0^2\, ds \right)^{1/2} \right). \tag{2.15}$$

Proof: As in the parabolic problem, we use the Ritz projector R_h to write the error $u - u_h$ as a sum

$$u - u_h = p + q \qquad \text{with} \qquad p := u - R_h u, \quad q := R_h u - u_h, \tag{2.16}$$

then estimate separately the norm of each part.

The regularity hypotheses and Lemma 5.6 imply that

$$\|p(t)\|_0 \leq c\,h^2 \|u(t)\|_2 \quad \text{and} \quad \|\nabla p(t)\|_0 \leq c\,h\,\|u(t)\|_2$$

for some constant $c > 0$. Hence

$$\|p(t)\|_0 + h\,|p(t)|_1 \leq c\,h^2 \|u(t)\|_2. \tag{2.17}$$

Similarly we have

$$\|p'(t)\|_0 \leq c\,h^2 \|u'(t)\|_2 \quad \text{and} \quad \|p''(t)\|_0 \leq c\,h^2 \|u''(t)\|_2, \tag{2.18}$$

where p' and p'' denote temporal derivatives.

Now we turn to estimating q. Since u and u_h satisfy the variational equations (2.2) and (2.9) respectively, and $V_h \subset V$, one obtains the Galerkin orthogonality relation

$$(u'' - u_h'', w_h) + a(u - u_h, w_h) = 0 \qquad \text{for all} \quad w_h \in V_h. \qquad (2.19)$$

The definition of the Ritz projection yields

$$a(q, w_h) = a(R_h u - u_h, w_h) = a(u - u_h, w_h) \qquad \text{for all} \quad w_h \in V_h. \quad (2.20)$$

Furthermore,

$$(q'', w_h) = ((R_h u)'' - u_h'', w_h) = ((R_h u - u)'' + u'' - u_h'', w_h)$$
$$= (p'', w_h) + (u'' - u_h'', w_h) \quad \text{for all } w_h \in V_h.$$

Combining this equation with (2.19) and (2.20), we see that q satisfies the variational equation

$$(q'', w_h) + a(q, w_h) = -(p'', w_h) \qquad \text{for all} \quad w_h \in V_h. \qquad (2.21)$$

To proceed further we use the superposition principle for linear problems. Set $q = \hat{q} + \bar{q}$, where the functions $\hat{q}$ and $\bar{q}$ in V_h are defined by the initial value problems

$$(\hat{q}'', w_h) + a(\hat{q}, w_h) = 0 \quad \text{for all } w_h \in V_h,$$
$$\hat{q}(0) = R_h u_h^0 - u^0, \quad \hat{q}'(0) = R_h v_h^0 - v^0 \qquad (2.22)$$

and

$$(\bar{q}'', w_h) + a(\bar{q}, w_h) = -(p'', w_h) \quad \text{for all } w_h \in V_h, \quad \bar{q}(0) = \bar{q}'(0) = 0. \quad (2.23)$$

By Lemma 5.31, the definition (2.22) of $\hat{q}$ implies that

$$|\hat{q}(t)|_1^2 + \|\hat{q}'(t)\|_0^2 = a(\hat{q}(t), \hat{q}) + (\hat{q}'(t), \hat{q}'(t))$$
$$= a(\hat{q}(0), \hat{q}(0)) + (\hat{q}'(0), \hat{q}'(0)) = |\hat{q}(0)|_1^2 + \|\hat{q}'(0)\|_0^2.$$

Now the equivalence of norms in $\mathbb{R}^2$ and the initial conditions of (2.22) give us

$$|\hat{q}(t)|_1 + \|\hat{q}'(t)\|_0 \le c \left(|u_h^0 - R_h u^0|_1 + \|v_h^0 - R_h v^0\|_0 \right). \qquad (2.24)$$

On taking $w_h = 2\bar{q}'(t)$ in (2.23), for each $t \in [0, T]$ we get

$$\frac{d}{dt} \left(a(\bar{q}(t), \bar{q}(t)) + (\bar{q}'(t), \bar{q}'(t)) \right) = -2 \left(p''(t), \bar{q}'(t) \right).$$

But $-2(p'', \bar{q}') \le \|p''\|_0^2 + \|\bar{q}'\|_0^2$. Integrating and recalling the homogeneous initial data of (2.23) leads to

$$a(\bar{q}(t), \bar{q}(t)) + (\bar{q}'(t), \bar{q}'(t)) \le \int_0^t \|p''(s)\|_0^2 \, ds + \int_0^t \|\bar{q}'(s)\|_0^2 \, ds. \qquad (2.25)$$

Taking into account the homogeneous initial conditions in (2.23), from Lemma 5.34 (Gronwall's inequality) and (2.18) it follows that

$$|\bar{q}(t)|_1^2 + \|\bar{q}'(t)\|^2 = a(\bar{q}(t), \bar{q}(t)) + (\bar{q}'(t), \bar{q}'(t)) \leq e^t \int_0^t \|p''(s)\|^2 ds \quad \forall\, t \in [0, T].$$

Hence

$$|\bar{q}(t)|_1 + \|\bar{q}'(t)\|_0 \leq c h^2 \left(\int_0^t \|u''(s)\|^2 ds \right)^{1/2} \qquad \text{for all} \quad t \in [0, T], \quad (2.26)$$

by (2.18). As $u - u_h = p + \hat{q} + \bar{q}$, the bounds (2.17), (2.24) and (2.26) and the triangle inequality deliver the desired estimate for two of the summands in (2.15).The third summand can be handled similarly ∎

5.2.3 Temporal Discretization

The initial value problem (2.9) for the system of ordinary differential equations that is generated by the spatial semi-discretization must now be discretized in time for its numerical implementation. In Section 5.1.3 this task was discussed in detail for parabolic problems. For wave-type problems like (2.1), the important attributes that the temporal discretization should possess include not only consistency and stability—as was shown for parabolic problems in Section 5.1.3—but also conservation of energy in the discrete problem. An example of such a discretization is given below by (2.27).

For constant step size $\tau := T/M$ with $M \in \mathbb{N}$, we define the equidistant grid $t_k := k\tau$, $k = 0, 1, \ldots, M$, which subdivides $[0, T]$. Let $u_h^k \in V_h$ denote the approximation of the spatially semi-discrete solution $u_h(t_k)$. Set $t_{k+1/2} = \frac{1}{2}(t_k + t_{k+1})$ and $u_h^{k+1/2} = \frac{1}{2}(u_h^k + u_h^{k+1})$. As an example of a suitable finite difference method we study the scheme (see [LT03])

$$(D_\tau^+ D_\tau^- u_h^k, w_h) + a\left(\tfrac{1}{2}(u_h^{k+1/2} + u_h^{k-1/2}), w_h \right) = (f(t_k), w_h) \tag{2.27}$$
$$\text{for all } w_h \in V_h, \ k = 1, \ldots, M - 1.$$

This scheme defines iteratively approximations $u_h^k \in V_h$ of $u_h(t_k)$, provided that one has initial discrete functions $u_h^0, u_h^1 \in V_h$.

Recall (see Chapter 2) that under sufficient regularity assumptions one has

$$D_\tau^+ D_\tau^- u_h(t) = \frac{1}{\tau^2} \left(u_h(t+\tau) - 2u_h(t) + u_h(t-\tau) \right) = u_h(t) + O(\tau^2).$$

The scheme (2.27) is explicitly

$$\tau^{-2}\left(u_h^{k+1} - 2u_h^k + u_h^{k-1}, w_h\right) + a\left(\tfrac{1}{2}(u_h^{k+1/2} + u_h^{k-1/2}), w_h\right)$$
$$= (f(t_k), w_h) \qquad \text{for all } w_h \in V_h, \ k = 1, \dots, M - 1. \tag{2.28}$$

The bilinearity of both components of the left-hand side enables us to rewrite this as

$$\tau^{-2}\left(u_h^{k+1}, w_h\right) + \tfrac{1}{4}a(u_h^{k+1}, w_h) = (b_h^k, w_h)$$
$$\text{for all } w_h \in V_h, \ k = 1, \dots, M - 1. \tag{2.29}$$

Here $b_h^k \in V_h^*$ is defined by

$$(b_h^k, w_h) := (f(t_k), w_h) + \tau^{-2}(2u_h^k - u_h^{k-1}, w_h) - a\left(\tfrac{1}{4}(2u_h^k + u_h^{k-1}), w_h\right)$$
$$\text{for all } w_h \in V_h. \tag{2.30}$$

As in the preceding section let $\{\varphi_j\}$ be a basis for V_h, set $\hat{u}^{k+1} = (u_j(t_{k+1}))$ where $u_h^{k+1} = \sum_{j=1}^{N} u_j(t_{k+1})\varphi_j$, and let $\hat{b}^k$ be the corresponding right-hand side of (2.30); then in the notation of that section we can write (2.29) in the matrix-vector form

$$\left(\frac{1}{4} A + \tau^{-2} D\right) \hat{u}^{k+1} = \hat{b}^k, \quad k = 1, \dots, M - 1. \tag{2.31}$$

This system can be solved to find iteratively $\hat{u}^{k+1}$ (and hence also u_h^{k+1}) from the discrete initial values $\hat{u}^0$ and $\hat{u}^1$. The matrix of the linear system (2.31) is symmetric and positive definite, and is independent of the index k of the recursion. It is typically of high dimension, but in finite element methods the number of non-zero entries is relatively small. Thus the sparse system (2.31) requires specific numerical methods for its efficient treatment; we shall discuss this in Chapter 8.

Analogously to (2.7), the following discrete conservation of energy principle is valid.

Lemma 5.36. *Assume that $f \equiv 0$. Then the discrete solutions $u_h^k \in V_h$ iteratively defined by (2.27) satisfy*

$$a(u_h^{k+1/2}, u_h^{k+1/2}) + (D_\tau^+ u_h^k, D_\tau^+ u_h^k) = a(u_h^{1/2}, u_h^{1/2}) + (D_\tau^+ u_h^0, D_\tau^+ u_h^0)$$
$$\text{for } k = 1, \dots, M - 1. \tag{2.32}$$

Proof: In (2.27), as test function choose $w_h = \frac{1}{2\tau}(u_h^{k+1} - u_h^{k-1})$. Then

$$w_h = \frac{1}{2}\left(D_\tau^+ u_h^k + D_\tau^+ u_h^{k-1}\right) = \frac{1}{\tau}\left(u_h^{k+1/2} - u_h^{k-1/2}\right)$$

and

$$(D_\tau^+ D_\tau^- u_h^k, w_h) = \frac{1}{2\tau} \left(D_\tau^+ u_h^k - D_\tau^+ u_h^{k-1}, D_\tau^+ u_h^k + D_\tau^+ u_h^{k-1} \right)$$

$$= \frac{1}{2} D_\tau^- \| D_\tau^+ u_h^k \|^2. \tag{2.33}$$

We also have

$$a\left(\frac{1}{2}(u_h^{k+1/2} + u_h^{k-1/2}), w_h \right) = \frac{1}{2\tau} a\left(u_h^{k+1/2} + u_h^{k-1/2}, u_h^{k+1/2} - u_h^{k-1/2} \right).$$

The bilinearity and symmetry of $a(\cdot, \cdot)$ reduce this to

$$a\left(\frac{1}{2}(u_h^{k+1/2} + u_h^{k-1/2}), w_h \right) = \frac{1}{2} D_\tau^- a(u_h^{k+1/2}, u_h^{k+1/2}).$$

Adding this identity to (2.33), then setting $f \equiv 0$ in (2.27), we get

$$D_\tau^- \left(a(u_h^{k+1/2}, u_h^{k+1/2}) + (D_\tau^+ u_h^k, D_\tau^+ u_h^k) \right) = 0, \quad k = 1, \ldots, M - 1.$$

The iterative application of this formula yields (2.32). ∎

Now we turn to the convergence analysis of the completely discretized method (2.27).

Theorem 5.37. *Let the solution u of (2.1) satisfy the regularity assumptions made in Theorem 5.35. Let V_h be the space of C^0 piecewise linear elements. Assume that $\tau \le 1$. Then there exists a constant $c > 0$ such that the solution $\{u_h^k\}$ of the discrete problem (2.27) satisfies:*

$$\| u_h^{k+1/2} - u(t_{k+1/2}) \|_0 + h \, | u_h^{k+1/2} - u(t_{k+1/2}) |_1 + \| D_\tau^+ u_h^k - u_t(t_{k+1/2}) \|_0$$

$$\le c \left(| R_h u(t_0) - u_h^0 |_1 + | R_h u(t_1) - u_h^1 |_1 + \| D_\tau^+ (R_h u(t_0) - u_h^0) \|_0 \right)$$

$$+ c(h^2 + \tau^2), \quad \text{for } k = 1, \ldots, M - 1. \tag{2.34}$$

Proof: Like the semi-discrete case, we use the Ritz projection $R_h : V \to V_h$ to decompose the error $u(t_k) - u_h^k$ as the sum

$$u(t_k) - u_h^k = p^k + q^k \tag{2.35}$$

where

$$p^k := u(t_k) - R_h u(t_k) \quad \text{and} \quad q^k := R_h u(t_k) - u_h^k. \tag{2.36}$$

We estimate separately the norms of p^k and q^k. As in the proof of Theorem 5.35, for p^k we have

$$\| p^k \|_0 + h \, | p^k |_1 \le c h^2 \| u(t_k) \|_2 \tag{2.37}$$

and

$$\|(p^k)''\|_0 \le c h^2 \|u''(t_k)\|_2 \quad \text{and} \quad \|(p^k)'\|_0 \le c h^2 \|u'(t_k)\|_2. \quad (2.38)$$

Next, consider q^k. At each time level t_k, since u is a solution of (2.1) it satisfies

$$(u_{tt}(t_k), w_h) + a(u(t_k), w_h) = (f(t_k), w_h) \quad \text{for all } w_h \in V_h.$$

Thus

$$([R_h u]''(t_k), w_h) + a([R_h u](t_k), w_h) = (f(t_k), w_h) + (([R_h u]'' - u_{tt})(t_k), w_h)$$
$$= (f(t_k), w_h) - ((p^k)'', w_h)$$
$$\text{for all } w_h \in V_h.$$

Under suitable regularity assumptions on the t-derivatives of u, this gives us

$$(D_\tau^+ D_\tau^- [R_h u](t_k), w_h) + \tfrac{1}{2} a([R_h u](t_{k+1/2}), w_h) + \tfrac{1}{2} a([R_h u](t_{k-1/2}), w_h)$$
$$= (f(t_k), w_h) - ((p^k)'', w_h) + (r^k, w_h) \quad \text{for all } w_h \in V_h,$$

with

$$\|r^k\|_0 \le c\tau^2. \quad (2.39)$$

Hence, recalling the discretization (2.27) and setting $q^{k+1/2} = R_h u(t_{k+1/2}) - u_h^{k+1/2}$, one gets

$$(D_\tau^+ D_\tau^- q^k), w_h) + a(\tfrac{1}{2} q^{k+1/2} + \tfrac{1}{2} q^{k-1/2}, w_h)$$
$$= -((p^k)'', w_h) + (r^k, w_h) \quad \text{for all } w_h \in V_h. \quad (2.40)$$

As in the proof of Theorem 5.35 we now apply a superposition principle, writing $q^k = \hat{q}^k + \bar{q}^k$ and $q^{k+1/2} = \hat{q}^{k+1/2} + \bar{q}^{k+1/2}$ where these terms are defined by

$$(D_\tau^+ D_\tau^- \hat{q}^k), w_h) + a(\tfrac{1}{2} \hat{q}^{k+1/2} + \tfrac{1}{2} \hat{q}^{k-1/2}, w_h) = 0 \quad \text{for all } w_h \in V_h,$$
$$\hat{q}^0 = R_h u(t_0) - u_h^0, \quad \hat{q}^1 = R_h u(t_1) - u_h^1, \quad (2.41)$$

and

$$(D_\tau^+ D_\tau^- \bar{q}^k), w_h) + a(\tfrac{1}{2} \bar{q}^{k+1/2} + \tfrac{1}{2} \bar{q}^{k-1/2}, w_h)$$
$$= -((p^k)'', w_h) + (r^k, w_h) \quad \text{for all } w_h \in V_h, \quad (2.42)$$
$$\bar{q}^0 = \bar{q}^1 = 0,$$

with $\hat{q}_h^{k+1/2} = \tfrac{1}{2}(\hat{q}_h^k + \hat{q}_h^{k+1})$ and $\bar{q}_h^{k+1/2} = \tfrac{1}{2}(\bar{q}_h^k + \bar{q}_h^{k+1})$. An application of Lemma 5.36 to (2.41) yields

$$|\hat{q}^{k+1/2}|_1^2 + \|D_\tau^+\hat{q}^k\|_0^2 = |\hat{q}^{1/2}|_1^2 + \|D_\tau^+\hat{q}^0\|_0^2, \quad k = 1,\ldots,M-1.$$

From the equivalence of norms in finite-dimensional spaces and the initial conditions in (2.41), there exists a constant $c > 0$ such that

$$|\hat{q}^{k+1/2}|_1 + \|D_\tau^+\hat{q}^k\|_0 \leq c \left(|R_h u(t_0) - u_h^0|_1 + |R_h u(t_1) - u_h^1|_1 \right. \tag{2.43}$$
$$\left. + \|D_\tau^+(R_h u(t_0) - u_h^0)\|_0\right), \quad k = 1,\ldots,M-1.$$

To estimate $\bar{q}$, as in the proof of Lemma 5.36 we choose the particular test function

$$w_h = \frac{1}{2}(D_\tau^+\bar{q}^k + D_\tau^+\bar{q}^{k-1}) = \frac{1}{\tau}(\bar{q}^{k+1/2} - \bar{q}^{k-1/2})$$

in (2.42). This gives

$$D_\tau^- \left(|\bar{q}^{k+1/2}|_1^2 + \|D_\tau^+\bar{q}^k\|^2\right) = (r^k, D_\tau^+\bar{q}^k) + (r^k, D_\tau^+\bar{q}^{k-1})$$
$$- ((p^k)'', D_\tau^+\bar{q}^k) - ((p^k)'', D_\tau^+\bar{q}^{k-1}). \tag{2.44}$$

Set

$$\alpha_k = |\bar{q}^{k+1/2}|_1^2 + \|D_\tau^+\bar{q}^k\|_0^2$$

for all k. Then by the Cauchy-Schwarz inequality and $ab \leq a^2 + b^2/4$,

$$\alpha_k - \alpha_{k-1} \leq \tau \left(\|(p^k)''\|_0^2 + \|r^k\|_0^2\right) + \frac{\tau}{2}\alpha_k + \frac{\tau}{2}\alpha_{k-1}, \quad k = 1,\ldots,M.$$

That is,

$$\left(1 - \frac{\tau}{2}\right)\alpha_k \leq \left(1 + \frac{\tau}{2}\right)\alpha_{k-1} + \tau \left(\|(p^k)''\|_0^2 + \|r^k\|_0^2\right), \quad k = 1,\ldots,M. \tag{2.45}$$

As $\tau \leq 1$ we can choose a number δ such that

$$\left(1 - \frac{\tau}{2}\right)\delta \geq \|(p^k)''\|_0^2 + \|r^k\|_0^2, \quad k = 1,\ldots,M.$$

Then (2.45) implies that

$$\alpha_k \leq \beta\alpha_{k-1} + \tau\delta, \quad k = 1,\ldots,M \tag{2.46}$$

with $\beta := (1 + \frac{\tau}{2})/(1 - \frac{\tau}{2})$. By induction we obtain

$$\alpha_k \leq \beta^k\alpha_0 + \tau\delta\sum_{j=0}^{k-1}\beta^j.$$

But

$$\sum_{j=0}^{k-1}\beta^j = \frac{\beta^k - 1}{\beta - 1}$$

and

$$\beta^k = \left(\frac{1 + \frac{\tau}{2}}{1 - \frac{\tau}{2}}\right)^k = \left(1 + \frac{\tau}{1 - \frac{\tau}{2}}\right)^k \leq \exp\left(\frac{\tau}{1 - \frac{\tau}{2}} k\right)$$

so

$$\alpha_k \leq \exp\left(\frac{\tau}{1 - \frac{\tau}{2}} k\right) \alpha_0 + \delta\left(1 - \frac{\tau}{2}\right)\left[\exp\left(\frac{\tau}{1 - \frac{\tau}{2}} k\right) - 1\right].$$

As $k\tau \leq T$ for $k = 0, 1, \ldots, M$ and $\tau \leq 1$, there exists a constant $c > 0$ such that

$$\alpha_k \leq c\left[\alpha_0 + \delta\left(1 - \frac{\tau}{2}\right)\right], \quad k = 1, \ldots, M. \tag{2.47}$$

Hence, recalling the definitions of α_k and δ and the bounds of (2.38) and (2.39) for $\|(p^k)''\|_0$ and $\|r^k\|_0$, the equivalence of norms on finite-dimensional spaces guarantees that some $c > 0$ exists with

$$|\bar{q}^{k+1/2}|_1 + \|D_\tau^+ \bar{q}^k\|_0 \leq c \left(|R_h u(t_0) - u_h^0|_1 + |R_h u(t_1) - u_h^1|_1 \right.$$
$$\left. + \|D_\tau^+(R_h u(t_0) - u_h^0)\|_0 + \tau^2 + h^2\right), \quad k = 1, \ldots, M - 1. \tag{2.48}$$

Finally, a triangle inequality together with (2.36) and (2.43) gives the desired bound on the first terms in (2.34).

A bound on $\|D_\tau^+ u_h^k - u_t(t_{k+1/2})\|_0$ can be shown similarly using the assumed regularity of u. ∎

Remark 5.38. To obtain the optimal bound of convergence of $O(h^2 + \tau^2)$ for this method, the discrete initial values u_h^0, u_h^1 have to be chosen in such a way that one has

$$|R_h u(t_0) - u_h^0|_1 + |R_h u(t_1) - u_h^1|_1 + \|D_\tau^+(R_h u(t_0) - u_h^0)\|_0 = O(h^2 + \tau^2). \tag{2.49}$$

A constructive way to ensure this condition is described in Exercise 5.42.   □

5.2.4 Rothe's Method for Hyperbolic Problems

The classical Rothe method (see Section 5.1.5) is a semi-discretization in time of initial-boundary value problems, i.e., the temporal derivatives are discretized but not the spatial derivatives. It is simply the application of the implicit Euler scheme to a first-order system of differential equations in an appropriate function space. In the present subsection we examine first a generalized Rothe's method based on the temporal discretization studied in Section 5.2.3; later we move on to the classical Rothe method.

Our starting point is once again the weak formulation of the wave equation (2.1):

$$\left(\frac{d^2u}{dt^2}, w\right) + a(u, w) = (f, w) \qquad \text{for all} \quad w \in V,$$

$$u(0) = u^0, \quad \frac{du}{dt}(0) = v^0.$$

(2.50)

Here $u^0 \in V = H_0^1(\Omega)$ and $v^0 \in H = L_2(\Omega)$ are given.

We select a fixed time step size $\tau := T/M$ with $M \in \mathbb{N}$ and define an equidistant grid $t_k := k\tau$, $k = 0, 1, \ldots, M$ over $[0, T]$. Let $u^k \in V$ denote the computed numerical approximation of the desired solution $u(t_k)$ at each grid point t_k. Once again set $t_{k+1/2} = \frac{1}{2}(t_k + t_{k+1})$ and $u^{k+1/2} = \frac{1}{2}(u^k + u^{k+1})$. We analyse the following finite difference method [LT03]:

$$(D_\tau^+ D_\tau^- u^k, w) + a(\tfrac{1}{2}(u^{k+1/2} + u^{k-1/2}), w) = (f(t_k), w)$$

$$\text{for all} \ w \in V, \ k = 1, \ldots, M - 1.$$

(2.51)

Starting from two initial functions u^0, $u^1 \in V$ this scheme defines iteratively an approximation $u^k \in V$ of $u(t_k)$ for each k. The variant (2.51) of Rothe's method is the spatially continuous analogue of (2.27). Once $u^{k-1} \in V$ and $u^k \in V$ are known, $u^{k+1} \in V$ is defined by the elliptic boundary value problem

$$\tau^{-2}(u^{k+1}, w) + \tfrac{1}{4}a(u^{k+1}, w) = (b^k, w)$$

$$\text{for all} \ w \in V, \ k = 1, \ldots, M - 1.$$

(2.52)

Here $b^k \in V^*$ is given by

$$(b^k, w) := (f(t_k), w) + \tau^{-2}(2u^k - u^{k-1}, w) - a\left(\tfrac{1}{4}(2u^k + u^{k-1}), w\right)$$

$$\text{for all} \ w \in V.$$

(2.53)

Lax-Milgram's lemma guarantees that the variational equation (2.52) has a unique solution, but it should be noted that as τ is small, (2.52) is a singularly perturbed reaction-diffusion problem (compare Chapter 6).

The convergence analysis of (2.51) is very similar to the analysis of the completely discretized method in Section 5.2.3. This can be seen for instance in the proof of the next lemma.

Lemma 5.39. *Assume that $f \equiv 0$. Then the functions $u^k \in V$ defined iteratively by (2.51) satisfy the identity*

$$a(u^{k+1/2}, u^{k+1/2}) + (D_\tau^+ u^k, D_\tau^+ u^k) = a(u^{1/2}, u^{1/2}) + (D_\tau^+ u^0, D_\tau^+ u^0),$$

$$k = 1, \ldots, M - 1.$$

(2.54)

Proof: In (2.51) choose as test function $w = \frac{1}{2\tau}(u^{k+1} - u^{k-1})$. We also have

$$w = \frac{1}{2}\left(D_\tau^+ u^k + D_\tau^+ u^{k-1}\right) = \frac{1}{\tau}\left(u^{k+1/2} - u^{k-1/2}\right).$$

Then the bilinearity and the symmetry of the scalar product imply that

$$(D_\tau^+ D_\tau^- u^k, w) = \frac{1}{2\tau} \left(D_\tau^+ u^k - D_\tau^+ u^{k-1}, D_\tau^+ u^k + D_\tau^+ u^{k-1} \right)$$

$$= \frac{1}{2} D_\tau^- \| D_\tau^+ u^k \|_0^2. \tag{2.55}$$

Next,

$$a\left(\frac{1}{2}(u^{k+1/2} + u^{k-1/2}), w \right) = \frac{1}{2\tau} a\left(u^{k+1/2} + u^{k-1/2}, u^{k+1/2} - u^{k-1/2} \right).$$

The bilinearity and symmetry of $a(\cdot, \cdot)$ therefore give

$$a\left(\frac{1}{2}(u^{k+1/2} + u^{k-1/2}), w \right) = \frac{1}{2} D_\tau^- a(u^{k+1/2}, u^{k+1/2})$$

Substituting this identity and (2.55) into (2.51) with $f \equiv 0$, we get

$$D_\tau^- \left(a(u^{k+1/2}, u^{k+1/2}) + (D_\tau^+ u^k, D_\tau^+ u^k) \right) = 0, \quad k = 1, \ldots, M - 1.$$

Finally, the iterative application of this relation proves (2.54). ∎

Theorem 5.40. *Assume that the solution u of (2.50) is sufficiently regular. Then there exists a constant $c > 0$ such that the solution $\{u^k\}$ of the semidiscrete problem (2.51) generated by the generalized Rothe's method satisfies the following bound:*

$$\|u^{k+1/2} - u(t_{k+1/2})\|_0 + |u^{k+1/2} - u(t_{k+1/2})|_1 + \|D_\tau^+ u^k - u_t(t_{k+1/2})\|_0$$

$$\leq c \left(\|D_\tau^+ (u(t_0) - u^0)\|_0 + \tau^2 \right), \quad k = 1, \ldots, M - 1. \tag{2.56}$$

The proof of this theorem is similar to the proof of Theorem 5.37, but shorter because Rothe's method does not include a spatial discretization. In particular no intermediate Ritz projection is required and the error $u^k - u(t_k)$ can be estimated by Lemma 5.39, like our handling of the summand q^k in the proof of Theorem 5.37.

Now we turn to the classical Rothe's method (compare Section 5.1.5). This is simply the application of the implicit Euler scheme to the temporal derivative in parabolic initial-boundary problems. To invoke this approach in the context of second-order hyperbolic problems, we consider instead of (2.50) the formulation (2.4) as a system of first-order (in time) differential equations. Unlike the preceding analysis we allow variable step sizes $\tau_{k+1} := t_{k+1} - t_k$. When the implicit Euler scheme is applied to

$$\begin{aligned}
(u_t, z) - (v, z) &= 0 && \text{for all} \quad z \in H, \\
(v_t, w) + a(u, w) &= (f, w) && \text{for all} \quad w \in V,
\end{aligned} \tag{2.57}$$

we obtain the semi-discrete problems

$$(D^-_{\tau_{k+1}} u^{k+1}, z) - (v^{k+1}, z) = 0 \qquad \text{for all} \quad z \in H,$$
$$(D^-_{\tau_{k+1}} v^{k+1}, w) + a(u^{k+1}, w) = (f(t_{k+1}), w) \quad \text{for all} \quad w \in V. \tag{2.58}$$

This corresponds to the classical formulation

$$\frac{1}{\tau_{k+1}}(u^{k+1} - u^k) - v^{k+1} = 0,$$
$$\frac{1}{\tau_{k+1}}(v^{k+1} - v^k) - \Delta u^{k+1} = f(t_{k+1}), \quad k = 1, \dots, M-1.$$

Eliminating v^k and v^{k+1} gives the scheme

$$\frac{1}{\tau_{k+1}}\left(\frac{1}{\tau_{k+1}}(u^{k+1} - u^k) - \frac{1}{\tau_k}(u^k - u^{k-1})\right) - \Delta u^{k+1} = f(t_{k+1}),$$
$$k = 1, \dots, M-1.$$

The continuous embedding $V \hookrightarrow H$ means that (2.58) implies also the weak formulation

$$\left(D^+_{\tau_{k+1}} D^-_{\tau_k} u^k, w\right) + a(u^{k+1}, w) = (f(t_{k+1}), w) \quad \text{for all} \quad w \in V, \tag{2.59}$$

i.e.,

$$(u^{k+1}, w) + \tau^2_{k+1} a(u^{k+1}, w) = \left(u^k + \frac{\tau_{k+1}}{\tau_k}(u^k - u^{k-1}), w\right) \tag{2.60}$$
$$\text{for all} \quad w \in V, \quad k = 1, \dots, M-1.$$

Given $u^{k-1}, u^k \in V$, the Lax-Milgram lemma guarantees that (2.60) has a unique solution $u^{k+1} \in V$. Thus, given u^0 and $u^1 \in V$, the semi-discrete method (2.60) defines the approximating functions $u^k \in V$ for $k = 2, \dots, M$. This scheme has an important stability property—see the following energy estimate—that is similar to the one stated in Lemma 5.39 for the generalized Rothe's scheme.

Lemma 5.41. *Assume that $f \equiv 0$. Then the functions $u^k \in V$ iteratively defined by (2.60) satisfy the following bound:*

$$\|D^+_{\tau_{k+1}} u^k\|^2 + a(u^{k+1}, u^{k+1}) \leq \|D^+_{\tau_1} u^0\|^2 + a(u^1, u^1), \tag{2.61}$$
$$k = 1, \dots, M-1.$$

Proof: At the kth time step choose $w = u^{k+1} - u^k$ as test function. Then the variational equation (2.59), which is equivalent to (2.60), implies the identity

$$\frac{1}{\tau_{k+1}}\left(\frac{1}{\tau_{k+1}}(u^{k+1} - u^k) - \frac{1}{\tau_k}(u^k - u^{k-1}), u^{k+1} - u^k\right)$$
$$+ a(u^{k+1}, u^{k+1} - u^k) = 0.$$

That is,

$$(D^+_{\tau_{k+1}} u^k, D^+_{\tau_{k+1}} u^k) + a(u^{k+1}, u^{k+1}) = (D^+_{\tau_{k+1}} u^k, D^+_{\tau_k} u^{k-1}) + a(u^{k+1}, u^k).$$

Hence, by the Cauchy-Schwarz inequality,

$$\begin{aligned}
(D^+_{\tau_{k+1}} u^k, D^+_{\tau_{k+1}} u^k) + a(u^{k+1}, u^{k+1}) &\leq \|D^+_{\tau_{k+1}} u^k\| \, \|D^+_{\tau_k} u^{k-1}\| \\
&\quad + a(u^{k+1}, u^{k+1})^{1/2} a(u^k, u^k)^{1/2} \\
&\leq \frac{1}{2} \left(\|D^+_{\tau_{k+1}} u^k\|^2 + \|D^+_{\tau_k} u^{k-1}\|^2 \right) \\
&\quad + \frac{1}{2} \left[a(u^{k+1}, u^{k+1}) + a(u^k, u^k) \right].
\end{aligned}$$

Rearranging, this inequality becomes

$$\|D^+_{\tau_{k+1}} u^k\|^2 + a(u^{k+1}, u^{k+1}) \leq \|D^+_{\tau_k} u^{k-1}\|^2 + a(u^k, u^k),$$
$$k = 1, \ldots, M - 1. \tag{2.62}$$

Finally, an iterative application of (2.62) proves (2.61). ∎

For a complete convergence analysis of the method (2.60) that includes adaptive grid generation see [15].

5.2.5 Remarks on Error Control

For simplicity in the preceding sections, with the exception of the analysis of (2.60), we assumed that the temporal grid was equidistant and the spatial discretization was fixed independently of time. But one can use appropriate error indicators to generate a complete discretization of the hyperbolic problem (2.2) that is adaptive in both space and time. This adaptivity permits a non-equidistant temporal grid and different finite element discretizations can be used at different time levels. That is, the fixed discrete space $V_h \subset V$ is replaced by spaces $V_k \subset V$ for $k = 2, \ldots, M$, e.g., spaces generated by piecewise linear C^0 elements over varying triangulations $\mathcal{T}_k$ of $\overline{\Omega}$. If one starts from the underlying temporal semi-discretization (2.59), then spatial discretization generates the variational equations

$$\left(D^+_{\tau_{k+1}} D^-_{\tau_k} u^k, w_h \right) + a(u^{k+1}, w_h) = (f(t_{k+1}), w_h) \quad \text{for all } w_h \in V_{k+1}. \tag{2.63}$$

Since $V_{k+1} \subset V$, the Lax-Milgram lemma guarantees that each problem (2.63) has a unique solution $u^{k+1} \in V_{k+1}$. Error indicators that can be used to control independently the spatial and temporal grids are described in [15] and analysed in detail. See also [BR03].

Exercise 5.42. Show that if the solution u and the initial functions u^0 and v^0 are sufficiently regular, then when solving the problem (2.1), the choices

$$u_h^0 := R_h u^0, \qquad u_h^1 := R_h \left(u^0 + \tau v^0 + \frac{1}{2}((\Delta u)(0) + f(0)) \right)$$

ensure that the requirement (2.49) is fulfilled.

Exercise 5.43. Prove the bound (2.56) by invoking Lemma 5.39.

6

Singular Perturbations

In this chapter we consider linear boundary value problems for differential equations of the form

$$-\varepsilon \triangle u + b\nabla u + cu = f$$

and initial-boundary value problems for differential equations of the form

$$\frac{\partial u}{\partial t} - \varepsilon \triangle u + b\nabla u + cu = f,$$

where the data are scaled in such a way that $\|f\|_\infty$, $\|c\|_\infty$ and $\|b\|_\infty$ are all $O(1)$ while $0 < \varepsilon \ll 1$. Problems such as these are said to be *singularly perturbed* because as $\varepsilon \to 0$, in general their solutions $u = u(\cdot, \varepsilon)$ do not converge pointwise to the solution of the related problem obtained when $\varepsilon = 0$ in the differential equation (combined with some appropriate subset of the boundary or initial-boundary conditions).

Difficulties arise in discretizing singularly perturbed problems because the stabilizing term $-\varepsilon \triangle u$ becomes less influential as $\varepsilon \to 0$. Moreover, the structure of the solution $u(\cdot, \varepsilon)$ causes the consistency errors of difference schemes and the approximation error of finite element schemes on standard meshes to increase as $\varepsilon \to 0$.

In this chapter we begin with one-dimensional boundary value problems, then move on to parabolic problems in one space dimension, and finally examine elliptic and parabolic problems in several space dimensions. To understand why it is difficult to solve singularly perturbed problems accurately we shall discuss the asymptotic structure of the solution and the types of layers that may occur in the solution of the problem. Our survey includes finite difference, finite volume and finite element methods.

Of course in a single chapter we can give only an introduction to the numerical solution of singularly perturbed problems. The monograph [RST96] presents a comprehensive survey of research in this area; a revised and extended second edition of this book is planned for 2007.

6.1 Two-Point Boundary Value Problems

6.1.1 Analytical Behaviour of the Solution

Consider the boundary value problem

$$Lu := -\varepsilon u'' + b(x)u' + c(x)u = f(x) \text{ on } (0,1), \quad u(0) = u(1) = 0, \quad (1.1)$$

with $0 < \varepsilon \ll 1$. Assume that $c(x) \geq 0$, which ensures that (1.1) has a unique solution u. Although $u = u(x, \varepsilon)$, we shall often write $u(x)$ for brevity and u' denotes differentiation with respect to x.

Even very simple examples show how standard discretization methods have difficulty in solving (1.1). The exact solution of the boundary value problem

$$-\varepsilon u'' - u' = 0, \quad u(0) = 0, \quad u(1) = 1,$$

is

$$u(x, \varepsilon) = \left(1 - \exp\left(-\frac{x}{\varepsilon}\right)\right) \Big/ \left(1 - \exp\left(-\frac{1}{\varepsilon}\right)\right).$$

Apply central differencing on an equidistant mesh of diameter h:

$$-\varepsilon D^+ D^- u_i - D^0 u_i = 0, \quad u_0 = 0, \; u_N = 1.$$

Then the computed solution is

$$u_i = \frac{1 - r^i}{1 - r^N} \quad \text{where} \quad r = \frac{2\varepsilon - h}{2\varepsilon + h}.$$

If ε is very small relative to the mesh so that $h > 2\varepsilon$, then it follows from the formula for u_i that the discrete solution oscillates. But even if one takes very small step sizes (which is impractical in higher dimensions) that satisfy $h < 2\varepsilon$, strange things can happen. For instance, if $\varepsilon = h$, then

$$\lim_{h \to 0} u_1 = \frac{2}{3} \neq \lim_{h \to 0} u(x_1) = 1 - \frac{1}{e}.$$

This surprising behaviour occurs because of the structure of the solution $u(x, \varepsilon)$. For

$$\lim_{\varepsilon \to 0} u(x, \varepsilon) = 1 \text{ for any arbitrary but fixed } x \in (0, 1], \text{ but } u(0, \varepsilon) = 0.$$

Thus the solution $u(x, \varepsilon)$ has a *boundary layer* at $x = 0$; this is a narrow region in which the solution changes rapidly if ε is very small. See Figure 6.1. Moreover, the derivatives of the solution are unbounded as $\varepsilon \to 0$: in our example, near $x = 0$ one gets

$$u^{(k)} \sim C\varepsilon^{-k}.$$

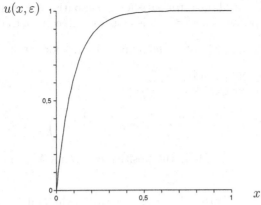

$u(x, \varepsilon)$

Figure 6.1 Boundary layer at $x = 0$ for $\varepsilon = 0.1$

Next we show that the behaviour of the solution outlined for our example is in fact typical of the class of boundary value problems (1.1) when $b(x) \neq 0$ for all $x \in [0, 1]$. We assume for simplicity that b, c and f are sufficiently smooth.

The following *localization rule* describes the location of the boundary layer in the solution of (1.1):

If b is positive, there is in general a layer at $x = 1$.

If b is negative, there is in general a layer at $x = 0$.

To verify this rule we consider the case $b(x) \geq \beta > 0$. Let u_0 be the solution of the *reduced problem*

$$b(x)u'(x) + c(x)u(x) = f(x), \quad u(0) = 0. \tag{1.2}$$

The reduced problem comprises the reduced differential equation (which is got by setting $\varepsilon = 0$ in the original differential equation) and a suitable subset of the original boundary conditions. To choose this subset correctly is critical in defining the reduced problem. For several classes of problems—including our problem (1.1)—localization rules can be specified. When using the correct definition of the reduced problem, one has

$$\lim_{\varepsilon \to 0} u(x, \varepsilon) = u_0(x) \tag{1.3}$$

for almost all x lying in the original domain. Neighbourhoods of points (or two- and three-dimensional manifolds in later problems) in the domain at which (1.3) does not hold are called *layer regions*.

Lemma 6.1. *Let $b(x) \geq \beta > 0$ on $[0, 1]$. Let $x_0 \in (0, 1)$ be arbitrary but fixed. Then for all $x \in [0, x_0]$ one has*

$$\lim_{\varepsilon \to 0} u(x, \varepsilon) = u_0(x).$$

Proof: Our differential operator satisfies a comparison principle on $[0,1]$. Set $v_1(x) = \gamma \exp(\beta x)$, where $\gamma > 0$ is constant. Recall that $c(x) \geq 0$. Then

$$Lv_1(x) \geq \gamma(-\beta^2 \varepsilon + b\beta) \exp(\beta x) \geq 1 \quad \text{for } x \in [0,1]$$

provided γ is chosen suitably.
Moreover, if $v_2(x) := \exp(-\beta(1-x)/\varepsilon)$ then

$$Lv_2(x) \geq \frac{\beta}{\varepsilon}(b - \beta) \exp\left(-\beta\frac{1-x}{\varepsilon}\right) \geq 0.$$

Set $v = M_1 \varepsilon v_1 + M_2 v_2$, where the positive constants M_1 and M_2 are chosen such that

$$Lv(x) \geq M_1 \varepsilon \geq |L(u - u_0)| = \varepsilon|u_0''(x)|,$$
$$v(0) \geq |(u - u_0)(0)| = 0,$$
$$\text{and } v(1) = M_1 \varepsilon v_1(1) + M_2 \geq |(u - u_0)(1)| = |u_0(1)|.$$

By the comparison principle,

$$|(u - u_0)(x)| \leq v(x) = M_1 \varepsilon + M_2 \exp\left(-\beta\frac{1-x}{\varepsilon}\right).$$

The assertion of the Lemma follows. Moreover we see that

$$|(u - u_0)(x)| \leq M^* \varepsilon \quad \text{for} \quad x \in [0, x_0],$$

for some constant M^*. That is, the solution u_0 of the reduced problem is a good approximation of the exact solution u outside the layer region. ∎

Looking again at the proof of Lemma 6.1, one discovers why for positive b the layer is located at $x = 1$. Suppose that for $b(x) \geq \beta > 0$ we try to prove a similar result in $[x_0, 1]$ with the reduced problem required to satisfy $u(1) = 0$. Then one needs the barrier function $v_2^*(x) := \exp(-\beta x/\varepsilon)$, but we have the wrong sign in

$$Lv_2^*(x) \geq -\frac{\beta}{\varepsilon}(b + \beta) \exp\left(-\frac{\beta x}{\varepsilon}\right),$$

so something is amiss!
To improve the outcome of Lemma 6.1 by approximating $u(x, \varepsilon)$ pointwise for all $x \in [0,1]$, one must add a correction term to the solution u_0 of the reduced problem. Now u_0 satisfies

$$Lu_0 = f - \varepsilon u_0'' \sim f,$$
$$u_0(0) = 0, \quad u_0(1) = u_0(1),$$

so the correction v_0 should satisfy the homogeneous equation $Lv_0 = 0$, the boundary condition $v_0(1) = -u_0(1)$ and decrease exponentially in the interior of $[0,1]$. We call v_0 a *boundary layer correction*.

If a layer is located at $x = x_1$ say, then in the theory of matched asymptotic expansions one introduces local coordinates to describe the layer correction: define $\xi = \pm(x - x_1)/\varepsilon^\alpha$, where $\alpha > 0$ is a constant. The parameter α is chosen such that in the new coordinates the main part of the transformed differential operator L^* has (exponentially, in general) decreasing solutions as $\xi \to \infty$. Here "main" means the term(s) with the largest coefficient(s) when ε is small.

In our example the choice $\xi = (1 - x)/\varepsilon^\alpha$ leads to

$$L^* = -\varepsilon^{1-2\alpha}\frac{d^2}{d\xi^2} - \varepsilon^{-\alpha}b(1 - \varepsilon^\alpha\xi)\frac{d}{d\xi} + c(1 - \varepsilon^\alpha\xi).$$

Hence the main part of the transformed operator is as follows:

$$L^* \sim -\varepsilon^{-1}\left(\frac{d^2}{d\xi^2} + b(1)\frac{d}{d\xi}\right) \quad \text{for} \quad \alpha = 1,$$

$$L^* \sim -\varepsilon^{1-2\alpha}\frac{d^2}{d\xi^2} \quad \text{for} \quad \alpha > 1,$$

$$L^* \sim -\varepsilon^{-\alpha}b(1)\frac{d}{d\xi} \quad \text{for} \quad 0 < \alpha < 1.$$

Exponentially decreasing solutions are possible only in the first case here so we choose $\alpha = 1$. The solution of the layer correction problem

$$\frac{d^2v}{d\xi^2} + b(1)\frac{dv}{d\xi} = 0, \quad v|_{\xi=0} = -u_0(1), \quad v|_{\xi=\infty} = 0$$

is

$$v_0(x) = -u_0(1)\exp\left(-b(1)\frac{1-x}{\varepsilon}\right).$$

Such a layer is called an *exponential* boundary layer.

Similarly to Lemma 6.1, one can prove:

Lemma 6.2. *Assume that $b(x) \geq \beta > 0$ on $[0, 1]$. Then there exists a constant C, which is independent of x and ε, such that the solution of the boundary value problem (1.1) satisfies*

$$\left|u(x, \varepsilon) - \left[u_0(x) - u_0(1)\exp\left(-b(1)\frac{1-x}{\varepsilon}\right)\right]\right| \leq C\varepsilon. \tag{1.4}$$

For the analysis of numerical methods it is important to have precise information about the behaviour of derivatives of the solution of (1.1).

Lemma 6.3. *Let $b(x) \geq \beta > 0$ on $[0, 1]$. Then the first-order derivative of the solution of (1.1) satisfies the bound*

$$|u'(x, \varepsilon)| \leq C\left(1 + \varepsilon^{-1}\exp\left(-\beta\frac{1-x}{\varepsilon}\right)\right) \quad \text{for } 0 \leq x \leq 1. \tag{1.5}$$

Here C is once again a generic constant independent of x and ε.

Proof: Apply integration by parts to

$$-\varepsilon u'' + bu' = h := f - cu.$$

Setting $B(x) = \int_0^x b$, one obtains (for some constants of integration K_1, K_2)

$$u(x) = u_p(x) + K_1 + K_2 \int_x^1 \exp[-\varepsilon^{-1}(B(1) - B(t))]\, dt,$$

where

$$u_p(x) := -\int_x^1 z(t)\, dt, \quad z(x) := \int_x^1 \varepsilon^{-1} h(t) \exp[-\varepsilon^{-1}(B(t) - B(x))]\, dt.$$

Since $u'(1) = -K_2$ we must estimate K_2.
The boundary conditions imply that $K_1 = 0$ and

$$K_2 \int_0^1 \exp[-\varepsilon^{-1}(B(1) - B(t))]\, dt = -u_p(0).$$

Clearly

$$|u_p(0)| \le \max_{[0,1]} |z(x)|.$$

One can use a comparison principle to show that $|u(x)| \le Cx \le C$ and it follows that

$$|z(x)| \le C\varepsilon^{-1} \int_x^1 \exp[-\varepsilon^{-1}(B(t) - B(x))]\, dt.$$

Now $B(t) - B(x) = \int_x^t b(\tau)d\tau$, so

$$\exp[-\varepsilon^{-1}(B(t) - B(x))] \le \exp[-\beta\varepsilon^{-1}(t - x)] \quad \text{for} \quad x \le t.$$

Consequently

$$|z(x)| \le C\varepsilon^{-1} \int_x^1 \exp(-\beta\varepsilon^{-1}(t - x))\, dt \le C,$$

which yields $|u_p(0)| \le C$. Also

$$\int_0^1 \exp(-\varepsilon^{-1}(B(1) - B(t)))\, dt \ge C\varepsilon$$

and it follows that

$$|K_2| = |u'(1)| \le C\varepsilon^{-1}.$$

Finally,

$$u'(x) = z(x) + K_2 \exp[-\varepsilon^{-1}(B(1) - B(x))]$$

gives the desired estimate

$$|u'(x)| \le C\left(1 + \varepsilon^{-1} \exp\left(-\beta \frac{1 - x}{\varepsilon}\right)\right). \quad \blacksquare$$

Corollary 6.4. *Assume that $b(x) \geq \beta > 0$ on $[0,1]$. Then for the solution of (1.1) one has*

$$\int_0^1 |u'(x)|\, dx \leq C$$

for some constant C that is independent of ε.

This inequality is significant because the L_1 norm plays an important role in stability estimates.

One can use mathematical induction to prove the following extension of Lemma 6.3:

Lemma 6.5. *Assume that $b(x) \geq \beta > 0$ on $[0,1]$. Then for $0 \leq x \leq 1$, the derivatives of the solution of the boundary value problem (1.1) satisfy*

$$|u^{(i)}(x,\varepsilon)| \leq C[1 + \varepsilon^{-i}\exp(-\beta\varepsilon^{-1}(1-x))]$$

$$\text{for } i = 0, 1, ..., q \text{ and } 0 \leq x \leq 1, \tag{1.6}$$

where q depends on the regularity of the data of the problem.

Remark 6.6. In the analysis of numerical methods it is sometimes advantageous to use, instead of Lemma 6.5, the following decomposition of the solution (which is in fact equivalent to Lemma 6.5):
The solution u of the boundary value problem (1.1) can be decomposed as

$$u = S + E, \tag{1.7a}$$

where S is the smooth part and E is a layer component, with

$$|S^{(k)}(x)| \leq C, \quad |E^{(k)}(x)| \leq C\varepsilon^{-k}\exp(-\beta(1-x)/\varepsilon) \tag{1.7b}$$

for $k = 0, 1, ..., q$ and $0 \leq x \leq 1$, where q depends on the regularity of the data of (1.1); moreover,

$$LS = f \quad \text{and} \quad LE = 0 \tag{1.7c}$$

(see [85]). We call such a decomposition an *S-decomposition* because it was introduced by Shishkin to analyse upwind finite difference schemes.

Remark 6.7. A strong layer is characterized by a first-order derivative that is not bounded as $\varepsilon \to 0$. If the first-order derivative happens to be bounded, we have a *weak layer* if the second-order derivative is not bounded as $\varepsilon \to 0$. Here the types of boundary conditions play an essential role.
For instance, assuming $b(x) > 0$ on $[0,1]$, the conditions

$$u(0) = 0 \quad \text{and} \quad u'(1) = \alpha$$

imply that the layer component E satisfies $|E'(x)| \le C$ on $[0,1]$, but $|E''(x)|$ is not uniformly bounded in ε. With the boundary conditions

$$u(0) = 0 \quad \text{and} \quad b(1)u'(1) = f(1),$$

the situation is different: the first-order and second-order derivatives are uniformly bounded in ε and the layer is very weak.

To summarize: the strength of a layer depends on the boundary conditions!
□

The structure of the solution becomes more complicated if the coefficient b vanishes at any point in $[0,1]$. Let $x_0 \in (0,1)$ be a zero of b. Then x_0 is called a *turning point*. We assume that b has only one turning point.

In this case, the localization role for layers tells us:

If $b'(x_0) > 0$ (implying $b(0) < 0$ and $b(1) > 0$), then, in general, there are boundary layers at both $x = 0$ and $x = 1$.

If $b'(x_0) < 0$ (implying $b(0) > 0$ and $b(1) < 0$), then layers at $x = 0$ or $x = 1$ are impossible.

Consequently, the solution u_0 of the reduced problem is defined as follows:

Case A: $b'(x_0) > 0$: u_0 is the solution of
$$b(x)u' + c(x)u = f \qquad \text{without boundary conditions!}$$

Case B: $b'(x_0) < 0$: u_0 is the solution of
$$b(x)u' + c(x)u = f \text{ on } (0, x_0 - \delta) \text{ with } u_0(0) = 0,$$
$$b(x)u' + c(x)u = f \text{ on } (x_0 + \delta, 1) \text{ with } u_0(1) = 0, \text{ for some } \delta > 0.$$

In Case A, u_0 is the smooth solution of $bu' + cu = f$ that exists if $c(x_0) \ne 0$. If one assumes $c(x_0) \ne 0$, then an analogue of Lemma 6.1 is valid in $(0,1)$. Adding two boundary corrections at $x = 0$ and $x = 1$ yields analogues of Lemmas 6.2 and 6.3.

Case B is more difficult than case A. Let us consider the particular but typical example $b(x) = bx$ with b and c constant and $f(x) = bx^k$, $x_0 = 0$. Then

$$u_0(x) = \frac{1}{\rho - k} \begin{cases} -x^\rho + x^k & \text{for } x > 0, \\ -((-x)^\rho) + (-x)^k & \text{for } x < 0, \end{cases}$$

if $\rho := c/b$ is not equal to k. The solution u_0 of the reduced problem is continuous but not continuous differentiable; u has a cusp layer.

Finally we consider the case where $b(x) = (x - x_0)b^*(x)$ with $b^*(x) \ne 0$ and $\rho = c(0)/b^*(0)$ is an integer. Consider the example

$$-\varepsilon u'' - xu' = x \quad \text{in} \quad (-1,1),$$
$$u(-1) = u(1) = 0.$$

Then the solution of the reduced problem is discontinuous:

$$u_0(x) = \begin{cases} 1 - x \text{ in } (\delta, 1), \\ -1 - x \text{ in } (-1, -\delta). \end{cases}$$

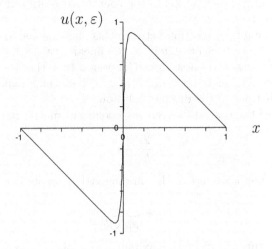

Figure 6.2 Interior layer with $\varepsilon = 0.01$

The solution u has an *interior layer* at $x = 0$. See Figure 6.2. This is clearly demonstrated analytically by the behaviour of the derivative of the exact solution:

$$u'(x) = \alpha(\varepsilon)e^{-x^2/(2\varepsilon)} - 1$$

with

$$\alpha(\varepsilon) = \left(\int_0^1 e^{-y^2/(2\varepsilon)} dy \right)^{-1} \geq \left(\int_0^\infty e^{-y^2/(2\varepsilon)} dy \right)^{-1} = \sqrt{\frac{2}{\pi\varepsilon}},$$

so $u'(0) \to \infty$ for $\varepsilon \to 0$.

Many books on the analysis of singular perturbation problems discuss turning point problems in detail; see for instance [dJF96].

6.1.2 Discretization on Standard Meshes

We first discuss finite difference methods, then later finite element methods for solving the boundary value problem (1.1):

$$-\varepsilon u'' + b(x)u' + c(x)u = f(x), \quad u(0) = u(1) = 0,$$

while assuming that $b(x) > \beta > 0$. In this subsection we discuss the behaviour of several methods on standard meshes; for simplicity only equidistant meshes with mesh size h are examined.

We saw already in an example in Section 6.1.1 that the central difference method

$$-\varepsilon D^+ D^- u_i + b_i D^0 u_i + c_i u_i = f_i, \quad u_0 = u_N = 0,$$

can yield unrealistic oscillations in the computed solution if the mesh size is not small enough.

There is a simple connection between this phenomenon and the M-matrix properties of the coefficient matrix of the linear system that corresponds to central differencing. We know from Chapter 2 that the M-matrix property guarantees stability and the validity of a discrete maximum principle; this excludes oscillations in the discrete solution.

Each row of the central differencing coefficient matrix has the form

$$0 \cdots 0 \quad -\frac{\varepsilon}{h^2} - \frac{b_i}{2h} \quad \frac{2\varepsilon}{h^2} + c_i \quad -\frac{\varepsilon}{h^2} + \frac{b_i}{2h} \quad 0 \cdots 0.$$

Consequently nonpositivity of the off-diagonal elements is equivalent to the condition

$$h \leq \frac{2\varepsilon}{\max |b|}. \tag{1.8}$$

One can show that the discrete maximum principle is indeed violated if (1.8) does not hold. But if ε is extremely small, it is impractical to satisfy condition (1.8) in numerical computation in 2D or 3D.

To ensure the validity of the discrete maximum principle, a very simple old idea is to replace the central difference quotient for the first-order derivative by a one-sided difference quotient in order to generate the correct sign in the off-diagonal elements. That means, depending on the sign of b, we choose

$$D^+ u_i = \frac{u_{i+1} - u_i}{h} \quad \text{if} \quad b < 0,$$

$$D^- u_i = \frac{u_i - u_{i-1}}{h} \quad \text{if} \quad b > 0.$$

This *upwind method* corresponds to our rule for the localization of layers: in the case $\varepsilon = 0$ and $b > 0$, we discretize

$$bu' + cu = f, \quad u(0) = 0$$

with

$$bD^- u_i + c_i u_i = f_i, \quad u_0 = 0$$

so that we move with the flow towards the layer.

Summarizing: The simple *upwind method* for discretizing (1.1) when $b > 0$ is

$$-\varepsilon D^+ D^- u_i + b_i D^- u_i + c_i u_i = f_i, \quad u_0 = u_N = 0. \tag{1.9}$$

With this method a typical row of the associated coefficient matrix looks like

$$0 \cdots 0 \quad -\frac{\varepsilon}{h^2} - \frac{b_i}{h} \quad \frac{2\varepsilon}{h^2} + \frac{b_i}{h} + c_i \quad -\frac{\varepsilon}{h^2} \quad 0 \cdots 0,$$

so we have a L-Matrix *without any restriction on the mesh size*. Moreover, the matrix is irreducibly diagonally dominant and hence an M-matrix.

Lemma 6.8. *The simple upwind finite difference scheme is stable, uniformly with respect to ε, i.e.,*

$$||u_h||_{\infty,h} \leq C||f_h||_{\infty,h}$$

with a stability constant C that is independent of ε.

Proof: We apply the M-criterion.
Let us define $e(x) := (1+x)/2$, and let the vector $e := (e(x_1), ..., e(x_{N-1}))$ be the restriction of $e(x)$ to the mesh points. Then, writing L_h for the discrete linear operator defined by the left-hand side of (1.9), we have

$$(L_h e)_i \geq \beta/2.$$

It follows that the coefficient matrix A of the linear system associated with (1.9) satisfies

$$||A^{-1}|| \leq \frac{1}{\beta/2}.$$

This bound is independent of ε. ∎

A numerical method is said to be *uniformly convergent* of order p with respect to the perturbation parameter ε in the L_∞ norm $||\cdot||_\infty$, if one has a bound of the form

$$||u - u_h||_\infty \leq Ch^p \quad \text{(constant } p > 0) \tag{1.10}$$

with a constant C independent of ε. Here u is the solution of the boundary value problem and u_h the computed solution, while h is the mesh diameter. In the case of a finite difference method, the maximum norm should be replaced by its discrete analogue.
Although the simple upwind scheme is uniformly stable, it is *not* uniformly convergent (for any constant $p > 0$) on an equidistant mesh. To show this, we consider the example

$$-\varepsilon u'' - u' = 0, \quad u(0) = 0, \quad u(1) = 1.$$

The discrete solution using simple upwinding on an equidistant mesh is given by

$$u_i = \frac{1 - r^i}{1 - r^N} \quad \text{with} \quad r = \frac{\varepsilon}{\varepsilon + h}.$$

It follows that for $\varepsilon = h$

$$\lim_{h \to 0} u_1 = \frac{1}{2} \neq \lim_{h \to 0} u(x_1) = 1 - \frac{1}{e}.$$

The missing adaption to the behaviour of the exact solution causes the scheme to be not uniformly convergent.

If one desires uniformly convergent or robust schemes that work well for all values of ε, there are two principal means of achieving this aim: one can use a standard method on a special mesh, or use a standard mesh and look for a special method (i.e., an adapted scheme). First we discuss adapted schemes on equidistant meshes and in Section 6.1.3 special meshes will be considered.

We choose central differencing as the starting point for our discussion. One could instead begin with the simple upwind scheme (1.9).

If $b(x) \equiv b = $ constant and $c \equiv 0$, then integration of the differential equation over the interval (x_i, x_{i+1}) yields

$$\varepsilon = b \frac{u(x_{i+1}) - u(x_i)}{u'(x_{i+1}) - u'(x_i)}.$$

Every numerical method will approximate the right-hand side of this equation, so instead of ε some approximation of it is generated. Conversely, one can hope to generate a numerical method with good properties by replacing ε by some clever approximation, followed by a discretization using a standard scheme. This is one way of motivating schemes of the form

$$-\varepsilon\sigma_i D^+ D^- u_i + b_i D^0 u_i + c_i u_i = f_i, \qquad (1.11)$$

which have *artificial diffusion*. It turns out that it is fruitful to choose $\sigma_i = \sigma(\rho_i)$ with $\rho_i = b_i h/(2\varepsilon)$. Now $D^+ D_- = \frac{1}{h}(D^+ - D^-)$ and $D^0 = \frac{1}{2}(D^+ + D^-)$, so the scheme (1.11) can be written as

$$-\varepsilon D^+ D^- u_i + b_i[(1/2 - \alpha_i)D^+ + (1/2 + \alpha_i)D^-]u_i + c_i u_i = f_i, \qquad (1.12)$$

where

$$\alpha_i := \frac{\sigma(\rho_i) - 1}{2\rho_i}. \qquad (1.13)$$

This shows that the choice $\sigma(\rho) = 1 + \rho$ generates the simple upwind scheme (1.9).

The coefficients in (1.12) make it natural to require $0 \le \alpha_i \le \frac{1}{2}$; this implies the condition

$$1 \le \sigma(\rho) \le 1 + \rho. \qquad (1.14)$$

Now let us consider the coefficient matrix of the linear system corresponding to (1.11). The ith row has the form

$$0 \cdots 0 \qquad -\frac{\varepsilon}{h^2}\sigma_i - \frac{b_i}{2h} \qquad \frac{2\varepsilon}{h^2}\sigma_i \qquad -\frac{\varepsilon}{h^2}\sigma_i + \frac{b_i}{2h} \qquad 0 \cdots 0.$$

Therefore, one can prove the M-matrix property if

$$\sigma(\rho) > \sigma \quad \text{or equivalently} \quad \alpha_i > \frac{1}{2} - \frac{1}{2\rho_i}. \qquad (1.15)$$

Analogously to Lemma 6.8, the scheme (1.11) with artificial diffusion is uniformly stable when the condition (1.15) is satisfied.

So far we need only satisfy the conditions $1 + \rho \geq \sigma(\rho) > \rho$. What is the best way now to choose $\sigma(\rho)$?

If one wants to generate a uniformly convergent scheme, then additional *necessary convergence conditions* can be derived. Fix $h/\varepsilon = \rho^*$, fix i, and let $h \to 0$. Then Lemma 6.2 implies that

$$\lim_{h \to 0} u(1 - ih) = u_0(1) - u_0(1) \exp(-ib(1)\rho^*). \tag{1.16}$$

The scheme (1.11) can be written as

$$-\frac{\sigma(\rho^* b_i/2)}{\rho^*}(u_{i-1} - 2u_i + u_{i+1}) + \frac{1}{2}(u_{i+1} - u_{i-1})b_i = h(f_i - c_i u_i).$$

We wish to calculate the limit approached by this equation as $h \to 0$. The assumption of uniform convergence justifies our replacing each u_j by the true solution u evaluated at the same mesh point; then, replacing i by $N - i$ and recalling (1.16), we obtain

$$\lim_{h \to 0} \frac{\sigma(\rho^* b_{N-i}/2)}{\rho^*} = \frac{1}{2}b(1) \coth \frac{1}{2}\rho^* b(1). \tag{1.17}$$

The choice

$$\sigma(\rho) = \rho \coth \rho \tag{1.18}$$

clearly satisfies (1.17), and in fact also satisfies the conditions (1.14) and (1.15)—see Figure 6.3.
The resulting scheme,

$$-\frac{h}{2}b_i \coth\left(\frac{h}{2\varepsilon}b_i\right)D^+D^-u_i + b_i D^0 u_i + c_i u_i = f_i, \tag{1.19}$$

is the famous *Iljin scheme* or Iljin-Allen-Southwell scheme.

The literature contains many different techniques for analysing the Iljin and related schemes for two-point boundary value problems. Among these, the most important are (cf. [RST96], [101]):

- a classical finite difference analysis, based on uniform stability and consistency, that may also use an asymptotic expansion of the exact solution
- the approximation of the exact difference scheme for the given problem
- the construction of so-called compact difference schemes that require the scheme to be exact for certain polynomials and for exponential functions related to the layer structure of the solution u
- collocation with exponential splines
- approximation of the given problem by a related problem with piecewise polynomial coefficients, then the exact solution of this new problem
- Petrov-Galerkin methods with exponential splines.

In the following we sketch first the classical finite difference analysis; later we shall discuss in detail Petrov-Galerkin methods with exponential splines.

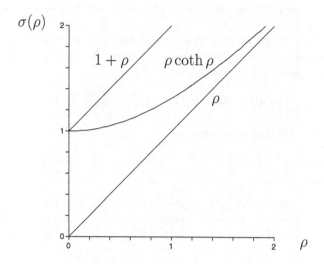

Figure 6.3 $\sigma(\rho)$ for the Iljin scheme

Theorem 6.9. *The Iljin scheme is first-order uniformly convergent in the discrete maximum norm, i.e.,*

$$\max_i |u(x_i) - u_i| \leq Ch \qquad (1.20)$$

with a constant C that is independent of ε and h.

Proof: Using the bounds $\|u^{(k)}\|_\infty \leq C\varepsilon^{-k}$, a standard consistency analysis combined with uniform stability (cf. Lemma 6.8) yields after some calculation

$$|u(x_i) - u_i| \leq C\Big(\frac{h^2}{\varepsilon^3} + \frac{h^3}{\varepsilon^4}\Big).$$

In the case $\varepsilon \geq h^{1/3}$ it follows immediately that

$$|u(x_i) - u_i| \leq Ch.$$

Thus consider the case $\varepsilon \leq h^{1/3}$. One can construct an asymptotic approximation $\psi = u_0 + \varepsilon u_1 + \varepsilon^2 u_2 + v_0 + \varepsilon v_1 + \varepsilon^2 v_2$ (cf. Lemma 6.2) with the property that

$$|u - \psi| \leq C\varepsilon^3.$$

Here u_0, u_1, u_2 are smooth in the sense that their derivatives are bounded, uniformly in ε, and v_0, v_1, v_2 are explicitly known functions that can be written as products of polynomials and decaying exponentials. Carefully estimating the "consistency error" $L_h(\psi_i - u_i)$ and invoking uniform stability leads to

$$|u_i - \psi(x_i)| \leq Ch.$$

But $\varepsilon^3 \leq h$, so the triangle inequality

$$|u(x_i) - u_i| \leq |u(x_i) - \psi(x_i)| + |u_i - \psi(x_i)|$$

yields the desired estimate. ∎

Remark 6.10. In the significant paper [77] the reader will find a precise analysis of three schemes: the simple upwind scheme, the modified upwind scheme of Samarskij with $\sigma(\rho) = 1 + \rho^2/(1+\rho)$, and the Iljin scheme. The analysis begins by using the estimates of Lemma 6.5 to investigate the consistency error. The next step uses special discrete barrier functions to yield the following: For the simple upwind scheme (1.9) one has

$$|u(x_i) - u_i| \leq \begin{cases} Ch[1 + \varepsilon^{-1}\exp(-\bar{\beta}\varepsilon^{-1}(1 - x_i))] & \text{for } h \leq \varepsilon, \\ Ch[h + \exp(-\beta(1 - x_i)/(\beta h + \varepsilon))] & \text{for } h \geq \varepsilon; \end{cases}$$

here $\bar{\beta}$ is a constant depending on β such that $\exp(\bar{\beta}t) \leq 1 + \beta t$ for $t \in [0, 1]$. For the Iljin scheme, it is shown in [77] that the error satisfies

$$|u(x_i) - u_i| \leq C\left[\frac{h^2}{h + \varepsilon} + \frac{h^2}{\varepsilon}\exp(-\beta(1 - x_i)/\varepsilon)\right].$$

The estimates for simple upwinding show that the method is not uniformly convergent in the layer region, as we expect. In the interval $[0, 1 - \delta]$ for any fixed $\delta > 0$, however, one has uniform convergence of order 1.
For the Iljin scheme one has convergence of order 2 if $\varepsilon \geq \varepsilon_0 > 0$; the order of uniform convergence is 1, and this is true in the entire interval $[0, 1]$! It is not difficult to verify that this is true for the example

$$-\varepsilon u'' + u' = x, \quad u(0) = u(1) = 0$$

and its discretization by the Iljin scheme.   □

Next we move on to finite element techniques on equidistant meshes. Naturally the practical arguments in favour of using finite elements are stronger in two or three space dimensions, but when finite element techniques are applied to singularly perturbed problems in 1D, we gain valuable information about the difficulties that arise if the singular perturbation parameter is extremely small.

Standard versions of finite elements on equidistant meshes have the same difficulties as standard finite difference methods: a lack of stability can generate oscillations in the discrete solution. For instance, using linear finite elements to discretize (1.1) on the mesh $\{x_i : i = 0, \ldots, N\}$ and evaluating the integrals by the midpoint quadrature rule generates the scheme

$$-\varepsilon D^+D^-u_i + 1/2(b(x_{i+1/2})D^+u_i + b(x_{i-1/2})D^-u_i)$$
$$+(1/2)(c(x_{i-1/2}) + c(x_{i+1/2}))u_i = (1/2)(f(x_{i-1/2}) + f(x_{i+1/2})).$$

We see the similarity to central differencing, and as in central differencing the discrete solution oscillates strongly (see Figure 6.13 for the two-dimensional case).

Is there some finite element analogue of the simple upwind scheme? An old idea from the 1970s is to generate an upwind effect by choosing the test functions differently from the ansatz functions. Then the finite element method is said to be of Petrov-Galerkin type. For instance, let us consider linear ansatz functions combined with test functions of the type

$$\psi_i(x) = \phi_i(x) + \alpha_{i-\frac{1}{2}}\sigma_{i-\frac{1}{2}}(x) - \alpha_{i+\frac{1}{2}}\sigma_{i+\frac{1}{2}}(x)$$

where

$$\phi_i(x) = \begin{cases} (x - x_{i-1})/h & x \in [x_{i-1}, x_i], \\ (x_{i+1} - x)/h & x \in [x_i, x_{i+1}], \\ 0 & \text{otherwise,} \end{cases}$$

with quadratic bubble functions $\sigma_{i+\frac{1}{2}}$; the parameters $\alpha_{i+\frac{1}{2}}$ are not yet chosen. This Petrov-Galerkin method generates the scheme (again applying the midpoint rule)

$$-\varepsilon D^+ D^- u_i + \bar{b}_{i+1/2} D^+ u_i + \bar{b}_{i-1/2} D^- u_i + [\bar{c}_{i+1/2} + \bar{c}_{i-1/2}] u_i$$
$$= \bar{f}_{i+1/2} + \bar{f}_{i-1/2}$$

with $\bar{q}_{i\pm1/2} := (1/2 \mp \alpha_{i\pm1/2}) q_{i\pm1/2}$ for $q = b, c, f$.
This bears some resemblance to the upwind scheme (1.9): for constant coefficients and $\alpha_{i\pm1/2} = 1/2$ the schemes coincide. In general the parameters $\alpha_{i\pm1/2}$ can be chosen using criteria similar to those used for difference schemes.

Petrov-Galerkin methods of this type are no longer used for 2D problems; alternative techniques are used to generate an upwind effect.

To take a more general view, let us assume that the ansatz functions are given and ask: which test functions are then optimal? To study this question, consider the abstract variational problem: find $u \in S$ such that

$$a(u, v) = (f, v) \quad \text{for all } v \in T.$$

Its Petrov-Galerkin discretization is: find $u_h \in S_h \subset S$ (so S_h is the space of ansatz functions) such that

$$a(u_h, v_h) = (f, v_h)$$

for all $v_h \in T_h \subset T$ (i.e., T_h is the test space).
Given a point x^* in the domain of u, define the Green's function $G \in T$ by

$$a(w, G) = w(x^*) \quad \text{for all } w \in S.$$

Let us for a moment assume that $G \in T_h$. Then the error of the Petrov-Galerkin method at the point x^* is given by

$$(u - u_h)(x^*) = a(u - u_h, G) = a(u, G) - a(u_h, G)$$
$$= (f, G) - (f, G) = 0$$

because $G \in T_h$. That is, when $G \in T_h$, the error at the point x^* is zero and the method is exact there! This is an attractive property—but in general we cannot guarantee that $G \in T_h$. Nevertheless there is one nontrivial situation where this occurs: for the problem $-u'' = f$ with $u(0) = u(1) = 0$, using linear elements and $S_h = T_h$, if x^* is a mesh point then $G \in T_h$. Consequently, the finite element solution of this problem is exact at all mesh points (assuming all integrals are evaluated exactly).

Now let $\bar{a}(\cdot, \cdot)$ be a bilinear form that approximates $a(\cdot, \cdot)$. Using this form we define $G \in T$ by

$$\bar{a}(w, G) = w(x^*) \quad \text{for all } w \in S.$$

Furthermore, assume that the numerical approximation u_h is a solution of

$$\bar{a}(u_h, v_h) = (f, v_h) \quad \text{for all } v_h \in T_h.$$

If we again assume that $G \in T_h$, then we obtain

$$(u - u_h)(x^*) = \bar{a}(u - u_h, G)$$
$$= (\bar{a} - a)(u, G) + (f, G) - \bar{a}(u_h, G),$$

i.e.,

$$(u - u_h)(x^*) = (\bar{a} - a)(u, G). \tag{1.21}$$

Therefore we conclude:
It is desirable to choose a test space that contains Green's functions for an approximating adjoint problem.

Remark 6.11. For arbitrary $v_h \in T_h$ one has

$$(u - u_h)(x^*) = a(u - u_h, G) = a(u - u_h, G - v_h).$$

Hence

$$|(u - u_h)(x^*)| \le \inf_{v_h \in T_h} |a(u - u_h, G - v_h)|.$$

Again this reveals the importance of choosing a test space that is related to some Green's function. □

Now we apply this idea to our singularly perturbed boundary value problem (1.1), whose bilinear form is (writing $(\cdot, \cdot)$ for the $L_2(0, 1)$ inner product)

$$a(u, v) := \varepsilon(u', v') + (bu' + cu, v). \tag{1.22}$$

We seek $u \in S := H_0^1(0, 1)$ such that $a(u, v) = (f, v)$ for all $v \in S$. Define

$$\bar{a}(u, v) := \varepsilon(u', v') + (\bar{b}u' + \bar{c}u, v), \tag{1.23}$$

where $\bar{b}$ and $\bar{c}$ are piecewise constant approximations of b and c.

Let $G_j \in S$ be the Green's function related to $\bar{a}$ and associated with the mesh point x_j. Then

$$\varepsilon(w', G'_j) + (\bar{b}w' + \bar{c}w, G_j) = w(x_j) \quad \text{for all } w \in S. \tag{1.24}$$

Alternatively, one can characterize G_j in the classical way:

(a) on the interior of each mesh interval G_j satisfies

$$-\varepsilon G''_j - \bar{b}G'_j + \bar{c}G_j = 0 \quad \text{and} \quad G_j(0) = G_j(1) = 0.$$

(b) G_j is continuous on $[0, 1]$.

(c) $\displaystyle\lim_{x \to x_i - 0}(\varepsilon G'_j - \bar{b}G_j) - \lim_{x \to x_i + 0}(\varepsilon G'_j - \bar{b}G_j) = -\delta_{ij} \quad \text{for} \quad i = 1, \dots, N-1$

—here δ_{ij} is the Kronecker delta.

The typical jump condition (c) for a Green's function follows from (a) and (1.24).

This motivates the following choice of test space: T_h is spanned by the $N-1$ functions ψ_k that satisfy

$$-\varepsilon \psi''_k - \bar{b}\psi'_k + \bar{c}\psi_k = 0 \quad \text{on each mesh interval,} \tag{1.25a}$$
$$\psi_k(x_j) = \delta_{kj}. \tag{1.25b}$$

It is obvious that each basis function ψ_k is non-zero only on the interval (x_{k-1}, x_{k+1}). These basis functions are *exponential splines*.

Next, choose some space of linearly independent ansatz functions ϕ_i with the property that $\phi_i(x_j) = \delta_{ij}$ and look for a numerical approximation of the given boundary value problem in the form

$$u_h(x) = \sum_{i=1}^{N-1} u_i\phi_i(x) \quad \text{for } x \in [0, 1]. \tag{1.26}$$

The Petrov-Galerkin method with exponential splines as test functions is: find $u_h \in V_h$ with

$$\bar{a}(u_h, \psi) = (\bar{f}, \psi) \quad \text{for all } \psi \in T_h. \tag{1.27}$$

This is a linear system of equations in the unknowns $\{u_i\}$. We shall compute the elements of the coefficient matrix of this linear system and verify that $G_j \in T_h$. Consider the contribution of $u_{k-1}\phi_{k-1}(x)$ to (1.27). Setting $\psi = \psi_k$ and invoking the property (1.25), the coefficient of u_{k-1} is seen to be

$$\int_{x_{k-1}}^{x_k} (\varepsilon\phi'_{k-1}\psi'_k + b_{k-1}\phi'_{k-1}\psi_k + c_{k-1}\phi_{k-1}\psi_k)dx = -\varepsilon\psi'_k(x_{k-1}).$$

Analogously, the coefficient of u_{k+1} is $\varepsilon\psi'_k(x_{k+1})$, while the coefficient of u_k is $\varepsilon[\psi'_k(x_k - 0) - \psi'_k(x_k + 0)]$. Now the definition of ψ_k implies that

$$\psi'_k(x_{k-1}) > 0, \ \ \psi'_k(x_k - 0) > 0, \ \ \psi'_k(x_k + 0) < 0, \ \ \psi'_k(x_{k+1}) < 0;$$

let us verify, for instance, that $\psi'_k(x_{k-1}) > 0$. For if $\psi'_k(x_{k-1}) < 0$, then ψ_k has a local minimum at some point $\tilde{x}$, at which one must have $\psi'_k(\tilde{x}) = 0$, $\psi_k(\tilde{x}) < 0$, $\psi''_k(\tilde{x}) \geq 0$, and this contradicts (1.25).

Thus the coefficient matrix of our linear system is an M-matrix. Hence the discrete problem has a unique solution.

To check that $G_j \in T_h$, we must show that parameters $\{\alpha_k\}$ can be chosen such that

$$G_j = \sum_{k=1}^{N-1} \alpha_k \psi_k.$$

For any choice of the α_k one has properties (a) and (b), so only the jump condition (c) needs to be satisfied. A calculation shows that this is equivalent to a linear system whose coefficient matrix is the transpose of the matrix just studied, and is therefore an M-matrix. Consequently we do have $G_j \in T_h$.

It is now straightforward to prove

Theorem 6.12. *If $\bar{b}, \bar{c}, \bar{f}$ are piecewise constant first-order approximations of b, c, f, then the Petrov-Galerkin method with exponential test functions (1.27) has a unique solution. Independently of the choice of the ansatz functions, the error in the mesh points satisfies*

$$\max_i |u(x_i) - u_i| \leq Ch \tag{1.28}$$

with a constant C that is independent of ε and h.

Proof: It remains only to prove (1.28). The properties of G_j and the definition of $\bar{a}$ imply that

$$(u - u_h)(x_j) = (f - \bar{f}, G_j) + (u', (-b + \bar{b})G_j) + (u, (-c + \bar{c})G_j).$$

Since G_j is uniformly bounded and moreover $\int |u'| \leq C$ by Corollary 6.4, it follows that

$$|(u - u_h)(x_j)| \leq Ch. \ \ \blacksquare$$

We close this subsection with some remarks related to Theorem 6.12.

Remark 6.13. The choice

$$\bar{q} = [q(x_{i-1}) + q(x_i)]/2 \ \ \text{for } q = b, c, f \text{ in each interval } (x_{i-1}, x_i)$$

implies second-order uniform convergence at each mesh point:

$$|u(x_i) - u_i| \leq Ch^2 \text{ for all } i.$$

See [RST96]. $\square$

Remark 6.14. The definition of the test functions ψ_k via (1.25) is called *complete exponential fitting.* More simply, one can define the basis functions by

$$-\psi_k'' - \bar{b}\psi_k' = 0 \text{ on each mesh interval}, \quad \psi_k(x_j) = \delta_{kj}.$$

Then gives a method with *exponential fitting* which can be further simplified using lumping: define

$$\bar{a}(v, \psi_i) := (\bar{b}v', \varepsilon\psi_i' + \bar{\psi}_i) + h\, c(x_i)v(x_i).$$

It should be noticed that the analysis of this scheme is technically delicate.
□

What influence do the ansatz functions have on the method? If one is satisfied with uniform convergence at the mesh points, then standard piecewise linear ansatz functions are fine. The stronger requirement that one obtain uniform convergence for all x in $[0, 1]$, i.e., that $\|u - u_h\|_\infty \leq Ch$, forces the use of exponential splines as ansatz functions—but in contrast to the test functions, these exponentials are related to the original differential operator and not to its adjoint. For instance, the following basis functions ϕ_k for the trial space are suitable:

$$-\varepsilon\phi_k'' + \bar{b}\phi_k' = 0, \quad \phi_k(x_i) = \delta_{ik}.$$

A study of the interpolation error for these exponential splines shows that the uniform $O(h)$ estimate is optimal.

6.1.3 Layer-adapted Meshes

Now we consider the alternative approach where standard discretization methods are used on special meshes that yield robust convergence with respect to the singular perturbation parameter. There is a wide range of publications that deal with layer adapted meshes. Since in the thesis [Lin07] a rather complete overview is given we refer the interested reader to it and omit widely to cite the original papers.

To simplify the notation we assume that the layer is located at $x = 0$ (if the layer is at $x = 1$, make the change of variable $x \mapsto 1 - x$) and study the boundary value problem

$$Lu := -\varepsilon u'' - bu' + cu = f, \quad u(0) = u(1) = 0, \tag{1.29}$$

with $b(x) > \beta > 0$. Assume that b, c and f are sufficiently smooth and $c(x) \geq 0$.

As early as 1969, Bakhvalov proposed a special mesh $0 = x_0 < x_1 < \cdots < x_N = 1$ that is adapted to the nature of the boundary layer. He chooses mesh points near $x = 0$ that satisfy

$$q\left[1 - \exp\left(-\frac{\beta x_i}{\sigma \varepsilon}\right)\right] = \xi_i := \frac{i}{N}.$$

Here $q \in (0, 1)$ and $\sigma > 0$ are parameters as yet undefined; q determines how many mesh points are in the layer and σ controls how fine the mesh is there. Outside the layer an equidistant mesh is used.

To be precise, Bakhvalov's mesh is specified by $x_i = \phi(i/N)$, $i = 0, 1, \ldots, N$, where

$$\phi(\xi) = \begin{cases} \chi(\xi) := -\frac{\sigma \varepsilon}{\beta} \ln(1 - \xi/q) & \text{for } \xi \in [0, \tau], \\ \chi(\tau) + \chi'(\tau)(\xi - \tau) & \text{for } \xi \in [\tau, 1]. \end{cases}$$

and τ is a transition point between the fine and coarse meshes. This point τ is a solution of the nonlinear equation

$$\chi(\tau) + \chi'(\tau)(1 - \tau) = 1, \tag{1.30}$$

which ensures that the mesh generating function ϕ lies in $C^1[0, 1]$ in this original *B-mesh*. An analysis shows that

$$\tau = \frac{\gamma \varepsilon}{\beta} |\ln \varepsilon| \tag{1.31}$$

for some γ that is bounded by a constant independent of ε. Much later it emerged that the C^1 property of ϕ is unnecessary. Instead of solving (1.30), one can choose a value for γ then define τ by (1.31) and describe the layer-adapted part of the mesh by

$$\phi(\xi) = -\frac{\gamma \varepsilon}{\beta} \ln(1 - 2(1 - \varepsilon)\xi) \quad \text{for } \xi = i/N, \ i = 0, 1, \ldots, N/2.$$

At the transition point τ for these *B-type meshes* one has

$$\exp\left(-\frac{\beta x}{\varepsilon}\right)\Big|_{x=\tau} = \varepsilon^\gamma. \tag{1.32}$$

From the numerical point of view when choosing the transition point, it seems better to replace smallness of the layer term with respect to ε by smallness with respect to the discretization error. If we desire a discretization error proportional to $N^{-\sigma}$, then the equation

$$\exp\left(-\frac{\beta x}{\varepsilon}\right)\Big|_{x=\tau} = N^{-\sigma} \tag{1.33}$$

leads to the choice $\tau = (\sigma \varepsilon / \beta) \ln N$ for the transition point between the fine and the coarse mesh. We call a mesh an *S-type mesh* if it is generated by

$$\phi(\xi) = \begin{cases} \frac{\sigma \varepsilon}{\beta} \hat{\phi}(\xi) & \text{with } \hat{\phi}(1/2) = \ln N \quad \text{for } \xi \in [0, 1/2] \\ 1 - (1 - \frac{\sigma \varepsilon}{\beta} \ln N)2(1 - \xi) & \text{for } \xi \in [1/2, 1]. \end{cases}$$

In particular when $\hat{\phi}(\xi) = 2(\ln N)\xi$, the mesh generated is piecewise equidistant; this *S-mesh* was introduced by Shishkin in 1988.

Remark 6.15. At first sight a comparison of B-type and S-type meshes favours B-type meshes because it seems reasonable to choose the transition point from the coarse to the fine mesh depending only on the boundary layer and independently of N. But S-meshes have a much simpler structure, and if one uses a piecewise equidistant B-type mesh, then one has a logarithmic factor $\ln \varepsilon$ in the L_∞ error estimate and therefore no uniform convergence. Consequently, in what follows we concentrate on S-meshes. It is in most cases straightforward to generalize the results to S-type meshes. For S-type meshes, special mesh-generating functions allow us to avoid the factor $\ln N$ in the error estimates.   □

If one wants to include also the non-singularly perturbed case, one defines the transition point for S-type meshes by

$$\tau = \min \left\{ q, (\sigma \varepsilon / \beta) \ln N \right\}. \tag{1.34}$$

We assume, for simplicity, that convection dominates and use the choice $\tau = (\sigma \varepsilon / \beta) \ln N$ for the transition point of our S-meshes.

Let us begin by analysing simple upwinding on an S-mesh.

Theorem 6.16. *The error of simple upwinding on an S-mesh with $\sigma = 1$ satisfies*

$$\max_i |u(x_i) - u_i^N| \le C N^{-1} \ln N, \tag{1.35}$$

where C is a constant independent of ε and N.

Proof: Recall the S-decomposition of the exact solution from Remark 6.6 and decompose the discrete solution analogously, viz.,

$$u_i^N = S_i^N + E_i^N$$

with (for $i = 1, \ldots, N - 1$)

$$L^N S_i^N = f_i, \ S_0^N = S(0), \ S_N^N = S(1) \quad \text{and}$$
$$L^N E_i^N = 0, \ E_0^N = E(0), \ S_N^N = E(1).$$

For the smooth part S, uniform stability and uniform consistency lead immediately to

$$\max_i |S(x_i) - S_i^N| \le C N^{-1}.$$

The consistency error of the layer part is in general *not* bounded pointwise uniformly in ε: one gets

$$|L^N (E(x_i) - E_i^N)| \le C \varepsilon^{-1} (N^{-1} \ln N) e^{-\beta x_i / \varepsilon}. \tag{1.36}$$

Nevertheless one can prove (1.35). There are several ways of doing this. We shall use a discrete maximum principle and a special barrier function. Let us define

$$w_i = C \prod_{k=1}^{i} \left(1 + \frac{\beta h_k}{\varepsilon}\right)^{-1} \quad \text{for } i = 0, \ldots, N.$$

After a short computation one finds that one can choose C such that $L^N w_i \geq |L^N E_i^N|$ for $i = 1, \ldots, N-1$ with $w_0 \geq |E_0|$ and $w_N \geq |E_N|$. Hence, in the region outside the layer,

$$|E_i^N| \leq w_i \leq w_{N/2} \leq C N^{-1} \quad \text{for} \quad i \geq N/2,$$

where the bound on $w_{N/2}$ can be deduced using $h_k = (\varepsilon N^{-1}/\beta) \ln N$ for $k = 1, \ldots, N/2$. One also has $|E_i| \leq C N^{-1}$ for $i \geq N/2$ from Remark 6.6.

Inside the layer region, one combines (1.36) with a discrete maximum principle and the barrier function

$$C N^{-1} + w_i (N^{-1} \ln N)$$

to get

$$|E(x_i) - E_i^N| \leq C N^{-1} + w_i (N^{-1} \ln N) \leq C N^{-1} \ln N. \quad \blacksquare$$

Remark 6.17. Alternatively, one can convert the consistency error bound to a convergence result by applying (L_∞, L_1) stability or the $(L_\infty, W_{-1,\infty})$ stability of the upwind operator which is valid on any mesh. Note that the consistency error on an S-mesh is uniformly bounded in the L_1 norm by $C N^{-1} \ln N$.

The improved stability properties of the upwind operator can be inferred from a detailed study of the related Green's function; see [86]. □

Similarly, for simple upwinding on the more sophisticated B-mesh one obtains

$$|u(x_i) - u_i^N| \leq C N^{-1} \quad \text{for all } i.$$

What about central differencing on a layer-adapted mesh?

The behaviour of the solution on such a mesh is not as bad as on an equidistant mesh: while oscillations are present, they are small and decrease as N increases. Under special assumptions on the mesh, the central differencing operator is uniformly stable in (L_∞, L_1). This leads to the following bounds for central differencing:

$$\max_i |u(x_i) - u_i^N| \leq \begin{cases} C N^{-2} & \text{on B-type meshes,} \\ C (N^{-1} \ln N)^2 & \text{on S-meshes.} \end{cases}$$

Next we analyse linear finite elements on an S-mesh with $\sigma = 2$. First we consider the interpolation error $u^I - u$. On each subinterval (x_{i-1}, x_i) one has

$$(u^I - u)(x) = \frac{x_i - x}{h_i} \int_{x_i}^{x_{i-1}} u''(\xi)(x_{i-1} - \xi) \, d\xi - \int_{x_i}^{x} u''(\xi)(x - \xi) \, d\xi.$$

This implies that

$$|(u^I - u)(x)| \leq 2 \int_{x_{i-1}}^{x_i} |u''(\xi)|(\xi - x_{i-1})\, d\xi.$$

Now any positive, monotonically decreasing function g satisfies the inequality

$$\int_{x_{i-1}}^{x_i} g(\xi)(\xi - x_{i-1})\, d\xi \leq \frac{1}{2}\left\{\int_{x_{i-1}}^{x_i} g^{1/2}(\xi)\, d\xi\right\}^2,$$

as can be seen by regarding both sides as a function of x_i, then differentiating. Hence

$$|(u^I - u)(x)| \leq C \left\{\int_{x_{i-1}}^{x_i} (1 + \varepsilon^{-1} \exp(-\beta\xi/(2\varepsilon))\, d\xi\right\}^2.$$

For an S-mesh it follows immediately that

$$|(u^I - u)(x)| \leq \begin{cases} C\, N^{-2} & \text{in } [\tau, 1], \\ C\, (N^{-1} \ln N)^2 & \text{in } [0, \tau]. \end{cases} \tag{1.37}$$

As well as pointwise estimates of the interpolation error, we also need estimates in the H^1 semi-norm. Integration by parts yields

$$|u^I - u|_1^2 = -\int_0^1 (u^I - u)u''dx \leq C\varepsilon^{-1}\|u^I - u\|_\infty,$$

so

$$\varepsilon^{1/2}|u^I - u|_1^2 \leq C\, N^{-1} \ln N. \tag{1.38}$$

Since $|u|_1$ is not bounded uniformly with respect to ε, it seems appropriate to analyse the error in the ε-weighted norm

$$\|v\|_\varepsilon := \sqrt{\varepsilon|v|_1^2 + |v|_0^2}.$$

Let us assume that $c + b'/2 \geq \alpha > 0$. As previously stated, this can be ensured by a change of variable using our assumption that $b(x) > \beta > 0$. Then the bilinear form

$$a(v, w) := \varepsilon(v', w') + (cv - bv', w)$$

is, with respect to ε, uniformly V-elliptic in the norm $\|\cdot\|_\varepsilon$ one does *not* have

$$|a(v, w)| \leq C\, \|v\|_\varepsilon\|w\|_\varepsilon$$

with a constant C that is independent of ε. Consequently it is impossible to deduce an error bound in $\|\cdot\|_\varepsilon$ directly from the above interpolation error estimates.

We therefore take the structure of our mesh into consideration and estimate separately on the fine mesh on $[0, \tau]$ and the coarse mesh on $[\tau, 1]$. Setting $\gamma = \min\{1, \alpha\}$ and $\tilde{e} = u^I - u^N$, we obtain

$$
\begin{aligned}
\gamma \|\tilde{e}\|_\varepsilon = \gamma \|u^I - u^N\|_\varepsilon &\leq a(u^I - u^N, u^I - u^N) = a(u^I - u, u^I - u^N) \\
&= \varepsilon((u^I - u)', (u^I - u^N)') + (b(u^I - u), (u^I - u^N)') \\
&\quad + ((c + b')(u^I - u), u^I - u^N) \\
&\leq C\|u^I - u\|_\varepsilon \|u^I - u^N\|_\varepsilon + C\left[\|u^I - u\|_{\infty,(0,\tau)} \|(u^I - u^N)'\|_{L_1(0,\tau)} \right. \\
&\quad \left. + \|u^I - u\|_{0,(0,\tau)} \|(u^I - u^N)'\|_{0,(0,\tau)}\right].
\end{aligned}
$$

On the coarse mesh we apply an inverse inequality, and on the fine mesh a Cauchy-Schwarz inequality yields

$$
\|(u^I - u^N)'\|_{L_1(0,\tau)} \leq C (\ln N)^{1/2} \|u^I - u^N\|_\varepsilon.
$$

Combining these estimates we have

$$
\|u^I - u^N\|_\varepsilon \leq C \left\{ \|u^I - u\|_\varepsilon + (\ln N)^{1/2} \|u^I - u\|_{\infty,(0,\tau)} + N \|u^I - u\|_{0,(\tau,1)} \right\}.
$$

This inequality, our earlier bounds and a triangle inequality yield the following result:

Theorem 6.18. *Consider the discretization of the boundary value problem (1.29) using linear finite elements on an S-mesh with $\sigma = 2$. Then the error satisfies*

$$
\|u - u^N\|_\varepsilon \leq C N^{-1} \ln N. \tag{1.39}
$$

Higher-order finite elements can be analysed in a similar way provided that one has adequate information about the behaviour of the requisite derivatives of the smooth and layer parts of the solution.

6.2 Parabolic Problems, One-dimensional in Space

6.2.1 The Analytical Behaviour of the Solution

Let us consider the initial-boundary value problem

$$
\frac{\partial u}{\partial t} - \varepsilon \frac{\partial^2 u}{\partial x^2} + b(x) \frac{\partial u}{\partial x} = f(x, t) \quad \text{in} \quad Q = (0, 1) \times (0, T), \tag{2.1a}
$$

$$
u(0, t) = u(1, t) = 0, \tag{2.1b}
$$

$$
u(x, 0) = g(x). \tag{2.1c}
$$

In general this problem has a unique solution whose smoothness depends on the smoothness of the data and on compatibility conditions at the corners of the domain.

The asymptotic behaviour of the solution of (2.1) as $\varepsilon \to 0$ depends strongly on the properties of the convection coefficient b. Let us define the function $X(t; \xi, \tau)$ to be the solution of

$$\frac{dX}{dt} = b(X, t) \quad \text{with} \quad X(\tau; \xi, \tau) = \xi.$$

The reduced equation is

$$\frac{\partial u}{\partial t} + b(x)\frac{\partial u}{\partial x} = f(x, t).$$

Let Q^* be the domain comprising all points (x, t) with $t \geq 0$ that lie on the characteristics of the reduced operator and pass through the side $t = 0$ of $\bar{Q}$. Then the solution of the reduced problem

$$\frac{\partial u_0}{\partial t} + b(x)\frac{\partial u_0}{\partial x} = f(x, t), \tag{2.2a}$$

$$u_0(x, 0) = g(x) \quad \text{for } x \in (0, 1), \tag{2.2b}$$

can be written in terms of the function X:

$$u_0(x, t) = g(X(0; x, t)) + \int_0^t f(X(\sigma; x, t), \sigma)d\sigma.$$

Now we have essentially three different cases (see Figure 6.4):

Case A: $b(0) < 0$, $b(1) > 0$ (the characteristics leave the domain).
Then $Q^* \not\supseteq Q$.

Case B: $b(0) > 0$, $b(1) < 0$ (the characteristics enter the domain).
Then $Q^* \subsetneq Q$.

Case C: $b(0) = 0$, $b(1) = 0$ (characteristics tangential to the boundary).
Then $Q^* = Q$.

In case B one can define the reduced problem by the reduced equation and all the initial-boundary conditions. Difficulties may arise along the characteristics through the points $(0, 0)$ and $(1, 0)$; depending on the compatibility of the data, the solution of the reduced problem (or its derivatives) can be discontinuous.

In case A one expects layers at $x = 0$ and $x = 1$ similar to those appearing in stationary 1D problems—*exponential boundary layers*. In case C the layer correction turns out to be the solution of an associated parabolic problem, so the layer is called a *parabolic boundary layer*.

As in Section 6.1 let us assume in general that b satisfies $b(x) > 0$. Then it is appropriate to define the reduced problem by

$$\frac{\partial u_0}{\partial t} + b(x)\frac{\partial u_0}{\partial x} = f(x, t), \tag{2.3a}$$

$$u_0(x, 0) = g(x) \quad \text{for} \quad x \in (0, 1). \tag{2.3b}$$

$$u_0(0, t) = 0. \tag{2.3c}$$

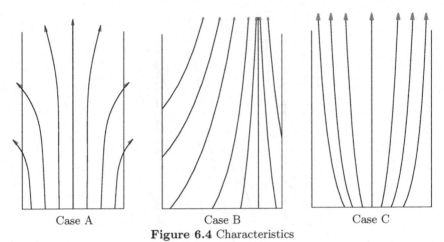

Case A   Case B   Case C

Figure 6.4 Characteristics

At the boundary $x = 1$ the solution to (2.1) has an exponential boundary layer. Setting $\zeta = (1 - x)/\varepsilon$, the layer correction $v(\zeta)$ associated with the point $(1, t)$ is defined by

$$\frac{d^2v}{d\zeta^2} + b(1)\frac{dv}{d\zeta} = 0 \quad \text{with} \quad v(0) = u_0(1, t).$$

One expects that $u_0(x, t) + v(\zeta, t)$ is a uniform asymptotic approximation of the solution of the given problem. Indeed, if the second-order derivatives of u_0 are bounded then the maximum principle allows us to prove that

$$|u(x, t) - [u_0(x, t) + v(\zeta, t)]| \leq C\varepsilon \quad \text{on } Q.$$

Assuming still more compatibility conditions, pointwise estimates for derivative approximations can also be derived.

Without compatibility one can only prove (using a weak maximum principle) that

$$|u(x, t) - [u_0(x, t) + v(\zeta, t)]| \leq C\varepsilon^{1/2}.$$

In this case it is difficult to prove pointwise bounds for derivatives.

In summary, the parabolic case, though one-dimensional in space, turns out to be more complicated than stationary one-dimensional problems: parabolic boundary layers may be present, and even in the case $b > 0$ the interior layer that may arise along the characteristic through the corner $(0, 0)$ is a source of difficulty.

6.2.2 Discretization

First we discuss a standard difference scheme for the discretization of the initial-boundary value problem (2.1). On the mesh nodes $\{(x_i, t^k)\}$ given by $x_i = ih$, $t^k = k\tau$, consider the difference scheme

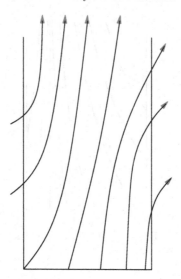

Figure 6.5 the case $b(x) > 0$

$$\frac{u_i^{k+1} - u_i^k}{\tau} + (1-\theta)\left[-\varepsilon\sigma_i\frac{u_{i-1}^k - 2u_i^k + u_{i+1}^k}{h^2} + b_i\frac{u_{i+1}^k - u_{i-1}^k}{2h}\right] +$$

$$+\theta\left[-\varepsilon\sigma_i\frac{u_{i-1}^{k+1} - 2u_i^{k+1} + u_{i+1}^{k+1}}{h^2} + b_i\frac{u_{i+1}^{k+1} - u_{i-1}^{k+1}}{2h}\right] \tag{2.4}$$

$$= (1-\theta)f_i^k + \theta f_i^{k+1},$$

where the user-chosen parameter θ lies in $[0,1]$ and u_i^k is the solution computed at (x_i, t^k). The value of θ determines whether the scheme is explicit ($\theta = 0$) or implicit ($\theta > 0$). The user-chosen parameter σ_i allows us to introduce artificial diffusion. Rearranging, we have

$$u_i^{k+1}(1 + 2\theta\tau\mu_i) + u_{i-1}^{k+1}\theta\tau\left(-\frac{b_i}{2h} - \mu_i\right) + u_{i+1}^{k+1}\theta\tau\left(\frac{b_i}{2h} - \mu_i\right)$$

$$= u_i^k(1 - 2(1-\theta)\tau\mu_i) + u_{i-1}^k(1-\theta)\tau\left(\frac{b_i}{2h} + \mu_i\right) +$$

$$u_{i+1}^k(1-\theta)\tau\left(-\frac{b_i}{2h} + \mu_i\right) + \tau[(1-\theta)f_i^k + \theta f_i^{k+1}]$$

with $\mu_i := \varepsilon\sigma_i/h^2$.

The stability of this scheme in the discrete maximum norm is easily analysed. Assuming $b(x) > \beta > 0$, one can establish

Lemma 6.19. *Set $\sigma_i = \sigma(\rho_i)$, where $\rho_i := h\,b_i/(2\varepsilon)$. Assume that $\sigma(\rho) > \rho$. Then*

(i) the implicit scheme (2.4) with $\theta = 1$ is stable in the discrete maximum norm, uniformly with respect to ε;
(ii) if $2\tau\varepsilon\sigma/h^2 \leq 1$, then the explicit scheme with $\theta = 0$ is also uniformly stable.

For the implicit scheme, central differencing is again unsuitable because stability requires

$$\rho < 1, \quad \text{i.e.,} \quad h < \frac{2}{\varepsilon b_i}.$$

On the other hand, the implicit scheme combined with (for instance) simple upwinding, so $\sigma(\rho) = 1 + \rho$, restricts neither the spatial mesh width h nor the time step τ.

In what follows, at first we continue the discussion of the case $b > 0$, then at the end of this subsection we sketch some difficulties associated with parabolic boundary layers.

When $b > 0$ we expect an exponential boundary layer at $x = 1$. Therefore, on standard (e.g., equidistant) meshes in space we can expect uniform convergence only if we choose σ carefully. Assuming for the moment *constant coefficients* (i.e., constant b) and equidistant mesh sizes in space and time, let us study the 6-point scheme

$$\sum_{m=0,1} \sum_{n=-1,0,1} \alpha_{nm} u_{i+n}^{j+m} = h f_i^j.$$

Using an asymptotic approximation of the solution, we can derive necessary conditions for uniform convergence just as in the stationary case—recall (1.17). In this way one obtains

$$\sum_m \sum_n \alpha_{nm} = 0 \qquad (2.5)$$

and, setting $\rho = bh/(2\varepsilon)$,

$$(\alpha_{-1,0} + \alpha_{-1,1})\exp(2\rho) + (\alpha_{0,0} + \alpha_{0,1}) + (\alpha_{1,0} + \alpha_{1,1})\exp(-2\rho) = 0. \quad (2.6)$$

Now our special scheme (2.4) has

$$\alpha_{-1,0} = \tau(1-\theta)(-b/(2h) - \mu), \quad \alpha_{-1,1} = \tau\theta(-b/(2h) - \mu),$$
$$\alpha_{0,0} = -1 + 2\tau(1-\theta)\mu, \qquad \alpha_{0,1} = 1 + 2\tau\theta\mu,$$
$$\alpha_{1,0} = \tau(1-\theta)(b/(2h) - \mu), \qquad \alpha_{1,1} = \tau\theta(b/(2h) - \mu).$$

Clearly (2.5) is automatically satisfied. Condition (2.6) places no restriction on θ and τ and yields the formula

$$\sigma(\rho) = \rho\coth(\rho).$$

Recalling Lemma 6.19, we conclude that our favorite scheme on equidistant meshes in the class (2.4) is an implicit scheme that is of Iljin type in space, i.e., is characterized by

$$\sigma_i = \rho_i \coth(\rho_i) \quad \text{with} \quad \rho_i = hb_i/(2\varepsilon).$$

If one now assumes sufficient compatibility of the data at the corners of Q, then one can prove that this scheme is first-order uniformly convergent; without compatibility the analysis is difficult.

Alternatively—again assuming sufficient compatibility—it is possible to generate and analyse a related scheme by exponential fitting in a Petrov-Galerkin framework. Let us start from the weak formulation of the initial-boundary value problem (2.1):

$$(u_t, v) + \varepsilon(u_x, v_x) + (bu_x, v) = (f, v),$$

where $(\cdot, \cdot)$ is the $L_2(Q)$ inner product. For each $k > 0$, we seek a numerical solution u_h at time $t = t^k$, setting

$$u_h(x, t^k) = \sum_i u_i^k \phi_i(x, t^k).$$

As in Section 6.1, approximate b by a piecewise constant $\bar{b}$ on every subinterval of $[0, 1]$ and define a modified bilinear form $\bar{a}$ by

$$\bar{a}(v, w) := \varepsilon(v_x, w_x) + (\bar{b}v_x, w).$$

Then implicit discretization in time combined with a Petrov-Galerkin method in space can be described as follows:

$$\frac{u_i^{k+1} - u_i^k}{\tau} + \bar{a}\big(u_h(\cdot, t^{k+1}), \psi_i\big) = (f^{k+1}, \psi_i).$$

As in Section 6.1 we choose exponential splines ψ_i as test functions; these test functions are the solutions of

$$-\varepsilon\psi_i'' - \bar{b}\psi_i' = 0 \quad \text{on every mesh interval,}$$
$$\psi_i(x_j) = \delta_{i,j}.$$

More details are in [RST96]. Assuming compatibility, the error satisfies the uniform estimate

$$\max_{i,j} |u(x_i, t^k) - u_i^k| \le C(h + \tau)$$

at the mesh points.

Instead of using exponential splines on an equidistant mesh, one could use standard splines on a mesh that is layer-adapted in the x direction.

Next we give some hints regarding the treatment of parabolic boundary layers, which are an important phenomenon. These appear in reaction-diffusion problems of the following type:

$$u_t - \varepsilon u_{xx} + c(x,t)u = f(x,t) \quad \text{in } Q = (0,1) \times (0,T), \qquad (2.7a)$$
$$u(0,t) = u(1,t) = 0, \quad u(x,0) = g(x). \qquad (2.7b)$$

Without any restriction we can assume that $c(x,t) \geq c_0 > 0$.

Here the solution u_0 of the reduced problem satisfies only the initial condition, and in general boundary layers are present along the sides $x = 0$ and $x = 1$ of Q. Let us study the structure of the solution at $x = 0$: introduce the local variable $\zeta = x/\varepsilon^{1/2}$. Then at $x = 0$ the local layer correction $v(\zeta,t)$ satisfies

$$v_t - v_{\zeta\zeta} + c(0,t)v = 0, \quad v(\zeta,0) = 0, \quad v(0,t) = -u_0(0,t).$$

This is a parabolic differential operator, which explains the descriptive terminology "parabolic boundary layer" for v. One can estimate v and its derivatives using an explicit but complicated representation of v in terms of the complementary error function, or one can proceed indirectly using maximum principles and carefully chosen barrier functions. Again assuming sufficient compatibility, a non-trivial analysis yields

$$\left| \frac{\partial^{l+m}}{\partial x^l \partial t^m} v(x,t) \right| \leq C \varepsilon^{-l/2} e^{-\gamma x/\varepsilon^{1/2}}, \tag{2.8}$$

where the constant $\gamma \in (0, c_0)$ is arbitrary. This estimate might lead us to hope that one could successfully apply some form of exponential fitting on equidistant meshes to deal with parabolic boundary layers, but blocking this road is the following remarkable result of Shishkin [111]:

> For problems with parabolic boundary layers there does not exist a scheme on standard meshes whose solution can be guaranteed to converge to the solution u in the maximum norm, uniformly with respect to the perturbation parameter ε.

We try to give a heuristic explanation of this deep result. First, in certain cases the boundary layer problem has a relatively simple explicit solution. For instance, the solution of

$$v_t - v_{\zeta\zeta} = 0, \quad v(\zeta,0) = 0, \quad v(0,t) = t^2,$$

is given by

$$v(\zeta,t) = 2 \int_0^t \int_0^\tau \operatorname{erfc}(\zeta/(2\mu^{1/2})) \, d\mu \, d\tau, \tag{2.9}$$

where erfc is the complementary error function. There are analogous formulae when instead one has $v(0,t) = t^m$ for integer $m > 2$. Now for problems with constant coefficients the necessary conditions for uniform convergence turn out to be of the form

$$AV_{i-1} + BV_i + DV_{i+1} = 0.$$

Such conditions are not satisfied by functions of the form e^{-di^2}, but the layer functions appearing in (2.9) and its analogues for $v(0,t) = t^m$ are of exactly this type, as can be seen from properties of the complementary error function.

Therefore we conclude:

For a problem with parabolic boundary layers, if one wants to construct a scheme which is guaranteed to converge uniformly in the maximum norm, then one has to use layer-adapted meshes! Furthermore, this is true not only for reaction-diffusion problems such as (2.7) but also for convection-diffusion problems whose solutions have parabolic boundary layers.

It is possible to define a suitable S-mesh based on the estimates (2.8). Set

$$\rho = \min\left\{1/4, \frac{\varepsilon^{1/2}}{\gamma}\ln N\right\}.$$

Subdivide the three subintervals $[0, \rho]$, $[\rho, 1 - \rho]$ and $[1 - \rho, 1]$ of $[0, 1]$ by equidistant meshes containing $N/4$, $N/2$ and $N/4$ subintervals respectively. Use an equidistant mesh in time. On this space-time mesh we discretize (2.7) implicitly:

$$\frac{u_i^{k+1} - u_i^k}{\tau} - \varepsilon D^+ D^- u_i^{k+1} + c(x_i, t^{k+1})u_i^{k+1} = f_i^{k+1}.$$

Then (see [Shi92]) one can prove the uniform error estimate

$$|u(x_i, t^{k+1}) - u_i^{k+1}| \le C\left((N^{-1}\ln N)^2 + \tau\right). \tag{2.10}$$

Similar results are valid for problems of the type (2.1) where convection is present. The order of convergence with respect to the time step can be improved by defect correction or by a time discretization of Crank-Nicolson type [78].

6.3 Convection-Diffusion Problems in Several Dimensions

6.3.1 Analysis of Elliptic Convection-Diffusion Problems

Convection-diffusion equations of the form

$$\frac{\partial u}{\partial t} - \nu \triangle u + b \cdot \nabla u + cu = f$$

play an important role in many applications such as fluid mechanics. We mention explicitly

- the computation of temperature in compressible flows;
- the equations for the concentration of pollutants in fluids;
- the momentum relation of the Navier-Stokes equations
  (here three difficulties overlap: the convective character, the nonlinearity and the divergence-free condition on the flow).

Numerical methods for singularly perturbed problems of the form

$$-\varepsilon u'' + bu' + cu = f$$

with $0 < \varepsilon \ll 1$ are frequently motivated by referring to the Navier-Stokes equations in the regime of large Reynolds numbers, but it is a long road from the analysis of discretization methods for one-dimensional linear problems to the Navier-Stokes equations. Even today there are many open problems concerning the numerical solution of the Navier-Stokes equations for large Reynolds numbers.

In this subsection we discuss stationary problems of the form

$$-\varepsilon \Delta u + b \cdot \nabla u + cu = f \quad \text{in} \quad \Omega \subset \mathbb{R}^2, \tag{3.1a}$$
$$u = 0 \quad \text{on} \quad \Gamma := \partial\Omega, \tag{3.1b}$$

with $0 < \varepsilon \ll 1$. We restrict ourselves to the two-dimensional case and assume, for simplicity, homogeneous Dirichlet boundary conditions. In (3.1) the quantity b is, of course, a vector function. If there is no possibility of misunderstanding we write $b\nabla$ to mean the scalar product $b \cdot \nabla$.

First we consider the asymptotic behaviour of the solution of the boundary value problem (3.1) as $\varepsilon \to 0$. Typically we suppose that $c \geq 0$ or $c - (1/2) \operatorname{div} b \geq \omega > 0$ so that a unique classical or weak solution exists. Setting $\varepsilon = 0$ in the differential equation yields the reduced equation

$$L_0 u := b\nabla u + cu = f.$$

From the one-dimensional case we know already that one cannot satisfy the boundary conditions on the whole boundary. Therefore, we try to define the reduced problem by

$$L_0 u = f, \tag{3.2a}$$
$$u|_\Sigma = 0, \tag{3.2b}$$

and ask for which subset Σ of the boundary Γ the problem is well posed. It is clear that the characteristics of (3.2) should play an important role. These are the solutions of

$$\frac{d\xi}{d\tau} = b(\xi(\tau)). \tag{3.3}$$

To define Σ, the behaviour of the characteristics near Γ is critical. Assume that we can define a function F in a strip near the boundary in such a way that $F|_\Gamma = 0$ and $F < 0$ on the interior of Ω. Then define

$$\Gamma_+ = \{x \in \Gamma \mid b\nabla F > 0\} \quad \text{“outflow boundary”}$$
$$\Gamma_- = \{x \in \Gamma \mid b\nabla F < 0\} \quad \text{“inflow boundary”}$$
$$\Gamma_0 = \{x \in \Gamma \mid b\nabla F = 0\} \quad \text{“characteristic boundary”}.$$

On $\Gamma_- \cup \Gamma_+$ the characteristics of (3.2) intersect the boundary, while at points in Γ_0 the characteristic is a tangent to the boundary. If Γ is smooth then $\Gamma = \Gamma_+ \cup \Gamma_- \cup \Gamma_0$.

The standard method of characteristics tells us that we obtain a well-posed problem on setting $\Sigma = \bar{\Gamma}_+$ or $\Sigma = \bar{\Gamma}_-$. To decide which of these to use in (3.2), one must find (as in the one-dimensional case) which choice permits decreasing solutions for the local layer correction. It turns out—compare the 1D situation—that the correct choice is $\Sigma = \bar{\Gamma}_-$. Thus the reduced problem is defined by

$$L_0 u = f, \tag{3.4a}$$
$$u|_{\bar{\Gamma}_-} = 0. \tag{3.4b}$$

We expect a boundary layer in the neighbourhood of the outflow boundary Γ_+. Let us classify that layer. Near Γ_+ we introduce local variables by setting

$$x_1 = x_1(\rho, \phi),$$
$$x_2 = x_2(\rho, \phi),$$

where $\rho(x) = \mathrm{dist}\,(x, \Gamma)$, $0 < \rho < \rho_0$ (some positive constant ρ_0) and ϕ is associated with position along the boundary. Applying the transformation $\zeta = \rho/\varepsilon$ and expanding in ε, it turns out that the ordinary differential equation (with coefficients $B_0(0, \phi)$, $A_{2,0}(0, \phi)$) determining the layer correction has the solution

$$v = -u_0|_{\Gamma_+} \exp\left(-\frac{B_0(0, \phi)}{A_{2,0}(0, \phi)}\zeta\right)$$

with The ellipticity of (3.1) enables us to show that $A_{2,0}$ is negative, and $B_0(0, \phi) = b\nabla\rho|_{\rho=0} < 0$ from the definition of Γ_+. We recognize an exponential boundary layer.

When the geometrical behaviour of the characteristics is complicated the solution of the reduced problem may also be complicated. In fact, even in geometrically simple situations the solution of the reduced problem may still fail to be straightforward.

Figure 6.6 shows a situation which often occurs in practice and corresponds to a channel flow. The solution has exponential layers on Γ_+ and parabolic layers on Γ_0. In the neighbourhood of the points A and B there may be additional difficulties.

In Figure 6.7 no parabolic boundary layers appear in the solution. But, depending on the boundary conditions, there could be an interior parabolic layer along the characteristic through the point P owing to a discontinuity in the solution of the reduced problem.

The characteristics are quite complicated in the example

$$-\varepsilon\Delta u + (x^2 - 1)u_x + (y^2 - 1)u_y + cu = f \quad \text{in} \quad \Omega = \{(x, y) : x^2 + y^2 < 3\},$$
$$u = 0 \quad \text{on} \quad \Gamma = \partial\Omega;$$

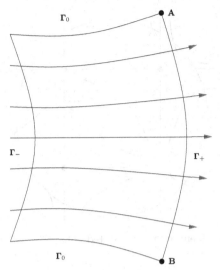

Figure 6.6 Standard situation

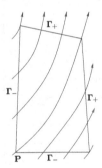

Figure 6.7 A problem without parabolic boundary layers

see Figure 6.8.

Finally, if the domain is not simply connected, then a simple structure to the characteristics does not necessarily imply a simple asymptotic structure to the solution. In Figure 6.9, for example, interior layers will arise along CD and EF. For a detailed description of this complicated topic see [GFL+83]; many other textbooks on singular perturbation problems give at least a discussion of simple situations.

In a nutshell, singularly perturbed convection-diffusion problems in 2D are much more complicated than in 1D. As a consequence, it is also more complicated to generate and analyse robust discretization methods.

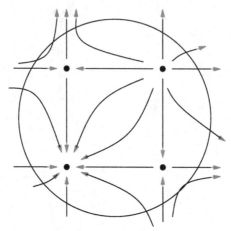

Figure 6.8 A problem with complicated behaviour of the characteristics

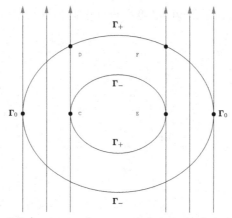

Figure 6.9 A non-simply connected domain: interior layers

As in the one-dimensional case, it is extremely helpful for the numerical analyst to have a decomposition of the solution into a sum of a smooth part (where the required derivatives are uniformly bounded with respect to ε) and layer parts—without the usual remainder term of an asymptotic expansion. Unfortunately, so far there exist complete proofs of the existence of such decompositions in only a few cases.

Let us consider a standard problem with exponential layers, namely

$$-\varepsilon\Delta u + b \cdot \nabla u + cu = f \quad \text{in} \quad \Omega = (0,1)^2, \tag{3.5a}$$

$$u = 0 \quad \text{on} \quad \partial\Omega, \tag{3.5b}$$

assuming $b_1(x,y) > \beta_1 > 0$, $b_2(x,y) > \beta_2 > 0$. If we assume some smoothness of the coefficients and moreover the compatibility condition

$$f(0,0) = f(1,0) = f(0,1) = f(1,1) = 0,$$

then this problem has a unique classical solution $u \in C^{3,\alpha}(\bar{\Omega})$. This solution has exponential layers along the sides $x = 1$ and $y = 1$ of Ω. One can prove the S-type decomposition [87]

$$u = S + E_1 + E_2 + E_3 \tag{3.6}$$

with

$$\left| \frac{\partial^{i+j}}{\partial x^i \partial y^j} S \right| \leq C, \tag{3.7a}$$

$$\left| \frac{\partial^{i+j}}{\partial x^i \partial y^j} E_1(x,y) \right| \leq C\varepsilon^{-i} e^{-\beta_1(1-x)/\varepsilon}, \tag{3.7b}$$

$$\left| \frac{\partial^{i+j}}{\partial x^i \partial y^j} E_2(x,y) \right| \leq C\varepsilon^{-j} e^{-\beta_2(1-y)/\varepsilon}, \tag{3.7c}$$

$$\left| \frac{\partial^{i+j}}{\partial x^i \partial y^j} E_3(x,y) \right| \leq C\varepsilon^{-(i+j)} e^{-(\beta_1(1-x)/\varepsilon + \beta_2(1-y)/\varepsilon)}. \tag{3.7d}$$

These estimates are valid for $i + j \leq 1$ without any additional assumptions; to extend them to higher-order derivatives, sufficient conditions on the data are given in [87].

Remark 6.20. For reaction-diffusion problems Melenk's book [Mel02] presents a complete theory for solution decompositions, including corner singularities, in polygonal domains. Transport phenomena mean that the corresponding convection-diffusion problems are more complicated. Some results for problems with parabolic boundary layers can be found in [Shi92], [76].   □

It is very instructive to look at a graph of the Green's function of a convection-diffusion problem in 2D because the difficulties encountered in its numerical solution are then obvious. Let us consider problem (3.5) with constant coefficients $b_1 > 0$, $b_2 > 0$, $c > 0$ and let G be the Green's function associated with an arbitrary but fixed point $(x,y) \in \Omega$ for this problem. If $\tilde{G}$ is the Green's function associated with (x,y) and the same differential operator but on the domain $\mathbb{R}^2$ (the "free-space Green's function"), then a maximum principle implies that

$$0 \leq G \leq \tilde{G}.$$

Now $\tilde{G}$ can be computed explicitly, using e.g. Fourier transforms. One gets

$$\tilde{G}(x,y;\xi,\eta) = \frac{1}{2\pi\varepsilon} \exp\left[(b_1(\xi - x) + b_2(\eta - y)/(2\varepsilon)) \right] K_0(\lambda r)$$

where

$$r^2 = (\xi - x)^2 + (\eta - y)^2, \qquad (2\varepsilon\lambda)^2 = b_1^2 + b_2^2 + 4\varepsilon c,$$

and K_0 is a modified Bessel function. Of course, $\tilde{G}$ has a logarithmic singularity at (x, y). Figure 6.10 shows the typical behaviour of $\tilde{G}$ for small ε. It shows convincingly how upwind data strongly influences the solution at (x, y). A direct estimate yields

$$\|G\|_{L_p} \leq C\,\varepsilon^{-(p-1)/p} \quad \text{for} \quad 1 \leq p < \infty.$$

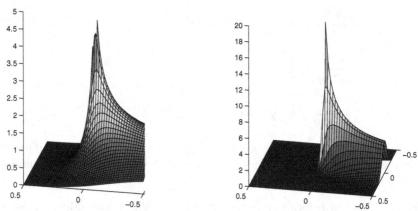

Figure 6.10 The Green's function (excluding the neighbourhood of the singularity) for $x = y = c = 0$, $b_1 = 1$, $b_2 = 0$ with $\varepsilon = 0.1$ (left) and $\varepsilon = 0.01$ (right)

6.3.2 Discretization on Standard Meshes

When discretizing the given convection-diffusion problem it is natural to ask: what qualities do we expect in our discretization method? Some possible aims are:

(V_1): Simple methods that work for $h \leq h_0$ with h_0 independent of ε, are stable and produce good results in those subdomains of Ω where one can neglect the influence of layers and singularities;

(V_2): Higher-order methods having the same properties as the methods of (V_1);

(V_3): Methods that are robust with respect to the parameter ε and can resolve layers; for instance, methods that are uniformly convergent with respect to a sufficiently strong norm.

When discussing uniform convergence, the choice of norm to measure the error is of fundamental importance. For instance, the layer term $e^{-x/\varepsilon}$ is $O(1)$ when measured in the maximum norm, but if one uses the L_1 norm then this term is only $O(\varepsilon)$ and consequently extremely small for small ε.

Methods typical of the class V_1 are *upwind difference schemes* and the corresponding upwind methods used in finite volume methods. Important methods belonging to class V_2 are the *streamline diffusion finite element method*, methods with *edge stabilization* (see [28]) and the *discontinuous Galerkin method*. We know already from the one-dimensional case that we can try to achieve uniform convergence using *exponential fitting* or *layer-adapted meshes*; these meshes will be discussed in the next subsection.

We begin with upwind finite difference schemes. Consider the model problem (see Figure 6.10)

$$-\varepsilon \Delta u + b\nabla u + cu = f \quad \text{in} \quad \Omega = (0,1) \times (0,1), \tag{3.8a}$$

$$u = 0 \quad \text{on} \quad \Gamma, \tag{3.8b}$$

where we assume that

$$(a) \qquad b = (b_1, b_2) > (\beta_1, \beta_2) > 0, \tag{3.9a}$$

$$(b) \qquad c \geq 0. \tag{3.9b}$$

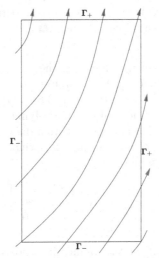

Figure 6.11 Characteristics of (3.8) and (3.9)

Both classical and weak solutions of (3.8) satisfy maximum principles: $f \geq 0$ implies $u \geq 0$. Often this property is very important: for example, the unknown u may correspond to a chemical concentration which must be non-negative. In such cases it is desirable to have the maximum principle for the discrete problem also. The methods of class (V_1) frequently satisfy a discrete maximum principle, in contrast to higher-order methods. As we know from Chapter 2, inverse-monotone matrices (in particular M-matrices) do satisfy maximum principles. This property of the matrix of the discrete problem also allows us to prove stability of the method in many cases.

It is not difficult to generalise the simple upwind difference scheme from one dimension to the two-dimensional case. For problem (3.8) on a equidistant square mesh with mesh spacing h in each coordinate direction, the simple upwind scheme (written as a difference stencil) is

$$-\frac{\varepsilon}{h^2} \begin{bmatrix} \cdot & 1 & \cdot \\ 1 & -4 & 1 \\ \cdot & 1 & \cdot \end{bmatrix} + \frac{1}{h} b_1^h \begin{bmatrix} \cdot & \cdot & \cdot \\ -1 & 1 & \cdot \\ \cdot & \cdot & \cdot \end{bmatrix} + \frac{1}{h} b_2^h \begin{bmatrix} \cdot & \cdot & \cdot \\ \cdot & 1 & \cdot \\ \cdot & -1 & \cdot \end{bmatrix} + c_h \begin{bmatrix} \cdot & \cdot & \cdot \\ \cdot & 1 & \cdot \\ \cdot & \cdot & \cdot \end{bmatrix} = f_h. \quad (3.10)$$

Here $b_i^h = b_i$ at each mesh point.

Lemma 6.21. *The upwind scheme (3.10) for the boundary value problem (3.8), (3.9) is inverse monotone and uniformly stable with respect to ε in the discrete maximum norm.*

Proof: It is easy to check that the coefficient matrix of the linear system associated with (3.10) is an L-matrix. Using $b_1 > 0$ and $c \geq 0$, the restriction of the function $e(x) = (1 + x)/2$ to the mesh yields a majorising element for the discrete problem. Now the usual M-criterion delivers both the M-matrix property and the desired stability estimate. ∎

When treating more general domains, the finite element method is more flexible than finite differencing. We therefore ask for an upwind version of the finite element method. Consider again (3.8),(3.9) and discretize this using linear finite elements. Assume that the underlying triangulation comes from an equidistant square mesh where one draws a diagonal in the same direction in each square. If b_1, b_2 are constants and $c \equiv 0$, then the finite element method generates the difference stencil

$$-\frac{\varepsilon}{h} \begin{bmatrix} \cdot & 1 & \cdot \\ 1 & -4 & 1 \\ \cdot & 1 & \cdot \end{bmatrix} + \frac{b_1}{6} \begin{bmatrix} \cdot & -1 & 1 \\ -2 & \cdot & 2 \\ -1 & 1 & \cdot \end{bmatrix} + \frac{b_2}{6} \begin{bmatrix} \cdot & 2 & 1 \\ 1 & \cdot & -1 \\ -1 & -2 & \cdot \end{bmatrix}.$$

It is clearly impossible to have the M-matrix property here. In the 1970s several strategies were proposed to modify this approach to yield M-matrices; for instance, Petrov-Galerkin techniques with linear ansatz functions and quadratic test functions were used—see [RST96]. Eventually it turned out that in the framework of finite elements there is no true analogue of the simple upwind finite difference method!

Among finite element methods for convection-diffusion problems, the most frequently used methods today are streamline diffusion and its modifications and the discontinuous Galerkin methods that we met already in Chapter 4. If one desires inverse monotonicity in the discrete problem, this can be got using finite volume methods.

To describe the upwind finite volume method, we leave the model problem (3.8),(3.9) and instead consider the more general boundary value problem

$$-\varepsilon \triangle u + b\nabla u + cu = f \quad \text{in} \quad \Omega \subset \mathbb{R}^2, \tag{3.11a}$$

$$u = 0 \quad \text{on} \quad \Gamma \tag{3.11b}$$

in a polygonal domain. Assume that

$$c - \frac{1}{2}\operatorname{div} b \geq 0. \tag{3.12}$$

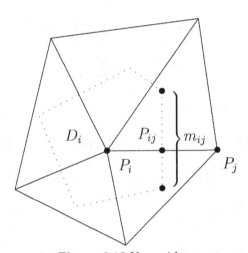

Figure 6.12 Voronoi box

We start from an admissible decomposition of the given domain into weakly acute triangles. Next, construct a dual mesh based on Voronoi boxes (recall Chapter 2); see Figure 6.12. We shall use the notation of Figure 6.12 and also set $l_{ij} = |P_i P_j|$.

As discussed in Chapter 2 and Chapter 4, the discretization of $-\triangle u$ on this mesh using either the finite volume method or linear finite elements generates the same difference scheme, namely

$$\sum_{j \in \Lambda_i} \frac{m_{ij}}{l_{ij}}(u(P_i) - u(P_j)). \tag{3.13}$$

The corresponding matrix is an M-matrix as it is an L-matrix, irreducible and diagonally dominant.

The discretization of the convective term is important in preserving the M-matrix property. While remaining in the finite volume framework, we describe the discretization in finite element terms. Start from the splitting

$$(b\nabla u_h, w_h) = (\text{div}(u_h b), w_h) - ((\text{div } b)u_h, w_h).$$

Approximating the first term here,

$$(\text{div}(u_h b), w_h) = \sum_i \int_{D_i} \text{div}(u_h b) w_h \, dx$$

$$\approx \sum_i w_h(P_i) \int_{D_i} \text{div}(u_h b) \, dx,$$

and Gauss's integral theorem yields

$$(\text{div}(u_h b), w_h) \approx \sum_i w_h(P_i) \int_{\partial D_i} (b \cdot \nu) u_h \, d\Gamma_i$$

$$= \sum_i w_h(P_i) \sum_{j \in \Lambda_i} \int_{\Gamma_{ij}} (b \cdot \nu_{ij}) u_h \, d\Gamma_{ij}.$$

Next we apply the quadrature rule

$$\int_{\Gamma_{ij}} (b \cdot \nu_{ij}) u_h d\Gamma_{ij} = (b(P_{ij}) \cdot \nu_{ij})(\text{meas } \Gamma_{ij})[\lambda_{ij} u_h(P_i) + (1 - \lambda_{ij}) u_h(P_j)]$$

where the weights λ_{ij} are not yet specified. We now have

$$(\text{div}(u_h b), w_h) \approx \sum_i w_h(P_i) \sum_{j \in \Lambda_i} (b(P_{ij}) \cdot \nu_{ij}) m_{ij} [\lambda_{ij} u_h(P_i) + (1 - \lambda_{ij}) u_h(P_j)].$$

The second term is approximated in a similar manner:

$$((\text{div } b)u_h, w_h) = \sum_i \int_{D_i} (\text{div } b) u_h w_h \, dx$$

$$\approx \sum_i u_h(P_i) w_h(P_i) \int_{D_i} \text{div } b \, dx,$$

so

$$((\text{div } b)u_h, w_h) \approx \sum_i u_h(P_i) w_h(P_i) \sum_{j \in \Lambda_i} (b(P_{ij}) \cdot \nu_{ij}) m_{ij}.$$

Collecting terms, we have derived the following discretization of the convective term:

$$(b\nabla u_h, w_h) \approx \sum_i w_h(P_i) \sum_{j \in \Lambda_i} (b(P_{ij}) \cdot \nu_{ij}) m_{ij} [(\lambda_{ij} - 1) u_h(P_i) + (1 - \lambda_{ij}) u_h(P_j)].$$

The corresponding matrix B_h has the following entries:

$$(B_h)_{kk} = \sum_{j \in \Lambda_k} (b(P_{kj}) \cdot \nu_{kj}) m_{kj} (\lambda_{kj} - 1),$$

$$(B_h)_{kl} = (b(P_{kl}) \cdot \nu_{kl}) m_{kl} (1 - \lambda_{kl}) \quad \text{if} \quad l \in \Lambda_k,$$

$$(B_h)_{kl} = 0 \quad \text{otherwise.}$$

Therefore we conclude that the choice

$$\lambda_{kl} = \begin{cases} 1 \text{ if } & b(P_{kl}) \cdot \nu_{kl} \geq 0, \\ 0 \text{ if } & b(P_{kl}) \cdot \nu_{kl} < 0, \end{cases} \tag{3.14}$$

will guarantee that B_h is an L-matrix.

Remark 6.22. The choice of λ_{kl} in (3.14) is the natural extension of the upwind strategy from the one-dimensional to the two-dimensional case. On an equidistant mesh, the discretization of the model problem (3.8),(3.9) then yields the upwind scheme (3.10). □

To complete the discretization, we discretize $cu - f$ analogously to the second term in the splitting above; this technique was already described in Chapter 4, Section 4.5.4.

An examination of the matrix generated by the full discretization shows that the choice (3.14) is sufficient but not necessary for an M-matrix; there is still some freedom in choosing λ_{ij}.

Let us write the full discrete problem as

$$a_l(u_h, v_h) = (f, v_h)_l \quad \text{for all} \quad v_h \in V_h.$$

To analyse the method we introduce the norm

$$\|v_h\|_\varepsilon := \left\{ \varepsilon |v_h|_1^2 + \|v_h\|_l^2 \right\}^{1/2}$$

with

$$\|v_h\|_l^2 = \sum_i v_h^2(P_i) \cdot (meas\, D_i).$$

Then from [5] we have the following result:

Theorem 6.23. *Assume that the triangulation is weakly acute. Then the upwind finite volume method has the following properties:*
(a) The discrete problem is inverse monotone for $h \leq h_0$, where h_0 is independent of the perturbation parameter ε.
(b) The error satisfies

$$\|u - u_h\|_\varepsilon \leq C\varepsilon^{-1/2} h(\|u\|_2 + \|f\|_{W_q^1}).$$

(c) Furthermore, $\|u - u_h\|_\varepsilon \leq Ch(\|u\|_2 + \|f\|_{W_q^1})$ if the triangulation is uniform.

Here we say that the triangulation is uniform if the mesh is three-directional.

The estimates of the theorem do not appear useful because in general $\|u\|_2$ is large when ε is small, but in [100] localized results tell us that in subdomains that exclude layers the method works well. Thus the upwind finite volume method belongs to the class V_1.

If one is interested in a higher-order method of type (V_2), then it is not realistic to insist on the inverse monotonicity of the discrete problem. It is well known that the discretization of $-\triangle$ by quadratic finite elements does not in general give inverse monotonicity; some special geometric situations are an exception. The very popular streamline diffusion finite element method (SDFEM)—also called the streamline upwind Petrov-Galerkin method (SUPG)—is a higher-order method that is not inverse monotone but has improved stability properties compared with the standard Galerkin method.

Consider again the boundary value problem (3.11) under the assumption (3.12). To introduce the SDFEM, we take the general Petrov-Galerkin method on a standard quasi-uniform mesh:
Find $u_h \in V_h$ such that

$$\varepsilon(\nabla u_h, \nabla w_h) + (b\nabla u_h + c u_h, w_h) = (f, w_h)$$

for all w_h from the test space W_h, where $(\cdot, \cdot)$ is the $L_2(\Omega)$ inner product. The basic idea of *streamline diffusion* is to choose test functions of the type

$$w_h := v_h + \beta\, b \nabla v_h \qquad \text{for} \quad v_h \in V_h$$

with a parameter β that can be specified element by element.

Example 6.24. The discretization of the convection-diffusion operator

$$-\varepsilon\triangle - p\frac{\partial}{\partial x} - q\frac{\partial}{\partial y}$$

with constant coefficients p, q on a uniform mesh of Friedrichs-Keller type (squares with one diagonal drawn) by linear finite elements and SDFEM generates the difference stencil

$$\varepsilon \begin{bmatrix} \cdot & -1 & \cdot \\ -1 & 4 & -1 \\ \cdot & -1 & \cdot \end{bmatrix} + \frac{h}{6} \begin{bmatrix} \cdot & -p+2q & p+q \\ -2p+q & \cdot & 2p-q \\ -(p+q) & p-2q & \cdot \end{bmatrix} +$$

$$+\beta \begin{bmatrix} \cdot & pq-q^2 & -pq \\ pq-q^2 & 2(p^2+q^2-pq) & pq-q^2 \\ -pq & pq-q^2 & \cdot \end{bmatrix}.$$

Is this related to stabilization by artificial diffusion?

In 2D it turns out that adding isotropic artificial diffusion by modifying ε to $\varepsilon + \gamma$ for some parameter γ yields poor results: layers are excessively smeared.

A better idea, based on the characteristics of the reduced equation, is to transform the differential operator to the form

$$-\varepsilon\Delta + \frac{\partial}{\partial\xi}$$

and then to add artificial diffusion *only in the flow direction*; that means, to replace

$$\varepsilon\frac{\partial^2}{\partial\xi^2} \quad \text{by} \quad (\varepsilon + \beta^*)\frac{\partial^2}{\partial\xi^2}.$$

Then transform back to the original variables and discretize the new differential operator by linear finite elements. The surprising result is: the difference stencil above is generated!

That is, the SDFEM is closely related to the addition of artificial diffusion in the streamline direction.   □

For C^0 elements one does not have

$$v_h + \beta b\nabla v_h \in H^1,$$

so the streamline diffusion method has a certain nonconforming character and it is necessary to define it carefully.

Let us define the bilinear form

$$A(w,v) := \varepsilon(\nabla w, \nabla v) + (b\nabla w + cw, v) + \sum_K \beta_K(-\varepsilon\Delta w + b\nabla w + cw, b\nabla v)_K,$$

where $(\cdot,\cdot)_K$ is the $L_2(K)$ inner product on each mesh element K and the parameter β_K is specified element-wise, and the linear form

$$F(v) := (f,v) + \sum_K \beta_K(f, b\nabla v)_K.$$

Then the SDFEM is defined by: Find $u_h \in V_h$ such that

$$A(u_h, v_h) = F(v_h) \quad \text{for all } v_h \in V_h. \tag{3.15}$$

If $\beta_K = 0$ for all K, the SDFEM reduces to the standard Galerkin method; for $\beta_K > 0$ the additional terms yield improved stability, as we shall see. This method is *consistent* if $u \in H^2(\Omega)$ because then

$$A(u, v_h) = F(v_h) \quad \text{for all } v_h \in V_h.$$

This implies the Galerkin-orthogonality property

$$A(u - u_h, v_h) = 0 \quad \text{for all } v_h \in V_h,$$

which is very useful in the error analysis of the SDFEM.

We now analyse the streamline diffusion finite element method for linear elements while assuming for simplicity that

$$\beta_K = \beta \quad \forall K, \quad c = \text{constant} > 0, \quad \text{div } b = 0. \tag{3.16}$$

Let us denote by $e := u - u_h$ the error of the method and by $\eta := \Pi_h u - u$ the difference between the exact solution and some projection of u into the finite element space V_h. Then $A(e, v_h) = 0$ for all $v_h \in V_h$ and it follows as usual that

$$A(e, e) = A(e, \eta). \tag{3.17}$$

The definition of the bilinear form yields

$$\begin{aligned} A(e, e) &= -\varepsilon(\triangle u, \beta \nabla e) + \varepsilon\|\nabla e\|^2 + \beta\|b\nabla e\|^2 + (1 + \beta c)(b\nabla e, e) + c(e, e) \\ &= -\varepsilon(\triangle u, \beta \nabla e) + \varepsilon\|\nabla e\|^2 + \beta\|b\nabla e\|^2 + c\|e\|^2 \end{aligned} \tag{3.18}$$

because $\text{div } b = 0$ implies that $(b\nabla e, e) = 0$. On the other hand we have

$$\begin{aligned} A(e, \eta) &= -\varepsilon(\triangle u, \beta b \nabla \eta) + \varepsilon(\nabla e, \nabla \eta) + (b\nabla e, \beta b \nabla \eta) + (b\nabla e, \eta) \\ &\quad + (ce, \beta b \nabla \eta) + (ce, \eta). \end{aligned}$$

Now we apply several times the inequality

$$ab \le \frac{\alpha}{2}a^2 + \frac{1}{2\alpha}b^2$$

with an appropriate chosen positive α:

$$\begin{aligned} A(e, \eta) &\le -\varepsilon(\triangle u, \beta b \nabla \eta) + \frac{\alpha_1}{2}\varepsilon\|\nabla e\|^2 + \frac{\alpha_2}{2}\beta\|b\nabla e\|^2 + \frac{\alpha_3}{2}\|b\nabla e\|^2 \\ &\quad + \frac{\alpha_4 + \alpha_5}{2}c\|e\|^2 + \frac{1}{2\alpha_1}\varepsilon\|\nabla \eta\|^2 + \frac{1}{2\alpha_2}\beta\|b\nabla \eta\|^2 + \frac{1}{2\alpha_3}\|\eta\|^2 \\ &\quad + \frac{\beta^2}{2\alpha_4}\|b\nabla \eta\|^2 + \frac{1}{2\alpha_5}c\|\eta\|^2. \end{aligned}$$

The choices $\alpha_1 = 1, \alpha_2 = 1/2, \alpha_3 = \beta/2$ and $\alpha_4 = \alpha_5 = 1/2$ yield in the interesting case $\beta \le 1$

$$\begin{aligned} A(e, \eta) &\le -\varepsilon(\triangle u, \beta b \nabla \eta) + \frac{1}{2}\Big(\varepsilon\|\nabla e\|^2 + \beta\|b\nabla e\|^2 + c\|e\|^2\Big) + \\ &\quad + \frac{\varepsilon}{2}\|\nabla \eta\|^2 + 2\beta\|b\nabla \eta\|^2 + \Big(c + \frac{1}{\beta}\Big)\|\eta\|^2. \end{aligned}$$

Recalling (3.17) and (3.18), we obtain

$$\begin{aligned} \varepsilon\|\nabla e\|^2 + \beta\|b\nabla e\|^2 + c\|e\|^2 &\le 2\varepsilon(\triangle u, \beta b \nabla e) - 2\varepsilon(\triangle u, \beta b \nabla \eta) + \\ &\quad + \varepsilon\|\nabla \eta\|^2 + 4\beta\|b\nabla \eta\|^2 + 2\Big(c + \frac{1}{\beta}\Big)\|\eta\|^2. \end{aligned}$$

Now assume that $u \in H^2(\Omega)$, so we can apply standard estimates for the interpolation error for linear elements:

$$\|\eta\| \le Ch^2\|u\|_2, \qquad \|\nabla\eta\| \le C\,h\|u\|_2.$$

This gives us

$$\varepsilon\|\nabla e\|^2 + \beta\|b\nabla e\|^2 + c\|e\|^2 \le C\Big[\varepsilon\beta\|u\|_2\|b\nabla e\| + \varepsilon\beta h\|u\|_2^2 +$$

$$+ \varepsilon h^2\|u\|_2^2 + 4\beta h^2\|u\|_2^2 + 2\Big(c + \frac{1}{\beta}\Big)h^4\|u\|_2^2\Big].$$

The estimate

$$\varepsilon\beta\|u\|_2\|b\nabla e\| \le \frac{\beta}{2}\|b\nabla e\|^2 + \frac{\beta}{2}\varepsilon^2\|u\|_2^2$$

leads finally to

$$\varepsilon\|\nabla e\|^2 + \beta\|b\nabla e\|^2 + c\|e\|^2 \le C\|u\|_2^2\Big[\beta\varepsilon(\varepsilon + h) + h^2(\varepsilon + \beta) + \Big(c + \frac{1}{\beta}\Big)h^4\Big].$$

Setting $\beta = \beta^* h$ for some constant β^*, we get

Theorem 6.25. *Assume that the solution of (3.11) lies in $H^2(\Omega)$ and that we are in the convection-dominated case $\varepsilon < Ch$.*
Then the choice $\beta = \beta^* h$ of the streamline diffusion parameter β leads to the following error estimates for the SDFEM with linear elements:

$$\|u - u_h\|_0 \le Ch^{3/2}\|u\|_2,$$

$$\varepsilon^{1/2}\|u - u_h\|_1 \le Ch^{3/2}\|u\|_2,$$

$$\|b\nabla(u - u_h)\|_0 \le Ch\|u\|_2.$$

Here the constants C are independent of ε and h.

These estimates clearly demonstrate the success of the SDFEM stabilization because it is impossible to prove such estimates, with constants independent of ε, for a Galerkin method on a standard mesh. A comparison with the upwind method (see Theorem 6.23) shows that the SDFEM has improved convergence rates for the derivative in the streamline direction. On the other hand, as already pointed out, the matrix of the discrete SDFEM problem is not inverse monotone.

Note that the results of Theorem 6.25 can be generalized to higher-order elements. If V_h contains all polynomials of degree k, then [RST96] one has

$$\|u - u_h\|_0 \le Ch^{k+\frac{1}{2}}\|u\|_{k+1},$$

$$\varepsilon^{1/2}\|u - u_h\|_1 \le Ch^{k+\frac{1}{2}}\|u\|_{k+1}$$

$$\text{and} \quad \|b\nabla(u - u_h)\|_0 \le Ch^k\|u\|_{k+1}.$$

These error estimates for SDFEM contain the factor $\|u\|_{k+1}$ which depends badly on ε in the sense that $\|u\|_{k+1} \to \infty$ as $\varepsilon \to 0$. Thus the method is not uniformly convergent. But a more sophisticated analysis yields local estimates of the form

$$\||u - u_h\||_{\Omega'} \le Ch^{k+\frac{1}{2}} \|u\|_{k+1,\Omega''}$$

on subdomains $\Omega' \subset \Omega''$ away from layers, with

$$\||w\||^2 := \|w\|_0^2 + \varepsilon|w|_1^2 + h\|b\nabla w\|_0^2.$$

These local estimates guarantee uniform convergence in subdomains on which certain derivatives of the exact solution are bounded, uniformly with respect to ε.

From the mathematical point of view it is interesting to ask: in the two-dimensional case, are there uniformly convergent methods (in the maximum norm) on standard meshes?
Just like the situation for parabolic problems that are 1D in space and have parabolic boundary layers in their solutions, for elliptic problems with parabolic boundary layers in their solutions it is *impossible* to devise a method on standard meshes that attains uniform convergence in the discrete maximum norm. Nevertheless, for some problems with exponential layers exponential fitting can be generalized to the two-dimensional case, as we now outline.

Let us consider the model problem

$$-\varepsilon\triangle u + b\nabla u + cu = f \quad \text{in} \quad \Omega = (0,1)^2, \tag{3.19a}$$
$$u = 0 \quad \text{on} \quad \Gamma, \tag{3.19b}$$

with constant coefficients $b = (b_1, b_2)$ and c, with $c > 0$. We choose a uniform mesh of squares with mesh width h and aim to achieve uniform convergence via a Petrov-Galerkin method. The crucial question is how to choose the ansatz and test functions.

As we observed in the one-dimensional case, it seems helpful to use as test functions exponential splines that solve some adjoint problem. Therefore we define $\phi_i(x)$ to be the solution of

$$-\varepsilon\phi_i'' - b_1\phi_i' = 0 \quad \text{in} (0,1) \quad \text{with the exception of the mesh points,}$$
$$\phi_i(x_j) = \delta_{i,j},$$

and define $\psi_k(y)$ to be the solution of

$$-\varepsilon\psi_k'' - b_2\psi_k' = 0 \quad \text{in} (0,1) \quad \text{with the exception of the mesh points,}$$
$$\psi_i(y_k) = \delta_{k,j}.$$

The test space is finally defined to be the span of the basis functions $\{\phi_i\psi_k\}$ obtained by multiplying the above 1-dimensional test functions. We call this test space the space of L^*-splines.

Remark 6.26. In the case of variable coefficients it is a little more complicated to define a suitable basis for the test space. Suppose that the south-west corner of each rectangular element $R_{i,j}$ has coordinates (x_i, y_j). We define one-dimensional exponential splines by freezing coefficients at the midpoints of edges. For instance, the spline $\psi_{i+1/2,j}$, for $x \in (x_i, x_{i+1})$ and $y = y_j$, is a solution of the equation

$$-\varepsilon\psi'' - b_1(x_{i+1/2}, y_j)\psi' + c(x_{i+1/2}, y_j)\psi = 0.$$

Using splines of this type, let the function $\psi^0_{i+1/2,j}$ satisfy $\psi(x_i) = 1$ and $\psi(x_{i+1}) = 0$, while the function $\psi^1_{i+1/2,j}$ satisfies $\psi(x_i) = 0$ and $\psi(x_{i+1}) = 1$. Then define the basis functions for our two-dimensional exponentially fitted spline space by

$$\varphi_{ij}(x, y) = \begin{cases} \psi^1_{i-1/2,j}(x)\psi^1_{i,j-1/2}(y) & \text{in } R_{i-1,j-1}, \\ \psi^0_{i+1/2,j}(x)\psi^1_{i,j-1/2}(y) & \text{in } R_{i,j-1}, \\ \psi^0_{i+1/2,j}(x)\psi^0_{i,j+1/2}(y) & \text{in } R_{i,j}, \\ \psi^1_{i-1/2,j}(x)\psi^0_{i,j+1/2}(y) & \text{in } R_{i-1,j}. \end{cases}$$

The construction of exponential splines on triangular meshes is, for instance, described in [104], [120].   □

Analogously to the above L^*-Splines, one can define L-splines using the original differential operator. Surprisingly, unlike the one-dimensional case, the analysis of the discrete problem for the Petrov-Galerkin method with L-splines as ansatz functions and L^*-Splines as test functions shows that the discrete problem is unstable. Thus possible choices of ansatz and test functions that may yield uniform convergence are the following:

ansatz functions	test functions
L-splines	L-splines
bilinear	L^*-splines

Equivalent combinations are (L^*-splines, L^*-splines) and (L^*-splines, bilinear elements). In [66] the reader can find the result of numerical experiments for these methods that indicate uniform convergence. But at present a complete analysis in a strong norm is known only for the Galerkin method with L-splines; the analysis of Petrov-Galerkin methods is more difficult and remains more or less open.

Let us sketch the derivation of an error estimate for the Galerkin method with exponential L-splines $\phi_i(x)\psi_j(y)$. We study the error in the ε-weighted norm

$$\|u\|_\varepsilon^2 := \varepsilon|u|_1^2 + \|u\|_0^2.$$

Set $V = H_0^1(\Omega)$. The bilinear form

$$a(u, v) := \varepsilon(\nabla u, \nabla v) + (b\nabla u + cu, v) \tag{3.20}$$

is uniformly V-elliptic with respect to this norm:

$$a(u, u) \geq \alpha \|u\|_\varepsilon^2 \qquad \text{(constant } \alpha > 0, \text{ independent of } \varepsilon\text{),} \qquad (3.21)$$

provided $c - (1/2) \operatorname{div} b > 0$ on $\bar\Omega$. For simplicity we take b to be constant in what follows.

As we saw in the one-dimensional case, the bilinear form is not uniformly bounded with respect to the ε-weighted norm. Consequently a convergence bound does not follow immediately from the interpolation error; a more sophisticated approach is needed.

We start from the inequality

$$\alpha \|u - u_h\|_\varepsilon^2 \leq a(u - u_h, u - u_h) = a(u - u_h, u - u_I). \qquad (3.22)$$

Here the nodal interpolant u_I is defined by

$$u_I(x, y) := \sum_i \sum_j u(x_i, y_j) \phi_i(x) \psi_j(y). \qquad (3.23)$$

To proceed further, we need estimates of the interpolation error; in fact one can prove

Lemma 6.27. *Assume that $f(0,0) = f(1,0) = f(0,1) = f(1,1) = 0$. Then the error between the solution u of (3.19) and its nodal interpolant by L-splines is bounded by*

$$\|u - u_I\|_\infty \leq Ch, \qquad (3.24a)$$

$$\|u - u_I\|_\varepsilon \leq Ch^{1/2}. \qquad (3.24b)$$

Sketch of the proof: First we prove the corresponding statement in the one-dimensional case where $-\varepsilon u'' + bu' + cu = f$. On the interval (x_{i-1}, x_i), let us define

$$Mz := -\varepsilon \frac{d^2 z}{dx^2} + b \frac{dz}{dx}.$$

Then $M\phi = 0$ for each of our splines, so

$$|M(u - u_I)| = |f - cu| \leq C$$

and

$$(u - u_I)|_{x_{i-1}} = (u - u_I)|_{x_i} = 0.$$

Hence, invoking a comparison principle with the barrier function $\Phi_1(x) = C(x - x_{i-1})$ and a sufficiently large C yields

$$|(u - u_I)(x)| \leq Ch \quad \text{for } x \in [x_{i-1}, x_i].$$

The more elaborate barrier function $\Phi_2(x) = C(x - x_{i-1})(1 - e^{-b(x_i - x)/\varepsilon})$ gives the sharper bound

$$|(u - u_I)(x)| \leq C(x - x_{i-1})(1 - e^{-b(x_i - x)/\varepsilon}).$$

To estimate the interpolation error in the $\| \cdot \|_\varepsilon$ norm we start from

$$\alpha \|u - u_I\|_\varepsilon^2 \leq a(u - u_I, u - u_I)$$
$$= \varepsilon((u - u_I)', (u - u_I)') + (b(u - u_I)', u - u_I)$$
$$+ (c(u - u_I), u - u_I).$$

After integration by parts, we have

$$\alpha \|u - u_I\|_\varepsilon^2 \leq \sum_i \int_{x_{i-1}}^{x_i} [-\varepsilon(u - u_I)'' + b(u - u_I)'](u - u_I) + (c(u - u_I), u - u_I).$$

But $-\varepsilon(u - u_I)'' + b(u - u_I)' = f - cu$ and $\|u - u_I\|_\infty \leq Ch$, so it follows that

$$\|u - u_I\|_\varepsilon \leq Ch^{1/2}.$$

The proof for the two-dimensional case is similar; see [94]. An extra difficulty here is that additional terms along the edges of rectangles arise from integration by parts and must be estimated. To do this one needs a priori information about the derivatives of the solution u of (3.19), and the compatibility conditions assumed enable us to derive this information. ∎

Remark 6.28. The above estimate for the interpolation error in the ε-weighted H^1 norm is of optimal order. For consider the problem

$$-\varepsilon z'' + bz' = w, \quad z(0) = z(1) = 0$$

with constant b and w. Then one can calculate the interpolation error exactly, obtaining

$$\|z - z_I\|_\varepsilon = \frac{w}{b}\varepsilon^{1/2}\left[\frac{bh}{2\varepsilon} \coth\left(\frac{bh}{2\varepsilon} - 1\right)\right]^{1/2}.$$

Set $\rho = bh/2\varepsilon$. For small ρ the interpolation error behaves like $\varepsilon^{1/2}\rho^{1/2}$, i.e., like $h^{1/2}$. □

To estimate the discretization error, we need still a technical result—an inverse inequality that relates the L_1 and L_2 norms of the derivative of any function from our discrete space of L-splines. This is given in [94]:

Lemma 6.29. *For every function $v_h \in V_h$ one has*

$$\left\|\frac{d}{dx}v_h\right\|_{L_1} \leq Ch^{-1/2}\varepsilon^{1/2}\left\|\frac{d}{dx}v_h\right\|_{L_2},$$

where the constant C is independent of v_h, h and ε.

All the ingredients for estimating the discretization error are now in our possession. We start from (3.22) and

$$a(u - u_h, u - u_I) = \varepsilon(\nabla(u - u_h), \nabla(u - u_I)) + (b\nabla(u - u_h), u - u_I)$$
$$+ (c(u - u_h), u - u_I).$$

The first and third terms on the right-hand side are estimated as follows. First we apply a Cauchy-Schwarz inequality and obtain, for instance,

$$|\varepsilon(\nabla(u - u_h), \nabla(u - u_I))| \le \varepsilon^{1/2}|u - u_h|_1 \, \varepsilon^{1/2}|u - u_I|_1 \le \varepsilon^{1/2}|u - u_h|_1 \, Ch^{1/2}.$$

Then the inequality

$$\alpha_1\alpha_2 \le \frac{\gamma_1\alpha_1^2}{2} + \frac{\alpha_2^2}{2\gamma_1}$$

gives

$$|\varepsilon(\nabla(u - u_h), \nabla(u - u_I))| \le C(\gamma, \alpha)h + \gamma\alpha\varepsilon|u - u_h|_1^2.$$

Provided γ is small enough, we can move the term $\varepsilon|u - u_h|_1^2$ to the left-hand side of (3.22).

It is more troublesome to estimate $(b\nabla(u - u_h), u - u_I)$. First we have

$$(b\nabla(u - u_h), u - u_I) = (b\nabla(u - u_I), u - u_I) + (b\nabla(u_I - u_h), u - u_I).$$

Integrating, one sees that $(b\nabla(u - u_I), u - u_I) = 0$, so

$$(b\nabla(u - u_h), u - u_I) = (b\nabla(u_I - u_h), u - u_I).$$

Now the application of Lemmas 6.27 and 6.29 yields

$$|(b\nabla(u - u_h), u - u_I)| \le Ch^{-1/2}\varepsilon^{1/2}\|u_I - u_h\|_1\|u - u_I\|_\infty$$
$$\le Ch^{1/2}\varepsilon^{1/2}(\|u_I - u\|_1 + \|u - u_h\|_1)$$
$$\le Ch + Ch^{1/2}\varepsilon^{1/2}\|u - u_h\|_1$$
$$\le Ch + C(\gamma, \alpha)h + \gamma\alpha\varepsilon\|u - u_h\|_1^2.$$

Assembling our calculations, we have proved

Theorem 6.30. *Let u be a solution of the boundary value problem (3.19) that has only exponential boundary layers, assuming the compatibility conditions of Lemma 6.27. Then the error of the Galerkin method with L-splines on a standard rectangular mesh satisfies the uniform convergence estimate*

$$\|u - u_h\|_\varepsilon \le Ch^{1/2},$$

where the constant C is independent of h and ε.

Detailed proofs for problems with variable coefficients and exponential boundary layers can be found in [102] and [Dör98]. In the latter paper there is also the estimate

$$\|u - u_h\|_\infty \le Ch,$$

but its derivation uses an inf-sup inequality whose proof is still open. This paper also describes some ideas for increasing the order of convergence of exponentially fitted methods in the two-dimensional case.

At present the analysis of exponentially fitted methods on triangular meshes is not as well developed as on rectangular meshes. As there is the additional drawback that for problems with strong parabolic boundary layers it is impossible to have a fitted method that is uniformly convergent in the maximum norm, we terminate our discussion of exponential fitting in 2D. From the practical point of view, it is more attractive to use adapted meshes and standard finite element spaces.

6.3.3 Layer-adapted Meshes

Consider the following model problem, which has exponential boundary layers at $x = 0$ and $y = 0$:

$$-\varepsilon \triangle u - b \cdot \nabla u + cu = f \quad \text{in } \Omega = (0,1)^2, \quad u = 0 \text{ on } \Gamma = \partial\Omega \qquad (3.25)$$

with

$$(b_1, b_2) > (\beta_1, \beta_2) > 0 \quad \text{and} \quad c + (\text{div } b)/2 \ge \omega > 0.$$

We assume that one can obtain an S-type decomposition (3.6) of u into smooth and exponential layer components that satisfy the bounds (3.7). Problems with parabolic boundary layers can be handled analogously if we assume the existence of an suitable S-type decomposition.

Let us use a piecewise constant layer-adapted S-mesh. As $\Omega = (0,1)^2$, its construction is simple: the mesh is a tensor product of one-dimensional S-meshes in the x and y-directions with transition points τ_x and τ_y defined by (1.34) with $\beta = \beta_1$ in the x-direction and $\beta = \beta_2$ in the y-direction.

Remark 6.31. If the geometry of the given domain is more complicated, then it is more difficult to construct layer-adapted meshes. But if the layer structure is known then the construction of layer-adapted meshes is possible; see [Mel02], for instance.   □

For the upwind difference scheme whose solution is $\{u_{ij}\}$ on an S-mesh with N subintervals in each coordinate direction, one can prove, similarly to the one-dimensional case, that

$$|u(x_i, y_j) - u_{ij}^N| \le C N^{-1} \ln N.$$

The same error bound is valid for the upwind finite volume method on an S-mesh [86].

We shall study in detail linear and bilinear finite elements on an S-mesh. First consider the interpolation error.

Lemma 6.32. *For the solution u of problem (3.25), the interpolation error for nodal linear or bilinear interpolants u^I on an S-mesh with $\sigma = 2$ in (1.34) satisfies*

$$\|u - u^I\|_\infty \le C(N^{-1}\ln N)^2, \qquad \|u - u^I\|_0 \le CN^{-2}, \qquad (3.26\text{a})$$
$$\varepsilon^{1/2}|u - u^I|_1 \le CN^{-1}\ln N. \qquad (3.26\text{b})$$

All constants here are independent of ε and N.

Proof: We use the solution decomposition (3.6) and study separately the interpolation error for the smooth and layer parts. Moreover, we split Ω into several subdomains: $\Omega_g := [\tau_x, 1] \times [\tau_y, 1]$ is characterized by a mesh width of order $O(N^{-1})$ in both coordinate directions, $\Omega_f := [0, \tau_x] \times [0, \tau_y]$ has the extremely small mesh width $O(\varepsilon N^{-1}\ln N)$ in both directions, and the remainder of the domain, $\Omega_a := \Omega \backslash (\Omega_g \cup \Omega_f)$, has highly anisotropic elements where the length/width ratio in each element is not uniformly bounded in ε.

On the anisotropic elements it is vital to use the anisotropic interpolation error estimates that we described in Chapter 4, Section 4.4:

$$\|w - w^I\|_{L_p(\tau)} \le C \left\{ h_x^2 \|w_{xx}\|_{L_p(\tau)} + h_x h_y \|w_{xy}\|_{L_p(\tau)} \right.$$

$$\left. + h_y^2 \|w_{yy}\|_{L_p(\tau)} \right\}, \qquad (3.27\text{a})$$

$$\|(w - w^I)_x\|_0 \le C \left\{ h_x \|w_{xx}\|_0 + h_y \|w_{xy}\|_0 \right\}. \qquad (3.27\text{b})$$

The estimation of the interpolation error is easy for the smooth part of the solution decomposition. Thus we take as an example one of the layer components, namely $E_1(x, y)$, which satisfies

$$\left| \frac{\partial^{i+j}}{\partial x^i \partial y^j} E_1(x, y) \right| \le C \varepsilon^{-i} \exp(-\beta_1 x / \varepsilon).$$

In the region $x \le \tau_x$ we apply (3.27a):

$$\|E_1 - E_1^I\|_\infty \le C \left\{ (\varepsilon N^{-1}\ln N)^2 \varepsilon^{-2} + (\varepsilon N^{-2}\ln N)\varepsilon^{-1} + N^{-2} \right\}$$
$$\le C(N^{-1}\ln N)^2,$$

$$\|E_1 - E_1^I\|_0 \le C \left\{ (\varepsilon N^{-1}\ln N)^2 \varepsilon^{-3/2} + (\varepsilon N^{-2}\ln N)\varepsilon^{-1/2} + N^{-2} \right\}$$
$$\le C N^{-2},$$

where we made the reasonable assumption that $\varepsilon \le N^{-1}$. For $x \ge \tau_x$, the choice $\sigma = 2$ ensures that the layer term E_1 is sufficiently small:

$$\|E_1 - E_1^I\|_0 \le \|E_1 - E_1^I\|_\infty \le 2\|E_1\|_\infty \le C N^{-2}.$$

The other layer components are handled analogously. The error in the $|\cdot|_1$ norm can be derived by imitating the analysis leading to (1.38). ∎

Given these interpolation error estimates, it is straightforward to transfer the ideas of the proof of Theorem 6.18 from the one-dimensional case to the two-dimensional case. One thereby obtains

Theorem 6.33. *Suppose that the boundary value problem (3.25) is discretized using a Galerkin finite element method with linear or bilinear finite elements on an S-mesh with N subintervals in each coordinate direction and $\sigma = 2$. Then one gets the ε-uniform error estimate*

$$\|u - u^N\|_\varepsilon \leq C\,N^{-1}\ln N,$$

where the constant C is independent of ε and N.

When treating parabolic boundary layers one can prove a similar result for an appropriate S-mesh.

Figure 6.13 shows the numerical solution for a test problem with exponential boundary layers. On the left is the solution produced by the Galerkin finite element method with bilinears on a standard mesh—large oscillations are clearly visible—and on the right we solved the same problem using the same method but on an S-mesh.

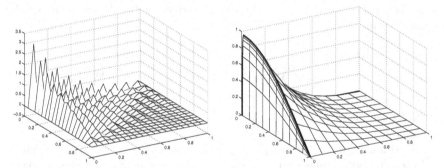

Figure 6.13 Numerical solution of a problem with exponential layers

Optimal L_∞ norm error estimates have not been proved for the linear and bilinear methods of Theorem 6.33. Numerical experiments support the conjecture that one obtains second-order convergence in the subdomain Ω_g defined in the proof of Lemma 6.32. But in the layer region the numerical convergence rate observed for bilinear elements (approximately second order) is twice the rate for linear elements! This surprising fact can be explained through superconvergence properties of bilinear elements [86].

We have proved uniform convergence of the Galerkin method on layer-adapted meshes. It is clear that layer-adapted meshes improve the stability properties of the discretization, compared with standard meshes. Nevertheless in general we do not recommend the use of the pure Galerkin method because the associated stiffness matrix has some undesirable properties: it has eigenvalues with very large imaginary parts so standard iterative methods for solving linear systems of equations have great difficulty in efficiently computing the discrete solution.

In principle, one can combine every stabilization method (e.g., streamline diffusion or discontinuous Galerkin) with layer-adapted meshes, but optimal error estimates for such combinations are few and far between.

6.3.4 Parabolic Problems, Higher-Dimensional in Space

To finish this Chapter on singularly perturbed problems we give a brief discussion of non-stationary problems where the space dimension is greater than one. Let $\Omega \subset \mathbb{R}^n$ with $n \geq 2$ and consider the initial-boundary value problem

$$\frac{\partial u}{\partial t} - \varepsilon \triangle u + b \nabla u + cu = f \quad \text{in} \quad \Omega \times (0, T), \tag{3.28a}$$

$$u = 0 \quad \text{on} \quad \partial \Omega \times (0, T), \tag{3.28b}$$

$$u = u_0(x) \quad \text{for} \quad t = 0. \tag{3.28c}$$

The streamline diffusion finite element method is often used to discretize problems like this. Writing $(\cdot, \cdot)$ for the inner product in $L_2(\Omega)$, let us define the bilinear form

$$A(w, v) : = (w_t, v) + \varepsilon(\nabla w, \nabla v) + (b \nabla w + cw, v)$$
$$+ \delta \sum_K (w_t - \varepsilon \triangle w + b \nabla w + cw, v_t + b \nabla v)_K$$

with, for simplicity, a globally constant streamline diffusion stabilization parameter δ.

Partition $[0, T]$ by the temporal grid $0 = t_0 < t_1 < \cdots < t_M = T$ where $t_m = m\tau$ for some time-step τ. A first possibility is to use continuous space-time elements as ansatz functions. Then the discrete problem is generated by

$$\int_{t_{m-1}}^{t_m} A(u, v) \, dt = \int_{t_{m-1}}^{t_m} (f, v + \delta(v_t + b \nabla v)) \, dt \quad \text{for } m = 1, 2, \dots, M.$$

It seems more natural to follow the standard philosophy of initial-value problems by time-stepping in the discretization of (3.28). The *discontinuous Galerkin method in time* allows to implement this strategy. Let V_h be a finite element space of functions defined on Ω and let U be a polynomial in t with values in V_h on the time interval (t_{m-1}, t_m). Then we can use streamline diffusion in space, combined with discontinuous Galerkin in time, as follows: Find U such that for all $v \in V_h$ and $m = 1, 2, \dots, M$ we have

$$\int_{t_{m-1}}^{t_m} A(U, v)\, dt + (U_+^{m-1}, v_+^{m-1}) =$$

$$= \int_{t_{m-1}}^{t_m} (f, v + \delta(v_t + b\nabla v))\, dt + (U_-^{m-1}, v_+^{m-1}). \tag{3.29}$$

where $w_-^{m-1}(w_+^{m-1})$ means the limit reached as we approach $t = t_{m-1}$ from below (above). When $m = 1$ one takes $U_-^{m-1} \equiv 0$ and U_+^{m-1} to be some approximation of the initial data u_0. This method allows solutions to be discontinuous across each time level t_m so it is possible to change the spatial mesh when moving from the strip (t_{m-1}, t_m) to the strip (t_m, t_{m+1}). Nävert [Näv82] proved error estimates for (3.29) like those for the stationary case.

The combination of a discontinuous Galerkin discretization in space with the above discontinuous Galerkin approach in time was recently analysed in [49].

Another popular method in the non-stationary case is based on the behaviour of the characteristics of the reduced problem. It discretizes the so-called total (first-order) derivative and is called the *Lagrange-Galerkin method*.

Let us define the vector field $X = X(x, s, t)$ by

$$\frac{dX(x, s, t)}{dt} = b(X(x, s, t), t), \qquad X(x, s, t)|_{t=s} = x. \tag{3.30}$$

Then setting $u^*(x, t) = u(X(x, s, t), t)$, the chain rule yields

$$\frac{\partial u^*}{\partial t} - \varepsilon \triangle u^* + c\, u^* = f.$$

Thus in the strip (t_{m-1}, t_m), with $x = X(x, t_m, t_m)$, it is reasonable to use the following approximation:

$$\frac{\partial u^*}{\partial t} \approx \frac{u^*(x, t_m) - u^*(x, t_{m-1})}{\tau} = \frac{1}{\tau}\left[u(x, t_m) - u(X(x, t_m, t_{m-1}), t_{m-1})\right].$$

Finally, the Lagrange-Galerkin method can be written as:
For $m = 1, 2, \ldots, M$, find $U^m \in V_h$ such that for all $v_h \in V_h$

$$\frac{1}{\tau}\left((U^m - U^{m-1}(X(\cdot, t_m, t_{m-1}), t_{m-1})), v_h\right)$$

$$+\varepsilon(\nabla U^m, \nabla v_h) + (cU^m, v_h) = (f^m, v_h).$$

Error estimates for this method can be found for example in [12]. It is possible to weaken the requirement that the system (3.30) must be solved exactly for the characteristics.

The analysis of discretization methods on layer-adapted meshes for the problem (3.28) is still in its infancy. So far results are available only for combinations of the upwind finite difference method with low-order methods in time that make it possible to apply a discrete maximum principle [Shi92].

Exercise 6.34. Consider the singularly perturbed boundary value problem

$$-\varepsilon u'' + b(x)u' + c(x)u = f(x) \quad \text{on } (0,1), \quad u(0) = u(1) = 0,$$

with $0 < \varepsilon \ll 1$, $b(x) > 0$.

On an equidistant mesh, for the discretization of u'' we use the standard difference stencil and for u' a one-sided stencil (which one?). Investigate stability and consistency while monitoring the dependence on the parameter ε of any constants appearing in the analysis.

Exercise 6.35. Consider the singularly perturbed boundary value problem

$$-\varepsilon u'' + bu' = f \quad \text{on } (0,1), \quad u(0) = u(1) = 0,$$

with constant b and f. Construct the exact difference scheme on an equidistant mesh and compare it with the Iljin scheme.

Exercise 6.36. Construct a three-point scheme of the form

$$r_- u_{i-1} + r_c u_i + r_+ u_{i+1} = f_i$$

on an equidistant mesh to discretize the singularly perturbed boundary value problem

$$-\varepsilon u'' + bu' = f \quad \text{on } (0,1), \quad u(0) = u(1) = 0,$$

with $b > 0$, subject to the requirement that the functions 1, x and $\exp(-b(1 - x)/\varepsilon)$ are discretized exactly. Compare your result with the Iljin scheme.

Exercise 6.37. Consider the singularly perturbed boundary value problem

$$-\varepsilon u'' + bu' = f \quad \text{on } (0,1), \quad u(0) = u(1) = 0,$$

with constant b and f. To discretize this problem use a Petrov-Galerkin method on an equidistant mesh with linear ansatz and quadratic test functions. Write down the discrete problem generated. Discuss the properties of the scheme.

Exercise 6.38. Apply a Petrov-Galerkin method on an equidistant mesh with L-splines as ansatz functions and L^*-splines as test functions to discretize the boundary value problem

$$-\varepsilon u'' + b(x)u' + c(x)u = f(x) \quad \text{on } (0,1), \quad u(0) = u(1) = 0.$$

Compute the discrete problem if the coefficients of the given problem are approximated by piecewise constants on your mesh.

Exercise 6.39. Consider the elliptic boundary value problem

$$-\varepsilon \triangle u + b\nabla u = f \quad \text{in } \quad \Omega \subset \mathbb{R}^2,$$
$$u = 0 \quad \text{on } \quad \Gamma.$$

Discretize the problem using a finite volume method based on Voronoi boxes dual to a given triangulation.

a) Let Ω be polygonal and the triangulation weakly acute. Under which conditions on the free parameters λ_{kl} of the quadrature rule is the coefficient matrix of the discrete problem an M-matrix?

b) Let $\Omega = (0,1)^2$. We triangulate Ω with a uniform three-directional mesh based on the straight lines $x = 0$, $y = 0$, $x = y$. Compare explicitly the discrete problem generated by the finite volume approach with upwind finite differencing.

Exercise 6.40. Consider the boundary value problem

$$-\varepsilon \Delta u - p u_x - q u_y = f \quad \text{in} \quad \Omega = (0,1)^2,$$
$$u = 0 \quad \text{on} \quad \Gamma,$$

and its discretization on an equidistant mesh of squares using the streamline diffusion method with bilinear elements. Here p and q are positive constants.

a) What difference star is generated?

b) Examine whether or not it is possible to choose the streamline diffusion parameter in such a way that the coefficient matrix of the discrete problem generated is an M-matrix.

Exercise 6.41. Discretize the boundary value problem

$$\varepsilon \Delta u + b \nabla u = f \quad \text{in} \quad \Omega = (0,1)^2,$$
$$u = 0 \quad \text{on} \quad \Gamma,$$

for constant $b = (b_1, b_2)$ on a uniform mesh of squares, using a Petrov-Galerkin method with L-Splines ϕ as the ansatz functions. The test functions ψ have the tensor-product form

$$\psi^{i,j}(x,y) = \psi^i(x)\psi_j(y)$$

with

$$\psi^i(x_j) = \delta_{ij}, \quad \psi_i(y_j) = \delta_{ij}$$

and

$$\text{supp}\,(\psi^{i,j}) = [x_{i-1}, x_i] \times [y_{j-1}, y_j]$$

but are otherwise arbitrary. Verify that the difference stencil generated has the form

$$\frac{\varepsilon}{h^2} \sum_{r=i-1}^{i+1} \sum_{q=j-1}^{j+1} \alpha_{rq} u_{rq}$$

with

$$[\alpha_{ij}] = \begin{bmatrix} R_x^- S_y^+ + R_y^+ S_x^- & R_x^c S_y^+ + R_y^+ S_x^c & R_x^+ S_y^+ + R_y^+ S_x^+ \\ R_x^- S_y^c + R_y^c S_x^- & R_x^c S_y^c + R_y^c S_x^c & R_x^+ S_y^c + R_y^c S_x^+ \\ R_x^- S_y^- + R_y^- S_x^- & R_x^c S_y^- + R_y^- S_x^c & R_x^+ S_y^- + R_y^- S_x^+ \end{bmatrix}$$

and
$$R_x^+ = \sigma(\rho_x), \; R_x^- = \sigma(-\rho_x), \quad R_x^c = -(R_x^- + R_x^+),$$
$$\rho_x = b_1 h/\varepsilon, \quad \rho_y = b_2 h/\varepsilon, \quad \sigma(x) = x/(1 - \exp(-x)),$$

with
$$hS_x^+ = (\phi^{i+1}, \psi^i), \; hS_x^c = (\phi^i, \psi^i), \; hS_x^- = (\phi^{i-1}, \psi^i),$$
$$hS_y^+ = (\phi_{j+1}, \psi_j), \; hS_y^c = (\phi_j, \psi_j), \; hS_y^- = (\phi_{j-1}, \psi_j).$$

Exercise 6.42. Discretize the boundary value problem

$$-\varepsilon u'' + bu' = f \quad \text{on } (0,1), \quad u(0) = u(1) = 0,$$

with constant b and f by central differencing on an S-mesh. Does the obtained numerical solution oscillate?

7

Numerical Methods for Variational Inequalities and Optimal Control

7.1 The Problem and its Analytical Properties

Weak formulations of partial differential equations can, as shown in Chapter 3, be written as variational equations. They also give necessary and sufficient optimality conditions for the minimization of convex functionals on some linear subspace or linear manifold in an appropriate function space. In models that lead to variational problems, the constraints that appear have often a structure that does not produce problems posed on subspaces. Then optimality conditions do not lead automatically to variational equations, but may instead be formulated as variational inequalities. In the present section, several important properties of variational inequalities are collected and we sketch their connection to the problems of convex analysis.

Let $(V, \|\cdot\|)$ be a real Hilbert space and G a nonempty closed convex subset of V. Suppose we are given a mapping $F : G \to V^*$, where V^* is the dual space of V. We consider the following abstract problem:

Find $u \in G$ such that

$$\langle Fu, v - u \rangle \geq 0 \qquad \text{for all } v \in G. \tag{1.1}$$

Here $\langle s, v \rangle$ denotes the value of the functional $s \in V^*$ applied to $v \in V$. The relation (1.1) is called a *variational inequality* and u is a solution of this variational inequality.

If G is a linear subspace of V then $v := u \pm z$ is in G for each $z \in G$. This fact and the linearity of $\langle Fu, \cdot \rangle$ together imply that the variational inequality (1.1) is then equivalent to

$$\langle Fu, z \rangle = 0 \qquad \text{for all } z \in G.$$

Thus variational equations can be considered as particular examples of variational inequalities and (1.1) generalizes the variational equation problem in a natural way.

In what follows we consider two simple examples of physical models that lead directly to variational inequalities.

Example 7.1. A membrane is stretched over some domain $\Omega \subset \mathbb{R}^2$ and is deflected by some force having pointwise density $f(x)$. At the boundary Γ the membrane is fixed and in the interior of Ω the deflection is assumed to be bounded from below by a given function g (an obstacle). Then the deflection $u = u(x)$ is the solution of the variational inequality

$$\text{find } u \in G \text{ such that } \int_\Omega \nabla u \nabla (v - u)\, dx \geq \int_\Omega f\,(v - u)\, dx \tag{1.2a}$$
$$\text{for all } v \in G$$

where

$$G := \{\, v \in H_0^1(\Omega) \, : \, v \geq g \quad \text{a.e. in } \Omega \,\}. \tag{1.2b}$$

In this example the operator $F : G \to V^*$ that appears in the abstract problem (1.1) is defined on the whole space $V = H_0^1(\Omega)$ by

$$\langle Fu, v \rangle = \int_\Omega \nabla u \nabla v\, dx - \int_\Omega fv\, dx \qquad \text{for all } u, v \in V. \tag{1.3}$$

Similarly to the relationship between differential equations and variational equations, (1.2) can be considered as a weak formulation of a system of differential inequalities. If the solution u of (1.2) has the additional regularity property that $u \in H^2(\Omega)$, then from (1.2) we can show that u satisfies the conditions

$$\left. \begin{array}{r} -\Delta u \geq f \\ u \geq g \\ (\Delta u + f)(u - g) = 0 \end{array} \right\} \quad \text{in } \Omega, \tag{1.4}$$
$$u|_\Gamma = 0.$$

Conversely, any function $u \in H^2(\Omega)$ that satisfies (1.4) is also a solution of the variational inequality problem (1.2). The system (1.4) can be interpreted as follows: depending on the solution u, the domain Ω can be partitioned into a subset D_1 in which the differential equation $-\Delta u = f$ holds and its complement D_2—the contact zone—in which the solution u touches the given obstacle g, i.e.,

$$D_2 = \{\, x \in \Omega \, : \, u(x) = g(x) \,\}.$$

Figure 7.1 illustrates this situation.

Under this interpretation the problem (1.2) is known in the literature as an *obstacle problem*. As a rule, the differential equation is valid only in some subdomain D_1 that depends upon the solution; the inner boundary $\Gamma_* := \partial D_1 \cap \partial \overline{D}_2$ and the conditions that u has to satisfy there are not known a priori. For this reason such problems are also called *free boundary value problems*. From the matching conditions for piecewise-defined functions we have the equations

$$u|_{\Gamma_*} = g|_{\Gamma_*} \quad \text{and} \quad \left. \frac{\partial u}{\partial n} \right|_{\Gamma_*} = \left. \frac{\partial g}{\partial n} \right|_{\Gamma_*},$$

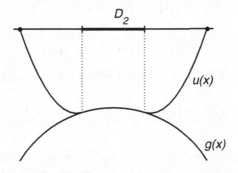

Figure 7.1 Obstacle problem

where n is a unit normal to Γ_*. A detailed and extensive analysis of various models that lead to free boundary value problems can be found in [Fri82].
□

Example 7.2. Once again we start from a physical model and give a weak formulation of the problem as a variational inequality. Consider a domain Ω with boundary Γ. In Ω let $u(x)$ be the difference between the concentration of some substance and a given reference value. Changes in the concentration in the interior of Ω are caused by sources (or sinks) with intensity f and at the boundary by diffusion. The boundary Γ is divided into two disjoint parts Γ_1 and Γ_2. On Γ_1 one has homogeneous Dirichlet boundary conditions while the portion Γ_2 of the boundary is assumed to be semi-permeable. This means that on Γ_2 only a flux directed from outside to inside Ω is possible and this flux occurs only if the concentration just inside the boundary is lower than the given reference concentration. If additional transition phenomena at the boundary are ignored, then this model leads to the following system (see [DL72]):

$$-\Delta u = f \text{ in } \Omega,$$

$$u = 0 \text{ on } \Gamma_1,$$

$$\left.\begin{array}{r} u \geq 0 \\[2mm] \dfrac{\partial u}{\partial n} \geq 0 \\[2mm] \dfrac{\partial u}{\partial n} u = 0 \end{array}\right\} \quad \text{on } \Gamma_2. \tag{1.5}$$

Here, as described above, we have $\Gamma = \Gamma_1 \cup \Gamma_2$ and $\Gamma_1 \cap \Gamma_2 = \emptyset$, while n is the outward-pointing unit normal to subdomain D_1, but the inward-pointing unit normal to D_2.

To formulate this model as a variational inequality, we choose the space

$$V = \{ v \in H^1(\Omega) \ : \ v|_{\Gamma_1} = 0 \} \tag{1.6}$$

and define

$$G = \{ v \in V \ : \ v|_{\Gamma_2} \geq 0 \}. \tag{1.7}$$

Any solution $u \in H^2(\Omega)$ of (1.5) will also satisfy the variational inequality

$$\int_\Omega \nabla u \, \nabla(v - u) \, dx \geq \int_\Omega f \, (v - u) \, dx \qquad \text{for all } v \in G. \tag{1.8}$$

On the other hand, problem (1.8) has a unique solution $u \in G$. Thus this variational inequality provides a weak formulation of (1.5). Problem (1.5), and its weak formulation (1.8), are known as *Signorini's problem*. Since this model can be characterized by an obstacle condition at the boundary it is often described as a problem with a "thin" obstacle. □

For our further abstract analysis we assume that the operator F of (1.1) has the following properties:

- The operator F is strongly monotone over G, i.e., there exists a constant $\gamma > 0$ such that

$$\gamma \|u - v\|^2 \leq \langle Fu - Fv, u - v \rangle \qquad \text{for all } u, v \in G; \tag{1.9}$$

- F is Lipschitz continuous in the sense that for some non-decreasing function $\nu : \mathbb{R}_+ \to \mathbb{R}_+$ one has

$$\|Fu - Fv\|_* \leq \nu(\delta)\|u - v\| \qquad \text{for all } u, v \in G_\delta, \tag{1.10a}$$

where G_δ is defined by

$$G_\delta := \{ v \in G \ : \ \|v\| \leq \delta \}. \tag{1.10b}$$

We make these relatively strong assumptions to simplify our presentation, but they could be relaxed in various ways. For the analysis of variational inequalities under weaker conditions we refer the reader to, e.g., [HHNL88, KS80]. Our aim here is to present the basic principles and ideas that are used in the analysis of (1.1); this will help us later to develop numerical methods and investigate their convergence properties.

First, some useful lemmas are given.

Lemma 7.3. *Under our previous assumptions, there exists a non-decreasing function* $\mu : R_+ \to R_+$ *such that*

$$| \langle Fu, v \rangle | \leq \mu(\|u\|)\|v\| \qquad \text{for all } u, v \in G. \tag{1.11}$$

Proof: Since $G \neq \emptyset$, we can choose $\tilde{v} \in G$. From (1.10) and the definition of the norm $\|\cdot\|_*$ in the dual space V^* it follows that

$$|\langle Fu - F\tilde{v}, v \rangle| \leq \nu(\max\{\|u\|, \|\tilde{v}\|\})\|u - \tilde{v}\| \, \|v\|.$$

Hence

$$|\langle Fu, v \rangle| \leq \nu(\max\{\|u\|, \|\tilde{v}\|\})(\|u\| + \|\tilde{v}\|)\|v\| + |\langle F\tilde{v}, v \rangle|.$$

Now set

$$\mu(s) := \nu(\max\{s, \|\tilde{v}\|\})(s + \|\tilde{v}\|) + \|F\tilde{v}\|_*.$$

This function satisfies the desired inequality (1.11). ∎

Let $(\cdot, \cdot)$ denote the inner product in V.

Lemma 7.4. *Let $Q \subset V$ be nonempty, convex and closed. Then for each $y \in V$ there exists a unique $u \in Q$ such that*

$$(u - y, v - u) \geq 0 \qquad \text{for all } v \in Q. \tag{1.12}$$

The associated projector $P : V \to Q$ defined by $Py := u$ is non-expansive, i.e.

$$\|Py - P\tilde{y}\| \leq \|y - \tilde{y}\| \qquad \text{for all } y, \tilde{y} \in V. \tag{1.13}$$

Proof: For fixed but arbitrary $y \in V$ we consider the variational problem

$$\min_{v \in Q} J(v) := (v - y, v - y). \tag{1.14}$$

Now

$$Q \neq \emptyset \quad \text{and} \quad J(v) \geq 0 \quad \text{for all } v \in V,$$

so $\inf_{v \in Q} J(v) \geq 0$. Let $\{\varepsilon_k\}$ be a monotone sequence that satisfies

$$\varepsilon_k > 0 \text{ for } k = 1, 2, \ldots \qquad \text{and} \qquad \lim_{k \to +\infty} \varepsilon_k = 0.$$

The definition of the infimum implies that we can choose a sequence $\{v^k\} \subset Q$ with

$$J(v^k) \leq \inf_{v \in Q} J(v) + \varepsilon_k, \qquad \text{for } k = 1, 2, \ldots \tag{1.15}$$

It follows that

$$\|v^k - y\|^2 \leq \|z - y\|^2 + \varepsilon_k, \qquad k = 1, 2, \ldots,$$

for any $z \in Q$. Thus the sequence $\{v^k\}$ is bounded. Now the reflexivity of V implies that $\{v^k\}$ is weakly compact. Without loss of generality we may suppose that the entire sequence $\{v^k\}$ is weakly convergent to some $\hat{v} \in V$. Since Q is convex and closed, Q is also weakly closed. Hence $\hat{v} \in Q$. The convexity and continuity of the functional $J(\cdot)$ imply that it is weakly lower semi-continuous (see, e.g., [Zei90]). Putting all these facts together, we obtain

$$J(\hat{v}) \le \liminf_{k \in \mathcal{K}} J(v^k)$$

where $\mathcal{K} = \{1, 2, \dots\}$. Recalling (1.15), we see that $\hat{v}$ is a solution of the variational problem (1.14). Taking into account the properties of the functionals $J(\cdot)$ leads to

$$(\hat{v} - y, v - \hat{v}) \ge 0 \qquad \text{for all } v \in Q, \tag{1.16}$$

which proves (1.12).

Next, we prove the uniqueness of the solution $\hat{v}$. Suppose that $\tilde{v} \in Q$ satisfies

$$(\tilde{v} - y, v - \tilde{v}) \ge 0 \qquad \text{for all } v \in Q.$$

Taking $v = \hat{v}$ here and $v = \tilde{v}$ in (1.16), then adding, we get

$$(\hat{v} - \tilde{v}, \tilde{v} - \hat{v}) \ge 0,$$

which implies that $\tilde{v} = \hat{v}$. That is, the mapping $P : V \to Q$ is uniquely defined.

Finally, let $y, \tilde{y} \in V$ be arbitrary. Then

$$(Py - y,\, v - Py) \ge 0 \qquad \text{for all } v \in Q,$$
$$(P\tilde{y} - \tilde{y},\, v - P\tilde{y}) \ge 0 \qquad \text{for all } v \in Q.$$

Taking $v = P\tilde{y}$ in the first inequality and $v = Py$ in the second, then adding, we arrive at

$$(Py - P\tilde{y} - (y - \tilde{y}), P\tilde{y} - Py) \ge 0.$$

That is,

$$(y - \tilde{y}, Py - P\tilde{y}) \ge (Py - P\tilde{y}, Py - P\tilde{y}).$$

A Cauchy-Schwarz inequality now gives the bound (1.13). ∎

Remark 7.5. The argument used in the proof of Lemma 7.4 can also be used to prove the Riesz representation theorem. See Exercise 7.13.   □

Lemma 7.6. *Let $Q \subset V$ be nonempty, convex and closed. Assume that the operator F is strongly monotone and Lipschitz continuous on Q, with monotonicity and Lipschitz constants $\gamma > 0$ and $L > 0$, respectively. Let $P_Q : V \to Q$ denote the projector onto Q and $j : V^* \to V$ the Riesz representation operator. Then for any fixed parameter r in the interval $\left(0, \dfrac{2\gamma}{L^2}\right)$, the mapping $T : V \to Q$ defined by*

$$T(v) := P_Q(I - r\, j\, F)v \quad \text{for all} \quad v \in V \tag{1.17}$$

is contractive on Q.

Proof: We have (compare the proof of the Lax-Milgram Lemma)

$$\|(I - rjF)v - (I - rjF)\tilde{v}\|^2 =$$
$$= (v - \tilde{v} - rj(Fv - F\tilde{v}), v - \tilde{v} - rj(Fv - F\tilde{v}))$$
$$= \|v - \tilde{v}\|^2 - 2r\langle Fv - F\tilde{v}, v - \tilde{v}\rangle + r^2 \|Fv - F\tilde{v}\|^2$$
$$\leq (1 - 2r\gamma + r^2 L^2) \|v - \tilde{v}\|^2 \qquad \text{for all } v, \tilde{v} \in Q.$$

This inequality (1.13) shows that T_Q is for $r \in \left(0, \dfrac{2\gamma}{L^2}\right)$ contractive on Q. ∎

Next we turn our attention to existence and uniqueness results for the solution of the variational inequality (1.1).

Theorem 7.7. *The variational inequality (1.1) has a unique solution $u \in G$, with*

$$\|u\| \leq \frac{1}{\gamma}\mu(\|\hat{v}\|) + \|\hat{v}\| \quad \text{for any } \hat{v} \in G,$$

where the function $\mu : R_+ \to R_+$ is defined in Lemma 7.3.

Proof: We shall use Lemma 7.6 to construct an suitable contraction mapping, but unlike the proof of the Lax-Milgram lemma, the mapping F may be not globally Lipschitz continuous. To deal with this inconvenience we introduce an additional restriction that depends on a parameter $\delta > 0$. Then using an a priori bound we shall show that this restriction is inessential and can be omitted provided that δ is sufficiently large.

Choose $\delta > 0$ such that the set G_δ defined by (1.10b) is nonempty. As it is the intersection of two closed and convex sets, G_δ is also convex and closed. By Lemma 7.6 the mapping $T_{G_\delta} : G_\delta \to G_\delta$ defined by (1.17) is contractive for each $r \in (2\gamma/\nu(\delta)^2)$. Hence Banach's fixed point theorem guarantees the existence of a unique $u_{r\delta} \in G_\delta$ that satisfies

$$u_{r\delta} = T_{G_\delta} u_{r\delta}. \tag{1.18}$$

The characterization (1.12) of projections given in Lemma 7.4, the definition (1.17) of T_{G_δ} and (1.18) together yield the inequality

$$(u_{r\delta} - (I - rjF)u_{r\delta}, v - u_{r\delta}) \geq 0 \quad \text{for all } v \in G_\delta.$$

That is,

$$r\,(jFu_{r\delta}, v - u_{r\delta}) \geq 0 \qquad \text{for all } v \in G_\delta.$$

But $j : V^* \to V$ is the Riesz representation operator and $r > 0$, so this is equivalent to

$$\langle Fu_{r\delta}, v - u_{r\delta}\rangle \geq 0 \qquad \text{for all } v \in G_\delta. \tag{1.19}$$

Thus $u_{r\delta} \in G_\delta$ is the solution of the variational inequality subject to the additional constraint $\|v\| \leq \delta$.

Let $\hat{v} \in G_\delta$ be arbitrary but fixed. Then (1.19) implies that

$$\langle Fu_{r\delta} - F\hat{v} + F\hat{v}, \hat{v} - u_{r\delta} \rangle \geq 0 \,.$$

Hence, invoking the strong monotonicity assumption (1.9) and Lemma 7.3, we get

$$\|u_{r\delta} - \hat{v}\| \leq \frac{1}{\gamma} \mu(\|\hat{v}\|) \,.$$

Consequently

$$\|u_{r\delta}\| \leq \frac{1}{\gamma} \mu(\|\hat{v}\|) + \|\hat{v}\| \,. \tag{1.20}$$

If the parameter δ is increased we can nevertheless use the same $\hat{v} \in G_\delta$. Thus $\delta > 0$ can be chosen such that

$$\delta > \frac{1}{\gamma} \mu(\|\hat{v}\|) + \|\hat{v}\| \,.$$

Now (1.20) gives

$$\|u_{r\delta}\| < \delta \,.$$

Next we show that this inequality and (1.19) together imply

$$\langle Fu_{r\delta}, v - u_{r\delta} \rangle \geq 0 \qquad \text{for all } v \in G \,. \tag{1.21}$$

For suppose that (1.21) does not hold. Then there exists $\overline{v} \in G$ such that

$$\langle Fu_{r\delta}, \overline{v} - u_{r\delta} \rangle < 0 \,. \tag{1.22}$$

Inequality (1.19) tells us that $\overline{v} \notin G_\delta$, i.e., that $\|\overline{v}\| > \delta$. We select

$$\tilde{v} := (1 - \lambda)u_{r\delta} + \lambda\overline{v} \tag{1.23}$$

where

$$\lambda := \frac{\delta - \|u_{r\delta}\|}{\|\overline{v}\| - \|u_{r\delta}\|} \,.$$

Now $\|u_{r\delta}\| < \delta < \|\overline{v}\|$ implies that $\lambda \in (0, 1)$. That is, $\tilde{v}$ lies on the line segment joining $u_{r\delta}$ and $\overline{v}$, and

$$\|\tilde{v}\| \leq (1 - \lambda)\|u_{r\delta}\| + \lambda\|\overline{v}\| = \delta \,.$$

The convexity of G then yields $\tilde{v} \in G_\delta$. But (1.22) and (1.23) give

$$\langle Fu_{r\delta}, \tilde{v} - u_{r\delta} \rangle = \lambda \langle Fu_{r\delta}, \overline{v} - u_{r\delta} \rangle < 0 \,,$$

which contradicts (1.19). This contradiction shows that (1.21) does hold, i.e., $u_{r\delta}$ is a solution of the variational inequality (1.1).

Uniqueness follows in the usual way from the monotonicity property (1.9). Finally, the a priori bound of the theorem has already been proved in (1.20).

The proof of Lemma 7.4 exploited the relationship between variational problems with constraints and variational inequalities. A generalization of this is also valid (see [KS80], [ET76]):

Lemma 7.8. *Assume that the functional* $J : G \to R$ *is Fréchet differentiable on* G. *If* $u \in G$ *is a solution of the variational problem*

$$\min_{v \in G} J(v), \tag{1.24}$$

then u *satisfies the variational inequality*

$$\langle J'(u), v - u \rangle \geq 0 \quad \text{for all } v \in G. \tag{1.25}$$

If in addition J *is convex, then (1.25) is a sufficient condition for* $u \in G$ *to be a solution of the variational problem (1.24).*

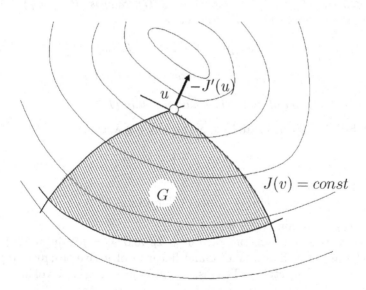

Figure 7.2 Characterization of the optimal solution u

In the finite-dimensional case, Lemma 7.8 has a geometric interpretation that is sketched in Figure 7.2.

In what follows, where in particular we seek to generate an approximation G_h of the set G, a further specification of the representation of G is required, e.g., in the form (1.2b) that occurs in obstacle problems. An equivalent description is given by

$$G = \left\{ v \in H_0^1(\Omega) : \int_\Omega vw\, dx \geq \int_\Omega gw\, dx \quad \text{for all } w \in L_2(\Omega),\ w \geq 0 \right\}. \quad (1.26)$$

Here in $L_2(\Omega)$ the natural (almost everywhere) partial ordering is applied.

Let W be another real Hilbert space. A set $\mathcal{K} \subset W$ is said to be a cone if and only if the implication

$$w \in \mathcal{K} \qquad \Longrightarrow \qquad \lambda w \in \mathcal{K} \quad \text{for all } \lambda \geq 0$$

is valid. Let $\mathcal{K} \subset W$ be a closed convex cone in W. Furthermore, let $b : V \times W \to \mathbb{R}$ be a continuous bilinear form. As a generalization of (1.26), let us assume that the set $G \subset V$ has the representation

$$G = \{ v \in V : b(v, w) \leq g(w) \quad \text{for all } w \in \mathcal{K} \} \qquad (1.27)$$

for some $g \in W^*$. As for mixed variational equations (compare Section 4.7) one has

Lemma 7.9. *Let $G \subset V$ be defined by (1.27). Suppose that $(u, p) \in V \times \mathcal{K}$ satisfies the mixed variational inequalities*

$$\langle Fu, v \rangle + b(v, p) \;=\; 0 \qquad\qquad \text{for all } v \in V, \qquad (1.28a)$$

$$b(u, w - p) \qquad \leq\; g(w - p) \quad \text{for all } w \in \mathcal{K}. \qquad (1.28b)$$

Then u is a solution of the variational inequality (1.1).

Proof: Since $\mathcal{K}$ is convex and $p \in \mathcal{K}$, for any $y \in \mathcal{K}$ we have

$$w := p + y \in \mathcal{K}.$$

Now (1.28b) gives
$$b(u, y) \leq g(y) \quad \text{for all } y \in \mathcal{K}.$$

By the representation (1.27) we get $u \in G$.

Since $\mathcal{K}$ is a cone, we can take $w = 2p$ and $w = p/2$ in (1.28b). With these choices, the bilinearity of b and linearity of g give $b(u, p) \leq g(p)$ and $b(u, p) \geq g(p)$, respectively. That is, $b(u, p) = g(p)$. Hence, invoking (1.28a), we obtain

$$\langle Fu, v - u \rangle + b(v, p) - g(p) = 0 \qquad \text{for all } v \in V.$$

Then $p \in \mathcal{K}$ and (1.27) deliver the inequality

$$\langle Fu, v - u \rangle \geq 0 \qquad \text{for all } v \in G.$$

We see that u is a solution of the variational inequality (1.1). ∎

The subsequent Theorem 7.11 that ensures the existence of a solution $(u, p) \in V \times \mathcal{K}$ of the mixed problem in the case of variational equalities rests upon an extension of the bipolar theorem (see [Goh02]) of closed convex cones. For a given cone $\mathcal{K} \subset W$ we apply the continuous bilinear form $b : V \times W \to \mathbb{R}$ instead of the usual scalar product to define the modified polar and bipolar cone by

$$\mathcal{K}^* := \{\, v \in V \ : \ b(v, w) \leq 0 \ \forall w \in \mathcal{K} \,\} \tag{1.29}$$

and

$$\mathcal{K}^{**} := \{\, w \in W \ : \ b(v, w) \leq 0 \ \forall v \in \mathcal{K}^* \,\}, \tag{1.30}$$

respectively. The definitions (1.29), (1.30) together with the continuity of the bilinear form $b(\cdot, \cdot)$ imply that $\mathcal{K}^{**}$ is a closed convex cone. Further we have

$$\mathcal{K} \subset \mathcal{K}^{**}. \tag{1.31}$$

To be able to prove $\mathcal{K} = \mathcal{K}^{**}$ we assume that the Babuška-Brezzi conditions are fulfilled. In particular, it is assumed that for some $\delta > 0$ we have

$$\sup_{v \in V} \frac{b(v, w)}{\|v\|} \geq \delta \|w\| \quad \text{for all } w \in W. \tag{1.32}$$

Lemma 7.10. *The bipolar cone $\mathcal{K}^{**}$ defined by (1.29), (1.30) equals the original cone $\mathcal{K}$ if and only if $\mathcal{K}$ is closed and convex.*

Proof: Since $\mathcal{K}^{**}$ is a closed convex cone and (1.31) holds we have only to show that for any closed convex cone $\mathcal{K} \subset W$ the inclusion $\mathcal{K}^{**} \subset \mathcal{K}$ holds.

Assume this is not true. Then there exists some element $w^{**} \in \mathcal{K}^{**} \backslash \mathcal{K}$. Because $\mathcal{K}$ is closed and convex it can be strictly separated from w^{**} by some hyperplane, i.e. there exists some $q \in W^*$ such that

$$q(w) \leq 0 \quad \text{for all } w \in \mathcal{K} \qquad \text{and} \qquad q(w^{**}) > 0. \tag{1.33}$$

The supposed Babuška-Brezzi condition (1.32) guarantee that some $\bar{v} \in V$ exists such that
$$b(\bar{v}, w) = q(w) \quad \text{for all } w \in W.$$

(see Theorem 3.48) Thus (1.33) is equivalent to

$$b(\bar{v}, w) \leq 0 \quad \text{for all } w \in \mathcal{K} \qquad \text{and} \qquad b(\bar{v}, w^{**}) > 0. \tag{1.34}$$

The first part implies
$$\bar{v} \in \mathcal{K}^*$$

Hence the second part yields $w^{**} \notin \mathcal{K}^{**}$ which contradicts our assumption. ∎

Theorem 7.11. *Let $G \neq \emptyset$ be defined by (1.27). Assume that (1.32) is satisfied. Then the mixed formulation (1.28) has a solution $(u, p) \in V \times \mathcal{K}$.*

Proof: From $G \neq \emptyset$ and Theorem 7.7, the variational inequality (1.1) has a unique solution $u \in G$. We set

$$Z = \{ v \in V : b(v, w) = 0 \quad \text{for all } w \in W \}.$$

Clearly $u + z \in G$ for all $z \in Z$. Since $Z \subset V$ is a linear subspace, the variational inequality (1.1) implies that

$$\langle Fu, z \rangle = 0 \quad \text{for all } z \in Z.$$

As the Babuška-Brezzi condition (1.32) is valid, there exists $p \in W$ such that (see Theorem 3.48, compare also [BF91])

$$\langle Fu, v \rangle + b(v, p) = 0 \quad \text{for all } v \in V. \tag{1.35}$$

Furthermore, there exists $\tilde{u} \in Z^{\perp}$ such that

$$b(\tilde{u}, w) = g(w) \quad \text{for all } w \in W. \tag{1.36}$$

With (1.29) we obtain

$$v \in G \quad \Longleftrightarrow \quad v - \tilde{u} \in \mathcal{K}^*. \tag{1.37}$$

In addition, (1.36) together with the definition (1.27) of G and $\mathcal{K} \subset W$ imply $\tilde{u} \in G$ and $2u - \tilde{u} \in G$. Now the variational inequality (1.1) with the choices $v = \tilde{u}$ and $v = 2u - \tilde{u}$ yields

$$\langle Fu, \tilde{u} - u \rangle \geq 0 \quad \text{and} \quad \langle Fu, u - \tilde{u} \rangle \geq 0.$$

Hence we have
$$\langle Fu, \tilde{u} - u \rangle = 0.$$

Taking $v = \tilde{u} - u$ in (1.35) now gives

$$b(\tilde{u} - u, p) = 0.$$

Then in (1.36) set $w = p$ and we get

$$b(\tilde{u}, p) = g(p). \tag{1.38}$$

Consequently, using $u \in G$ and (1.27), one has

$$b(u, w - p) \leq g(w - p) \quad \text{for all } w \in \mathcal{K}.$$

It remains only to show that $p \in \mathcal{K}$. From (1.27) and (1.38) we see that

$$b(v - \tilde{u}, p) \leq 0 \quad \text{for all } v \in G$$

and because of (1.37) we have $q \in \mathcal{K}^{**}$. Now, Lemma 7.10 completes the proof. ∎

Like mixed variational equations, saddle-point results can be derived for the problems considered here. If G is given by (1.27), then the related Lagrange functional can be defined as before by

$$L(v, w) := J(v) + b(v, w) - g(w) \qquad \text{for all } v \in V, \, w \in \mathcal{K}. \qquad (1.39)$$

Unlike the case of a variational equation, the cone $\mathcal{K}$ is used here instead of the whole space W. For this situation a pair $(u, p) \in V \times \mathcal{K}$ is called a *saddle point* of the Lagrange functional defined by (1.39) if

$$L(u, w) \leq L(u, p) \leq L(v, p) \qquad \text{for all } v \in V, \, w \in \mathcal{K}. \qquad (1.40)$$

By means of the arguments used in the proof of Lemma 7.9 we can prove, analogously to saddle-point results for variational equations, the following lemma:

Lemma 7.12. *Let $(u, p) \in V \times \mathcal{K}$ be a saddle point of the Lagrange functional $L(\cdot, \cdot)$. Then u is a solution of the variational problem (1.24).*

Proof: As Exercise 7.14.

Exercise 7.13. Use the arguments of Lemma 7.4 to prove the Riesz representation theorem: For each $f \in V^*$ there is a unique $g \in V$ such that

$$\langle f, v \rangle = (g, v) \qquad \text{for all} \quad v \in V.$$

Exercise 7.14. Prove Lemma 7.12.

7.2 Discretization of Variational Inequalities

In this section we shall discuss the discretization of variational inequalities by finite element methods. For the numerical treatment of variational inequalities by finite difference methods and an investigation of their convergence properties, the reader can consult [45], for example.

As a test problem we shall consider the variational inequality (1.1), where the general assumptions made in Section 7.1 are presumed to be satisfied.

Let V_h be a finite-dimensional subspace of V and let $G_h \subset V_h$ be nonempty, closed and convex. To discretize the problem (1.1) we consider the method: *Find some $u_h \in G_h$ such that*

$$\langle F u_h, v_h - u_h \rangle \geq 0 \qquad \text{for all } v_h \in G_h. \qquad (2.1)$$

At first sight this problem seems to be a conforming finite element discretization of (1.1), but a closer look shows that $V_h \subset V$ and $G_h \subset V_h$ do not necessarily imply $G_h \subset G$. In fact this property is inappropriate in many cases: for obstacle problems, the set G defined by (1.2b) will not in general be approximated well by

$$\tilde{G}_h := \{\, v_h \in V_h \;:\; v_h(x) \geq g(x) \;\text{ a.e. in } \Omega \,\}$$

—instead one should use

$$G_h := \{\, v_h \in V_h \;:\; v_h(p^i) \geq g(p^i)\,, \quad i = 1,\dots,N \,\}. \tag{2.2}$$

Here the p^i, $i = 1,\dots,N$, are the interior grid points in the partition of the domain Ω that is used in the finite element discretization V_h.

As $G_h \not\subset G$ is possible, the properties (1.9) and (1.10) are not automatically true on G_h. To compensate for this, one can place additional hypotheses on F. Suppose that $F : V \to V^*$, i.e., F is defined on the whole space V. We also assume property (1.10) in the extended form

$$\|Fu - Fv\|_* \leq \nu(\delta)\|u - v\| \quad \text{for all } u, v \in V \text{ with } \|u\| \leq \delta, \ \|v\| \leq \delta, \tag{2.3}$$

and we suppose that there exists a constant $\gamma_h > 0$ that is independent of the discretization and is such that

$$\gamma_h \|u_h - v_h\|^2 \leq \langle Fu_h - Fv_h, u_h - v_h \rangle \qquad \text{for all } u_h, v_h \in G_h. \tag{2.4}$$

Under all these assumptions Theorem 7.7 can be immediately extended to the discrete problem and we get

Theorem 7.15. *The discrete variational inequality (2.1) has a unique solution $u_h \in G_h$. Furthermore,*

$$\|u_h\| \leq \frac{1}{\gamma_h}\mu(\|\hat{v}_h\|) + \|\hat{v}_h\| \qquad \text{for all } \hat{v}_h \in G_h,$$

where the function μ is defined as in Lemma 7.3 with G replaced by G_h.

Before we proceed with the analysis of the variational inequality (2.1) and consider the convergence behaviour of the solutions u_h of the discrete problems (2.1) to the solution u of the continuous problem (1.1) as $h \to 0$, we study in more detail the structure of the finite-dimensional problems generated.

Let $V_h = \operatorname{span}\{\varphi_i\}_{i=1}^N$, i.e., each $v_h \in V_h$ has the representation

$$v_h(x) = \sum_{i=1}^N v_i \varphi_i(x).$$

In particular $u_h(x) = \sum_{j=1}^N u_j \varphi_j(x)$. Then (2.1) is equivalent to the system

$$\left\langle F\left(\sum_{j=1}^N u_j \varphi_j \right), \sum_{i=1}^n (v_i - u_i)\varphi_i \right\rangle \geq 0 \qquad \text{for all } v_h = \sum_{i=1}^N v_i \varphi_i \in G_h. \tag{2.5}$$

In the case of a linear mapping F and variational equations, the inequalities (2.5) are equivalent to the Galerkin equations, but in the context of variational inequalities a further characterization of G_h is needed to relate (2.5) to a system that can be handled efficiently.

The discretization (2.2) of the obstacle problem (1.2) yields the following complementarity problem:

$$\left.\begin{array}{r}\displaystyle\sum_{j=1}^{N} a(\varphi_j, \varphi_i)u_j \geq f_i, \\[2mm] u_i \geq g_i, \\[2mm] \displaystyle\left(\sum_{j=1}^{N} a(\varphi_j, \varphi_i)u_j - f_i\right)(u_i - g_i) = 0,\end{array}\right\} \text{ for } i = 1, \ldots, N, \qquad (2.6)$$

with $a(\varphi_j, \varphi_i) := \int_\Omega \nabla\varphi_j \nabla\varphi_i \, dx$, $f_i := \int_\Omega f\varphi_i \, dx$ and $g_i := g(p^i)$. We emphasize that complementarity problems require special algorithms for their efficient numerical treatment. In addition, the finite element discretization of variational inequalities generates large sparse matrices whose special structure should be taken into account when designing numerical algorithms for their solution.

If the variational inequality (1.1) can by Lemma 7.8 be considered as an optimality criterion for a certain constrained variational problem, then the discrete variational inequality (2.1) can similarly be interpreted as the optimality criterion

$$\langle J'(u_h), v_h - u_h\rangle \geq 0 \qquad \text{for all } v_h \in G_h \qquad (2.7)$$

for the discrete problem

$$\min_{v_h \in G_h} J(v_h). \qquad (2.8)$$

In the case of the obstacle problem (1.2) this becomes

$$\min J(v_h) := \frac{1}{2}a(v_h, v_h) - f(v_h), \qquad (2.9)$$

where the minimum is taken over the set of $v_h \in V_h$ such that $v_h(p^i) \geq g(p^i)$ for $i = 1, \ldots, N$. Writing $A_h := (a(\varphi_j, \varphi_i))$ for the stiffness matrix and setting $f_h := (f(\varphi_i))$, $\underline{g} := (g(p^i))$ and $\underline{v} := (v_1, \ldots, v_N)$, the problem (2.9) is equivalent to the quadratic programming problem

$$\min_{v_h \in G_h} z(\underline{v}) := \frac{1}{2}\underline{v}^T A_h \underline{v} - f_h^T \underline{v}, \quad \text{where } G_h := \{v \in \mathbb{R}^N, \underline{v} \geq \underline{g}\}. \qquad (2.10)$$

As in the case of the general complementarity problems previously considered, the discrete optimization problems generated by the finite element method have a high dimension and a special structure. Their numerical treatment requires special methods. Penalty methods are a means of approximating variational inequalities by variational equations, which can then be treated efficiently by special methods. This topic will be discussed in more detail in Section 7.3.

As an intermediate result for the present, we estimate the error $\|u - u_h\|$ for fixed discretizations. We have

Lemma 7.16. *Let the operator $F : V \to V^*$ satisfy the conditions (1.9), (1.10), (2.3) and (2.4). Also assume the overlapping monotonicity property*

$$\gamma \|v - v_h\|^2 \leq \langle Fv - Fv_h, v - v_h \rangle \qquad \text{for all } v \in G, \; v_h \in G_h. \qquad (2.11)$$

Then the variational inequalities (1.1) and (2.1) have solutions $u \in G$ and $u_h \in G_h$ respectively, and these solutions are unique. Furthermore, one has the estimate

$$\frac{\gamma}{2} \|u - u_h\|^2 \leq \inf_{v \in G} \langle Fu, v - u_h \rangle + \inf_{v_h \in G_h} \left\{ \langle Fu, v_h - u \rangle + \frac{\sigma^2}{2\gamma} \|v_h - u\|^2 \right\} \quad (2.12)$$

with $\sigma := \nu(\max\{\|u\|, \|u_h\|\})$.

Proof: Existence and uniqueness of the solutions u and u_h of the problems (1.1) and (2.1) are guaranteed by Theorems 7.7 and 7.15. Thus we need only estimate $\|u - u_h\|$. By (2.11) we have

$$\gamma \|u - u_h\|^2 \leq \langle Fu - Fu_h, u - u_h \rangle$$
$$= \langle Fu, u \rangle - \langle Fu, u_h \rangle + \langle Fu_h, u_h \rangle - \langle Fu_h, u \rangle$$
$$= \langle Fu, u - v \rangle - \langle Fu, u_h - v \rangle + \langle Fu_h, u_h - v_h \rangle$$
$$- \langle Fu_h, u - v_h \rangle \qquad \text{for all } v \in G, \; v_h \in G_h .$$

But u and u_h are solutions of the variational inequalities (1.1) and (2.1), respectively, and it follows that

$$\gamma \|u - u_h\|^2 \leq \langle Fu, v - u_h \rangle + \langle Fu, v_h - u \rangle + \langle Fu_h - Fu, v_h - u \rangle .$$

Hence, the definition of σ and property (2.3) give

$$\gamma \|u - u_h\|^2 \leq \langle Fu, v - u_h \rangle + \langle Fu, v_h - u \rangle + \sigma \|u_h - u\| \, \|v_h - u\| .$$

The standard inequality

$$\|u_h - u\| \, \|v_h - u\| \leq \frac{1}{2} \left(\frac{\gamma}{\sigma} \|u - u_h\|^2 + \frac{\sigma}{\gamma} \|v_h - u\|^2 \right)$$

then yields

$$\frac{\gamma}{2} \|u - u_h\|^2 \leq \langle Fu, v - u_h \rangle + \langle Fu, v_h - u \rangle + \frac{\sigma^2}{2\gamma} \|v_h - u\|^2 .$$

As $v \in G$ and $v_h \in G_h$ are arbitrary, this proves the desired estimate. ∎

Remark 7.17. If $G = V$, then condition (1.1) is equivalent to

$$\langle Fu, v \rangle = 0 \qquad \text{for all } v \in V.$$

Thus in this case (2.12) becomes the estimate

$$\|u - u_h\| \leq \frac{\sigma}{\gamma} \inf_{v_h \in V_h} \|v_h - u\| \tag{2.13}$$

and we see that Lemma 7.16 is a generalization of Cea's Lemma.   □

Next, we study the convergence of a sequence of solutions u_h for a family of discretizations (2.1). For this purpose we make the following assumptions about the approximability of G by G_h (compare [GLT81]):

- For each $v \in G$ there exists a sequence $\{v_h\}$, with each $v_h \in G_h$, such that

$$\lim_{h \to 0} v_h = v \,;$$

- $v_h \in G_h$ with $v_h \rightharpoonup v$ as $h \to 0$ implies that $v \in G$.

Theorem 7.18. *Make the same assumptions as in Lemma 7.16 and let the approximation G_h of the set G fulfill the two conditions stated above. Furthermore, assume that the property (2.4) holds uniformly, i.e., for some $\gamma_0 > 0$ one has $\gamma_h \geq \gamma_0$ for all $h > 0$. Then*

$$\lim_{h \to 0} \|u - u_h\| = 0 \,.$$

Proof: The first approximation property of G_h guarantees that

$$\lim_{h \to 0} \inf_{v_h \in G_h} \|v_h - u\| = 0 \,. \tag{2.14}$$

Hence there exists $\hat{v}_h \in G_h$ and $r > 0$ such that

$$\|\hat{v}_h\| \leq r \qquad \text{for all } h > 0. \tag{2.15}$$

Theorem 7.15, (2.15), and the assumption $\gamma_h \geq \gamma_0 > 0$ for all $h > 0$ give the estimate

$$\|u_h\| \leq \frac{1}{\gamma_0} \mu(r) + r \,.$$

It follows that the sequence $\{u_h\}$ is weakly compact. Let $\{\tilde{u}_h\} \subset \{u_h\}$ be a subsequence that converges weakly to some $\tilde{u} \in V$. Then the second approximation property of G_h yields $\tilde{u} \in G$. Thus

$$\inf_{v \in G} \langle Fu, v - \tilde{u}_h \rangle \leq \langle Fu, \tilde{u} - \tilde{u}_h \rangle \,.$$

But now applying this inequality, $\tilde{u}_h \rightharpoonup \tilde{u}$ and (2.12) in Lemma 7.16 gives

$$\lim_{h \to 0} \|\tilde{u} - \tilde{u}_h\| = 0.$$

As the solution $u \in G$ of (1.1) is unique and the sequence $\{u_h\}$ is weakly compact, the convergence of the full sequence $\{u_h\}$ to u follows. ∎

Before we start to investigate quantitative convergence estimates, it should be noted that solutions of variational inequalities have in general only limited regularity. It is thus unrealistic to assume higher-order regularity. Unlike the case of variational equations, the smoothness of the solution u does not depend solely on the underlying domain Ω and the regularity properties of the operator F. In the case of obstacle problems, one may have $u \in H^2(\Omega)$ under additional assumptions (see [KS80]), but in general no higher-order regularity can be expected. Consequently higher-order finite element methods cannot be recommended. In the case of second-order elliptic variational inequalities, appropriate convergence results (first-order convergence) can already be achieved using piecewise linear C^0 elements on triangular grids. In special situations the order of approximation can be improved to $O(h^{3/2})$ by the use of piecewise quadratic elements; see [26].

Recall (2.12). In this inequality we study the terms

$$\inf_{v \in G} \langle Fu, v - u_h \rangle \qquad \text{and} \qquad \inf_{v_h \in G_h} \langle Fu, v_h - u \rangle.$$

Set $V = H_0^1(\Omega)$. Then we have the continuous embedding

$$V = H_0^1(\Omega) \hookrightarrow L_2(\Omega) \hookrightarrow H^{-1}(\Omega) = V^*. \tag{2.16}$$

As an additional regularity hypothesis on the solution u of the variational inequality (1.1), we assume that there exists $\tilde{F}u \in L_2(\Omega)$ such that

$$\langle Fu, v \rangle = (\tilde{F}u, v)_{0, \Omega} \qquad \text{for all } v \in V. \tag{2.17}$$

In particular, this implies that

$$|\langle Fu, v - u_h \rangle| \leq \|\tilde{F}u\|_{0, \Omega} \|v - u_h\|_{0, \Omega} \tag{2.18}$$

and

$$|\langle Fu, v_h - u \rangle| \leq \|\tilde{F}u\|_{0, \Omega} \|v_h - u\|_{0, \Omega}. \tag{2.19}$$

Now approximation results in $L_2(\Omega)$, which yield a higher order of convergence than can be got in $H^1(\Omega)$, are feasible in the analysis.

Theorem 7.19. *Let $\Omega \subset \mathbb{R}^2$ be a polyhedron with boundary Γ. Assume that $f \in L_2(\Omega)$ and $g \in H^2(\Omega)$ with $g|_\Gamma \leq 0$. Furthermore, assume that the solution u of the associated obstacle problem (1.2) satisfies the regularity requirement $u \in H^2(\Omega)$. Suppose that we approximate this problem using piecewise linear C^0 elements and approximate solutions u_h defined by (2.2). Then there exists a constant $C > 0$, which is independent of h, such that*

$$\|u - u_h\| \leq C h.$$

Proof: The property $u \in H^2(\Omega)$ allows us to integrate by parts in (1.3), obtaining

$$\langle Fu, v \rangle = - \int_\Omega (\Delta u + f) v \, dx \qquad \text{for all } v \in V.$$

Thus (2.17) is satisfied with $\tilde{F}u = -(\Delta u + f)$ and the estimates (2.18) and (2.19) can be applied. Let $\Pi_h : V \to V_h$ denote the operator defined by interpolation at the grid points. Since $u \in G$, by (1.2b) and (2.2) we have $\Pi_h u \in G_h$. Interpolation theory yields

$$\|u - \Pi_h u\| \le C h \qquad \text{and} \qquad \|u - \Pi_h u\|_{0,\Omega} \le C h^2$$

for some $C > 0$. Consequently

$$\inf_{v_h \in G_h} \left\{ \langle Fu, v_h - u \rangle + \frac{\sigma^2}{2\gamma} \|v_h - u\|^2 \right\} \le C h^2$$

for the second summand in the right-hand side of (2.12).

Next, we estimate $\inf_{v \in G} \langle Fu, v - u_h \rangle$. Define pointwise

$$\tilde{u}_h := \max\{u_h, g\}.$$

It can be shown (see, e.g., [KS80]) that $\tilde{u}_h \in V$. By (1.2b) we clearly have $\tilde{u}_h \in G_h$. From (2.2) the property $u_h \in G_h$ holds if and only if $u_h \ge \Pi_h g$. We have only the two possibilities

$$\tilde{u}_h(x) = u_h(x) \qquad \text{or} \qquad \tilde{u}_h(x) = g(x).$$

In the second case with $u_h \ge \Pi_h g$ we have $\tilde{u}_h(x) - u_h(x) \le g(x) - (\Pi_h g)(x)$. Thus, we get

$$0 \le (\tilde{u}_h - u_h)(x) \le |(g - \Pi_h g)(x)| \qquad \text{for all } x \in \Omega.$$

Using (2.18), this yields

$$\inf_{v \in G} \langle Fu, v - u_h \rangle \le |\langle Fu, \tilde{u}_h - u_h \rangle|$$

$$\le \|\Delta u + f\|_{0,\Omega} \|\tilde{u}_h - u_h\|_{0,\Omega}$$

$$\le \|\Delta u + f\|_{0,\Omega} \|\tilde{g} - \Pi_h g\|_{0,\Omega}.$$

By general approximation results we therefore have

$$\inf_{v \in G} \langle Fu, v - u_h \rangle \le C h^2$$

since $g \in H^2(\Omega)$. Now an appeal to Lemma 7.16 completes the proof. ∎

Next, we consider a mixed finite element discretization of the variational inequality (1.1). Recall (1.27):

$$G = \{\, v \in V \,:\, b(v,w) \le g(w) \quad \text{for all } w \in \mathcal{K} \,\}$$

where W is a Hilbert space, $\mathcal{K} \subset W$ is a closed convex cone, $b : V \times W \to \mathbb{R}$ is a continuous bilinear form, and $g \in W^*$. We choose some finite-dimensional subspace $W_h \subset W$ and a closed convex cone $\mathcal{K}_h \subset W_h$. Then a natural discretization of G is

$$G_h := \{\, v_h \in V_h \,:\, b(v_h, w_h) \le g(w_h) \quad \text{for all } w_h \in \mathcal{K}_h \,\}. \tag{2.20}$$

If $\mathcal{K}_h$ is a convex polyhedral cone, i.e.,

$$\mathcal{K}_h = \Big\{\, v_h = \sum_{l=1}^{L} \lambda_l s_l \,:\, \lambda_l \ge 0,\ l = 1, \ldots, L \,\Big\} \tag{2.21}$$

where the elements $s_l \in W_h$, $l = 1, \ldots, L$ are finite in number, then the linearity of $g(\cdot)$ and the bilinearity of $b(\cdot, \cdot)$ imply that

$$v_h \in G_h \quad\Longleftrightarrow\quad \sum_{i=1}^{N} b(\varphi_i, s_l) v_i \le g(s_l), \quad l = 1, \ldots, L.$$

That is, the property that $v_h \in V_h$ lies in the set G_h is characterized by a system of a finite number of linear inequalities.

Lemmas 7.9 and 7.12 can be immediately extended to the case of general discrete problems (2.1) with feasible sets G_h defined by (2.20).

Lemma 7.20. *Let $G_h \subset V_h$ be defined by (2.20). Let $(u_h, p_h) \in V_h \times \mathcal{K}_h$ satisfy the mixed variational inequalities*

$$\begin{aligned}
\langle Fu_h, v_h \rangle + b(v_h, p_h) &= 0 && \text{for all } v_h \in V_h, \\
b(u_h, w_h - p_h) &\le g(w_h - p_h) && \text{for all } w_h \in \mathcal{K}_h.
\end{aligned} \tag{2.22}$$

Then u_h is a solution of the variational inequality (2.1).

Lemma 7.21. *Let the original problem be given as a restricted variational problem (1.24) with a feasible set G described by (1.27). If the pair $(u_h, p_h) \in V_h \times \mathcal{K}_h$ is a discrete saddle point of the Lagrange functional defined by (1.39), i.e.,*

$$L(u_h, w_h) \le L(u_h, p_h) \le L(v_h, p_h) \qquad \text{for all } v_h \in V_h,\ w_h \in \mathcal{K}_h, \tag{2.23}$$

then u_h is a solution of the finite-dimensional quadratic programming problem (2.10).

We now analyse the convergence of the mixed finite element discretization (2.22) of the variational inequality (1.1), following the approach developed in [63].

Theorem 7.22. *Consider the problem (1.1) with G defined by (1.27). Assume that the operator $F : V \to V^*$ satisfies (2.3) and is strongly monotone on the whole space V, i.e., there exists a constant $\gamma > 0$ such that*

$$\gamma \|y - v\|^2 \leq \langle Fy - Fv, y - v \rangle \qquad \text{for all } y, v \in V. \qquad (2.24)$$

Furthermore, assume that the corresponding mixed formulation (1.28) has a solution $(u, p) \in V \times \mathcal{K}$.

Suppose that (1.28) is discretized by the mixed finite element method (2.22) with $V_h \subset V$, $W_h \subset W$ and $\mathcal{K}_h \subset \mathcal{K}$, where the discretization satisfies, uniformly in h, the Babuška-Brezzi condition

$$\delta \|w_h\| \leq \sup_{v_h \in V_h} \frac{b(v_h, w_h)}{\|v_h\|} \qquad \text{for all } w_h \in W_h \qquad (2.25)$$

—here $\delta > 0$ is some constant. Suppose that there exists a constant $c_0 > 0$ such that for each $h > 0$ there is a $\tilde{v}_h \in G_h$ with $\|\tilde{v}_h\| \leq c_0$. Then the discrete mixed formulation (2.22) has a solution $(u_h, p_h) \in V_h \times \mathcal{K}_h$ for each $h > 0$. Furthermore, the following estimates are valid:

$$\|u - u_h\|^2 \leq c_1 \{\|u - v_h\|^2 + \|p - w_h\|^2\}$$
$$+ c_2(g(w_h - p) - b(u, w_h - p)) \qquad (2.26)$$
$$\text{for all } v_h \in V_h, \ w_h \in \mathcal{K}_h,$$

$$\|p - p_h\| \leq c(\|u - u_h\| + \|p - w_h\|)$$
$$\text{for all } w_h \in W_h, \qquad (2.27)$$

for some constants $c, c_1, c_2 > 0$ that are independent of h.

Proof: By hypothesis there exists $\tilde{v}_h \in G_h$ with $\|\tilde{v}_h\| \leq c_0$. We know that the set G_h is convex and closed. Thus by Theorem 7.7 the discrete variational inequality

$$\langle Fu_h, v_h - u_h \rangle \geq 0 \qquad \text{for all } v_h \in G_h$$

has a unique solution $u_h \in G_h$, and u_h is uniformly bounded as $h \to 0$. Invoking (2.25), it follows that the mixed formulation (2.22) has a solution $(u_h, p_h) \in G_h \times \mathcal{K}_h$. From (1.28a) and (2.22) we have

$$\langle Fu, v \rangle \ + \ b(v, p) \ = 0 \qquad \text{for all } v \in V,$$
$$\langle Fu_h, v_h \rangle + b(v_h, p_h) = 0 \qquad \text{for all } v_h \in V_h \subset V. \qquad (2.28)$$

Now for arbitrary $v_h \in V_h$ and $w_h \in W_h$ one gets

$$b(v_h, w_h - p_h) = b(v_h, w_h) - b(v_h, p_h) = b(v_h, w_h) + \langle Fu_h, v_h \rangle$$
$$= b(v_h, w_h) + \langle Fu_h, v_h \rangle - \langle Fu, v_h \rangle - b(v_h, p)$$
$$= b(v_h, w_h - p) + \langle Fu_h - Fu, v_h \rangle$$
$$\leq (\beta \|w_h - p\| + \|Fu_h - Fu\|_*) \|v_h\|,$$

where β denotes the continuity constant of the bilinear form b. Again appealing to (2.25), we see that

$$\delta \, \|w_h - p_h\| \leq \beta \, \|w_h - p\| + \|Fu_h - Fu\|_*. \tag{2.29}$$

Now the bound (2.27) follows from the uniform boundedness of u_h combined with the local Lipschitz continuity of the operator F given in (2.3) and a triangle inequality.

Next, we turn to the proof of (2.26). Clearly (1.28b), (2.22) and $\mathcal{K}_h \subset K$ give

$$b(u, p_h - p) \leq g(p_h - p) \quad \text{and} \quad b(u_h, w_h - p_h) \leq g(w_h - p_h) \quad \text{for all } w_h \in \mathcal{K}_h.$$

This implies

$$b(u - u_h, p_h - p) \leq g(w_h - p) - b(u_h, w_h - p) \qquad \text{for all } w_h \in \mathcal{K}_h. \tag{2.30}$$

On the other hand, setting $v = u - u_h$ and $v_h = u_h - u + u$ in (2.28) yields

$$\langle Fu - Fu_h, u - u_h \rangle + b(u - u_h, p - p_h) + \langle Fu_h, u \rangle + b(u, p_h) = 0.$$

From this inequality and (2.24), (2.28) and (2.30) by straight forward calculations we get

$$\begin{aligned}
\gamma \, \|u - u_h\|^2 &\leq g(w_h - p) - b(u, w_h - p) + b(u - u_h, w_h - p) \\
&\quad - \langle Fu - Fu_h, u - v_h \rangle - b(u - v_h, p - p_h) \\
&\leq g(w_h - p) - b(u, w_h - p) + \beta \, \|u - u_h\| \, \|w_h - p\| \\
&\quad + L \, \|u - u_h\| \, \|u - v_h\| + \beta \, \|u - v_h\| \, \|p - p_h\|.
\end{aligned}$$

for all $v_h \in V_h$ and $w_h \in \mathcal{K}_h$. By (2.27) this implies

$$\begin{aligned}
\gamma \, \|u - u_h\|^2 &\leq g(w_h - p) - b(u, w_h - p) + \beta \, \|u - u_h\| \, \|w_h - p\| \\
&\quad + (L + c\beta) \, \|u - u_h\| \, \|u - v_h\| + c\beta \, \|u - v_h\| \, \|w_h - p\|.
\end{aligned}$$

Using the standard inequality $2st \leq \varepsilon s^2 + \frac{1}{\varepsilon} t^2$ for arbitrary $\varepsilon > 0$ and $s, t \in \mathbb{R}$, we obtain

$$\begin{aligned}
\gamma \, \|u - u_h\|^2 &\leq g(w_h - p) - b(u, w_h - p) + \tfrac{\varepsilon}{2}((1 + c)\beta + L) \, \|u - u_h\|^2 \\
&\quad + \tfrac{\beta}{2\varepsilon} \, \|w_h - p\|^2 + \tfrac{L + c\beta}{2\varepsilon} \, \|u - v_h\|^2 + \tfrac{c\beta}{2} (\|u - v_h\|^2 + \|w_h - p\|^2)
\end{aligned}$$

for all $v_h \in V_h$, $w_h \in \mathcal{K}_h$. If $\varepsilon > 0$ is chosen sufficiently small then (2.26) follows. ∎

Remark 7.23. The assumption of Theorem 7.22—that for each $h > 0$ there exists $\tilde{v}_h \in G_h$ with $\|\tilde{v}_h\| \le c_0$—can be justified in several special models that lead to variational inequalities. For instance, in the case of the obstacle problem (1.2) with $g \le 0$, one can make the trivial choice $\tilde{v}_h = 0$ independently of the discretization. Instead of (2.25) it is also possible to use the condition

$$\delta \|v_h\| \le \sup_{w_h \in W_h} \frac{b(v_h, w_h)}{\|w_h\|} \qquad \text{for all } v_h \in V_h \qquad (2.31)$$

for some $\delta > 0$. This inequality is known in the context of mixed variational equations (see [BF91]) and in a sense complements (2.25). $\square$

7.3 Penalty Methods and the Generalized Lagrange Functional for Variational Inequalities

7.3.1 Basic Concept of Penalty Methods

The *penalty method*, which uses penalty functions, is a standard technique in constrained nonlinear programming (see [GLT81], [56]). Its basic principle is to guarantee—in an asymptotic sense—the fulfillment of constraints by including in the objective function an additional term (the penalty) that acts against the optimization goal if constraints are violated. In this way the given constrained programming problem is embedded in a family of variational problems that depend upon some parameter appearing in the penalty term and contain no restrictions, i.e., are unconstrained.

If the given variational inequality (1.1) is associated via Lemma 7.4 with the optimization problem (1.24) as an optimality condition, then an appropriate penalty method can be applied directly to (1.24). The unconstrained auxiliary problems that are generated can be discretized by, e.g., a finite element method. A similar discrete problem is obtained if initially a finite element method is applied to problem (1.24) to generate a finite-dimensional optimization problem that is then treated using a penalty method.

If the given problem (1.1) is not derived from a variational problem (1.24), then an analogue of the penalty method of Section 4.7 can be applied to the regularization of the mixed variational formulation (1.28) and its discretization (2.22). We now consider this approach in more detail. Suppose we are given a mapping $F : V \to V^*$ and a continuous bilinear form $b : V \times W \to \mathbb{R}$. Let the set $G \subset V$ be defined by (1.27) for some $g \in W^*$. Corresponding to the mixed variational formulation (1.28), we consider the following problem: Find $(u_\rho, p_\rho) \in V \times \mathcal{K}$ such that

$$\langle Fu_\rho, v \rangle + b(v, p_\rho) = 0 \qquad\qquad \text{for all } v \in V, \qquad (3.1a)$$

$$b(u_\rho, w - p_\rho) - \frac{1}{\rho}(p_\rho, w - p_\rho) \le g(w - p_\rho) \qquad \text{for all } w \in \mathcal{K}. \qquad (3.1b)$$

Here $(\cdot, \cdot)$ denotes the scalar product in W and $\rho > 0$ is some fixed parameter. To simplify the notation we again identify W^* with W. Let $B : V \to W^*$ be defined by

$$(Bv, w) = b(v, w) \qquad \text{for all } v \in V,\ w \in W. \qquad (3.2)$$

Then (3.1b) can be described equivalently by

$$(p_\rho - \rho(Bu_\rho - g), w - p_\rho) \geq 0 \qquad \text{for all } w \in \mathcal{K}. \qquad (3.3)$$

Let $P_\mathcal{K} : W \to \mathcal{K}$ denote the projection defined in Lemma 7.4 from W onto the closed convex cone $\mathcal{K} \subset W$. Using the property that $\mathcal{K}$ is a cone, inequality (3.3) and the representation of $P_\mathcal{K}$ given in Lemma 7.4 imply that

$$p_\rho = \rho\, P_\mathcal{K}(Bu_\rho - g). \qquad (3.4)$$

Thus (3.1) is equivalent to (3.4) and

$$\langle Fu_\rho, v \rangle + \rho\,(P_\mathcal{K}(Bu_\rho - g), Bv) = 0 \qquad \text{for all } v \in V. \qquad (3.5)$$

This forms a penalty problem that corresponds to (1.1) and (1.27). Thus we have transformed the original variational inequality (1.1) to an auxiliary problem in the form of a variational equation. The positive number ρ is called the penalty parameter. Later we shall discuss the existence and the convergence behaviour of u_ρ as $\rho \to +\infty$, but we begin by discussing a simple example to illustrate the use of (3.5).

Example 7.24. Consider the obstacle problem (1.2). Setting $W = L_2(\Omega)$, the cone $\mathcal{K}$ here has the particular form

$$\mathcal{K} = \{\, w \in L_2(\Omega) :\ w \geq 0 \,\}.$$

From the definition of G in (1.2b), the mapping $B : V = H_0^1(\Omega) \to W = L_2(\Omega)$ is the identity operator. Let $[\cdot]_+ : W \to \mathcal{K}$ be defined by

$$[w]_+(x) := \max\{w(x), 0\} \qquad \text{a.e. in } \Omega. \qquad (3.6)$$

Then the projection operator can be described by

$$P_\mathcal{K} w = [w]_+ \qquad \text{for all } w \in W.$$

Hence the penalty problem (3.5) has the form (see, e.g., [GLT81])

$$\int_\Omega \nabla u_\rho \nabla v\, dx - \rho \int_\Omega [g - u_\rho]_+ v\, dx = \int_\Omega f v\, dx \qquad \text{for all } v \in H_0^1(\Omega). \qquad \square \ (3.7)$$

The next lemma gives an existence result for problem (3.5).

Lemma 7.25. *Assume that the mapping $F : V \to V^*$ is strongly monotone on the full space V. Then for each $\rho > 0$, the penalty problem (3.5) has a unique solution $u_\rho \in V$.*

Proof: We define an operator $S : V \to V^*$ by

$$\langle Su, v \rangle := (P_{\mathcal{K}}(Bu - g), Bv) \qquad \text{for all } u, v \in V. \tag{3.8}$$

Then the penalty problem (3.5) can be written as

$$\langle (F + \rho S)u_\rho, v \rangle = 0 \qquad \text{for all } v \in V. \tag{3.9}$$

First we show that S is Lipschitz continuous. From (3.8) it is clear that

$$\langle Su - S\tilde{u}, v \rangle = (P_{\mathcal{K}}(Bu - g) - P_{\mathcal{K}}(B\tilde{u} - g), Bv) \qquad \text{for all } u, \tilde{u}, v \in V.$$

Now the Cauchy-Schwarz inequality, the linearity of B and Lemma 7.4 give

$$|\langle Su - S\tilde{u}, v \rangle| \le \|P_{\mathcal{K}}(Bu - g) - P_{\mathcal{K}}(B\tilde{u} - g)\| \, \|Bv\|$$
$$\le \|B(u - \tilde{u})\| \, \|Bv\| \qquad \text{for all } u, \tilde{u}, v \in V.$$

By hypothesis the bilinear form $b(\cdot, \cdot)$ is continuous, so (3.2) implies the existence of some constant $\beta > 0$ such that $\|Bv\| \le \beta \|v\|$ for all $v \in V$. It follows that

$$|\langle Su - S\tilde{u}, v \rangle| \le \beta^2 \|u - \tilde{u}\| \, \|v\| \qquad \text{for all } u, \tilde{u}, v \in V,$$

which yields

$$\|Su - S\tilde{u}\|_* \le \beta^2 \|u - \tilde{u}\| \qquad \text{for all } u, \tilde{u} \in V,$$

i.e., the mapping S is Lipschitz continuous.

Next, we investigate the monotonicity properties of S. The definition of the projector $P_{\mathcal{K}}$ in (1.12) clearly implies that

$$(P_{\mathcal{K}}(a + c) - (a + c), w - P_{\mathcal{K}}(a + c)) \ge 0 \qquad \text{for all } w \in \mathcal{K},$$
$$(P_{\mathcal{K}}(b + c) - (b + c), w - P_{\mathcal{K}}(b + c)) \ge 0 \qquad \text{for all } w \in \mathcal{K},$$

for any a, b, $c \in W$. Choose $w = P_{\mathcal{K}}(b + c)$ in the first inequality and $w = P_{\mathcal{K}}(a + c)$ in the second, then add; one then obtains

$$(P_{\mathcal{K}}(a + c) - P_{\mathcal{K}}(b + c) - (a - b), P_{\mathcal{K}}(b + c) - P_{\mathcal{K}}(a + c)) \ge 0.$$

Rearranging,

$$\begin{aligned} P_{\mathcal{K}}(a + c) - P_{\mathcal{K}}(b + c), a - b) &\ge \\ &\ge (P_{\mathcal{K}}(b + c) - P_{\mathcal{K}}(a + c), P_{\mathcal{K}}(b + c) - P_{\mathcal{K}}(a + c)) \ge 0. \end{aligned} \tag{3.10}$$

Let $u, \tilde{u} \in V$ be arbitrary. Taking $a = Bu$, $b = B\tilde{u}$ and $c = -g$ in (3.10), then (3.8) gives

$$\langle Su - S\tilde{u}, u - \tilde{u} \rangle \geq 0.$$

Recalling (2.24), we conclude that

$$\langle (F + \rho S)u - (F + \rho S)\tilde{u}, u - \tilde{u} \rangle \geq \gamma \|u - \tilde{u}\|^2 \qquad \text{for all } u, \tilde{u} \in V. \quad (3.11)$$

Then Theorem 7.7 (with $G = V$) guarantees that (3.5) has a unique solution.
∎

Now we investigate the convergence behaviour of the penalty method (3.5) applied to variational inequalities. In this study we assume that the Babuška-Brezzi conditions (1.32) are satisfied; we shall see later that alternative criteria could be applied to penalty methods applied to optimization problems, but it is then difficult to derive sharp bounds for the rate of convergence of u_ρ independently of the penalty parameter $\rho > 0$.

Theorem 7.26. *Let G be nonempty. Assume that the Babuška-Brezzi conditions (1.32) are satisfied. Let $F : V \to V^*$ be strongly monotone. Then the solution u_ρ of the penalty problem (3.5) converges as $\rho \to +\infty$ to the solution u of the original problem. Furthermore, as $\rho \to +\infty$ the elements $p_\rho \in \mathcal{K}$ defined by (3.1) converge to the component p of the solution of the mixed formulation (1.28). There exists a constant $c > 0$ such that*

$$\|u - u_\rho\| \leq c\rho^{-1} \qquad and \qquad \|p - p_\rho\| \leq c\rho^{-1}. \quad (3.12)$$

Proof: First we show that the u_ρ are bounded for all $\rho > 0$. Our hypotheses justify the application of Lemma 7.25, which tells us that for each $\rho > 0$ the penalty problem has a unique solution u_ρ. By (3.5) we have

$$\langle Fu_\rho, v \rangle + \rho \left(P_\mathcal{K}(Bu_\rho - g), Bv \right) = 0 \qquad \text{for all } v \in V.$$

From Theorem 7.11 it follows that there exists $p \in \mathcal{K}$ such that

$$\langle Fu, v \rangle + b(v, p) = 0 \qquad \text{for all } v \in V.$$

Recalling the definition of $B : V \to W^*$ in (3.2), subtraction yields

$$\langle Fu_\rho - Fu, v \rangle + (\rho P_\mathcal{K}(Bu_\rho - g) - p, Bv) = 0 \qquad \text{for all } v \in V.$$

Setting $v = u_\rho - u$, we obtain

$$\langle Fu_\rho - Fu, u_\rho - u \rangle + \rho(P_\mathcal{K}(Bu_\rho - g) - P_\mathcal{K}(Bu - g), B(u_\rho - u))$$
$$= (p - \rho P_\mathcal{K}(Bu - g), B(u_\rho - u)). \quad (3.13)$$

Now $u \in G$, so we have

$$b(u, w) \leq g(w) \qquad \text{for all } w \in \mathcal{K}$$

and consequently

$$(0 - (Bu - g), w - 0) \geq 0 \qquad \text{for all } w \in \mathcal{K}.$$

Hence Lemma 7.4 implies that $P_{\mathcal{K}}(Bu - g) = 0$. Substituting this into (3.13) and invoking the monotonicity inequality (3.11) that was derived during the proof of Lemma 7.25, we arrive at the inequality

$$\gamma \|u_\rho - u\|^2 \leq \beta \|p\| \|u_\rho - u\|.$$

That is,

$$\|u_\rho - u\| \leq \frac{\beta}{\gamma} \|p\|.$$

It follows that the family $\{u_\rho\}_{\rho>0}$ is bounded.

Next,

$$\langle Fu_\rho, v \rangle + b(v, p_\rho) = 0 \qquad \text{for all } v \in V$$

and the Babuška-Brezzi condition (1.32) then gives

$$\delta \|p_\rho\| \leq \sup_{v \in V} \frac{b(v, p_\rho)}{\|v\|} = \|Fu_\rho\|_*.$$

The boundedness of $\{u_\rho\}_{\rho>0}$ and the continuity of F now show that the set $\{p_\rho\}_{\rho>0}$ is also bounded.

After these preliminary results we turn to the convergence estimates. From

$$\langle Fu_\rho, v \rangle + b(v, p_\rho) = 0 \qquad \text{for all } v \in V,$$

$$\langle Fu, v \rangle + b(v, p) = 0 \qquad \text{for all } v \in V,$$

the linearity of $b(v, \cdot)$ gives

$$\langle Fu - Fu_\rho, v \rangle + b(v, p - p_\rho) = 0 \qquad \text{for all } v \in V. \qquad (3.14)$$

Taking $v = u - u_\rho$, we get

$$\langle Fu - Fu_\rho, u - u_\rho \rangle = b(u_\rho - u, p - p_\rho). \qquad (3.15)$$

Inequalities (3.1b) and (1.28) give

$$b(u_\rho, w - p_\rho) - \frac{1}{\rho}(p_\rho, w - p_\rho) \leq g(w - p_\rho) \qquad \text{for all } w \in \mathcal{K}$$

and

$$b(u, w - p) \leq g(w - p) \qquad \text{for all } w \in \mathcal{K}.$$

Choosing $w = p$ in the first inequality and $w = p_\rho$ in the second, then adding, we obtain

$$b(u_\rho - u, p - p_\rho) \leq \frac{1}{\rho}(p_\rho, p - p_\rho).$$

Then the boundedness of $\{p_\rho\}_{\rho>0}$ implies that

$$b(u_\rho - u, p - p_\rho) \le c\rho^{-1} \|p - p_\rho\|;$$

here and subsequently c denotes a generic positive constant which can take different values in different places. Now the strong monotonicity of F and (3.15) give

$$\gamma \|u - u_\rho\|^2 \le c\rho^{-1} \|p - p_\rho\|. \tag{3.16}$$

Applying the Babuška-Brezzi condition (1.32) to (3.14), we get

$$\|p - p_\rho\| \le \frac{1}{\delta} \|Fu - Fu_\rho\|_*.$$

As $\{u_\rho\}_{\rho>0}$ is bounded, we infer from (1.10) that

$$\|p - p_\rho\| \le c \|u - u_\rho\|.$$

This inequality and (3.16) yield the bounds

$$\|u - u_\rho\| \le c\rho^{-1}$$

and

$$\|p - p_\rho\| \le c\rho^{-1}$$

for some constant $c > 0$. ■

Remark 7.27. The Babuška-Brezzi condition (1.32) guarantees that the mixed variational inequality system (1.28) has a unique solution and ensures its stability. This condition was crucial in the proof of Theorem 7.26. Under weaker assumptions we shall now study the convergence of penalty methods for variational inequalities that come from convex optimization problems.   □

Now we transfer our attention to the convergence analysis of a penalty method for the approximate solution of the constrained variational problem (1.24). The objective functional $J : V \to \mathbb{R}$ is assumed to be convex and continuous. In particular this implies that there exist $q_0 \in \mathbb{R}$ and $q \in V^*$ such that

$$J(v) \ge q_0 + \langle q, v \rangle \qquad \text{for all } v \in V. \tag{3.17}$$

To deal with the constraint $v \in G$, we introduce a continuous convex penalty functional $\Psi : V \to \mathbb{R}$ that increases an augmented objective functional whenever the constraint is violated. The (exterior) penalty property is described by

$$\Psi(v) \begin{cases} > 0 & \text{if } v \notin G, \\ = 0 & \text{if } v \in G. \end{cases} \tag{3.18}$$

In the penalty method the original problem (1.24) is replaced by an unconstrained auxiliary problem of the form

$$\min_{v \in V} [J(v) + \rho \Psi(v)]. \tag{3.19}$$

Here $\rho > 0$ is a fixed penalty parameter, just like in (3.5).

Theorem 7.28. *Let $\{u_\rho\}_{\rho>0}$ be a family of solutions of the augmented problems (3.19). Then any weak accumulation point of this family as $\rho \to +\infty$ is a solution of the original problem (1.24).*

Proof: The optimality of u_ρ for the augmented problem (3.19) and the property (3.18) imply that for each $\rho > 0$ we have

$$J(u_\rho) \le J(u_\rho) + \rho \Psi(u_\rho) \le J(v) + \rho \Psi(v) = J(v) \qquad \text{for all } v \in G. \tag{3.20}$$

The functional J is convex and continuous, and consequently weakly lower semi-continuous; see, e.g., [Zei90]. Hence (3.20) implies that

$$J(\bar{u}) \le \inf_{v \in V} J(v) \tag{3.21}$$

for any weak accumulation point $\bar{u}$ of $\{u_\rho\}_{\rho>0}$.

From (3.17) and (3.20) we have

$$q_0 + \langle q, u_\rho \rangle + \rho \Psi(u_\rho) \le J(v) \qquad \text{for all } v \in G \text{ and } \rho > 0. \tag{3.22}$$

Let $\{u_{\rho_k}\} \subset \{u_\rho\}$ be some subsequence that converges weakly to $\bar{u}$. As G is nonempty and $\lim_{k \to \infty} \rho_k = \infty$, we see from (3.22) that

$$\lim_{k \to \infty} \Psi(u_{\rho_k}) = 0.$$

But Ψ is convex and continuous, so $u_{\rho_k} \rightharpoonup \bar{u}$ as $k \to \infty$ implies that

$$\Psi(\bar{u}) \le 0.$$

Now (3.18) tells us that $\bar{u} \in G$. Recalling (3.21), we are done. ∎

Remark 7.29. To show existence and boundedness of the sequence $\{u_\rho\}_{\rho>0}$, additional assumptions are needed. In the case of the weakly coercive Signorini problem (1.5), (1.8), this can be guaranteed by assuming for example that $f < 0$. An alternative possibility is to regularize the original problem in an appropriate way: see [KT94]. □

Remark 7.30. In order to implement a penalty method, one must specify the penalty functional Ψ. In the case of an obstacle condition of the form

$$G = \{v \in H_0^1(\Omega) : v \ge g\},$$

the functional used in (3.7) could be used. The simple type of the considered constraints allow to express this functional also by

$$\Psi(v) = \frac{\rho}{2} \int_{\Omega} [g - v]_+^2(x) \, dx. \tag{3.23}$$

For constraints of the type (1.7), the analogous penalty functional

$$\Psi(v) = \frac{\rho}{2} \int_{\Gamma_2} [-v]_+^2(s) \, ds \tag{3.24}$$

can be applied. Finally, the pure penalty property (3.18) can be relaxed by introducing an asymptotic dependence on the penalty parameter as $\rho \to \infty$; an example of this technique will be presented in Section 7.4. □

Next we derive an L_∞ convergence estimate for the penalty method (3.5) applied to the obstacle problem (1.2).

Theorem 7.31. *Let $\Omega \subset R^2$ be a convex polyhedron with boundary Γ. Let $f \in L_\infty(\Omega)$ and $g \in W_\infty^2(\Omega)$ with $g|_\Gamma \leq 0$ in the sense of traces. Consider the obstacle problem (1.2), viz., find $u \in G$ such that*

$$\int_{\Omega} \nabla u \nabla (v - u) \, dx \geq \int_{\Omega} f(v - u) \, dx \quad \text{for all } v \in G \tag{3.25a}$$

where

$$G := \{ v \in H_0^1(\Omega) : v \geq g \quad a.e. \text{ in } \Omega \}. \tag{3.25b}$$

Suppose that this problem is treated by the following penalty method: find $u_\rho \in H_0^1(\Omega)$ such that

$$\int_{\Omega} \nabla u_\rho \nabla v \, dx - \rho \int_{\Omega} [g - u_\rho]_+ v \, dx = \int_{\Omega} f v \, dx \quad \text{for all } v \in H_0^1(\Omega). \tag{3.26}$$

Then the variational equation (3.26) has a unique solution $u_\rho \in H_0^1(\Omega)$ for each $\rho > 0$ and

$$\|u - u_\rho\|_{0,\infty} \leq \rho^{-1}(\|g\|_{2,\infty} + \|f\|_{0,\infty}). \tag{3.27}$$

Proof: Our hypotheses imply that the given problem (3.25) has a unique solution $u \in H_0^1(\Omega) \cap H^2(\Omega)$; see, e.g., [KS80]. Let

$$\Omega_0 := \{ x \in \Omega : u(x) = g(x) \}$$

be the contact zone of this solution. Then the optimal Lagrange multiplier p of (3.25) is given by

$$p(x) = \begin{cases} -\Delta g - f & \text{a.e. in } \Omega_0, \\ 0 & \text{elsewhere.} \end{cases} \tag{3.28}$$

As $g \in W^2_\infty(\Omega)$ and $f \in L_\infty(\Omega)$ we see that $p \in L_\infty(\Omega)$. We define a bilinear form by

$$a(u, v) = \int_\Omega \nabla u \nabla v \, dx \qquad \text{for all } u, v \in H^1(\Omega)$$

and write $(\cdot, \cdot)$ for the scalar product in $L_2(\Omega)$. The particular form of the constraint (3.25b), (3.28) and the mixed formulation equation (1.28a) yield

$$a(u, v) - (f, v) = (p, v) \qquad \text{for all } v \in H^1_0(\Omega)$$

where u is the solution of (3.25). Now $p > 0$ (compare (1.4)) so we obtain

$$a(u, v) - (f, v) \geq 0 \qquad \text{for all } v \in H^1_0(\Omega) \text{ such that } v \geq 0.$$

But $u \in G$ means that this inequality is equivalent to

$$a(u, v) - (f, v) - \rho \int_\Omega [g - u]_+ v \, dx \geq 0 \qquad \text{for all } v \in H^1_0(\Omega) \text{ such that } v \geq 0.$$

On the other hand, from (3.26) we have

$$a(u_\rho, v) - (f, v) - \rho \int_\Omega [g - u_\rho]_+ v \, dx = 0 \qquad \text{for all } v \in H^1_0(\Omega). \tag{3.29}$$

Hence, as $u|_\Gamma = u_\rho|_\Gamma = 0$, a comparison theorem (compare Section 3.2) based on the weak boundary maximum principle implies that

$$u_\rho \leq u \qquad \text{a.e. in } \Omega. \tag{3.30}$$

We continue by constructing a lower bound for the solution u_ρ of the auxiliary problem (3.26). Towards this aim let us set

$$\underline{u}(x) = u(x) - \delta \qquad \text{a.e. in } \Omega \tag{3.31}$$

for some $\delta > 0$. The definition of $a(\cdot, \cdot)$ gives immediately

$$a(\underline{u}, v) = a(u, v) \qquad \text{for all } v \in H^1_0(\Omega).$$

Thus

$$a(\underline{u}, v) - (f, v) - \rho \int_\Omega [g - \underline{u}]_+ v \, dx =$$

$$= a(u, v) - (f, v) - \rho \int_\Omega [g - \underline{u}]_+ v \, dx$$

$$= (p, v) - \rho \int_\Omega [g - \underline{u}]_+ v \, dx$$

$$\leq \int_{\Omega_0} (p - \rho\delta) v \, dx \qquad \text{for all } v \in H^1_0(\Omega) \text{ with } v \geq 0.$$

Now take $\delta = \|p\|_{0,\infty}\,\rho^{-1}$. Then we get

$$a(\underline{u}, v) - (f, v) - \rho \int_\Omega [g - \underline{u}]_+ v\,dx \le 0 \qquad \text{for all } v \in H_0^1(\Omega) \text{ with } v \ge 0.$$

By (3.29) and the previously-mentioned comparison theorem it follows that

$$\underline{u} \le u_\rho \qquad \text{a.e. in } \Omega\,.$$

Recalling (3.30) and (3.31), we have now shown that $\|u - u_\rho\|_{0,\infty} \le \delta$. Our choice $\delta = \|p\|_{0,\infty}\,\rho^{-1}$ and (3.28) then give (3.27). ∎

A well-known drawback of penalty methods is that the auxiliary problem generated is often ill-conditioned when the penalty parameter ρ is large. The selection of this parameter should be in harmony with the mesh size h of the discretization. For example, for certain finite element discretizations and specific problems one can make a tuned parameter selection such that the overall conditioning of the discrete penalty problem has the same order as the related discrete elliptic variational equation. For this we refer to Section 7.4.

An alternative way of relating unconstrained auxiliary problems to variational inequalities is through augmented Lagrangians. The basic principle here, as in the case of variational equations, can be obtained from a formulation based on PROX-regularization. If the mixed formulation (1.28) is used, then the associated augmented Lagrangian method is: *find $u^k \in V$ and $p^k \in W$ such that*

$$\langle Fu^k, v \rangle + b(v, p^k) = 0 \quad \text{for all } v \in V, \tag{3.32a}$$

$$b(u^k, w - p^k) - \frac{1}{\rho}(p^k - p^{k-1}, w - p^k) \le g(w - p^k) \quad \text{for all } w \in K, \tag{3.32b}$$

for $k = 1, 2, \ldots$ Here ρ is some fixed positive parameter and $p^0 \in K$ is a user-chosen starting iterate. Write (3.32b) in the form

$$(p^k - (p^{k-1} + \rho(Bu^k - g)), w - p^k) \ge 0 \qquad \text{for all } w \in K, \tag{3.33}$$

where, as in (3.2), we set $(Bv, w) = b(v, w)$ for all $v \in V$ and $w \in W$. Then, using the projector P_K defined in Lemma 7.4, equation (3.32a) can be formulated as: *Find $u^k \in V$ such that*

$$\langle Fu^k, v \rangle + (P_K[p^{k-1} + \rho(Bu^k - g)], Bv) = 0 \quad \text{for all } v \in V.$$

Thus, given $p^{k-1} \in K$, the associated component $u^k \in V$ is obtained as the solution of an unconstrained variational equation. Then the next iterate $p^k \in K \subset W$ is got from

$$p^k = P_K[p^{k-1} + \rho(Bu^k - g)]\,. \tag{3.34}$$

In the particular case of the obstacle problem (1.2), one then sees that (compare Example 7.24) the projector $P_{\mathcal{K}} : W \to \mathcal{K}$ is simply

$$P_{\mathcal{K}} w = [w]_+ \qquad \text{for all } w \in W,$$

and the iterative scheme of the augmented Lagrange method has the form: *Determine* $u^k \in V := H_0^1(\Omega)$ *such that*

$$\int_\Omega \nabla u^k \nabla v \, dx - \int_\Omega [p^{k-1} + \rho(g - u^k)]_+ v \, dx = \int_\Omega f v \, dx \qquad \text{for all } v \in V$$

holds, then define $p^k \in \mathcal{K}$ *by*

$$p^k = [p^{k-1} + \rho(g - u^k)]_+ \, .$$

A convergence analysis for the augmented Lagrangian method (3.32) is our next target. For this we have

Theorem 7.32. *Assume that the mixed variational formulation (1.28) has a solution* $(u, p) \in V \times \mathcal{K}$ *and that the mapping* $F : V \to V^*$ *is strongly monotone. Then for any initial iterate* $p^0 \in \mathcal{K}$, *the scheme (3.32) is well defined and*

$$\lim_{k \to \infty} \|u^k - u\| = 0.$$

Under the additional assumption that the Babuška-Brezzi condition (1.32) is satisfied, we also have

$$\lim_{k \to \infty} \|p^k - p\| = 0$$

and the estimates

$$\|p^k - p\| \leq \frac{1}{1 + c_1 \rho} \|p^{k-1} - p\|,$$

$$\|u^k - u\| \leq c_2 \, \rho^{-1} \, \|p^{k-1} - p\| \tag{3.35}$$

are valid for $k = 1, 2, \ldots$ *with some positive constants* c_1 *and* c_2.

Proof: Recall the mixed formulation (1.28):

$$\langle Fu, v \rangle + b(v, p) = 0 \qquad \text{for all } v \in V,$$

$$b(u, w - p) \qquad \leq g(w - p) \qquad \text{for all } w \in \mathcal{K}.$$

From this and (3.32) we infer that

$$\langle Fu^k - Fu, v \rangle + b(v, p^k - p) = 0 \qquad \text{for all } v \in V \tag{3.36a}$$

and

$$b(u^k - u, p - p^k) - \frac{1}{\rho} (p^k - p^{k-1}, p - p^k) \leq 0. \tag{3.36b}$$

Choosing $v = u^k - u$ in (3.36) and invoking (2.24), we get

$$\gamma \|u^k - u\|^2 \le \frac{1}{\rho} (p^k - p^{k-1}, p - p^k) = \frac{1}{\rho} [(p^k - p, p - p^k) + (p - p^{k-1}, p - p^k)].$$

Rearranging, then appealing to the Cauchy-Schwarz inequality, one has

$$\gamma \rho \|u^k - u\|^2 + \|p^k - p\|^2 \le \|p^k - p\| \|p^{k-1} - p\|. \tag{3.37}$$

In particular,
$$\|p^k - p\| \le \|p^{k-1} - p\| \quad \text{for } k = 1, 2, \ldots$$

Hence the sequence of non-negative real numbers $\{\|p^k - p\|\}$ must be convergent, and from (3.37) it then follows that

$$\lim_{k \to \infty} \|u^k - u\| = 0. \tag{3.38}$$

If the Babuška-Brezzi condition (1.32) is satisfied, then (3.36a) implies that

$$\|p^k - p\| \le \frac{1}{\delta} \|Fu^k - Fu\|. \tag{3.39}$$

Hence, since F is continuous, (3.38) forces convergence of $\{p^k\}$ to p. We see from (3.38) that the sequence $\{u^k\}$ is bounded. Thus the local Lipschitz continuity of F stated in (1.10) yields

$$\|Fu^k - Fu\| \le L \|u^k - u\| \quad \text{for } k = 1, 2, \ldots \tag{3.40}$$

for some constant L. Invoking (3.39) and (3.40) in (3.37), we get

$$\frac{\delta \gamma \rho}{L} \|u^k - u\| + \|p^k - p\| \le \|p^{k-1} - p\| \quad \text{for } k = 1, 2, \ldots \tag{3.41}$$

Using (3.39) and (3.40) again yields

$$\left(\frac{\delta^2 \gamma \rho}{L^2} + 1 \right) \|p^k - p\| \le \|p^{k-1} - p\| \quad \text{for } k = 1, 2, \ldots$$

This proves the first bound in (3.35). The second inequality there is immediate from (3.41). ∎

The penalty methods and augmented Lagrangian methods considered up to this point can be interpreted as regularizations of the second component (1.28b) of the mixed variational formulation. We shall now apply these regularization techniques directly to the original problem with the aim of embedding each convex variational problem into a family of strongly convex problems.

In *Tychonoff regularization* the basic idea is to associate the given variational problem

$$\min_{v \in G} J(v) \tag{3.42}$$

with auxiliary problems of the form

$$\min_{v \in G} [J(v) + \varepsilon \|v\|^2] \tag{3.43}$$

where $\varepsilon > 0$ is some fixed regularization parameter. As before, we assume that the feasible set $G \subset V$ of the original problem (3.42) is nonempty, convex and bounded. We also assume that (3.42) has an optimal solution that is not necessarily unique. The objective functional of the auxiliary problem (3.43) is strongly convex because $\varepsilon > 0$ and J is convex. Hence for each $\varepsilon > 0$ there exists a unique solution $u_\varepsilon \in G$ of (3.43). Convergence of the Tychonoff regularization is described by

Theorem 7.33. *Under the above assumptions, the solutions u_ε of the auxiliary problems (3.43) converge as $\varepsilon \to 0$ to the solution $\hat{u} \in G$ of (3.42) that satisfies*

$$\|\hat{u}\| \leq \|u\| \qquad \text{for all } u \in G_{opt},$$

where G_{opt} is the set of all optimal solutions of (3.42).

Proof: For each $\varepsilon > 0$, the functional $J(\cdot) + \varepsilon \| \cdot \|^2$ is continuous and strongly convex and the nonempty set G is convex and bounded; consequently there exists a unique $u_\varepsilon \in G$ such that

$$J(u_\varepsilon) + \varepsilon \|u_\varepsilon\|^2 \leq J(v) + \varepsilon \|v\|^2 \qquad \text{for all } v \in G. \tag{3.44}$$

By our hypotheses the original problem (3.42) has a solution. Choose $u \in G_{opt}$. Then u is an arbitrary solution of the problem (3.42) and we have

$$J(u) \leq J(v) \qquad \text{for all } v \in G. \tag{3.45}$$

Taking $v = u$ and $v = u_\varepsilon$ in (3.44) and (3.45) respectively, we get

$$\|u_\varepsilon\| \leq \|u\| \qquad \text{for all } u \in G_{opt}, \quad \varepsilon > 0. \tag{3.46}$$

Thus $\{u_\varepsilon\}_{\varepsilon > 0}$ is bounded and consequently weakly compact in the Hilbert space V. Hence one can choose a sequence $\{\varepsilon_k\}$ with $\lim_{k \to \infty} \varepsilon_k = 0$ and $\{u_{\varepsilon_k}\}$ weakly convergent to $\hat{u}$ for some $\hat{u} \in V$. But $u_{\varepsilon_k} \in G$ for $k = 1, 2, \ldots$ and G is convex and closed so we must have $\hat{u} \in G$. The convexity and continuity of the mapping $v \mapsto \|v\|$ implies its weak lower semi-continuity. From the inequality (3.46) we therefore have

$$\|\hat{u}\| \leq \|u\| \qquad \text{for all } u \in G_{opt}. \tag{3.47}$$

Again appealing to (3.44), we obtain

$$J(u_{\varepsilon_k}) \leq J(u) + \varepsilon_k \|u\|^2 \qquad \text{for all } u \in G_{opt}.$$

The continuity and convexity of J and $\lim_{k \to \infty} \varepsilon_k = 0$ then give

$$J(\hat{u}) \leq J(u) \qquad \text{for all } u \in G_{opt}.$$

That is, $\hat{u}$ is a solution of (3.42). In addition, from (3.47) we see that $\hat{u}$ is an element of the solution set G_{opt} that has minimal norm—which, since G_{opt} is closed and convex, implies that $\hat{u}$ is uniquely determined. The weak compactness of $\{u_\varepsilon\}_{\varepsilon>0}$ now implies the weak convergence property $u_{\varepsilon_k} \rightharpoonup \hat{u}$ as $k \to \infty$ for *any* sequence $\{\varepsilon_k\}$ with $\varepsilon_k \to 0$. The weak lower semi-continuity of the functional $\| \cdot \|$ then gives

$$\liminf_{\varepsilon \to 0} \|u_\varepsilon\| \geq \|\hat{u}\|,$$

and recalling (3.46) we get finally

$$\lim_{\varepsilon \to 0} \|u_\varepsilon\| = \|\hat{u}\|. \tag{3.48}$$

Exploiting the scalar product, we write

$$\|u_\varepsilon - \hat{u}\|^2 = (u_\varepsilon - \hat{u}, u_\varepsilon + \hat{u} - 2\hat{u}) = \|u_\varepsilon\|^2 - \|\hat{u}\|^2 - 2(\hat{u}, u_\varepsilon - \hat{u}).$$

The weak convergence property $u_\varepsilon \rightharpoonup \hat{u}$ and (3.48) imply that $\lim_{\varepsilon \to 0} \|u_\varepsilon - \hat{u}\| = 0$. That is, as $\varepsilon \to 0$, u_ε converges to that solution $\hat{u}$ of (3.42) that has minimal norm. ∎

By means of Tychonoff regularization one can embed both weakly elliptic variational inequalities and variational equations in a family of V-elliptic problems, but as the regularization parameter ε approaches 0, the positive ellipticity constant $\gamma = \gamma(\varepsilon)$ also tends to zero. To cope with this drawback when solving the regularized problems numerically, the regularization parameter and the discretization step size have to be adjusted together in some appropriate way.

An alternative form of sequential regularization is provided by *PROX-regularization*. This is based on the application of augmented Lagrangian methods to the original problem (3.42). It leads to a sequence of auxiliary problems

$$\min_{v \in G} \left[J(v) + \varepsilon \|v - u^{k-1}\|^2 \right] \quad \text{for } k = 1, 2, \ldots \tag{3.49}$$

Here one must select an initial guess $u^0 \in G$ and a fixed $\varepsilon > 0$, then each u^k is iteratively defined as the unique solution of the variational problem (3.49). For a discussion of the convergence of this method we refer to, e.g., [Mar70], [58].

These regularization methods can be combined with penalty methods or with augmented Lagrangian methods; see [KT94].

To close this section we study finite element discretizations of penalty problems and of penalty techniques for discrete elliptic variational inequalities. In both cases one obtains parameter-dependent finite-dimensional variational equations, which can be treated by efficient numerical methods (see Chapter 5). Note that the type of problem generated depends on the order in which one applies the discretization and the penalization.

We begin with conforming discretizations of the variational equations (3.5) generated by penalty methods, i.e., the discretization of the problems

$$\langle Fu_\rho, v \rangle + \rho\left(P_K(Bu_\rho - g), Bv\right) = 0 \qquad \text{for all } v \in V. \tag{3.50}$$

Define the operator $T_\rho : V \to V^*$ by

$$\langle T_\rho u, v \rangle := \langle Fu, v \rangle + \rho\left(P_K(Bu - g), Bv\right) \qquad \text{for all } u,\, v \in V, \tag{3.51}$$

where $\rho > 0$ is a fixed parameter. Under the assumptions of Lemma 7.25, by (3.11) we have

$$\gamma \|u - v\|^2 \leq \langle T_\rho u - T_\rho v, u - v \rangle \qquad \text{for all } u,\, v \in V \tag{3.52}$$

and

$$\|T_\rho u - T_\rho v\|_* \leq \left(L(\sigma) + \rho\beta^2\right)\|u - v\| \tag{3.53}$$
$$\text{for all } u,\, v \in V,\ \|u\| \leq \sigma,\ \|v\| \leq \sigma.$$

Here $L(\cdot)$ is the local Lipschitz constant of F and β is the boundedness constant of the bilinear form $b(\cdot, \cdot)$. The operator T_ρ is also V-elliptic and bounded. Thus all the standard conditions for the application of finite element methods are fulfilled.

Let V_h be a finite-dimensional subspace of V. Then (3.50) is discretized by the finite-dimensional problem

Find $u_{h\rho} \in V_h$ such that

$$\langle Fu_{h\rho}, v_h \rangle + \rho\left(P_K(Bu_{h\rho} - g), Bv_h\right) = 0 \qquad \text{for all } v_h \in V_h. \tag{3.54}$$

We write this problem as

$$\langle T_{h\rho} u_{h\rho}, v_h \rangle = 0 \qquad \text{for all } v_h \in V_h;$$

this notation is used in Section 7.3.2. The convergence of this discrete penalty method is dealt with in the next theorem.

Theorem 7.34. *Let the assumptions of Theorem 7.26 be satisfied. Then for each $\rho > 0$ the discrete penalty problem (3.54) has a unique solution $u_{h\rho} \in V_h$, and*

$$\|u_\rho - u_{h\rho}\| \leq (c_1 + c_2\rho) \inf_{v_h \in V_h} \|u_\rho - v_h\| \tag{3.55}$$

for some positive constants c_1 and c_2. Here $u_\rho \in V$ denotes the solution of the underlying continuous penalty problem (3.5).

Proof: The properties (3.52) and (3.53) allow us to apply Lemma 7.16 with G and G_h replaced by V and V_h. The conforming nature of the discretization (i.e., $V_h \subset V$) means that the conclusion of Lemma 7.16 simplifies to (2.13) by Remark 7.17. This gives (3.55) immediately. ∎

The bound of Theorem 7.34 illustrates a well-known phenomenon that occurs when finite element methods are applied to penalty methods, namely that the order of convergence of the finite element method can be reduced when ρ is large. The loss of accuracy is caused by the asymptotic singularity of the penalty problem as $\rho \to +\infty$. This theoretical drawback of penalty methods is also observed in computational experiments. It can be avoided by using a penalty method obtained from the regularization of a discrete mixed formulation that satisfies a uniform Babuška-Brezzi condition, i.e., with a constant $\delta > 0$ that is independent of the discretization parameter $h > 0$. Such a penalty method leads to the following discrete problem:

Find $(u_{h\rho}, p_{h\rho}) \in V_h \times \mathcal{K}_h$ such that

$$\langle F u_{h\rho}, v_h \rangle + b(v_h, p_{h\rho}) = 0 \quad \text{for all } v_h \in V_h, \quad (3.56a)$$

$$b(u_{h\rho}, w_h - p_{h\rho}) - \frac{1}{\rho}(p_{h\rho}, w_h - p_{h\rho}) \le g(w_h - p_{h\rho}) \quad (3.56b)$$

$$\text{for all } w_h \in \mathcal{K}_h.$$

Like the continuous problem (3.1), inequality (3.56b) can be represented by a projector $P_{\mathcal{K}_h} : W_h \to \mathcal{K}_h$ that is defined using Lemma 7.4. Then (3.56) yields the penalty method

$$\langle F u_{h\rho}, v_h \rangle + \rho \left(P_{\mathcal{K}_h}(B u_{h\rho} - g), B v_h \right) = 0 \quad \text{for all } v_h \in V_h. \quad (3.57)$$

To keep the notation simple we have written $u_{h\rho}$ for the solutions of both (3.57) and (3.54), but in general these functions are not identical.

The convergence of the method (3.57) can be shown using convergence results for mixed finite elements for variational equations that appear in [BF91].

Theorem 7.35. *Let $u_{h\rho} \in V_h$ be the solution of (3.57). Assume that the mixed finite element discretizations satisfy the Babuška-Brezzi condition uniformly, i.e., that there exists $\delta > 0$, which is independent of h, such that*

$$\sup_{v_h \in V_h} \frac{b(v_h, w_h)}{\|v_h\|} \ge \delta \|w_h\| \quad \text{for all } w_h \in W_h. \quad (3.58)$$

Then there is a positive constant c such that

$$\|u_h - u_{h\rho}\| \le c\,\rho^{-1}. \quad (3.59)$$

Here $u_h \in G_h$ denotes the solution of the discrete problem (1.28).

Remark 7.36. Unlike the penalty method (3.54), in the mixed finite element penalty method (3.57) the penalty parameter $\rho = \rho(h) > 0$ can be selected as a function of the discretization mesh size $h > 0$ in such a way that the overall method retains the same order of convergence as the underlying mixed finite element discretization (see, e.g., Theorem 7.32).   □

Remark 7.37. The evaluation of the projection $P_{\mathcal{K}_h}$ can be simplified if a non-conforming implementation of (3.56) such as mass lumping (see Section 4.6) is used. In this case we obtain

$$\langle Fu_{h\rho}, v_h \rangle + b_h(v_h, p_{h\rho}) = 0 \quad \text{for all } v_h \in V_h, \quad (3.60a)$$

$$b_h(u_{h\rho}, w_h - p_{h\rho}) - \frac{1}{\rho}(p_{h\rho}, w_h - p_{h\rho})_h \le g_h(w_h - p_{h\rho}) \quad (3.60b)$$

$$\text{for all } w_h \in \mathcal{K}_h$$

instead of (3.56). If the obstacle problem (1.2) is discretized with piecewise linear C^0 elements for V_h and piecewise constant elements for W_h, then with the mass lumping

$$(w_h, z_h) := \sum_{j=1}^{N} \text{meas}(D_j) w_j z_j$$

and setting

$$b_h(v_h, z_h) - g_h(z_h) := \sum_{j=1}^{N} \text{meas}(D_j)(g_j - v_j) z_j$$

we obtain the mixed finite element penalty problem

$$\sum_{j=1}^{N} a(\varphi_i, \varphi_j) u_j - f_i - \rho \, \text{meas}(D_i) \max\{0, g_i - u_i\} = 0, \quad i = 1, \ldots, N. \quad (3.61)$$

Here the D_j are the elements of the dual subdivision (see Section 4.6) and we set $u_{h\rho} = (u_j)_{j=1}^{N}$, $f_i = \int_{\Omega} f\varphi \, dx$ and $g_i = g(x_i)$. Our weighting of each ansatz function φ_i by the measure of the associated dual subdomain D_i is reasonable. The method (3.61) can be interpreted as the application of the well-known quadratic loss penalty function to the discrete problem (2.6).   □

7.3.2 Adjustment of Penalty and Discretization Parameters

When variational inequalities are discretized using penalty methods then, as we have seen, the following problems are relevant (the arrows in the diagram show the effects of penalization and discretization):

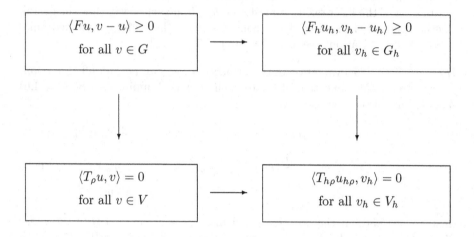

The penalization and finite element discretization each provide only an approximation of the original problem. Furthermore, in these methods the amount of computational work becomes greater as one increases the penalty parameter ρ and refines the mesh in the discretization. The first of these effects is caused by the ill-conditioning of penalty problems when ρ is large. Thus the parameters in the discretization scheme and penalty method should be selected together in such a way that

- the order of convergence of the discretization method is preserved in the full numerical method;
- the conditioning of the penalty problem is asymptotically no worse than the conditioning of a comparable discrete variational equation.

These goals can be achieved by the parameter adjustment strategy derived in [57] for a specific penalty method, which we now describe.

Consider the obstacle problem (1.2) with piecewise linear C^0 elements and with the discretization (2.2) for G_h. Unlike [57], we shall proceed here via the discretization of a continuous penalty problem. The given obstacle problem (1.2) leads first to the associated unconstrained variational problem

$$\min_{v \in V := H_0^1(\Omega)} J_\rho(v) \tag{3.62}$$

where

$$J_\rho(v) := \frac{1}{2}a(v,v) - (f,v) + s \int_\Omega \left[g(x) - v(x) + \sqrt{(g(x) - v(x))^2 + \rho^{-1}} \right] dx.$$

Here $a(\cdot, \cdot)$ and $(f, \cdot)$ are again defined by

$$a(u,v) = \int_{\Omega} \nabla u \nabla v \, dx \qquad \text{and} \qquad (f,v) = \int_{\Omega} f v \, dx.$$

The quantities $s > 0$ and $\rho > 0$ are parameters of the specific penalty method used in (3.62). Our study of the convergence behaviour of the method will be for an appropriately chosen but fixed $s > 0$ and for $\rho \to +\infty$, so we write the solution as u_ρ (i.e., we do not make explicit the dependence on s).

The strong convexity and continuity of $J_\rho(\cdot)$ imply that for each $\rho > 0$ the auxiliary problem (3.62) has a unique solution $u_\rho \in V$. This solution is characterized by the necessary and sufficient condition

$$\langle T_\rho u_\rho, v \rangle := a(u_\rho, v) - (f,v) - s \int_{\Omega} \left(1 + \frac{g - u_\rho}{\sqrt{(g - u_\rho)^2 + \rho^{-1}}} \right) v \, dx \tag{3.63}$$
$$= 0 \qquad \forall v \in V.$$

Let V_h be a finite element space based on piecewise linear C^0 elements over some quasi-uniform triangulation of $\Omega \subset \mathbb{R}^2$. To avoid additional boundary approximations, for simplicity in our analysis we assume that the domain Ω is polygonal. Assume also that the triangulation is of weakly acute type, i.e., all inner angles of the triangles of the subdivision do not exceed $\pi/2$.

Using partial mass lumping we discretize the variational equation (3.63) by

$$\langle T_{h\rho} u_{h\rho}, v_h \rangle = 0 \qquad \text{for all } v_h \in V_h. \tag{3.64}$$

Here the operator $T_{h\rho} u_{h\rho}$ is defined by

$$\langle T_{h\rho} y_h, v_h \rangle := a(y_h, v_h) - (f, v_h) - s \sum_{i=1}^{N} \text{meas}(D_i) \left(1 + \frac{g_i - y_i}{\sqrt{(g_i - y_i)^2 + \rho^{-1}}} \right) v_i$$

with

$$y_i := y_h(x_i), \ v_i := v_h(x_i), \ g_i := g(x_i) \ \text{and} \ u_i^{h\rho} := u_{h\rho}(x_i) \ \text{for } i = 1, \ldots, N.$$

Here the x_i, for $i = 1, \ldots, N$, are the inner grid points of the mesh (i.e., triangle vertices that lie inside Ω), and each D_i is the dual subdomain associated with x_i (see Section 4.6). The Ritz-Galerkin equations equivalent to (3.64) are

$$\sum_{j=1}^{N} a_{ij} u_j^{h\rho} - f_i - \text{meas}(D_i) \left(1 + \frac{g_i - u_i^{h\rho}}{\sqrt{(g_i - u_i^{h\rho})^2 + \rho^{-1}}} \right) = 0, \tag{3.65}$$
$$i = 1, \ldots, N,$$

with $a_{ij} := a(\varphi_j, \varphi_i)$ and $f_i := (f, \varphi_i)$. The variational equations (3.64) and the Ritz-Galerkin equations (3.65) each give a necessary and sufficient optimality criterion for the variational problem

$$\min_{v_h \in V_h} J_{h\rho}(v_h) \tag{3.66}$$

with

$$J_{h\rho}(v_h) := J(v_h) + s \sum_{i=1}^{N} \text{meas}(D_i)\big(g_i - v_i + \sqrt{(g_i - v_i)^2 + \rho^{-1}}\big). \quad (3.67)$$

We now analyse the convergence of the discretization (3.65), while assuming that all the hypotheses of Theorem 7.19 for the problem (1.2) are satisfied.

Lemma 7.38. *There exists a constant $c_0 > 0$, which is independent of the discretization, such that $u_{h\rho} \in G_h$ for*

$$s \geq c_0 > 0 \qquad and \qquad \rho > 0.$$

Proof: First, recall the discrete problem (2.6) associated with (1.2). Using the discrete solution $u_h(x) = \sum_{j=1}^{N} u_j \varphi_j(x)$ we define

$$p_i := \frac{1}{\text{meas}(D_i)}[a(u_h, \varphi_i) - (f, \varphi_i)], \quad i = 1, \ldots, N. \quad (3.68)$$

From the complementarity condition in (2.6) we get

$$p_i = 0 \text{ when } u_i > g_i.$$

Now consider any vertex x_i with $u_i = g_i$, i.e., where the obstacle is active. We have $u_j \geq g_j$ for all j by (2.6), and $a(\varphi_i, \varphi_j) \leq 0$ for $i \neq j$ follows from the assumption that the triangulation is weakly acute, so

$$0 \leq p_i \leq \frac{1}{\text{meas}(D_i)}[a(g_h, \varphi_i) - (f, \varphi_i)],$$

where g_h is the piecewise linear interpolant of the obstacle function g over the triangulation. Rewrite this inequality as

$$0 \leq p_i \leq \frac{1}{\text{meas}(D_i)}[a(g, \varphi_i) - (f, \varphi_i) + a(g_h - g, \varphi_i)].$$

By hypothesis $g \in W_\infty^2(\Omega)$ and $f \in L_\infty(\Omega)$. Because of $\text{meas}(\text{supp}\varphi_i) = O(h^2)$ consequently there exists a constant $c_1 > 0$ such that

$$|a(g, \varphi_i) - (f, \varphi_i)| \leq c_1 h^2, \quad i = 1, \ldots, N,$$

where h is the mesh diameter. Furthermore, for an arbitrary triangle Δ of the triangulation with $\Delta \in \text{supp } \varphi_i$, we have

$$\left| \int_\Delta \nabla(g_h - g) \nabla \varphi_i \, dx \right| = \left| \int_{\partial \Delta} (g_h - g) \frac{\partial \varphi_i}{\partial n} \, ds \right| \leq c_2 h^{-1} \int_{\partial \Delta} |g_h - g| \, ds$$

for some constant $c_2 > 0$. Hence

$$|a(g_h - g, \varphi_i)| \leq c_3 h^2$$

for some constant $c_3 > 0$. But $\operatorname{meas}(D_i) \geq c_4 h^2$ for some positive constant c_4, and combining our bounds we obtain

$$0 \leq p_i \leq c_0, \quad i = 1, \dots, N,$$

for some constant $c_0 > 0$ that is independent of the triangulation.

Suppose that $s \geq c_0$. The definition (3.68) of p_i gives

$$a(u_h, \varphi_i) - (f, \varphi_i) = \operatorname{meas}(D_i) \, p_i, \quad i = 1, \dots, N.$$

Using the notation of (3.65) and recalling that $g_i \leq u_i$ for all i, then this yields

$$\sum_{j=1}^{N} a_{ij} u_j - f_i - s \operatorname{meas}(D_i) \left(1 + \frac{g_i - u_i}{\sqrt{(g_i - u_i)^2 + \rho^{-1}}} \right) \tag{3.69}$$
$$\leq \operatorname{meas}(D_i) \, (p_i - s) \leq 0, \quad i = 1, \dots, N,$$

where we used $p_i \leq c_0 \leq s$. As we mentioned already, $a_{ij} \leq 0$ for $i \neq j$. The function

$$t \mapsto s \operatorname{meas}(D_i) \left(1 + \frac{t}{\sqrt{(g_i - t)^2 + \rho^{-1}}} \right), \quad i = 1, \dots, N$$

is monotonically increasing for $t \in R$. Thus a discrete comparison principle can be applied and (3.69) this yields

$$u_j \leq u_j^{h\rho}, \quad j = 1, \dots, N.$$

But $u_h = (u_j)_{j=1}^{N} \in G_h$ and it follows that $u_{h\rho} \in G_h$ also. ∎

Lemma 7.39. *Suppose that the parameter s is chosen such that $s \geq c_0$ as in Lemma 7.38. Let u_h be the solution of the discrete problem (2.6). Then*

$$\|u_h - u_{h\rho}\| \leq c \rho^{-1/4},$$

where the constant c is independent of the discretization.

Proof: We have

$$J_{h\rho}(u_{h\rho}) \leq J_{h\rho}(u_h)$$

since the function $u_{h\rho}$ is the optimal solution of the auxiliary problem (3.66). From (3.67) it is clear that $J_{h\rho}(v_h) \geq J(v_h)$ for all $v_h \in V_h$ and

$$J_{h\rho}(v_h) \leq J(v_h) + s \operatorname{meas}(\Omega) \rho^{-1/2} \quad \text{for all } v_h \in G_h.$$

Taking $v_h = u_h$ and invoking the previous inequalities, we deduce that

$$J(u_{h\rho}) \leq J(u_h) + s \operatorname{meas}(\Omega) \rho^{-1/2} \quad \text{for all } v_h \in G_h. \tag{3.70}$$

On the other hand, the optimality of u_h implies that

$$\langle J'(u_h), v_h - u_h \rangle \geq 0 \quad \text{for all } v_h \in G_h.$$

Lemma 7.38 enables us to take $v_h = u_{h\rho}$ here and the strong convexity of J then yields

$$J(u_{h\rho}) \geq J(u_h) + \langle J'(u_h), u_{h\rho} - u_h \rangle + \gamma \|u_{h\rho} - u_h\|^2 \geq J(u_h) + \gamma \|u_{h\rho} - u_h\|^2.$$

Recalling (3.70), we infer that

$$\gamma \|u_{h\rho} - u_h\|^2 \leq s \operatorname{meas}(\Omega) \rho^{-1/2}.$$

The desired result follows. ∎

To ensure that the $O(h)$ rate of convergence demonstrated in Theorem 7.19 is retained for the complete method, one should select the parameter s according to Lemma 7.38 and set

$$\rho = \rho(h) = h^{-4}. \tag{3.71}$$

We are also interested in the conditioning of the discretized penalty problem. For this purpose we introduce the mapping $\Phi_{h\rho} : \mathbb{R}^N \to \mathbb{R}$ defined by

$$\Phi_{h\rho}(\underline{v}) := s \sum_{i=1}^{N} \operatorname{meas}(D_i) \left(g_i - v_i + \sqrt{(g_i - V_i)^2 + \rho^{-1}} \right) \quad \text{for all } \underline{v} = (v_i)_{i=1}^{N}.$$

Theorem 7.40. *Let $s \geq c_0$, where c_0 is the constant of Lemma 7.38. Choose the penalty parameter $\rho = \rho(h)$ according to (3.71). Then there exists a constant $c > 0$ such that*

$$\|u - u_{h\rho}\| \leq ch,$$

where u denotes the solution of the original problem (1.2). Furthermore, the matrices $A_h + \Phi''(\underline{u}_{h\rho})$, with $\underline{u}_{h\rho} = (u_i^{h\rho})_{i=1}^{N}$, have asymptotically the same conditioning behaviour as the stiffness matrix A_h of the elliptic problem without the obstacle condition.

Proof: By Theorem 7.19, Lemmas 7.38 and 7.39, and (3.71) we have

$$\|u - u_{h\rho}\| \leq \|u - u_h\| + \|u_h - u_{h\rho}\| \leq ch$$

for some constant $c > 0$. Thus we need only analyse the conditioning of $A_h + \Phi''(\underline{u}_{h\rho})$. Using Rayleigh quotients to compute the extremal eigenvalues of this matrix, one has

$$\lambda_{min}(A_h + \Phi''(\underline{u}_{h\rho})) = \min_{z \in \mathbb{R}^n,\, z \neq 0} \frac{z^T(A_h + \Phi''(\underline{u}_{h\rho}))z}{z^T z}$$

$$\geq \min_{z \in \mathbb{R}^n,\, z \neq 0} \frac{z^T A_h z}{z^T z} + \min_{z \in \mathbb{R}^n,\, z \neq 0} \frac{z^T \Phi''(\underline{u}_{h\rho})z}{z^T z}$$

$$\geq \min_{z \in \mathbb{R}^n,\, z \neq 0} \frac{z^T A_h z}{z^T z} \geq m\,h^2$$

for some constant $m > 0$. Similarly,

$$\lambda_{max}(A_h + \Phi''(\underline{u}_{h\rho})) = \max_{z \in \mathbb{R}^n,\, z \neq 0} \frac{z^T(A_h + \Phi''(\underline{u}_{h\rho}))z}{z^T z}$$

$$\leq \max_{z \in \mathbb{R}^n,\, z \neq 0} \frac{z^T A_h z}{z^T z} + \max_{z \in \mathbb{R}^n,\, z \neq 0} \frac{z^T \Phi''(\underline{u}_{h\rho})z}{z^T z}$$

$$\leq \max_{z \in \mathbb{R}^n,\, z \neq 0} \frac{z^T A_h z}{z^T z}$$

$$\leq M + \max_{1 \leq i \leq N} \{ s\,\mathrm{meas}(D_i)\,\rho^{1/2} \}$$

for some constant $M > 0$. Hence, for our choice of the parameters s and ρ we get

$$\mathrm{cond}(A_h + \Phi''(\underline{u}_{h\rho})) = \frac{\lambda_{max}(A_h + \Phi''(\underline{u}_{h\rho}))}{\lambda_{min}(A_h + \Phi''(\underline{u}_{h\rho}))} \leq c\,h^{-2}$$

for some $c > 0$. ∎

In many application it is important to find not only an approximate solution $u_{h\rho} \in V_h$ but also an approximation of the contact zone

$$\Omega_0 := \{ x \in \overline{\Omega} : u(x) = g(x) \}.$$

We shall discuss perturbations of Ω_0. For this we need the notation

$$\Omega[\tau] := \{ x \in \overline{\Omega} : u(x) \leq g(x) + \tau \}$$

for each $\tau \geq 0$, and the set $\Omega_h^\beta \subset \mathbb{R}^2$ defined for any fixed parameter $\sigma > 0$ by

$$\Omega_h^\beta := \{ x \in \overline{\Omega} : u_{h\rho}(x) \leq g_h(x) + \sigma h^\beta \}$$

for $\beta \in (0, 1)$. Finally, let

$$d(A, B) := \max \left\{ \sup_{y \in A} \inf_{x \in B} \|x - y\|, \ \sup_{y \in B} \inf_{x \in A} \|x - y\| \right\}$$

denote the Hausdorff distance from the set A to the set B, where both sets lie in $\mathbb{R}^2$. The next result, which is taken from [57], gives some idea of how well Ω_0 is approximated.

Theorem 7.41. *Under the hypotheses of Theorem 7.40, there exist constants* $c > 0$ *and* $\overline{h} > 0$ *such that*

$$d(\Omega_h^\beta, \Omega_0) \leq \max \{ h; d(\Omega[ch^\beta], \Omega_0) \} \qquad \text{for all } h \in (0, \overline{h}].$$

To conclude this section, we look at the example of elastic-plastic torsion of a cylindrical rod; see [GLT81], [Fri82]. Consider a cylindrical rod made from an ideal elastoplastic material with cross-section $\Omega \subset \mathbb{R}^2$. If the constant angle of torsion per unit length is denoted by C, then this model leads to the variational problem

$$\min J(v) := \tfrac{1}{2}a(v,v) - C \int_{\Omega} v(x)\,dx$$
$$\text{where} \qquad v \in H_0^1(\Omega) \ \text{ and } \ |\nabla v| \leq 1 \ \text{ a.e. in } \Omega. \tag{3.72}$$

It can be shown that this problem is equivalent to

$$\min J(v) := \tfrac{1}{2}a(v,v) - C \int_{\Omega} v(x)\,dx$$
$$\text{where} \qquad v \in H_0^1(\Omega) \ \text{ and } \ |v(x)| \leq \delta(x, \partial\Omega) \ \text{ in } \Omega; \tag{3.73}$$

here $\delta(x, \partial\Omega)$ denotes the distance from the point $x \in \Omega$ to the boundary $\partial\Omega$. In the case $C < 0$ we obtain an obstacle problem of the type (1.2) and our previous analysis can be applied under some mild weakening of the assumptions made on the obstacle function g that are described in [57]. Numerical approximations of the boundaries of the elastoplastic behaviour of the rod are given in [57] for the sets

$$\Omega = (0,1) \times (0,1) \qquad \text{and} \qquad \Omega = ((0,1) \times (0.5,1)) \cup ((0.25, 0.75) \times (0,1)).$$

7.4 Optimal Control of Partial Differential Equations

The availability of highly efficient numerical methods for the approximate solution of partial differential equations has lead increasingly to the consideration of problems of optimal control whose state equations are partial differential equations. Among examples of this type of problem are optimal control of temperature ([115], [92]), optimization in fluid dynamics ([51], [43]) and the construction of optimal design ([64], [70]). As well as numerous publications in scientific journals, monographs such as [Lio71] and [NT94] study various aspects of the analysis and numerical solution of optimal control problems based on partial differential equations. Now in Tröltzsch's recent book [Trö05] we have an advanced textbook that succinctly discusses the principal problems of the theory of these problems, their applications, and numerical methods for their solution. As this topic is today the subject of much research and has a close mathematical relationship to the variational inequalities we have already discussed, we now present a brief introduction to optimal control with partial differential equations and the efficient solution of these problems by finite element methods.

7.4.1 Analysis of an Elliptic Model Problem

Let $\Omega \subset \mathbb{R}^n$ be a bounded domain with a regular boundary Γ. Let $V := H_0^1(\Omega)$ be the usual Sobolev space and set $U = L_2(\Omega)$. Let

$$a(w, v) := \int_{\Omega} \nabla w \cdot \nabla v \quad \text{for all } w, v \in V$$

be the bilinear form associated with the Laplace operator $-\Delta$. In the current section we often use the L_2 norm; thus we shall write $\| \cdot \|$ for this norm, instead of the notation $\| \cdot \|_0$ that appears widely in the rest of this book.

As a model problem we consider

$$\min_{w,u} J(w, u) := \frac{1}{2}\|w - z\|^2 + \frac{\rho}{2}\|u\|^2$$

where $\quad w \in V, \quad a(w, v) = (u, v) \quad$ for all $v \in V,$ $\qquad$ (4.1)

and $\quad u \in U_{ad} := \{u \in U : \alpha \le u \le \beta\}.$

Here $z \in L_2(\Omega)$ is a given target state and $\alpha, \beta \in \mathbb{R}$ are given constants with $\alpha < \beta$. These constants are bounds on the control u. The semi-ordering "$\le$" in (4.1) is the almost-everywhere pointwise one on Ω and the bounds defining U_{ad} are identified with the constant functions $\alpha, \beta \in L_2(\Omega)$. Finally, $\rho > 0$ is some regularization constant or cost factor for the control.

In the model problem (4.1) the control u acts as a source term in the entire domain Ω; this is called a problem with a *distributed control*. Often instead problems with *boundary control* are considered (see [Trö05]) where, unlike (4.1), the control u acts only via a boundary condition.

The classical formulation of (4.1) is

$$\min_{w,u} J(w, u) := \frac{1}{2} \int_{\Omega} [w(x) - z(x)]^2 \, dx + \frac{\rho}{2} \int_{\Omega} u(x)^2 \, dx$$

where $\quad -\Delta w = u \quad$ in $\Omega \quad$ and $w = 0 \quad$ on $\Gamma,$ $\qquad$ (4.2)

$$u \in L_2(\Omega), \quad \alpha \le u \le \beta \quad \text{in } \Omega.$$

In the optimal control literature it is standard to use u to denote the control. We follow this practice in this section and (unlike the rest of the book) write w for the solution of the variational equation that appears in (4.1) and the solution of the Poisson equation in (4.2). The occurring Poisson equation is called the *state equation* of the original control problem and its solution is briefly called the state. By the Lax-Milgram Lemma, for each $u \in U$ there exists a unique $w \in V$ such that

$$a(w, v) = (u, v) \qquad \text{for all } v \in V$$

and for some $c > 0$ one has $\|w\| \le c\|u\|$. Hence the definition

$$a(Su, v) := (u, v) \qquad \text{for all } v \in V \qquad (4.3)$$

specifies a continuous linear operator $S : U \to V$. In terms of this operator and the scalar product $(\cdot, \cdot)$ in $U = L_2(\Omega)$, problem (4.1) can be written in the form

$$\min_u J(u) := \frac{1}{2}(Su - z, Su - z) + \frac{\rho}{2}(u, u) \quad \text{where} \quad u \in G := U_{ad}. \quad (4.4)$$

Here we have identified G with U_{ad} to simplify the notation and to underline the connection with the preceding section. Because there is no danger of confusion we use the same notation J for the objective functionals in (4.4) and (4.1).

The problem (4.4) is called the reduced problem associated with (4.1).

Lemma 7.42. *The functional J defined in (4.4) is continuously differentiable and strongly convex on U. One has*

$$\langle J'(u), d \rangle = (Su - z, Sd) + \rho(u, d) \quad \text{for all } u, d \in U. \quad (4.5)$$

Proof: By expanding $J(u + d)$ we get

$$J(u + d) = J(u) + (Su - z, Sd) + \rho(u, d) + \frac{1}{2}(Sd, Sd) + \frac{\rho}{2}(d, d) \quad (4.6)$$
$$\text{for all } u, d \in U.$$

Now S is linear and continuous and hence there exists a constant $c_S > 0$ such that

$$\| Sd \| \le c_S \|d\| \quad \text{for all } d \in U. \quad (4.7)$$

Consequently

$$\frac{1}{2}(Sd, Sd) + \frac{\rho}{2}(d, d) = \frac{1}{2}\|Sd\|^2 + \frac{\rho}{2}\|d\|^2 \le \frac{1}{2}(c_S^2 + \rho)\|d\|^2 \quad \text{for all } d \in U.$$

This estimate and the identity (4.6) imply that J is continuously differentiable and one has the representation (4.5). Furthermore, (4.6) yields

$$J(u+d) = J(u) + \langle J'(u), d \rangle + \frac{1}{2}(Sd, Sd) + \frac{\rho}{2}(d, d) \ge J(u) + \langle J'(u), d \rangle + \frac{\rho}{2}\|d\|^2$$

It follows that J is strongly convex. ∎

Let $S^* : U \to V$ denote the operator adjoint to S, i.e.,

$$(u, S^*v) = (Su, v) \quad \text{for all} \quad u \in U, v \in V.$$

Then the Fréchet derivative $J'(u)$ of the cost functional can be represented as an element in $U = L_2(\Omega)$ by

$$J'(u) = S^*(Su - z) + \rho u. \quad (4.8)$$

Theorem 7.43. *The problem (4.4) has a unique optimal solution $\bar{u}$. The condition*

$$\langle J'(\bar{u}), u - \bar{u} \rangle \ge 0 \quad \text{for all } u \in G \quad (4.9)$$

is a necessary and sufficient criterion for $\bar{u} \in G$ to be the optimal solution of (4.4).

Proof: The set $G = U_{ad} \subset U$ is convex and bounded (see Exercise 7.66), and since $\alpha < \beta$ it is clear that $G \neq \emptyset$. The proof of existence of an optimal solution of (4.4) resembles the proof of Lemma 7.4 (recall problem (1.14)): to show that J is bounded from below, from (4.4) and (4.7) we have

$$J(u) = \frac{1}{2}(Su, Su) - (Su, z) + \frac{1}{2}(z, z) + \frac{\rho}{2}(u, u)$$

$$\geq \frac{1}{2}\|z\|^2 - c_S\|u\|\,\|z\| + \frac{\rho}{2}\|u\|^2$$

$$= \frac{1}{2}\|z\|^2 - \frac{c_S^2}{2\rho}\|z\|^2 + \left(\frac{c_S}{\sqrt{2\rho}}\|z\| - \sqrt{\frac{\rho}{2}}\|u\|\right)^2$$

$$\geq \frac{1}{2}(1 - c_S^2)\|z\|^2 \qquad \text{for all } u \in U,$$

provided that $\rho \geq 1$, which is not a restriction. Lemma 7.42 establishes the strong convexity of J. Hence the optimal solution is unique and from Lemma 7.8 the optimality criterion (4.9) is valid. ∎

Let us return to the original form (4.1) of the control problem. This is an abstract optimization problem in a function space with inequality constraints, and these constraints are fairly simple because they restrict only the controls. In what follows we shall retain these inequalities as constraints in the problem and do not include them in the Lagrange functional. That is, to treat the state equation we apply the Lagrange functional defined by

$$L(w, u, v) := J(w, u) - a(w, v) + (u, v) \qquad \text{for all} \quad w, v \in V, \ u \in U. \quad (4.10)$$

The structure of the problem (4.1) ensures the existence of all partial derivatives of L. These derivatives have the representations

$$\langle L_w(w, u, v), \psi \rangle = (w - z, \psi) - a(\psi, v) \qquad \text{for all} \quad \psi \in V, \quad (4.11)$$

$$\langle L_v(w, u, v), \xi \rangle = a(w, \xi) - (u, \xi) \qquad \text{for all} \quad \xi \in V, \quad (4.12)$$

$$\langle L_u(w, u, v), \mu \rangle = \rho(u, \mu) + (v, \mu) \qquad \text{for all} \quad \mu \in U, \quad (4.13)$$

for all $w, v \in V$ and $u \in U$. We then get the following result:

Theorem 7.44. *A pair $(\bar{w}, \bar{u}) \in V \times G$ is a solution of (4.1) if and only if there exists $\bar{v} \in V$ such that*

$$\langle L_w(\bar{w}, \bar{u}, \bar{v}), w \rangle = 0 \qquad \text{for all} \quad w \in V, \quad (4.14)$$

$$\langle L_v(\bar{w}, \bar{u}, \bar{v}), v \rangle = 0 \qquad \text{for all} \quad v \in V, \quad (4.15)$$

$$\langle L_u(\bar{w}, \bar{u}, \bar{v}), u - \bar{u} \rangle \geq 0 \qquad \text{for all} \quad u \in G. \quad (4.16)$$

Proof: We shall use (4.11)–(4.13) to show that (4.14)–(4.16) are equivalent to (4.9). Then by Theorem 7.43 we are done.

First, assume that (4.14)–(4.16) are valid for some triple $(\bar{w}, \bar{u}, \bar{v})$. Now (4.12) and (4.15) imply that

$$\bar{w} = S\bar{u}.$$

From (4.11) and (4.14) we then have

$$(S\bar{u} - z, w) = a(w, \bar{v}) \qquad \text{for all} \quad w \in V.$$

Choosing $w = Sd$ here for an arbitrary $d \in U$, the definition (4.3) of the operator S yields

$$(S\bar{u} - z, Sd) = a(Sd, \bar{v}) = (d, \bar{v}) \qquad \text{for all} \quad d \in U.$$

Hence $\bar{v} = S^*(S\bar{u} - z)$. But then (4.8), (4.13) and (4.16) give

$$\langle J'(\bar{u}), u - \bar{u} \rangle = (S^*(S\bar{u} - z) + \rho\bar{u}, u - \bar{u}) \geq 0 \qquad \text{for all} \quad u \in G.$$

That is, (4.9) is satisfied.

The converse argument is similar; one chooses $\bar{w} := S\bar{u}$ and $\bar{v} := S^*(S\bar{u} - z)$. ∎

Remark 7.45. The relations that constitute the optimality conditions of Theorem 7.44 can be characterized as follows: the identity (4.15) is simply the given state equation, the relation (4.14) is the associated adjoint equation and (4.16) is the feasible direction characterization for the objective functional J that was established in Theorem 7.43.   □

Remark 7.46. The proofs above reveal that the derivative $J'(u) \in U$ has the representation

$$J'(u) = v + \rho u,$$

where $v \in V$ is defined iteratively by the variational equations

$$a(w, \xi) = (u, \xi) \qquad \text{for all} \quad \xi \in V, \qquad (4.17\text{a})$$

$$a(\psi, v) = (w - z, \psi) \qquad \text{for all} \quad \psi \in V. \qquad (4.17\text{b})$$

As we already stated, (4.17a) is a weak formulation of the state equation and (4.17b) is the related adjoint equation. The Lax-Milgram lemma guarantees existence and uniqueness of a solution of (4.17a) and of (4.17b), i.e., to the state and adjoint state equations, respectively. Written as a boundary value problem, (4.17a) describes the classical state equation

$$-\Delta w = u \quad \text{in } \Omega, \qquad w = 0 \quad \text{on } \Gamma.$$

Similarly (4.17b) corresponds to the classical adjoint equation

$$-\Delta v = w - z \quad \text{in } \Omega, \qquad w = 0 \quad \text{on } \Gamma.$$

In this example the differential operator with its homogeneous boundary condition is selfadjoint. Consequently the gradient of the objective functional can be evaluated by solving two Poisson equations. If the underlying differential operator is not selfadjoint then of course the adjoint equation will not coincide with the state equation. $\square$

The set $G \subset U$ is nonempty, convex and bounded. Thus, given $u \in U$, we can define the L_2 projector $P : U \to G$ onto the feasible set G by $Pu := \tilde{u} \in G$ where

$$(\tilde{u} - u, \tilde{u} - u) \leq (v - u, v - u) \qquad \text{for all} \quad v \in G.$$

By Lemma 7.4 the operator P is non-expansive, i.e.,

$$\|Pu - Pv\| \leq \|u - v\| \qquad \text{for all} \quad u, v \in U.$$

The optimality condition can be reformulated in terms of the projector P:

Lemma 7.47. *For each $\sigma > 0$ the optimality condition (4.9) is equivalent to the fixed point equation*

$$\bar{u} = P\left(\bar{u} - \sigma J'(\bar{u})\right). \tag{4.18}$$

In particular if $\sigma = 1/\rho$ then the element $\bar{u} \in G$ minimizes J in (4.4) if and only if, with the adjoint state $\bar{v}$ defined by (4.17), one has

$$\bar{u} = P\left(-\frac{1}{\rho}\bar{v}\right). \tag{4.19}$$

Proof: This is left to the reader as Exercise 7.67. ∎

Remark 7.48. In the case $G = U_{ad}$ that we have considered, it is easy to evaluate the projector P. Indeed, the following representation holds almost everywhere in Ω:

$$[Pu](x) = \begin{cases} \beta & \text{if} \quad u(x) > \beta, \\ u(x) & \text{if} \quad \alpha \leq u(x) \leq \beta, \\ \alpha & \text{if} \quad u(x) < \alpha. \end{cases} \tag{4.20}$$

$\square$

Lemma 7.49. *Consider the case of functions α, β and assume that $\alpha, \beta \in H^1(\Omega)$. Then the optimal solution $\bar{u}$ also lies in the space $H^1(\Omega)$.*

Proof: The optimal adjoint state $\bar{v}$ automatically lies in $V = H_0^1(\Omega)$. Now (4.19), the representation (4.20) of the projector P, and the property (see [KS80]) that

$$\psi, \zeta \in H^1(\Omega) \qquad \Longrightarrow \qquad \max\{\psi, \zeta\} \in H^1(\Omega)$$

together prove the assertion. ∎

Remark 7.50. In the case $\rho = 0$ which we excluded, where no regularization is made, then $\bar{u} \in H^1(\Omega)$ cannot be guaranteed. Moreover, the characterization (4.19) is then no longer valid. When $\rho = 0$, one typically encounters discontinuous optimal solutions (the bang-bang principle; see [Trö05]).   □

Lemma 7.47 tells us that the problem (4.4) can be reformulated as a fixed point problem. Then one can generate a sequence $\{u^k\}_{k=1}^{\infty} \subset G$ via the fixed point iteration

$$u^{k+1} = T(u^k) := P\left(u^k - \sigma J'(u^k)\right), \quad k = 0, 1, \dots \qquad (4.21)$$

with some step-size parameter $\sigma > 0$ and any starting element $u^0 \in V$. This is the *method of projected gradients* and for it we have the following convergence result:

Theorem 7.51. *Consider the projected gradient method defined by (4.21) for a sufficiently small step size parameter $\sigma > 0$ and an arbitrary $u^0 \in V$. Then the method generates a sequence $\{u^k\}_{k=1}^{\infty} \subset G$ that converges to the optimal solution $\bar{u}$ of problem (4.4).*

Proof: Since J is strongly convex, the operator J' is strongly monotone. Furthermore, the quadratic structure of J ensures that J' is Lipschitz continuous. Lemma 7.6 (cf. Exercise 7.68) implies that for any sufficiently small parameter $\sigma > 0$ the sequence $\{u^k\} \subset G$ converges to the fixed point $\bar{u}$ of the operator T defined by (4.21). Then, by Lemma 7.47, the element $\bar{u}$ is an optimal solution of (4.4). ∎

Remark 7.52. To accelerate the convergence of the projected gradient method, the step size should be changed adaptively at each iteration. That is, (4.21) is modified to

$$u^{k+1} = T(u^k) := P\left(u^k - \sigma_k J'(u^k)\right), \quad k = 0, 1, \dots$$

with appropriate $\sigma_k > 0$.   □

The projected gradient method (4.21) is defined directly in the space U. If the Fréchet derivative J' is expressed via the adjoint state and the simple evaluation (4.20) of the projection P is taken into account, then each step of (4.21) can be implemented in the following way:

- Determine $w^k \in V$ as the solution of the state equation

$$a(w^k, v) = (u^k, v) \qquad \text{for all} \quad v \in V.$$

- Determine $v^k \in V$ as the solution of the adjoint equation

$$a(v, v^k) = (v, w^k - z) \qquad \text{for all} \quad v \in V.$$

- Set $u^{k+1/2} = u^k - \sigma(v^k + \rho u^k)$ then compute $u^{k+1} \in G$ by projection into G, i.e., according to the rule

$$u^{k+1}(x) := \begin{cases} \beta & \text{if} \quad u^{k+1/2}(x) > \beta, \\ u^{k+1/2}(x) & \text{if} \quad \alpha \leq u^{k+1/2}(x) \leq \beta, \\ \alpha & \text{if} \quad u^{k+1/2}(x) < \alpha. \end{cases} \qquad \text{for } x \in \Omega$$

The first two steps here require the solution of elliptic boundary value problems, which will be done numerically by some appropriate discretization. We will consider this aspect in Section 7.4.2.

In the Lagrange functional (4.10) we included the state equations but omitted the restrictions that define the set U_{ad} of admissible states—these constraints were treated directly. We now study briefly an extended form of the Lagrange functional that includes the control restrictions as well as the state equations. For this purpose, the underlying semi-ordering will be defined by means of the closed convex cone

$$K := \{ u \in U : u \geq 0 \quad \text{a.e. in } \Omega \}. \tag{4.22}$$

The set K coincides with its dual cone, viz.,

$$K = K^+ := \{ u \in U : (u, z) \geq 0 \quad \forall z \in K \}.$$

Now the set of admissible controls can be described equivalently by

$$U_{ad} = \{ u \in U : (\alpha, z) \leq (u, z) \leq (\beta, z) \quad \forall z \in K \}. \tag{4.23}$$

Recalling (4.10), we arrive at the extended Lagrange functional

$$L(w, u, v, \zeta, \eta) := J(w, u) - a(w, v) + (u, v)$$
$$+ (\alpha - u, \zeta) + (u - \beta, \eta) \tag{4.24}$$
$$\text{for all} \quad w, v \in V, \quad \zeta, \eta \in K, \ u \in U$$

that also includes terms from the control constraints. We use our previous notation L for this extended Lagrange functional also as there is no danger of confusion.

Analogously to Theorem 7.44, we have the following characterization (see also Exercise 7.69):

Theorem 7.53. *A pair $(\bar{w}, \bar{u}) \in V \times G$ is a solution of problem (4.1) if and only if there exist $\bar{v} \in V$ and $\bar{\zeta}, \bar{\eta} \in K$ such that*

$$\langle L_w(\bar{w}, \bar{u}, \bar{v}, \bar{\eta}, \bar{\zeta}), w \rangle = 0 \qquad \text{for all} \quad w \in V, \tag{4.25a}$$

$$\langle L_v(\bar{w}, \bar{u}, \bar{v}, \bar{v}, \bar{\eta}, \bar{\zeta}), v \rangle = 0 \qquad \text{for all} \quad v \in V, \tag{4.25b}$$

$$\langle L_u(\bar{w}, \bar{u}, \bar{v}, \bar{v}, \bar{\eta}, \bar{\zeta}), u \rangle = 0 \qquad \text{for all} \quad w \in U, \tag{4.25c}$$

$$\langle L_\eta(\bar{w}, \bar{u}, \bar{v}, \bar{v}, \bar{\eta}, \bar{\zeta}), \eta \rangle \leq 0 \qquad \text{for all} \quad \eta \in K, \tag{4.25d}$$

$$\langle L_\eta(\bar{w}, \bar{u}, \bar{v}, \bar{v}, \bar{\eta}, \bar{\zeta}), \bar{\eta} \rangle = 0 \tag{4.25e}$$

$$\langle L_\zeta(\bar{w}, \bar{u}, \bar{v}, \bar{v}, \bar{\eta}, \bar{\zeta}), \zeta \rangle \leq 0 \qquad \text{for all} \quad \zeta \in K, \tag{4.25f}$$

$$\langle L_\zeta(\bar{w}, \bar{u}, \bar{v}, \bar{v}, \bar{\eta}, \bar{\zeta}), \bar{\zeta} \rangle = 0. \tag{4.25g}$$

Remark 7.54. The system (4.25a)–(4.25d) are the Karush-Kuhn-Tucker conditions for the problem (4.1). Primal-dual methods, such as primal-dual penalty methods, can be constructed using these conditions. Furthermore, the system (4.25a)–(4.25d) gives a way of controlling active set strategies in algorithms of this type; see, e.g., [68]. □

At several points in the current section we have emphasized the close connection between variational inequalities and optimal control problems. Before we move on to study discretization techniques, an optimal control problem from [72] that contains elliptic variational inequalities as constraints will be considered. In particular, the definition of the feasible domain in this problem involves a variational inequality.

In addition to the previously-known data in the model problem (4.1), let us now be given a function $\Psi \in U$ such that at least one $v \in V$ exists with $v \leq \Psi$. In the case of functions Ψ that are continuous on $\overline{\Omega}$, this assumption is equivalent to $\Psi(x) \geq 0$ for all $x \in \Gamma$. Setting

$$Q := \{ v \in V : v \leq \Psi \}, \tag{4.26}$$

we consider the following optimal control problem with variational inequalities:

$$\min_{w,u} J(w, u) := \frac{1}{2} \|w - z\|^2 + \frac{\rho}{2} \|u\|^2 \tag{4.27a}$$

where $w \in Q,$ $a(w, v - w) \geq (u, v - w)$ for all $v \in Q,$ \qquad (4.27b)

$$u \in U_{ad} := \{ u \in U : a \leq u \leq b \}. \tag{4.27c}$$

The bilinear form $a(\cdot, \cdot)$ here is slightly simpler than the one appearing in [72]. Our bilinear form induces a mapping $F : V \to V^*$ defined by

$$\langle Fw, v \rangle = a(w, v) \qquad \text{for all} \quad w, v \in V.$$

Then F satisfies the assumptions of Section 7.1 and Theorem 7.7 guarantees that for each $u \in U$ there is a unique $w \in Q$ that satisfies the variational inequality (4.27b). Thus, taking $\tilde{S}u = w$, we can define an operator $\tilde{S} : U \to Q$ by

$$\tilde{S}u \in Q, \qquad a(\tilde{S}u, v - \tilde{S}u) \geq (u, v - \tilde{S}u) \quad \text{for all} \quad v \in Q. \tag{4.28}$$

Unlike the operator S of (4.3), $\tilde{S}$ is nonlinear. Nevertheless we have

Lemma 7.55. *The operator $\tilde{S}$ is Lipschitz continuous and*

$$(\tilde{S}u - \tilde{S}\tilde{u}, u - \tilde{u}) \geq 0 \qquad \text{for all} \quad u, \tilde{u} \in U. \tag{4.29}$$

Proof: Setting $w := \tilde{S}u$ and $\tilde{w} := \tilde{S}\tilde{u}$ in (4.28), we get

$$a(w, v - w) \geq (u, v - w) \quad \text{and} \quad a(\tilde{w}, v - \tilde{w}) \geq (u, v - \tilde{w}) \qquad \text{for all} \quad v \in Q.$$

Choose $v = \tilde{w}$ and $v = w$, respectively, in these inequalities then add them. The bilinearity of $a(\cdot, \cdot)$ yields

$$a(w - \tilde{w}, w - \tilde{w}) \leq (u - \tilde{u}, w - \tilde{w}). \tag{4.30}$$

But $a(v, v) \geq 0$ for all $v \in V$, and (4.29) follows. Applying the Cauchy-Schwarz inequality to (4.30), then invoking the ellipticity of $a(\cdot, \cdot)$ and the continuous embedding $V \hookrightarrow U$, we infer the Lipschitz continuity of $\tilde{S}$. ∎

Compared with (4.4), we now use $\tilde{S}$ instead of S and obtain the reduced control problem

$$\min_{u} \tilde{J}(u) := \frac{1}{2} (\tilde{S}u - z, \tilde{S}u - z) + \frac{\rho}{2} (u, u) \qquad \text{where} \quad u \in G. \tag{4.31}$$

One can prove results for this $\tilde{J}$ that are analogues of Lemma 7.42 and Theorem 7.43, and thereby obtain existence of an optimal solution.

Theorem 7.56. *The problem (4.31) has a unique optimal solution $\bar{u}$.*

Remark 7.57. Unlike the functional J of (4.4), in general the objective functional $\tilde{J}$ fails to be differentiable. Thus non-smooth methods are required for the numerical treatment of (4.31); see, e.g., [72] and [97]. □

7.4.2 Discretization by Finite Element Methods

In the model problem (4.1), the spaces $V = H_0^1(\Omega)$ and $U = L_2(\Omega)$ that occur naturally are infinite dimensional. In the numerical solution of optimal control problems with partial differential equations, a common approach is to discretize both spaces by, e.g., finite element methods and to approximate the set U_{ad} of admissible controls by a suitable set comprising finitely many conditions. One then has a *complete discretization*.

An alternative approach to discretization has been proposed by Hinze [69]. It is based directly on the optimality condition (4.19). Here the states, the state equation, the associated adjoint states and adjoint equation are discretized, but the space U that contains the controls is not. One can then achieve optimal-order convergence estimates because, unlike complete discretization, the discretization of control is not linked to a finite element space U_h that is chosen a priori. At the end of this section Hinze's technique will be briefly discussed, but we first describe complete discretization in detail and provide a convergence analysis for it.

Let $V_h \subset V$ be a conforming finite element space with a basis of functions $\varphi_j \in V$ where $j = 1, \ldots, N$, i.e.,

$$V_h = \text{span}\left\{\varphi_j\right\}_{j=1}^N.$$

We likewise select

$$U_H := \text{span}\left\{\psi_l\right\}_{l=1}^M$$

for linearly independent functions $\psi_l \in U$, $l = 1, \ldots, M$ lying in U. Here h, $H > 0$ denote the mesh sizes of the associated underlying grids (e.g., triangulations).

The choices of trial spaces $V_h \subset V$ and $U_H \subset U$ that we make in discretizing the states v_h and controls u_h could in principle be independent, but to obtain good error estimates these discretizations should be related. For instance, it is wasteful to solve the state equations to a very high degree of accuracy while simultaneously computing only a rough approximation U_H of U. In practical applications the same triangulation is often used for the construction of both V_h and U_H. Compared with the general case of independent discretizations of V and U, one then gets a simplified evaluation of the integrals that define $a(w, v)$ and (u, v). Here we shall restrict ourselves to the case where the grids coincide and index both discrete spaces by h, i.e., we write U_h instead of U_H.

We consider the following discretization of the model problem (4.1):

$$\min_{w_h, u_h} J(w_h, u_h) := \frac{1}{2}\|w_h - z\|^2 + \frac{\rho}{2}\|u_h\|^2 \tag{4.32a}$$

where $w_h \in V_h$, $a(w_h, v_h) = (u_h, v_h)$ for all $v_h \in V_h$ $\tag{4.32b}$

$$u_h \in G_h \subset U_h. \tag{4.32c}$$

Here $G_h := U_{h,ad}$ is some discretization of the set of admissible controls. In the literature the discretization

$$G_h := U_{h,ad} := \left\{ u_h \in U_h : \alpha \le u_h \le \beta \quad \text{a.e. in } \Omega \right\}$$

is frequently used, and if U_h comprises piecewise linear C^0 elements one then has

$$u_h \in U_{h,ad} \quad \Longleftrightarrow \quad \alpha \leq u(x_h) \leq \beta \quad \forall x_h \in \Omega_h,$$

where Ω_h denotes the set of all inner vertices of the triangulation. That is, in this case the set G_h is defined by a finite number of pointwise conditions.

The representation (4.23) inspires an alternative discretion of U_{ad} given by

$$G_h := U_{h,ad} := \{\, u_h \in U_h \,:\, (\alpha, z_h) \leq (u_h, z_h) \leq (\beta, z_h) \ \forall z_h \in U_h, \ z_h \geq 0 \,\}.$$

This is a type of mixed discretization where for simplicity we have used the same discrete space U_h as a dual space to approximate the constraints, i.e., the spaces of trial functions and test functions here are identical.

In what follows we concentrate on the case where $\Omega \subset \mathbb{R}^n$ is a polyhedral set and piecewise linear C^0 elements on a regular triangulation of Ω are used to discretize both state and adjoint equations. For the discretization of the controls we work with piecewise constant functions over the same triangulation. Let $\mathcal{T}_h = \{\Omega_j\}_{j=1}^M$ be our triangulation with mesh size $h > 0$. Then

$$V_h := \{\, v_h \in C(\overline{\Omega}) \,:\, v_h|_{\Omega_j} \in P_1(\Omega_j), \ j = 1, \ldots, M \,\},$$

$$U_h := \{\, u_h \in L_2(\Omega) \,:\, u_h|_{\Omega_j} \in P_0(\Omega_j), \ j = 1, \ldots, M \,\}.$$

We assume that all vertices of the polyhedron Ω are also vertices of our triangulation; thus no additional errors arise from boundary approximations. For the numerical treatment of more general domains Ω in optimal control problems, see [35] and [90].

Analogously to the solution operator $S : U \to V$ of the state equation in the original problem, we set $S_h u = w_h$ where $w_h \in V_h$ is the unique solution of the discrete variational equation

$$a(w_h, v_h) = (u, v_h) \qquad \text{for all } v_h \in V_h. \tag{4.33}$$

The operator $S_h : U \to V_h$ is linear and continuous. Clearly (4.33) is simply the application of the finite element method to the state equation (4.3). Since $V_h \subset V$ and the same bilinear form was used in both the continuous and discrete cases, (4.33) is a conforming discretization. Consequently the Lax-Milgram lemma guarantees existence and uniqueness of a solution $w_h \in V_h$ for each $u \in U$. Furthermore,

$$\| S_h d \| \leq c_S \| d \| \quad \text{for all } d \in U \tag{4.34}$$

where c_s is the same constant as in (4.7). In terms of the operator S_h, a reduced discrete control problem associated with (4.32) is given by

$$\min_{u_h} J_h(u_h) := \frac{1}{2} (S_h u_h - z, S_h u_h - z) + \frac{\rho}{2} (u_h, u_h) \quad \text{where } u_h \in G_h. \tag{4.35}$$

For this problem we have

Theorem 7.58. *The reduced discrete problem (4.35) has a unique optimal solution $\bar{u}_h \in G_h$. The condition*

$$\langle J'_h(\bar{u}_h), u_h - \bar{u}_h \rangle \geq 0 \qquad \text{for all } u_h \in G_h \tag{4.36}$$

is a necessary and sufficient criterion for $\bar{u}_h \in G_h$ to be a solution of (4.35).

Proof: The proof of Theorem 7.43 can easily be modified for the discrete case.

∎

Define the operator $S_h^* : U_h \to V_h$ by

$$(u_h, S_h^* v_h) = (S_h u_h, v_h) \qquad \text{for all } u_h \in U_h, \ v_h \in V_h,$$

i.e., S_h^* is the adjoint of S_h. Then the derivative $J'_h(u)$ of the discrete objective functional can be expressed as an element of U_h by

$$J'_h(u_h) = S_h^*(S_h u_h - z) + \rho \, u_h. \tag{4.37}$$

Similarly to Lemma 7.47, we have

Lemma 7.59. *Let $P_h : U \to G_h$ denote the $L_2(\Omega)$ projection onto G_h. For each $\sigma > 0$ the optimality condition (4.36) is equivalent to the fixed point equation*

$$\bar{u}_h = P_h \left(\bar{u}_h - \sigma J'_h(\bar{u}_h) \right). \tag{4.38}$$

In particular if $\sigma = 1/\rho$ the element $\bar{u}_h$ minimizes J_h in (4.35) if and only if, with the associated discrete adjoint state $\bar{v}_h$, one has

$$\bar{u}_h = P_h \left(-\frac{1}{\rho} \bar{v}_h \right). \tag{4.39}$$

Before proving convergence of the discrete optimal solution $\bar{u}_h$ to the solution $\bar{u}$ of the original problem, we briefly discuss the finite-dimensional representation of the discrete problem (4.35). In terms of our basis functions, elements $w_h \in V_h$ and $u_h \in U_h$ are given by

$$w_h = \sum_{j=1}^{N} \hat{w}_j \, \varphi_j \qquad \text{and} \qquad u_h = \sum_{j=1}^{M} \hat{u}_j \, \psi_j,$$

for some $\hat{w}_h = (w_j) \in \mathbb{R}^N$ and $\hat{u}_h = (u_j) \in \mathbb{R}^M$. Thus the discrete state equation (4.33) is equivalent to the Galerkin system of equations

$$\sum_{j=1}^{N} a(\varphi_j, \varphi_i) \, w_j = \sum_{j=1}^{M} (\psi_j, \varphi_i) \, u_j, \qquad i = 1, \ldots, N.$$

Defining the stiffness matrix $A_h = (a_{ij})$ where $a_{ij} := a(\varphi_j, \varphi_i)$ and the mass matrix $B_h = (b_{ij})$ where $b_{ij} := (\psi_j, \varphi_i)$, we can write the Galerkin system as

$$A_h \hat{w} = B_h \hat{u}. \tag{4.40}$$

Hence $\hat{w} = A_h^{-1} B_h \hat{u}$, but in practice computation of the matrix inverse must of course be avoided. Instead, $\hat{w} \in \mathbb{R}^N$ should be found using some efficient solution technique for the discrete elliptic system (4.40). To clarify this, observe that the finite-dimensional representation of the complete discrete system is

$$\min_{w_h, u_h} J_h(w_h, u_h) = \frac{1}{2} \hat{w}^T C_h \hat{w} - d_h^T \hat{w} + \frac{\rho}{2} \hat{u}^T E_h \hat{u} \tag{4.41a}$$

where $\qquad A_h \hat{w} = B_h \hat{u}, \qquad \alpha \leq u_j \leq \beta, \quad j = 1, \ldots, M, \tag{4.41b}$

and the matrices $C_h = (c_{ij})$ and $E_h = (e_{ij})$, and $d_h = (d_i) \in \mathbb{R}^N$, are defined by

$$c_{ij} := (\varphi_i, \varphi_j), \qquad e_{ij} := (\psi, \psi_j) \qquad \text{and} \qquad d_i := (z, \varphi_i).$$

Although the problem (4.41) is a quadratic optimization problem with rather simple constraints—namely box constraints—its numerical solution requires special solution techniques based on its special structure because of the very high dimensions of the discrete spaces involved. In this regard, (4.37) enables efficient evaluation of the gradient of the reduced objective function, which motivates the application of gradient-based minimization methods. Other numerical techniques that are suitable for this problem make successful use of active set strategies [13] or non-smooth Newton methods [116]. Also, the penalty methods discussed in Section 7 with a modified parameter-selection rule can be used to solve (4.41).

Next, we analyse the convergence of the solution $\bar{u}_h$ of the discrete problem (4.35) to the solution $\bar{u}$ of the continuous problem (4.4) as the mesh size h approaches zero. We shall use techniques developed in [35] for the semi-discrete case.

Let $\Pi_h : U \to U_h$ denote the orthogonal projector in $L_2(\Omega)$ defined by

$$\Pi_h u \in U_h \quad \text{with} \quad \|\Pi_h u - u\| \leq \|u_h - u\| \qquad \text{for all} \quad u_h \in U_h.$$

In our case, where U_h is the space of piecewise constant functions, this projector is explicitly given by

$$[\Pi_h u](x) = \frac{1}{\text{meas } \Omega_j} \int_{\Omega_j} u(\xi) \, d\xi \qquad \text{for all } x \in \Omega_j, \quad j = 1, \ldots, M \quad (4.42)$$

where meas Ω_j is the measure of the element Ω_j.

Lemma 7.60. *For the projector Π_h defined in (4.42), one has*

$$u \in G \qquad \Longrightarrow \qquad \Pi_h u \in G_h, \tag{4.43}$$

$$(\Pi_h u - u, v_h) = 0 \qquad \text{for all} \quad v_h \in U_h, \tag{4.44}$$

and there exists a constant $c > 0$ such that

$$\|\Pi_h v - v\| \leq ch \|v\|_1 \qquad \text{for all} \quad v \in H^1(\Omega). \tag{4.45}$$

Proof: From the definition of G we have

$$\alpha \leq u(x) \leq \beta \quad \text{for almost all } x \in \Omega.$$

By (4.42) this yields $\alpha \leq u_j \leq \beta$ for $j = 1, \ldots, M$, where $u_j :=$ $\dfrac{1}{\text{meas } \Omega_j} \int_{\Omega_j} u(\xi) \, d\xi$. Thus $u_h \in G_h$.

As U_h is a linear subspace of U, the identity (4.44) is a standard characterization of Π_h by means of necessary and sufficient conditions for attainment of the minimal distance in our definition.

Finally, the estimate (4.45) is a consequence of the Bramble-Hilbert Lemma. ∎

Theorem 7.61. *As $h \to 0$, the solution $\bar{u}_h \in G_h$ of the discrete problem (4.35) converges in the norm $\|\cdot\|$ to the solution $\bar{u} \in G$ of the original continuous problem (4.4). Furthermore, if Ω is convex, then there exists a constant $c > 0$ such that*

$$\|\bar{u}_h - \bar{u}\| \leq ch. \tag{4.46}$$

Proof: First, we show that $\|\bar{u}_h\|$ is bounded as $h \to 0$. The function $\hat{u}_h \equiv \alpha$ belongs to U_h and $\alpha \leq \hat{u}_h \leq \beta$. That is, $\hat{u}_h \in G_h$ for each $h > 0$. The optimality of $\bar{u}_h$ in the discrete problem (4.35) then implies that $J_h(\bar{u}_h) \leq J_h(\hat{u}_h)$. The properties of J_h yield

$$J_h(\bar{u}_h) \geq J_h(\hat{u}_h) + \langle J_h'(\hat{u}_h), \bar{u}_h - \hat{u}_h \rangle + \rho \|\hat{u}_h - \bar{u}_h\|^2$$

$$\geq J_h(\bar{u}_h) + \langle J_h'(\hat{u}_h), \bar{u}_h - \hat{u}_h \rangle + \rho \|\hat{u}_h - \bar{u}_h\|^2$$

and as $J_h'(\hat{u}_h) \in L_2(\Omega)$ one infers that

$$\|J_h'(\hat{u}_h)\| \, \|\bar{u}_h - \hat{u}_h\| \geq |\langle J_h'(\hat{u}_h), \bar{u}_h - \hat{u}_h \rangle| \geq \rho \|\hat{u}_h - \bar{u}_h\|^2.$$

Hence

$$\|\bar{u}_h\| \leq \|\hat{u}_h\| + \frac{1}{\rho} \|J_h'(\hat{u}_h)\|.$$

But $\hat{u}_h \equiv \alpha$ is constant so this bound on $\|\bar{u}_h\|$ is independent of h.

Now we study the convergence. From the representations of the gradients of J and J_h given in (4.8) and (4.37), and the optimality characterizations of Theorems 7.43 and 7.58, we see that

$$(S^*(S\bar{u} - z) + \rho \bar{u}, v - \bar{u}) \geq 0 \qquad \text{for all} \quad v \in G,$$

$$(S_h^*(S_h \bar{u}_h - z) + \rho \bar{u}_h, v_h - \bar{u}_h) \geq 0 \qquad \text{for all} \quad v_h \in G_h.$$

Choose here the particular test functions $v = \bar{u}_h$ and $v_h := \Pi_h \bar{u}$. (The former selection is permissible since $G_h \subset G$.) This yields

$$(S^*(S\bar{u} - z) + \rho\,\bar{u}, \bar{u}_h - \bar{u}) \geq 0,$$

$$(S_h^*(S_h\bar{u}_h - z) - (S^*(S\bar{u}_h - z), \Pi_h\bar{u} - \bar{u}_h) + \\ + ((S^*(S\bar{u}_h - z) + \rho\,\bar{u}_h, \Pi_h\bar{u} - \bar{u}_h) \geq 0.$$

Adding these inequalities and invoking the monotonicity of the operator S^*S yields

$$((S_h^*S_h - S^*S)\bar{u}_h, \Pi_h\bar{u} - \bar{u}_h) + ((S^* - S_h^*)z, \Pi_h\bar{u} - \bar{u}_h) \\ + ((S^*(S\bar{u}_h - z) + \rho\,\bar{u}_h, \Pi_h\bar{u} - \bar{u}) \\ \geq ((S^*S(\bar{u}_h - \bar{u}) + \rho\,(\bar{u}_h - \bar{u}), \bar{u}_h - \bar{u}) \geq \rho\,\|\bar{u}_h - \bar{u}\|^2. \qquad (4.47)$$

The convergence properties of the finite element method and of the projector Π_h ensure that

$$\lim_{h\to 0} \|S_h^*S_h - S^*S\| = 0, \quad \lim_{h\to 0} \|S_h^* - S^*\| = 0 \\ \text{and} \quad \lim_{h\to 0} \|\Pi_h\bar{u} - \bar{u}\| = 0. \qquad (4.48)$$

As we have already shown that $\|\bar{u}_h\|$ is bounded, the left-hand side of (4.47) tends to zero as $h \to 0$. Consequently

$$\lim_{h\to 0} \|\bar{u}_h - \bar{u}\| = 0,$$

as desired.

Suppose now that Ω is convex. Then $S\bar{u} \in V \cap H^2(\Omega)$ and there exists a constant $c > 0$ such that

$$\|(S_h - S)u\| \leq c\,h^2\,\|u\| \quad \text{and} \quad \|(S_h^* - S^*)u\| \leq c\,h^2\,\|u\| \quad \text{for all} \quad u \in U. \quad (4.49)$$

In particular

$$\|(S^* - S_h^*)\,z\| \leq c\,h^2\,\|z\| \qquad (4.50)$$

because $z \in L_2(\Omega)$ by hypothesis, and the boundedness of $\bar{u}_h$ and the estimate (4.49) yield with

$$S_h^*S_h - S^*S = (S_h^* - S^*)S_h + S^*(S_h - S)$$

the bound

$$\|(S_h^*S_h - S^*S)\bar{u}_h\| \leq c\,h^2. \qquad (4.51)$$

Hence we have

$$|((S_h^*S_h - S^*S)\bar{u}_h, \Pi_h\bar{u} - \bar{u}_h) + ((S^* - S_h^*)z, \Pi_h\bar{u} - \bar{u}_h)| \leq c\,h^2. \quad (4.52)$$

Finally, the third summand in the left-hand side of (4.47) will be estimated. First consider the identity

$$(S^*(S\bar{u}_h - z) + \rho\,\bar{u}_h, \Pi_h\bar{u} - \bar{u}) =$$
$$= (S^*(S\bar{u}_h - z) - \Pi_h S^*(S\bar{u}_h - z) + \Pi_h S^*(S\bar{u}_h - z) + \rho\,\bar{u}_h, \Pi_h\bar{u} - \bar{u}).$$

Here $\Pi_h S^*(S\bar{u}_h - z) + \rho\,\bar{u}_h \in U_h$ and the error orthogonality property (4.44) of the projector Π_h then gives immediately

$$(\Pi_h S^*(S\bar{u}_h - z) + \rho\,\bar{u}_h, \Pi_h\bar{u} - \bar{u}) = 0.$$

From the convexity of Ω we have $S^*(S\bar{u}_h - z) \in V \cap H^2(\Omega)$ and invoking Lemma 7.60 we get

$$|(S^*(S\bar{u}_h - z), \Pi_h\bar{u} - \bar{u})| \leq c\,h^2\,\|S^*(S\bar{u}_h - z)\|_2\,\|\bar{u}\|_2. \tag{4.53}$$

The convexity of Ω implies that for some constant $c > 0$ one has

$$\|S^*v\|_2 \leq c\,\|v\| \qquad \text{for all} \quad v \in L_2(\Omega).$$

Combining this with the previously-shown boundedness of $\bar{u}_h$, we see that the right-hand side of (4.53) is bounded by ch^2. Now recalling (4.52) and (4.47), we get

$$\rho\,\|\bar{u}_h - \bar{u}\|^2 \leq c\,h^2.$$

Then inequality (4.46) follows. ∎

Remark 7.62. Replacing the piecewise constant discretization of the control by a piecewise linear discretization will usually improve the order of convergence. Note however that the property (4.43), which was critical in our convergence proof, is no longer valid for the analogous projector $\tilde{\Pi}_h : U \to V_h$ in the piecewise linear case. For the complete discretization of states and controls with piecewise linear C^0 elements it can be shown that $\|\bar{u}_h - \bar{u}\| = O(h^{3/2})$; see [103].

Remark 7.63. As well as complete discretization (i.e., the discretization of the state equation, the adjoint equation and of the controls) semi-discretization techniques are also proposed and analysed in the literature. In [35] the controls are discretized but the state equation is not. Our proof of Theorem 7.61 is a modification of an argument from [35]. In that proof, when a complete discretization is used, some additional terms will appear in (4.50)–(4.52). In [69] Hinze proposes an alternative approach that discretizes the state and adjoint equations but not the controls. This simplifies the analysis substantially because then in both the continuous and semi-discrete cases one projects onto the same set of feasible controls. As a consequence of the projection that occurs in the optimality criterion, the optimal control usually has reduced smoothness compared with the solution of the state equation. Hence any discretization of the control that is chosen a priori can guarantee only a lower order of approximation. Unlike an explicit discretization, an implicit discretization via the projection can handle the specific behaviour of the piecewise representation of the projection. This leads to an overall discretization error that is derived only from the state equation and its adjoint, and thereby leads to an optimal order of convergence.

Analogously to the characterization (4.19), in the semi-discrete case a feasible control $\tilde{u}_h$ is optimal if and only if

$$\tilde{u}_h = P\left(-\frac{1}{\rho}\tilde{v}_h\right), \tag{4.54}$$

where $\tilde{v}_h \in V_h$ denotes the discrete adjoint state associated with $\tilde{u}_h \in U$. This is indeed a semi-discretization because no a priori selection of the discrete space U_h is required. It is shown in [69] under relatively mild assumptions that

$$\|\tilde{u}_h - \bar{u}\| \leq ch^2.$$

A fundamental property of the semi-discretization proposed in [69] is that, unlike in (4.39), the projector P appears but P_h does not. The method can be implemented numerically using the specific structure of $P\left(-\frac{1}{\rho}\tilde{v}_h\right)$. □

Remark 7.64. An alternative approach to improving the order of convergence appears in [90]. Here the optimal solution $\bar{u}_h$ obtained by full discretization is post-processed to yield an improved discretization $\hat{u}_h \in U$. Writing $\bar{v}_h$ for the discrete adjoint associated with $\bar{u}_h$, in a variant of (4.54) the new approximation $\hat{u}_h \in U$ is defined by

$$\hat{u}_h = P\left(-\frac{1}{\rho}\bar{v}_h\right).$$

One then has (see [90])

$$\|\hat{u}_h - \bar{u}\| \leq ch^2. □$$

Remark 7.65. The DWR method mentioned in Chapter 4 can also be used to adaptively solve control problems; see [BR03, Chapter 8].   □

In closing, let us remind the reader of the necessity of using suitably adapted solution techniques that exploit sparsity and special structure in any efficient numerical treatment of discrete control problems. As previously mentioned, active set strategies, non-smooth Newton methods or penalty methods can be applied. In the present book we shall not pursue these ideas but instead refer to the literature and point to the general aspects of efficient solution techniques for discrete variational equations that were presented in Chapter 8; some of these basic principles are relevant to control problems.

Exercise 7.66. Show that the set $U_{ad} \subset L_2(\Omega)$ defined in (4.1) is convex and closed.

Exercise 7.67. Prove Lemma 7.47.

Exercise 7.68. Use the norms of the operators S and S^* and properties of the bilinear form $a(\cdot, \cdot)$ to determine as large a constant $\sigma_{max} > 0$ as possible such that the projected gradient method converges for any $\sigma \in (0, \sigma_{max})$.

Exercise 7.69. Using the fact that K is a convex cone, show that the relations (4.25d), (4.25e) and (4.25f), (4.25g) are equivalent to the variational inequalities

$$\langle L_\eta(\bar{w}, \bar{u}, \bar{v}, \bar{v}, \bar{\eta}, \bar{\zeta}), \eta - \bar{\eta} \rangle \leq 0 \qquad \text{for all} \quad \eta \in K$$

and

$$\langle L_\zeta(\bar{w}, \bar{u}, \bar{v}, \bar{v}, \bar{\eta}, \bar{\zeta}), \zeta - \bar{\zeta} \rangle \leq 0 \qquad \text{for all} \quad \zeta \in K.$$

Exercise 7.70. Show by means of a simple example that for the L_2 projector $\tilde{\Pi}_h$ into the space of piecewise linear functions one does not always have (cf. Remark 7.62)

$$u \in G \quad \Longrightarrow \quad \tilde{\Pi}_h u \in G_h.$$

8

Numerical Methods for Discretized Problems

8.1 Some Particular Properties of the Problems

When a partial differential equation is discretized, the given infinite-dimensional problem —which implicitly determines a function—is transformed approximately into a finite-dimensional system of algebraic equations whose unknowns are the degrees of freedom of the discretization method. For example, the method of finite differences leads to a system of equations in the approximate values v_i of the desired continuous solution $v(\cdot)$ at the chosen grid points x_i, $i = 1, \ldots, N$. Similarly, the finite element method generates the Ritz-Galerkin system of equations whose unknowns are the coefficients of the finite-dimensional basis in the ansatz space. In both these techniques a linear differential equation is converted approximately into a finite-dimensional system of linear equations. This system typically has the following properties:

- an extremely high dimension;
- a sparse coefficient matrix;
- a bad condition number.

Because of these properties, standard methods of linear algebra such as Gaussian elimination are at best inefficient if they are applied to solve the linear systems that arise in the discretization of partial differential equations; in fact sometimes they cannot be used at all because of their excessive memory requirements or high computational time demands. For some linear systems that have a particular structure one can use instead appropriate fast direct solvers (see Section 8.2), while for general systems of equations special iterative methods can be recommended.

To illustrate the situation we consider as a sample problem a discretization of the Dirichlet problem

$$-\Delta u = f \quad \text{in } \Omega := (0,1)^2,$$
$$u|_\Gamma = 0,$$

$$(1.1)$$

by piecewise linear finite elements on a uniform mesh, as sketched in Figure 8.1.

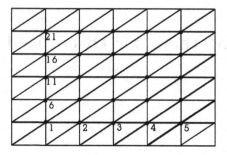

Figure 8.1 Uniform triangulation

Then (1.1) is replaced by the linear system

$$A_h v_h = f_h. \tag{1.2}$$

If we number the inner grid points row by row, as shown in Figure 8.1, then the induced system matrix A_h has the structure displayed in Figure 8.2 where all remaining entries of the matrix are zero.

$$
\left(
\begin{array}{ccccc|ccccc|ccccc|ccccc|ccccc}
4 & -1 & & & & -1 & & & & & & & & & & & & & & & & & & & \\
-1 & 4 & -1 & & & & -1 & & & & & & & & & & & & & & & & & & \\
 & -1 & 4 & -1 & & & & -1 & & & & & & & & & & & & & & & & & \\
 & & -1 & 4 & -1 & & & & -1 & & & & & & & & & & & & & & & & \\
 & & & -1 & 4 & & & & & -1 & & & & & & & & & & & & & & & \\
\hline
-1 & & & & & 4 & -1 & & & & -1 & & & & & & & & & & & & & & \\
 & -1 & & & & -1 & 4 & -1 & & & & -1 & & & & & & & & & & & & & \\
 & & -1 & & & & -1 & 4 & -1 & & & & -1 & & & & & & & & & & & & \\
 & & & -1 & & & & -1 & 4 & -1 & & & & -1 & & & & & & & & & & & \\
 & & & & -1 & & & & -1 & 4 & & & & & -1 & & & & & & & & & & \\
\hline
 & & & & & -1 & & & & & 4 & -1 & & & & -1 & & & & & & & & & \\
 & & & & & & -1 & & & & -1 & 4 & -1 & & & & -1 & & & & & & & & \\
 & & & & & & & -1 & & & & -1 & 4 & -1 & & & & -1 & & & & & & & \\
 & & & & & & & & -1 & & & & -1 & 4 & -1 & & & & -1 & & & & & & \\
 & & & & & & & & & -1 & & & & -1 & 4 & & & & & -1 & & & & & \\
\hline
 & & & & & & & & & & -1 & & & & & 4 & -1 & & & & -1 & & & & \\
 & & & & & & & & & & & -1 & & & & -1 & 4 & -1 & & & & -1 & & & \\
 & & & & & & & & & & & & -1 & & & & -1 & 4 & -1 & & & & -1 & & \\
 & & & & & & & & & & & & & -1 & & & & -1 & 4 & -1 & & & & -1 & \\
 & & & & & & & & & & & & & & -1 & & & & -1 & 4 & & & & & -1 \\
\hline
 & & & & & & & & & & & & & & & -1 & & & & & 4 & -1 & & & \\
 & & & & & & & & & & & & & & & & -1 & & & & -1 & 4 & -1 & & \\
 & & & & & & & & & & & & & & & & & -1 & & & & -1 & 4 & -1 & \\
 & & & & & & & & & & & & & & & & & & -1 & & & & -1 & 4 & -1 \\
 & & & & & & & & & & & & & & & & & & & -1 & & & & -1 & 4 \\
\end{array}
\right)
$$

Figure 8.2 Stiffness matrix

A discretization with step size $h = 1/6$ already yields a 25×25 matrix, yet this matrix contains only 105 non-zero entries. The uniformity of the triangulation and the systematic numbering of the grid points (and consequently of the components u_i of $u_h = (u_i)$) imply that the positions of the non-zero entries of the matrix A_h and their values can easily be determined. Consequently with this discretization the stiffness matrix A_h is not usually stored explicitly: instead, its entries are evaluated directly when they are needed in the calculation.

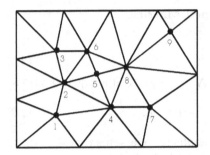

Figure 8.3 Irregular grid

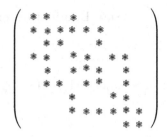

Figure 8.4 Stiffness matrix induced by the discretization of Fig. 8.3

The situation is very different if irregular grids are used. For example, let us consider the triangulation of Figure 8.3 with the numbering of the nodes given there. In this case a piecewise linear finite element discretization yields a stiffness matrix whose non-zero entries are denoted by $*$ in Figure 8.4; compare this with the regular pattern visible in Figure 8.2.

In the cases that we consider, the properties of the underlying continuous problem and of the triangulation generate a system matrix that is positive definite, weakly diagonally dominant and irreducible. Hence Gaussian elimination without pivoting can be applied to factor A_h into lower and upper triangular matrices L and U such that

$$A = LU. \tag{1.3}$$

To simplify the notation here and in what follows, the index h that is associated with the discretization is omitted when the properties of a fixed discretization are studied. The LU decomposition (1.3) of A generally produces non-zero entries at positions where the original matrix A had zero entries; this unwelcome property is called *fill-in*. It can be reduced by appropriate pivoting strategies—see, e.g., [HY81]. It is easy to see that without pivoting no fill-in occurs beyond the maximal bandwidth of the non-sparse entries of A. To reduce the computational effort, LU decomposition can be adapted to the banded structure of A (see Section 8.2).

An important attribute of the discretized problems is that from consistency one can deduce that asymptotically, as the mesh diameter tends to zero, the discrete problem inherits certain properties of the underlying boundary value problem. This attribute is the theoretical basis for multigrid methods and for the construction of optimal preconditioners; it leads to highly efficient solution techniques.

8.2 Direct Solution Techniques

8.2.1 Gaussian Elimination for Banded Matrices

Consider the system

$$A u = b \tag{2.1}$$

of linear equations where $A = (a_{ij})$ is an $N \times N$ matrix of bandwidth m, i.e.,

$$a_{ij} = 0 \qquad \text{for } |i - j| > m. \tag{2.2}$$

Suppose that $m \ll N$. Furthermore, for simplicity we assume that the system matrix A is strictly diagonally dominant, i.e., that

$$|a_{ii}| > \sum_{j \neq i} |a_{ij}|, \quad i = 1, \dots, N. \tag{2.3}$$

Under this condition, Gaussian elimination without pivoting shows that A can be factored as

$$A = L U \tag{2.4}$$

for some lower triangular matrix L and upper triangular matrix U. By matrix multiplication, taking into account the triangular structure of L and U, it follows that

$$\sum_{j=1}^{\min\{i,k\}} l_{ij} u_{jk} = a_{ik}, \qquad i, k = 1, \dots, N. \tag{2.5}$$

If the diagonal elements of either L or U are given (e.g., one could set $l_{ii} = 1$ for all i), then successive evaluations of (2.5) for $\min\{i, k\} = 1, \dots, N$ that alternate between the rows and columns of A can be solved easily to yield all entries of L and U. Here the triangular natures of L and U and the banded structure of A imply additionally that

$$l_{ij} = 0 \qquad \text{if } j > i \text{ or } j < i - m \tag{2.6}$$

and

$$u_{ij} = 0 \qquad \text{if } j < i \text{ or } j > i + m. \tag{2.7}$$

Combining (2.5)–(2.7), we get

$$a_{ik} = \sum_{j=\max\{i,k\}-m}^{\min\{i,k\}} l_{ij}u_{jk} \qquad \text{for } i,\, k = 1,\dots,N. \tag{2.8}$$

Thus when $l_{ii} = 1$ for all i, the previously-mentioned iterative elimination applied to (2.8) for $\min\{i,k\} = 1,\dots,N$ becomes

$$u_{ik} = a_{ik} - \sum_{j=\max\{1,k-m\}}^{i-1} l_{ij}\,u_{jk}, \quad k = i,\dots,i+m, \tag{2.9a}$$

$$l_{ik} = \frac{1}{u_{kk}}\left[a_{ik} - \sum_{j=\max\{1,i-m\}}^{k-1} l_{ij}\,u_{jk} \right], \quad i = k+1,\dots,k+m. \tag{2.9b}$$

If A is symmetric and positive definite, then a further reduction of memory requirements and computing time is made possible by using a symmetric triangular decomposition, which is obtained from the general formula by setting $l_{ii} = u_{ii}$ instead of $l_{ii} = 1$. Then the formulas (2.9a) and (2.9b) imply that $U = L^T$, i.e., we obtain

$$A = L L^T, \tag{2.10}$$

which is called the *Cholesky decomposition*. Positive definiteness of A (and consequently the invertibility of L) often follow from the properties of the underlying continuous problem. As an alternative to (2.10) one can use the modified form

$$A = L D L^T, \tag{2.11}$$

where D is a diagonal matrix and L is lower triangular with $l_{ii} = 1$, $i = 1,\dots,N$. Under additional conditions the splitting (2.11) can also be used when dealing with indefinite matrices.

The discretization of partial differential equations leads as a rule to systems of linear equations with a sparse system matrix A. The efficient implementation of Gaussian elimination in this setting requires a suitable management of memory storage for the relatively few non-zero entries present. In addition, special renumbering techniques (see, e.g., [HY81]) should be used a priori to reduce the amount of fill-in generated in the LU decomposition by Gaussian elimination. Another approach to the efficient treatment of the linear system is the construction of an approximate LU decomposition and its use as a preconditioner in iterative methods; see Section 8.4.

Finally we discuss a special case that often appears as a subproblem in the numerical treatment of discretizations of differential equations. If the system matrix A satisfies (2.2) with $m = 1$ (i.e., A is tridiagonal), then the elimination procedure (2.9) can be greatly simplified: set $u_{11} = a_{11}$ and $u_{NN} = a_{NN} - l_{N,N-1}u_{N-1,N}$, then compute

$$u_{ii} = a_{ii} - l_{i,i-1}u_{i-1,i}, \quad u_{i,i+1} = a_{i,i+1}, \quad l_{i+1,i} = a_{i+1,i}/u_{ii},$$
$$i = 2,\dots,N-1. \tag{2.12}$$

This special form of Gaussian elimination is often called the Thomas algorithm. It can be used very efficiently in the ADI-method (see Section 8.3) to solve the subproblems generated there.

Remark 8.1. In implementing the finite element method one might expect to assemble the entire stiffness matrix A before applying some solution method to the finite-dimensional system of linear equations, but in case of a fine mesh the linear system has such a high dimension that the amount of storage needed may exceed the available memory capacity. To reduce the quantity of data needed in memory, one can combine the assembling process with the generation of the LU decomposition of A. For if we have computed the first p rows and p columns of A, then by (2.5) the LU decomposition can already be started and the corresponding elements of L and U can be determined. This technique is called the *frontal solution method* (see [37], [48]). Its efficiency rests also on the evaluation of the stiffness matrix via element stiffness matrices. For consider some grid point x_i. If the element stiffness matrix associated with each subdomain that contains x_i is evaluated, then the sum of the entries in these matrices yields not only a_{ii} but also all remaining non-zero elements of the ith row and i^{th} column of the stiffness matrix A. In this way the LU decomposition can be carried out simultaneously with the generation of the stiffness matrix over successive subdomains. A simple example to demonstrate the frontal solution method appears in [GRT93].

The frontal solution technique presented in [47] is the basis for the program package MUMPS (MUltifrontal Massively Parallel sparse direct Solver), which gains further efficiency from parallelization; see http://graal.ens-lyon.fr/MUMPS   □

8.2.2 Fast Solution of Discrete Poisson Equations, FFT

Let A be a symmetric invertible $(N-1) \times (N-1)$ matrix and assume that we know a complete system $\{v^l\}_{l=1}^{N-1} \subset \mathbb{R}^{N-1}$ of eigenvectors of A that are mutually orthogonal with respect to some inner product $(\cdot, \cdot)$ - as a rule the usual Euclidean inner product. Denote the associated eigenvalues by $\{\lambda_l\}_{l=1}^{N-1}$. Consider the linear system

$$A u = b. \tag{2.13}$$

Then the solution u of (2.13) can be written as

$$u = \sum_{l=1}^{N-1} c_l v^l \quad \text{where} \quad c_l = \frac{1}{\lambda_l} \cdot \frac{(b, v^l)}{(v^l, v^l)}, \quad l = 1, \ldots, N-1. \tag{2.14}$$

In certain cases symmetries can be used in the evaluation of (2.14) to accelerate the calculation. This phenomenon will be illustrated in the following important one-dimensional example.

Suppose that (2.13) is generated by the discretization of the two-point boundary value problem

$$-u'' = f \quad \text{in } \Omega = (0,1), \qquad u(0) = u(1) = 0,$$

using the standard finite difference method on an equidistant grid with step size $h = 1/N$. That is, we consider the linear system

$$-u_{j-1} + 2u_j - u_{j+1} = h^2 f_j, \qquad j = 1, \ldots, N-1,$$
$$u_0 = u_N = 0. \tag{2.15}$$

The eigenvectors $v^l = (v_j^l)_{j=0}^N$ that correspond to this system matrix satisfy the homogeneous difference equation

$$-v_{j-1}^l + 2v_j^l - v_{j+1}^l = \lambda_l v_j^l, \qquad j = 1, \ldots, N-1,$$
$$v_0^l = v_N^l = 0. \tag{2.16}$$

For simplicity in the presentation of (2.15) and (2.16), the 0th and N^{th} components of the vectors have been included. Using the ansatz $v_j^l = e^{i\rho_l j}$, we obtain from the difference equation

$$2\left(1 - \cos \rho_l\right) e^{i\rho_l j} = \lambda_l e^{i\rho_l j}.$$

Linear combinations of such functions with the boundary conditions $v_0 = v_N = 0$ lead to

$$\sin(\rho_l N) = 0, \qquad l = 1, 2, \ldots$$

Hence $\rho_l = \pm l\pi/N$, $l = 1, 2, \ldots$ From above we then obtain

$$\lambda_l = 2\left(1 - \cos \frac{l\pi}{N}\right) = 4 \sin^2 \frac{l\pi}{2N}, \qquad l = 1, \ldots, N-1,$$

for the eigenvalues λ_l of (2.16). The related real eigenvectors $v^l \in \mathbb{R}^{N-1}$ have the form

$$v_j^l = -\frac{i}{2}(e^{i\rho_l j} - e^{-i\rho_l j}) = \sin(\rho_l j) = \sin \frac{l\pi j}{N}, \qquad l, j = 1, \ldots, N-1. \tag{2.17}$$

Finally one can verify

$$(v^l, v^m) = \begin{cases} 0 & \text{if } l \neq m, \\ \frac{N}{2} & \text{if } l = m. \end{cases}$$

Substituting these results into (2.14) yields

$$c_l = \left(2N \sin^2 \frac{l\pi}{2N}\right)^{-1} \sum_{j=1}^{N-1} b_j \sin \frac{l\pi j}{N}, \qquad l = 1, \ldots, N-1.$$

Thus one must evaluate sums of the form

$$c_l = \sum_{j=1}^{N-1} \tilde{b}_j \sin\frac{l\pi j}{N}, \qquad l = 1,\dots,N-1, \tag{2.18}$$

with $\tilde{b}_j := (2N \sin^2\frac{l\pi}{2N})^{-1} b_j$, to determine the coefficients c_l. A similar computation is needed to find the components u_j of the desired discrete solution $u_h = (u_j)_{j=1}^{N-1}$, because from (2.14) and (2.17) we have

$$u_j = \sum_{l=1}^{N-1} c_l \sin\frac{l\pi j}{N}, \qquad j = 1,\dots,N-1. \tag{2.19}$$

If one exploits the symmetries of the trigonometric functions and computes simultaneously all components u_j, $j = 1,\dots,N-1$, then the sums (2.18) and (2.19) can be calculated in a highly efficient manner by using the *fast Fourier transform (FFT)*. We now describe the basic principles of this technique by means of the complete discrete Fourier transform. The sine transform corresponding to (2.19) could be used instead but one then needs to distinguish four cases instead of the division into even and odd indices in our discussion below.

Setting $a := \exp(\frac{i2\pi}{N})$, we consider as a generic case the evaluation of the matrix-vector product whose individual components are sums of the form

$$z_j = \sum_{l=0}^{N-1} \beta_l a^{lj}, \qquad j = 0,1,\dots,N-1. \tag{2.20}$$

This apparently entails a total of N^2 arithmetic operations, but as we shall see the FFT significantly reduces that number. To simplify the analysis we assume that $N = 2^n$ for some positive integer n; for more general cases see, e.g., [107].

Clearly $a^N = 1$. Hence in (2.20) only linear combinations of $1, a, a^2, \dots, a^{N-1}$ can occur. Set $\tilde{N} = N/2$. We split and rewrite the sum (2.20) as

$$\begin{aligned} z_j &= \sum_{l=0}^{\tilde{N}-1} \beta_l a^{lj} + \sum_{l=\tilde{N}}^{N-1} \beta_l a^{lj} \\ &= \sum_{l=0}^{\tilde{N}-1} \left(\beta_l + \beta_{\tilde{N}+l} a^{\tilde{N}j}\right) a^{lj}, \quad j = 0,1,\dots,N-1. \end{aligned} \tag{2.21}$$

Recall that $1 = a^N = a^{2\tilde{N}}$. Set $\tilde{a} = a^2$. Then when $j = 2k$ is even we obtain

$$\begin{aligned} z_{2k} &= \sum_{l=0}^{\tilde{N}-1} \left(\beta_l + \beta_{\tilde{N}+l} a^{2\tilde{N}k}\right) a^{2lk} \\ &= \sum_{l=0}^{\tilde{N}-1} \left(\beta_l + \beta_{\tilde{N}+l}\right) \tilde{a}^{lk}, \quad k = 0,1,\dots,\tilde{N}-1. \end{aligned} \tag{2.22}$$

When $j = 2k + 1$ is odd, by (2.21) and $a^{\tilde{N}} = -1$ one has the representation

$$
\begin{aligned}
z_{2k+1} &= \sum_{l=0}^{\tilde{N}-1} \left(\beta_l + \beta_{\tilde{N}+l} \, a^{\tilde{N}\,(2k+1)} \right) a^{l\,(2k+1)} \\
&= \sum_{l=0}^{\tilde{N}-1} \left(\beta_l - \beta_{\tilde{N}+l} \right) a^l \, \tilde{a}^{lk}, \qquad k = 0, 1, \dots, \tilde{N} - 1.
\end{aligned}
\tag{2.23}
$$

Setting

$$
\begin{aligned}
\tilde{z}_k &= z_{2k}, & \tilde{\beta}_l &= \beta_l + \beta_{\tilde{N}+l}, \\
\hat{z}_k &= z_{2k+1}, & \hat{\beta}_l &= (\beta_l - \beta_{\tilde{N}+l})\,a^l,
\end{aligned}
\qquad k, l = 0, 1, \dots, \tilde{N} - 1, \tag{2.24}
$$

then for the even and odd indices we obtain the respective representations

$$
\tilde{z}_k = \sum_{l=0}^{\tilde{N}-1} \tilde{\beta}_l \, \tilde{a}^{lk} \qquad \text{and} \qquad \hat{z}_k = \sum_{l=0}^{\tilde{N}-1} \hat{\beta}_l \, \tilde{a}^{lk}, \qquad k = 0, 1, \dots, \tilde{N} - 1.
$$

In each case—this is the key observation—one has again the original form (2.20) but with $\tilde{a}^{\tilde{N}} = 1$ and only half the number of summands. We can use this idea recursively until only a single summand remains, since $N = 2^n$. That is, we obtain a fast realization of the discrete Fourier transform (2.20) where the evaluation of the desired matrix-vector product can be carried out using only $O(N \log N)$ operations instead of the N^2 that were needed in the initial formulation. Figure 8.5 sketches schematically the structure of a single recursive step of the algorithm.

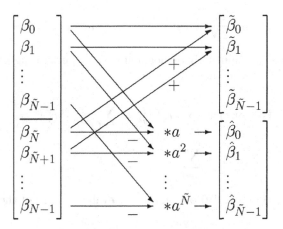

Figure 8.5 One step of an FFT scheme

In the representation (2.19) the discrete sine expansion leads to a similar method (see, e.g., [VL92]) for an efficient implementation of the discrete boundary value problem (2.15). We remark that algorithms for fast Fourier transforms are widely available as standard software (see MATLAB [Dav04] and the NAG Library [NAG05]) and are even supported on some computers by hardware components.

The Fourier transform can be extended to higher-dimensional cases provided that separation of variables can be applied to the given partial differential equation—then the above one-dimensional algorithm is applied separately in each coordinate direction.

Let us consider the model problem

$$-\Delta u = f \qquad \text{in } \Omega = (0,1) \times (0,1),$$
$$u|_\Gamma = 0.$$

Using an equidistant mesh with step size $h := 1/N$ in each coordinate direction, the standard discretization by finite differences (see Chapter 2) generates the discrete problem

$$\left.\begin{aligned}
4u_{jk} - u_{j-1,k} - u_{j,k-1} & \\
- u_{j+1,k} - u_{j,k+1} &= h^2 f_{jk} \\
u_{0k} = u_{Nk} = u_{j0} = u_{jN} &= 0
\end{aligned}\right\} \quad j, k = 1, \ldots, N-1. \qquad (2.25)$$

The associated eigenvectors $v^{lm} = \{v_{jk}^{lm}\}$ can be written in terms of the one-dimensional eigenvectors v^l, $v^k \in \mathbb{R}^{N-1}$ for $l, k = 1, \ldots, N-1$:

$$v_{jk}^{lm} = v_j^l v_k^m = \sin\frac{l\pi j}{N} \sin\frac{m\pi k}{N}. \qquad (2.26)$$

The eigenvalues are

$$\lambda_{lm} = 4\left(\sin^2\frac{l\pi}{2N} + \sin^2\frac{m\pi}{2N}\right), \quad l, m = 1, \ldots, N-1. \qquad (2.27)$$

The eigenvectors are mutually orthogonal with respect to the Euclidean inner product in $\mathbb{R}^{2(N-1)}$: one has

$$(v^{lm}, v^{rs}) = \sum_{j,k=1}^{N-1} v_{jk}^{lm} v_{jk}^{rs} = \sum_{j,k=1}^{N-1} v_j^l v_k^m v_j^r v_k^s =$$
$$= \sum_{j=1}^{N-1} v_j^l v_j^r \sum_{k=1}^{N-1} v_k^m v_k^s = \frac{N^2}{4}\delta_{lr}\,\delta_{ms}.$$

Instead of (2.18) and (2.19) one must now evaluate sums of the form

$$u_{jk} = \sum_{l,m=0}^{N-1} c_{lm} \sin\frac{l\pi j}{N} \sin\frac{m\pi k}{N}.$$

One can rewrite this as

$$u_{jk} = \sum_{m=0}^{N-1} \left(\sum_{l=0}^{N-1} c_{lm} \sin \frac{l\pi j}{N} \right) \sin \frac{m\pi k}{N},$$

which shows that once again we are dealing with one-dimensional problems that can be treated efficiently by FFT techniques.

Remark 8.2. The fast Fourier transform is directly applicable only to particular types of problem. Other fast solvers for discretizations of partial differential equations, such as the cyclic reduction method (see [SN89]), also suffer from restricted applicability. Domain decomposition techniques are a way of generating specific problems suited to fast solvers as subproblems of the original larger problem; in this way the range of applicability of the FFT can be extended. Another application of fast solvers is their use as preconditioners in iterative methods.   □

Remark 8.3. The treatment sketched above of a high-dimensional problem could be modified by using eigenfunctions only in some directions while in the remaining directions other discretization techniques such as finite element methods are applied. One important application of this approach is to problems posed in three-dimensional domains that have rotational symmetry, e.g., pistons; see [67].   □

Exercise 8.4. Let A be a block matrix of the form

$$A = \begin{pmatrix} A_{11} & A_{12} & \cdots & A_{1m} \\ A_{21} & A_{22} & \cdots & A_{2m} \\ & & \cdots & \\ A_{m1} & & \cdots & A_{mm} \end{pmatrix},$$

with $N_i \times N_j$ matrices $A_{ij} \in \mathcal{L}(\mathbb{R}^{N_j}, \mathbb{R}^{N_i})$, where the A_{ii} are invertible and $\sum_{i=1}^{m} N_i = N$. Assume that the condition (cf. 3.39) below

$$\|A_{ii}^{-1}\| \sum_{j \neq i} \|A_{ij}\| < 1, \qquad i = 1, \ldots, m,$$

is satisfied, where $\| \cdot \|$ denotes any matrix norm. Find a block LU decomposition analogue of standard LU decomposition and give a construction procedure for the sub-blocks occurring in it.

Exercise 8.5. Show that the vectors

$$v^{lm} = (v_{jk}^{lm})_{j,k=1}^{N-1} \in \mathbb{R}^{(N-1)^2}$$

defined by (2.26) are mutually orthogonal.

Exercise 8.6. Consider the problem

$$\frac{\partial^2}{\partial x^2} u(x,y) - \frac{\partial}{\partial y}(a(y)\frac{\partial}{\partial y} u(x,y)) = 1 \qquad \text{in } \Omega := (0,1) \times (0,1),$$
$$u|_\Gamma = 0 \qquad\qquad\qquad (2.28)$$

with some continuously differentiable function a. Discretize this problem using finite differences over an equidistant grid with step size $h = 1/N$ in both coordinate directions. Find a solution technique for the ensuing discrete problem that uses the discrete Fourier transform in the x-direction.

8.3 Classical Iterative Methods

8.3.1 Basic Structure and Convergence

Consider again the problem

$$Au = b \qquad\qquad (3.1)$$

where the $N \times N$ system matrix A is invertible and $b \in \mathbb{R}^N$ is given. For any invertible $N \times N$ matrix B, the problem (3.1) is equivalent to

$$Bu = (B - A)u + b.$$

This inspires the fixed-point iteration

$$Bu^{k+1} = (B - A)u^k + b, \qquad k = 0, 1, \ldots, \qquad (3.2)$$

which provides a general model for the iterative treatment of (3.1). In (3.2) the starting vector $u^0 \in \mathbb{R}^N$ has to be chosen. Defining the defect d^k in the kth iterate u^k by $d^k := b - Au^k$, the procedure (3.2) is equivalent to

$$u^{k+1} = u^k + B^{-1}d^k.$$

Introducing a step-size parameter $\alpha_k > 0$ to give us added flexibility, we examine the iteration

$$u^{k+1} = u^k + \alpha_k B^{-1}d^k, \qquad k = 0, 1, \ldots \qquad (3.3)$$

If $B = I$ this becomes the classical Richardson method. If we choose $B \neq I$ then (3.3) is a preconditioned version of Richardson's method and B is called the related preconditioner. When A is symmetric and positive definite, (3.3) is a preconditioned gradient method for the minimization of

$$F(u) := \frac{1}{2}u^T Au - b^T u.$$

In the case of constant step-size parameters, viz., $\alpha_k = \alpha > 0$, $k = 0, 1, \ldots$, the method (3.3) belongs to the general class of iterative methods of the type

$$u^{k+1} = T u^k + t, \quad k = 0, 1, \ldots,$$

where T is some matrix and the vector t lies in $\mathbb{R}^N$: for Richardson's method

$$T = T(\alpha) = I - \alpha B^{-1} A.$$

Let $\| \cdot \|$ an arbitrary vector norm in $\mathbb{R}^N$. To discuss the convergence of iterative methods for the solution of the finite-dimensional problem (3.1), one generally uses one of the following vector norms: (the notation here follows the traditional finite dimensional ones but differs from the previous chapters)

$$\|y\|_\infty := \max_{1 \leq i \leq N} |y_i| \quad (\text{ maximum norm })$$

$$\|y\|_2 := \left(\sum_{i=1}^N y_i^2 \right)^{1/2} \quad (\text{ Euclidean norm })$$

$$\|y\|_1 := \sum_{i=1}^N |y_i| \quad (\text{ discrete } L_1 \text{ norm })$$

$$\|y\|_A := \left(y^T A y \right)^{1/2} \quad (\text{ energy norm }).$$

In the case of the energy norm, it is assumed that the matrix A is symmetric and positive definite. When our analysis is independent of the norm used, we shall omit the subscript that identifies each norm in the list above. In finite-dimensional spaces all norms are equivalent so in theory one can work with any norm, but it is important to note that each change in the mesh size h of the discretization also changes the dimension of the finite-dimensional space where the computed solution lies and consequently the multiplicative factors that relate equivalent norms will usually change with h. Thus when convergence results are expressed in terms of h, the power of h can depend on the finite-dimensional norm used. Furthermore, the norms used in discretizations are often weighted to ensure consistency with continuous norms (cf. Section 2.1).

Whenever we do not explicitly specify the norm of an $N \times N$ matrix C, the reader can assume that we are using the norm

$$\|C\| := \sup_{y \in \mathbb{R}^N, y \neq 0} \frac{\|Cy\|}{\|y\|} \tag{3.4}$$

induced by the underlying vector norm $\| \cdot \|$. Clearly this induced matrix norm satisfies the standard inequality

$$\|Cy\| \leq \|C\| \|y\| \qquad \text{for all } y \in \mathbb{R}^N, \ C \in \mathcal{L}(\mathbb{R}^N),$$

where $\mathcal{L}(\mathbb{R}^N)$ denotes the space of $N \times N$ matrices.

The matrix B that one must select to implement the iterative method (3.2) should satisfy the following requirements:

- The linear systems

$$By = c \tag{3.5}$$

  are relatively easy to solve;
- For each $y \in \mathbb{R}^N$ the vector $c := (B - A)y + b$ can be easily evaluated;
- The norm $\|B^{-1}(B - A)\|$ is small.

These three conditions cannot all be satisfied simultaneously; one must find some compromise.

The convergence behaviour of (3.2) follows from Banach's fixed point theorem for the simpler iteration

$$u^{k+1} = Tu^k + t, \qquad k = 0, 1, \ldots, \tag{3.6}$$

with $T := B^{-1}(B - A)$ and $t := B^{-1}b$. For (3.6) we have

Lemma 8.7. *Assume that $T \in \mathcal{L}(\mathbb{R}^N)$ with $\|T\| < 1$. Then the fixed point problem*

$$u = Tu + t$$

has a unique solution $u \in \mathbb{R}^N$ and for any starting vector $u^0 \in \mathbb{R}^N$ and any $t \in \mathbb{R}^N$ the iteration (3.6) defines a sequence $\{u^k\} \subset \mathbb{R}^N$ that converges to u as $k \to \infty$. In fact

$$\|u^{k+1} - u\| \leq \|T\| \|u^k - u\|, \qquad k = 0, 1, \ldots$$

and

$$\|u^k - u\| \leq \frac{\|T\|^k}{1 - \|T\|} \|u^1 - u^0\|, \qquad k = 1, 2, \ldots . \tag{3.7}$$

Remark 8.8. Inequality (3.7) provides an a priori bound for the quality of the approximation of the desired solution u by the iterate u^k. Replacing u^0 and u^1 by u^{k-1} and u^k respectively, (3.7) also yields the a posteriori bound

$$\|u^k - u\| \leq \frac{\|T\|}{1 - \|T\|} \|u^k - u^{k-1}\|. \tag{3.8}$$

This bound is usually sharper than (3.7) but it is applicable only if u^{k-1} and u^k have already been evaluated.   □

The value of $\|T\|$ depends upon the norm chosen, so Lemma 8.7 gives a sufficient but not necessary condition for convergence of the iteration (3.6). Let

$$\rho(T) := \max_i \{ |\lambda_i| : \lambda_i \text{ an eigenvalue of } T \}$$

denote the *spectral radius* of the iteration matrix T. Then we obtain the following necessary and sufficient criterion:

Lemma 8.9. *The sequence $\{u^k\}$ defined by (3.6) converges for all $t \in \mathbb{R}^N$ and all $u^0 \in \mathbb{R}^N$ if and only if*

$$\rho(T) < 1. \tag{3.9}$$

Proof: We consider only the case of a symmetric matrix T; in the general case, complex eigenvalues and eigenvectors must also be taken into account.

Since T is symmetric there exists an orthogonal matrix C such that

$$CTC^T = diag(\lambda_i) =: \Lambda.$$

The property $C^TC = I$ implies that (3.6) is equivalent to the iteration

$$\tilde{u}^{k+1} = \Lambda\tilde{u}^k + \tilde{t}, \qquad k = 0, 1, \ldots, \tag{3.10}$$

with $\tilde{u}^k := Cu^k$, $\tilde{t} := Ct$. As Λ is a diagonal matrix the iteration (3.10) can be performed component by component, i.e., one can compute

$$\tilde{u}_i^{k+1} = \lambda_i\tilde{u}_i^k + \tilde{t}_i, \qquad i = 1, \ldots, N, \quad k = 0, 1, \ldots \tag{3.11}$$

By definition of the spectral radius $\rho(T)$, the condition (3.9) implies that the sequence $\{\tilde{u}^k\}$ converges as $k \to \infty$. But $\{u^k\}$ is related to $\{\tilde{u}^k\}$ by $u^k = C^T\tilde{u}^k$, and the convergence of the sequence $\{u^k\}$ follows.

If $\rho(T) \geq 1$ then (3.11) implies that the sequence $\{\tilde{u}^k\}$ is divergent, provided that for some index i with $|\lambda_i| \geq 1$ one has $\tilde{u}_i^0 \neq 0$ and $\tilde{t}_i \neq 0$. It follows that $\{u^k\}$ is also divergent. ∎

In the discretization of boundary value problems for partial differential equations, the dimension of the discrete system of equations and the properties of the matrices $A = A_h$ and $B = B_h$ both depend strongly upon the step size h of the discretization. Writing $\|\cdot\|_h$ for the mesh-dependent vector and matrix norms, the number of steps required to guarantee a given accuracy $\varepsilon > 0$ in the above stationary preconditioned Richardson iterative method is heavily influenced by the value

$$\sigma_h := \|T_h\|_h \qquad \text{where} \quad T_h := I_h - \alpha_h B_h^{-1} A_h. \tag{3.12}$$

Lemma 8.7 provides the a priori bound

$$\|u_h^k - u\|_h \leq \frac{(\sigma_h)^k}{1 - \sigma_h}\|u_h^1 - u_h^0\|_h, \quad k = 0, 1, \ldots$$

Hence the required accuracy ε is achieved after at most $k_h(\varepsilon)$ iteration steps, where

$$k_h(\varepsilon) := \min\left\{j \in \mathbb{N} : j \geq \frac{1}{\ln\sigma_h}\left[\ln\varepsilon + \ln(1 - \sigma_h) - \ln\|u_h^1 - u_h^0\|_h\right]\right\}. \tag{3.13}$$

In the next subsection we shall analyse this formula in detail in the context of the Gauss-Seidel and Jacobi iterative methods.

8.3.2 Jacobi and Gauss-Seidel Methods

In the original system (3.1), let us successively use the ith equation to eliminate x_i from the other equations for $i = 1, 2, \ldots, N$. This easily-implemented
approach yields the two iterative methods discussed in this subsection. If each
newly-obtained approximation of a variable is used immediately in the subsequent computation we obtain the *Gauss-Seidel scheme* (single step iteration),
while if the new iterates are only used after a full sweep has been performed,
this is the *Jacobi scheme* (complete step iteration). These schemes are simple
iterative methods for the numerical solution of (3.1). Consider the $(k + 1)^{\text{st}}$
sweep of the iterative method that evaluates $u^{k+1} \in \mathbb{R}^N$; in this sweep, the
i^{th} sub-step of the *Gauss-Seidel scheme* has the form

$$\sum_{j=1}^{i-1} a_{ij} u_j^{k+1} + a_{ii} u_i^{k+1} + \sum_{j=i+1}^{N} a_{ij} u_j^k = b_i, \quad i = 1, \ldots, N \qquad (3.14)$$

while for the *Jacobi scheme* it takes the form

$$\sum_{j=1}^{i-1} a_{ij} u_j^k + a_{ii} u_i^{k+1} + \sum_{j=i+1}^{N} a_{ij} u_j^k = b_i, \quad i = 1, \ldots, N. \qquad (3.15)$$

Let us split the matrix $A = (a_{ij})$ of (3.1) into the sum of its proper lower
triangular constituent L, its diagonal part D and its proper upper triangular
part R, viz.,

$$A = L + D + R \qquad (3.16)$$

with

$$L = \begin{pmatrix} 0 & \cdots & & 0 \\ a_{21} & 0 & \cdots & 0 \\ \cdot & & \cdot & \cdot \\ \cdot & & & \cdot \\ a_{N1} & \cdots & a_{N,N-1} & 0 \end{pmatrix}, \quad R = \begin{pmatrix} 0 & a_{12} & \cdots & a_{1N} \\ 0 & 0 & a_{23} & \cdot & a_{2N} \\ \cdot & & \cdot & & \cdot \\ \cdot & & & \cdot & a_{N-1,N} \\ 0 & \cdot & & \cdot & 0 \end{pmatrix}$$

and $D = diag(a_{ii})$. Then the methods (3.14) and (3.15) are the particular
cases of the general iterative scheme (3.2) obtained when

$$B = L + D \quad \text{and} \quad B = D, \qquad (3.17)$$

respectively. In both these cases the matrices B are either triangular or diagonal. Hence the system of equations (3.5) can be solved very simply.

We must still study the convergence behaviour of these two schemes by
estimating $\|B^{-1}(B - A)\|$ in each case. In the Jacobi scheme, the entries t_{ij}
of the iteration matrix $T := B^{-1}(B - A)$ are

$$t_{ij} = (\delta_{ij} - 1) \frac{a_{ij}}{a_{ii}}, \quad i, j = 1, \ldots, N, \qquad (3.18)$$

where δ_{ij} is the Kronecker delta. We shall use the maximum norm $\|\cdot\|_\infty$ as our vector norm. Then the matrix norm induced by (3.4) is well known to be

$$\|T\| = \max_{1\le i\le N} \sum_{j=1}^{N} |t_{ij}|.$$

Recalling (3.18), we see that

$$\|T\| = \max_{1\le i\le N} \mu_i \tag{3.19}$$

with

$$\mu_i := \frac{1}{|a_{ii}|} \sum_{j=1}^{N} |a_{ij}|, \qquad i = 1,\ldots,N. \tag{3.20}$$

We recall that A is strictly diagonally dominant (see Chapter 2) if one has

$$\mu_i < 1, \qquad i = 1,\ldots,N. \tag{3.21}$$

Lemma 8.10. *Let A be strictly diagonally dominant. Then for the Gauss-Seidel and Jacobi schemes we have the estimate*

$$\|B^{-1}(B-A)\| \le \max_{1\le i\le N} \mu_i < 1,$$

where the values μ_i are defined by (3.20) and $\|\cdot\|$ is induced by the vector norm $\|\cdot\|_\infty$.

Proof: Suppose that

$$Bz = (B-A)y$$

for some $y, z \in \mathbb{R}^N$. From (3.16) and (3.17) we get

$$z_i = -\frac{1}{a_{ii}} \left(\sum_{j=1}^{i-1} a_{ij} z_j + \sum_{j=i+1}^{N} a_{ij} y_j \right), \qquad i = 1,\ldots,N$$

and

$$z_i = -\frac{1}{a_{ii}} \left(\sum_{j=1}^{i-1} a_{ij} y_j + \sum_{j=i+1}^{N} a_{ij} y_j \right), \qquad i = 1,\ldots,N,$$

for the Gauss-Seidel and Jacobi schemes respectively. In the first of these, inequality (3.21) and the maximum norm can be used to prove inductively that

$$|z_i| \le \frac{1}{|a_{ii}|} \sum_{j\ne i} |a_{ij}| \, \|y\| < \|y\| \qquad \text{for} \quad i = 1,\ldots,N$$

and hence

$$\|z\| \le \max_{1\le i\le N} \mu_i \, \|y\|. \tag{3.22}$$

In the Jacobi case, (3.22) is obtained immediately.

The desired estimate now follows from the matrix norm definition (3.4). ∎

The discretization of elliptic problems like Poisson's equation often generates matrices A that are, however, only weakly diagonally dominant; that is, (see Chapter 2) the μ_i defined in (3.20) satisfy

$$\mu_i \leq 1, \qquad i = 1, \ldots, N. \tag{3.23}$$

The matrix A in such elliptic discretizations generally has the chain property (compare Chapter 2), viz., for each pair of indices $i, j \in \{1, \ldots, N\}$ with $i \neq j$ one can find $l = l(i,j) \in \{1, \ldots, N\}$ and a finite sequence $\{i_k\}_{k=0}^{l}$ of indices such that

$$
\begin{aligned}
&i_0 = i, \quad i_l = j; \\
&i_s \neq i_t && \text{for } s \neq t; \\
&a_{i_{k-1} i_k} \neq 0, && \text{for } k = 1, \ldots, l.
\end{aligned}
$$

In the language of graph theory, such a sequence of indices describes a directed path of length l from node i to node j, where two nodes are joined by an arc of the directed graph if the corresponding matrix entry is non-zero.

To keep the presentation simple we consider only the Jacobi scheme when we now analyse the convergence behaviour of our iterative method for weakly diagonal dominant matrices A.

Lemma 8.11. *Let A be a weakly diagonally dominant matrix that has the chain property. Assume also that for some index $m \in \{1, \ldots, N\}$ one has*

$$|a_{mm}| > \sum_{j \neq m} |a_{mj}|. \tag{3.24}$$

Then for the Jacobi scheme the estimate

$$\|(B^{-1}(B - A))^N\| < 1 \tag{3.25}$$

is valid in the maximum norm.

Proof: Let $y \in \mathbb{R}^N$, $y \neq 0$, be an arbitrary vector. Setting $y^0 = y$, compute

$$y^{k+1} := B^{-1}(B - A)\, y^k, \qquad k = 0, 1, \ldots$$

As we are dealing with the Jacobi scheme, this is simply

$$y_i^{k+1} = -\frac{1}{a_{ii}} \sum_{j \neq i} a_{ij}\, y_j^k, \qquad i = 1, \ldots, N, \quad k = 0, 1, \ldots \tag{3.26}$$

The weak diagonal dominance of A yields

$$|y_i^{k+1}| \leq \frac{1}{|a_{ii}|} \sum_{j \neq i} |a_{ij}|\, |y_j^k| \leq \|y^k\|, \qquad i = 1, \ldots, N, \quad k = 0, 1, \ldots,$$

and hence
$$\|y^{k+1}\| \leq \|y^k\| \leq \|y\|, \qquad k = 0,1,\ldots \tag{3.27}$$

Recalling the index m of (3.24), let us define

$$I^k := \{i: \text{ there exists a directed path of length } k \text{ from } i \text{ to } m\}$$

and set $I^0 = \{m\}$. We shall show by induction that

$$|y_i^{k+1}| < \|y\| \quad \text{ for all } i \in I^k, \; k = 0,1,\ldots \tag{3.28}$$

First, (3.24) implies that
$$|y_m^1| < \|y\|.$$

But $I^0 = \{m\}$, so we have verified (3.28) for $k = 0$. We move on to the inductive step "k true $\Rightarrow k+1$ true":
Let $i \in I^{k+1}\backslash\{m\}$. By the definition of I^{k+1}, there exists $s \in I^k$ with $a_{is} \neq 0$. From (3.26) we have

$$|y_i^{k+2}| \leq \frac{1}{a_{ii}} \sum_{j \neq i, \, j \neq s} |a_{ij}|\,|y_j^{k+1}| + \frac{|a_{is}|}{|a_{ii}|}\,|y_s^{k+1}|.$$

But $a_{is} \neq 0$, $s \in I^k$, and taking into account (3.27) and the inductive hypothesis (3.28) we deduce that

$$|y_i^{k+2}| < \left(\frac{1}{a_{ii}} \sum_{j \neq i} |a_{ij}| \right) \|y\| \leq \|y\|.$$

That is, (3.28) is valid, with k replaced by $k+1$, for all indices $j \in I^{k+1}\backslash\{m\}$. For the index m, from (3.24) and (3.27) we get

$$|y_m^{k+2}| \leq \left(\frac{1}{a_{mm}} \sum_{j \neq m} |a_{mj}| \right) \|y\| < \|y\|.$$

Thus the estimate (3.28) is valid in all cases.

The finite dimensionality of the problem (3.1) and the chain property of A imply that
$$I^{N-1} = \{1,\ldots,N\}.$$

The bound (3.25) now follows from (3.4) and (3.28). ∎

For the Gauss-Seidel scheme one has the following result (see [GR94]):

Lemma 8.12. *Let A be a symmetric and positive definite matrix. Then for the Gauss-Seidel scheme the estimate*

$$\|B^{-1}(B - A)\| < 1$$

is valid in the energy norm induced by A.

Lemmas 8.10 and 8.12 give sufficient conditions for

$$\|B^{-1}(B - A)\| < 1, \tag{3.29}$$

while Lemma 8.11 gives a sufficient condition for $\|(B^{-1}(B - A))^N\| < 1$. By Lemma 8.7 and a well-known generalization of it, either of these conditions is sufficient to guarantee the convergence of the associated iterative method (3.2). The smaller the value of $\|B^{-1}(B - A)\|$, the faster the convergence of the iteration. But for the schemes (3.14) and (3.15) one usually obtains

$$\lim_{h \to 0} \|B_h^{-1}(B_h - A_h)\| = 1,$$

where A_h and B_h denote the matrices associated with the discretization for mesh size h. For if for instance the standard method of finite differences with an equidistant grid is applied to

$$\begin{aligned} -\Delta u &= f & \text{in } \Omega = (0, 1) \times (0, 1), \\ u|_\Gamma &= 0, \end{aligned} \tag{3.30}$$

then, using a double indexing that is appropriate to the structure of the two-dimensional problem, the discrete problem $A_h u_h = f_h$ that is generated has the form

$$\left.\begin{aligned} 4u_{ii} - u_{i-1,j} - u_{i,j-1} & \\ -u_{i+1,j} - u_{i,j+1} &= h^2 f_{ij} \\ u_{0j} = u_{nj} = u_{i0} = u_{in} &= 0 \end{aligned}\right\} \quad i, j = 1, \ldots, n - 1. \tag{3.31}$$

Then the Gauss-Seidel scheme with a natural ordering of the sub-steps is

$$\left.\begin{aligned} 4u_{ii}^{k+1} - u_{i-1j}^{k+1} - u_{ij-1}^{k+1} & \\ -u_{i+1j}^k - u_{ij+1}^k &= h^2 f_{i,j} \\ u_{0j}^{k+1} = u_{nj}^{k+1} = u_{i0}^{k+1} = u_{in}^{k+1} &= 0 \end{aligned}\right\} \quad \begin{aligned} i, j &= 1, \ldots, n - 1, \\ k &= 0, 1, \ldots \end{aligned} \tag{3.32}$$

It turns out that for the matrices A_h and B_h associated with the Gauss-Seidel method (and also for the matrices in the Jacobi iteration) one has the asymptotic property (see, e.g., [SN89])

$$\|B_h^{-1}(B_h - A_h)\| = 1 - O(h^2). \tag{3.33}$$

In both methods the computational effort for a complete sweep is $O(N)$, where N is the total number of degrees of freedom. This is a consequence of the fact that each row of A_h contains a fixed number of non-zero entries (i.e., this number is independent of the mesh size h). Under mild additional assumptions, the same property holds for quite general discretizations and for problems posed in domains Ω that lie in any spatial dimension d.

From (3.33) we obtain the estimate

$$\|u_h^k - u_h\| \le \left(1 - O(h^2)\right)^k \|u_h^0 - u_h\|.$$

To ensure $\|u_h^k - u_h\| \le \varepsilon \|u_h^0 - u_h\|$ one takes

$$k \ge \frac{\ln(\varepsilon)}{\ln(1 - O(h^2))}$$

sweeps of the Gauss-Seidel or Jacobi method. With $\ln(1 - O(h^2)) = O(h^2)$ this leads to an expected number of $k = O(h^{-2}|\ln(\varepsilon)|)$ steps. Now with $N = O(h^{-d})$ we obtain the complexity

$$O\left(h^{-(2+d)}\,|\ln(\varepsilon)|\right). \tag{3.34}$$

of the Gauss-Seidel and Jacobi methods. In particular for problems posed in the plane the complexity is $O(N^2 \ln \varepsilon)$. Therefore when the mesh size h is small these methods require a large amount of computational effort to achieve a given accuracy; thus for very fine discretizations it is impractical to use the Gauss-Seidel or Jacobi method to solve the discrete problem. Nevertheless the good spatially local character of their iteration steps means that these schemes are very useful as smoothers in multigrid methods.

Before moving on to the investigation of some simple improvements of these iterative methods in the next section, we make a few further remarks.

The elimination process of the Gauss-Seidel iterative method in its natural ordering depends upon the chosen numbering of the variables and therefore upon the labelling of the grid points. If the Gauss-Seidel method (3.14) is applied to a problem (3.1) whose solution u has some symmetry property then the iteration will not in general preserve this symmetry even if the starting vector u^0 and the matrix A are symmetric. This undesirable behaviour may be eliminated or at least damped if methods that use a different ordering in the elimination are combined. If for example a forward version is coupled with a backward version, then one obtains the *symmetric Gauss-Seidel method* where each iteration uses two half-steps and has the form

$$(L + D)u^{k+1/2} + Ru^k = b,$$
$$Lu^{k+1/2} + (D + R)u^{k+1} = b. \tag{3.35}$$

On the other hand, the asymmetric behaviour of the standard Gauss-Seidel method makes it a suitable smoother in multi-grid methods for convection-diffusion problems if the order of elimination corresponds to the expected flow.

In the case of a two-dimensional grid, another way of constructing a symmetric version of the Gauss-Seidel method is through a chessboard-like elimination of the variables. This variant of the Gauss-Seidel method is called red-black iteration and will be discussed later as a particular block Gauss-Seidel method.

8.3.3 Block Iterative Methods

Often discretizations lead to a discrete system of linear equations (3.1) with a special *block structure*. Consider linear systems of the form

$$\sum_{j=1}^{m} A_{ij} u_j = b_i, \qquad i = 1, \ldots, m, \tag{3.36}$$

with

$$b_i, \ u_i \in \mathbb{R}^{N_i}, \quad A_{ij} \in \mathcal{L}(\mathbb{R}^{N_j}, \mathbb{R}^{N_i}), \quad \sum_{i=1}^{m} N_i = N,$$

and invertible matrices A_{ii}, where $\mathcal{L}(\mathbb{R}^{N_j}, \mathbb{R}^{N_i})$ is the space of $N_i \times N_j$ matrices. Then block-oriented versions of the Gauss-Seidel and Jacobi methods are described by

$$\sum_{j=1}^{i} A_{ij} u_j^{k+1} + \sum_{j=i+1}^{m} A_{ij} u_j^{k} = b_i, \qquad i = 1, \ldots, m \tag{3.37}$$

and

$$A_{ii} u_i^{k+1} + \sum_{j \neq i} A_{ij} u_j^{k} = b_i, \qquad i = 1, \ldots, m, \tag{3.38}$$

respectively. The convergence analysis of the original methods transfers readily to their block analogues. For example, the condition

$$\|A_{ii}^{-1}\| \sum_{j \neq i}^{m} \|A_{ij}\| < 1, \qquad i = 1, \ldots, m \tag{3.39}$$

is a sufficient condition for convergence of the iterative methods (3.37) and (3.38). This condition is a straightforward generalization of strong diagonal dominance.

To apply the block Gauss-Seidel method (3.37) to the discrete Poisson equation (3.31), we first collect the discrete variables u_{ij} in each column by setting

$$u_i = (u_{ij})_{j=1}^{n-1} \in \mathbb{R}^{n-1}, \qquad i = 1, \ldots, n-1.$$

Then the linear system (3.31), with $m = n - 1$ and $N_j = m$, $j = 1, \ldots, m$, can be written (see Figure 8.2) in the form

$$-A_{i,i-1} u_{i-1} + A_{ii} u_i - A_{i,i+1} u_{i+1} = h^2 f_i, \qquad i = 1, \ldots, m, \tag{3.40}$$

with

$$A_{ii} = \begin{pmatrix} 4 & -1 & 0 & & \cdot & 0 \\ -1 & 4 & -1 & & 0 & \cdot \\ 0 & -1 & 4 & -1 & 0 \\ & \cdot & \cdot & \cdot & \cdot & \cdot \\ 0 & & \cdot & & 0 & -1 & 4 \end{pmatrix}, \quad A_{i,i-1} = A_{i,i+1} = I, \quad f_i = (f_{ij})_{j=1}^{m}.$$

This leads to the following block version of the Gauss-Seidel method:

$$-A_{i,i-1}u_{i-1}^{k+1} + A_{ii}u_i^{k+1} - A_{i,i+1}u_{i+1}^k = h^2 f_i, \qquad i = 1,\ldots,n-1,$$
$$u_0^{k+1} = u_n^{k+1} = 0. \tag{3.41}$$

At the ith sub-step of the Gauss-Seidel method the vector u_i^{k+1} is defined to be the solution of the system of linear equations

$$A_{ii}\,u_i^{k+1} = b_i^k \qquad \text{with} \qquad b_i^k := h^2 f_i + A_{i,i-1}u_{i-1}^{k+1} + A_{i,i+1}u_{i+1}^k.$$

The system matrices A_{ii} here are strictly diagonally dominant and the fast Gauss elimination (the Thomas algorithm) of Section 8.2 can be applied to compute the solution $u_i^{k+1} \in \mathbb{R}^{n-1}$ of the linear system (3.41).

Alternatively, if in the discretization (3.31) the indices are split into two groups $I_\bullet$ and $I_\circ$ according to Figure 8.6, then the ith equation for any $i \in I_\bullet$ depends only on variables u_j with $j \in I_\circ$ and on u_i.

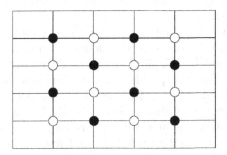

Figure 8.6 Red-black marking of the stiffness matrix

Thus we can eliminate independently all variables in each of the two groups while keeping all variables in the other group fixed. In this way parallel computation is possible and symmetries are preserved. The only drawback is that slowly decreasing oscillations may occur in components of the iterates in the case of certain starting vectors. The Gauss-Seidel method combined with this elimination strategy is called *chessboard iteration* or *red-black iteration*. If the variables are combined into two associated vectors $u_\bullet$ and $u_\circ$, then the original system $A u = b$ can be written as

$$\begin{pmatrix} A_{\bullet\bullet} & A_{\bullet\circ} \\ A_{\circ\bullet} & A_{\circ\circ} \end{pmatrix} \begin{pmatrix} u_\bullet \\ u_\circ \end{pmatrix} = \begin{pmatrix} b_\bullet \\ b_\circ \end{pmatrix} \tag{3.42}$$

and the red-black iteration has the form

$$A_{\bullet\bullet}u_\bullet^{k+1} = b_\bullet - A_{\bullet\circ}u_\circ^k, \qquad A_{\circ\circ}u_\circ^{k+1} = b_\circ - A_{\circ\bullet}u_\bullet^{k+1}. \tag{3.43}$$

It is thus a particular case of the block Gauss-Seidel method. Here $A_{\bullet\bullet}$ and $A_{\circ\circ}$ are diagonal matrices. If $u_\bullet^{k+1}$ is eliminated from the first equation in (3.43) then substituted into the second equation, we see that $u_\circ^{k+1}$ is defined by

$$A_{\circ\circ}u_\circ^{k+1} = b_\circ - A_{\circ\bullet}A_{\bullet\bullet}^{-1}b_\bullet + A_{\circ\bullet}A_{\bullet\bullet}^{-1}A_{\bullet\circ}u_\circ^k.$$

Similarly for $u_\bullet^{k+1}$ we obtain the iterative procedure

$$A_{\bullet\bullet}u_\bullet^{k+1} = b_\bullet - A_{\bullet\circ}A_{\circ\circ}^{-1}b_\circ + A_{\bullet\circ}A_{\circ\circ}^{-1}A_{\circ\bullet}u_\bullet^k.$$

These iterations coincide with the Picard iterative method for the solution of the associated reduced system

$$\left(A_{\circ\circ} - A_{\circ\bullet}A_{\bullet\bullet}^{-1}A_{\bullet\circ}\right)u_\circ = b_\circ - A_{\circ\bullet}A_{\bullet\bullet}^{-1}b_\bullet$$

and

$$\left(A_{\bullet\bullet} - A_{\bullet\circ}A_{\circ\circ}^{-1}A_{\circ\bullet}\right)u_\bullet = b_\bullet - A_{\bullet\circ}A_{\circ\circ}^{-1}b_\circ.$$

The matrices $A_{\circ\circ} - A_{\circ\bullet}A_{\bullet\bullet}^{-1}A_{\bullet\circ}$ and $A_{\bullet\bullet} - A_{\bullet\circ}A_{\circ\circ}^{-1}A_{\circ\bullet}$ that appear here are the Schur complements obtained from partial elimination of variables in the original linear system.

Similarly, by an appropriate subdivision of the indices into two complementary sets $I_\bullet$ and $I_\circ$, one can also define certain *domain decomposition methods (DD methods)*. We shall sketch this idea here for the particular case of the discrete Dirichlet problem (2.25). More general aspects of domain decomposition methods will be discussed in Section 8.6.

In one domain decomposition method the set $I_\circ$ contains all indices that refer to grid points along an interior boundary (see Figure 8.7) and $I_\bullet$ contains the remaining indices. Splitting the degrees of freedom in the same way, the linear subsystem

$$A_{\bullet\bullet}u_\bullet^{k+1} = b_\bullet - A_{\bullet\circ}u_\circ^k$$

is assembled from two independent discrete Poisson equations that are defined on the inner grid points on the opposite sides of the inner discrete boundary. Additionally, at each iteration one must solve the linear system

$$A_{\circ\circ}u_\circ^{k+1} = b_\circ - A_{\circ\bullet}u_\bullet^{k+1},$$

which couples the degrees of freedom on the discrete inner boundary with the Poisson subproblems mentioned above; this sub-step of the block Gauss-Seidel scheme requires only the solution of a tridiagonal linear system.

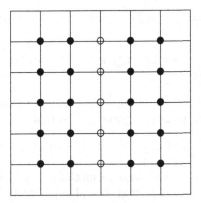

Figure 8.7 Discrete domain decomposition

The domain decomposition not only reduces the computational effort compared with a direct solution of the original discrete problem, but also permits the efficient use of parallel computing. The splitting of the domain in Figure 8.7 corresponds to the splitting of the matrix A given in Figure 8.8. The convergence analysis of the iterative method follows the approach used for standard Gauss-Seidel methods. The expected acceleration of the convergence of the block method, compared with the original pointwise method, will be discussed in the framework of more general domain decomposition methods in Section 8.6.

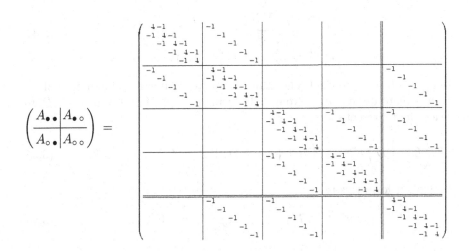

Figure 8.8 Rearranged matrix

Exercise 8.13. For the vector norms $\| \cdot \|_1$ and $\| \cdot \|_2$, show that the induced matrix norms defined by (3.4) can be given explicitly by

$$\|A\|_1 = \max_j \sum_i |a_{ij}| \qquad \text{and} \qquad \|A\|_2 = \rho(A^T A)^{1/2}.$$

In terms of $\|A\|_1$ find a simple criterion that is sufficient for convergence of the Gauss-Seidel method.

Exercise 8.14. Prove that in a finite-dimensional vector space all norms are equivalent. Find sharp constants $\underline{c}$ and $\bar{c}$ such that

$$\underline{c}\|u\|_\infty \leq \|u\|_2 \leq \bar{c}\|u\|_\infty \qquad \text{for all } u \in \mathbb{R}^N.$$

Exercise 8.15. Suppose that Jacobi's method is applied to the model problem (3.31). Use eigenvectors to find the value $\|B_h^{-1}(A_h - B_h)\|_2$ that characterizes the convergence behaviour of the method.

8.3.4 Relaxation and Splitting Methods

Simple iterative methods like Gauss-Seidel or Jacobi suffer severely from the asymptotically slow convergence implied by (3.33). It is possible to accelerate this convergence by means of relaxation, i.e., by modifying the method through the introduction of a user-chosen multiplier for the correction at each step of the iteration; cf. (3.3). Here we shall consider only the Gauss-Seidel method. Our analysis can easily be extended to the Jacobi method.

In the *relaxation method* the iterative step (3.14) is modified as follows:

$$\sum_{j=1}^{i-1} a_{ij} u_j^{k+1} + a_{ii}\tilde{u}_i^{k+1} + \sum_{j=i+1}^{N} a_{ij} u_j^k = b_i, \qquad i = 1, \ldots, N,$$

$$u_i^{k+1} = u_i^k + \omega\,(\tilde{u}_i^{k+1} - u_i^k). \tag{3.44}$$

Here $\omega > 0$ is a so-called relaxation parameter. After the elimination of the intermediate iterates $\tilde{u}_i^{k+1}$ from (3.44) using the splitting $A = L + D + R$, we reach the representation

$$\left(L + \frac{1}{\omega}D\right) u^{k+1} + \left(R + \left(1 - \frac{1}{\omega}\right)D\right) u^k = b, \qquad k = 0, 1, \ldots \tag{3.45}$$

In explicit form this is

$$u^{k+1} = T(\omega)u^k + t(\omega), \qquad k = 0, 1, \ldots, \tag{3.46}$$

where

$$T(\omega) := \left(L + \frac{1}{\omega}D\right)^{-1} \left(\left(L + \frac{1}{\omega}D\right) - A\right),$$

$$t(\omega) := \left(L + \frac{1}{\omega}D\right)^{-1} b. \tag{3.47}$$

Bearing Lemma 8.9 in mind, our goal is to choose the relaxation parameter $\omega > 0$ in such a way that the spectral radius $\rho(T(\omega))$ of the iteration matrix of (3.46) is minimized. As we shall show later, in many situations it is desirable to choose $\omega \in (1,2)$. Because $\omega > 1$ the method (3.45) is often called *successive over-relaxation (SOR)* .

Like the symmetric Gauss-Seidel method (3.35), in the SOR method alternating forward and backward sweeps can symmetrize the procedure. This leads to the following symmetric iteration which is known as the *SSOR method*:

$$\left(L + \tfrac{1}{\omega}D\right) u^{k+1/2} + \left(R + (1 - \tfrac{1}{\omega})D\right) u^k = b,$$

$$\left(L + (1 - \tfrac{1}{\omega})D\right) u^{k+1/2} + \left(R + \tfrac{1}{\omega}D\right) u^{k+1} = b, \qquad k = 0, 1, \dots \quad (3.48)$$

The next lemma relates the relaxation parameter ω to the spectral radius of $T(\omega)$.

Lemma 8.16. *For the iteration matrix $T(\omega)$ of the SOR method one has*

$$\rho(T(\omega)) \geq |1 - \omega| \quad \text{for all } \omega > 0.$$

Proof: The matrix $T(\rho)$ defined by (3.47) can be written in the form

$$T(\omega) = (I + \omega D^{-1}L)^{-1}((1 - \omega)I - \omega D^{-1}R).$$

As $D^{-1}L$ and $D^{-1}R$ are strictly lower and strictly upper triangular matrices, we have

$$\det(T(\omega)) = \det((I + \omega D^{-1}L)^{-1}) \det((1 - \omega)I - \omega D^{-1}R)$$
$$= 1 \cdot (1 - \omega)^N . \qquad (3.49)$$

The determinant of a matrix equals the product of its eigenvalues. Let the eigenvalues of $T(\omega)$ be λ_j, $j = 1, \dots, N$. From (3.49) we get

$$(1 - \omega)^N = \prod_{j=1}^{N} \lambda_j. \qquad (3.50)$$

By definition of the spectral radius, $|\lambda_j| \leq \rho(T(\omega))$ for $j = 1, \dots, N$. The assertion of the lemma now follows from (3.50). ∎

Lemmas 8.16 and 8.9 imply that the condition

$$\omega \in (0, 2)$$

is necessary for convergence of the SOR method.

Under additional assumptions on the system matrix A, the spectral radius $\rho(T(\omega))$ of the iteration matrix of the SOR method can be determined explicitly. We say that the system matrix of (3.1) has *property A* (see [Axe96]) if there exists a permutation matrix P such that (after a simultaneous exchange of rows and columns) one has

$$
P^T A P = \begin{pmatrix}
D_1 & R_1 & & & \\
L_2 & D_2 & R_2 & & \\
& \cdot & \cdot & \cdot & \\
& & \cdot & \cdot & R_{m-1} \\
& & & L_m & D_m
\end{pmatrix}, \tag{3.51}
$$

where the D_j, $j = 1, \ldots, m$, are invertible diagonal matrices. For matrices with property A, the SOR method is convergent; see [HY81], [Axe96].

Lemma 8.17. *Assume that the system matrix A has property A. Then the spectral radius of the iteration matrix $T(\omega)$ of the SOR method is given explicitly by*

$$
\rho(T(\omega)) = \begin{cases}
1 - \omega + \frac{\omega^2 \rho(J)^2}{2} + \omega \rho(J) \sqrt{\frac{\omega^2 \rho(J)^2}{4} - \omega + 1} & \text{if } \omega \in (0, \omega_{opt}], \\
\omega - 1 & \text{if } \omega \in (\omega_{opt}, 2),
\end{cases}
$$

where

$$
\omega_{opt} := \frac{2}{\rho(J)^2} \left(1 - \sqrt{1 - \rho(J)^2} \right) \tag{3.52}
$$

and J is the iteration matrix of the standard Jacobi method. In particular

$$
\rho(T(\omega_{opt})) \le \rho(T(\omega)) \qquad \text{for all } \omega \in (0, 2).
$$

If one uses the optimal value $\omega = \omega_{opt}$ of the relaxation parameter ω, then the SOR method applied to the model problem (3.31) has, as the mesh size h approaches zero, the asymptotic behaviour

$$
\rho(T_h(\omega_{opt})) = 1 - O(h).
$$

Remark 8.18. To determine ω_{opt} exactly from (3.52), one does need to know explicitly the spectral radius $\rho(J)$ of the standard Jacobi method. For any fixed relaxation parameter ω, an estimate $\tilde{\rho}(J)$ of $\rho(J)$ can be obtained from an approximation $\tilde{\rho}(T(\omega))$ of $\rho(T(\omega))$ via

$$
\tilde{\rho}(J) = \frac{\tilde{\rho}(T(\omega)) + \omega - 1}{\omega \, \tilde{\rho}(T(\omega))^{1/2}}.
$$

Then (3.52) will yield an approximation of the optimal relaxation parameter ω_{opt}.   □

For a detailed analysis of the SOR method we refer to [HY81].

Next, we discuss splitting methods. Consider the problem

$$A u = b \qquad (3.53)$$

for some positive definite system matrix A. Suppose that we have two positive definite matrices P and Q such that

$$A = P + Q. \qquad (3.54)$$

Then the system (3.53) is equivalent to either of the two formulations

$$(\sigma I + P)u = (\sigma I - Q)u + b$$

and

$$(\sigma I + Q)u = (\sigma I - P)u + b$$

where $\sigma > 0$ is an arbitrary real parameter. The basic idea of the *splitting method* is to exploit this equivalence by the two-stage iterative procedure

$$\begin{aligned}(\sigma I + P)u^{k+1/2} &= (\sigma I - Q)u^k + b, \\ (\sigma I + Q)u^{k+1} &= (\sigma I - P)u^{k+1/2} + b,\end{aligned} \qquad k = 0, 1, \ldots . \qquad (3.55)$$

After eliminating the intermediate step (which is sometimes called the half step) we obtain the iteration

$$u^{k+1} = T(\sigma)\, u^k + t(\sigma) \qquad (3.56)$$

with

$$T(\sigma) := (\sigma I + Q)^{-1}(\sigma I - P)(\sigma I + P)^{-1}(\sigma I - Q) \qquad (3.57)$$

and a corresponding vector $t(\sigma) \in \mathbb{R}^N$.

For the model problem (3.31), one can find suitable matrices P and Q in a natural way directly from the discretizations of $\dfrac{\partial^2 u}{\partial x^2}$ and $\dfrac{\partial^2 u}{\partial y^2}$. Substituting this choice into (3.55) yields the iteration scheme

$$(2 + \sigma)u_{ij}^{k+1/2} - u_{i-1,j}^{k+1/2} - u_{i+1,j}^{k+1/2} + (2 - \sigma)u_{ij}^k - u_{i,j-1}^k - u_{i,j+1}^k = h^2 f_{ij},$$

$$(2 - \sigma)u_{ij}^{k+1/2} - u_{i-1,j}^{k+1/2} - u_{i+1,j}^{k+1/2} + (2 + \sigma)u_{ij}^{k+1} - u_{i,j-1}^{k+1} - u_{i,j+1}^{k+1} = h^2 f_{ij},$$

for $i, j = 1, \ldots, n - 1$, $k = 0, 1, \ldots$ Peaceman and Rachford proposed this scheme in the context of implicit time discretizations of parabolic problems. Because of the alternation between discretization directions that appears in the iteration, this method is called the alternating direction implicit scheme or *ADI method*.

The convergence analysis of splitting methods is based on following lemma which is often called Kellogg's Lemma.

Lemma 8.19. *Let $C \in \mathcal{L}(\mathbb{R}^N)$ be a symmetric and positive definite matrix. Then for any $\sigma > 0$, the estimate*

$$\| (\sigma I - C)(\sigma I + C)^{-1} \| \leq \rho((\sigma I - C)(\sigma I + C)^{-1}) < 1 \qquad (3.58)$$

is valid in the Euclidean norm.

Proof: The matrix C is symmetric and consequently has an orthonormal system $\{v^i\}_{i=1}^N$ of eigenvectors with associated eigenvalues μ_i. Thus

$$(\sigma I - C)(\sigma I + C)^{-1} v^i = \frac{\sigma - \mu_i}{\sigma + \mu_i} v^i, \qquad i = 1, \dots, N. \qquad (3.59)$$

That is, the vectors $\{v^i\}_{i=1}^N$ are also an orthonormal eigensystem for the matrix $(\sigma I - C)(\sigma I + C)^{-1}$. As C is positive definite and $\sigma > 0$, one has

$$\rho((\sigma I - C)(\sigma I + C)^{-1}) = \max_{1 \leq i \leq N} \left| \frac{\sigma - \mu_i}{\sigma + \mu_i} \right| < 1. \qquad (3.60)$$

To prove the other inequality in (3.58), let $y \in \mathbb{R}^N$ be an arbitrary vector. Then y has the unique representation

$$y = \sum_{i=1}^N \eta_i v^i$$

in terms of the basis $\{v^i\}$. Define $z \in \mathbb{R}^n$ by

$$z := (\sigma I - C)(\sigma I + C)^{-1})y.$$

Then (3.59) implies that

$$z = \sum_{i=1}^N \frac{\sigma - \mu_i}{\sigma + \mu_i} \eta_i v^i.$$

The vectors $\{v^i\}$ are orthonormal and it follows that

$$\|z\|^2 = \sum_{i=1}^N \left(\frac{\sigma - \mu_i}{\sigma + \mu_i} \right)^2 \eta_i^2 \leq \left[\max_{1 \leq i \leq N} \left(\frac{\sigma - \mu_i}{\sigma + \mu_i} \right)^2 \right] \sum_{i=1}^N \eta_i^2$$

$$= \max_{1 \leq i \leq N} \left(\frac{\sigma - \mu_i}{\sigma + \mu_i} \right)^2 \|y\|^2.$$

Recalling (3.60), we therefore have

$$\|z\| \leq \rho((\sigma I - C)(\sigma I + C)^{-1}) \|y\|.$$

As $y \in \mathbb{R}^n$ was arbitrary, the definition (3.4) now implies that

$$\|(\sigma I - C)(\sigma I + C)^{-1}\| \leq \rho((\sigma I - C)(\sigma I + C)^{-1}). \qquad \blacksquare$$

Remark 8.20. If one is interested only in the norm estimate

$$\| (\sigma I - C)(\sigma I + C)^{-1} \| < 1,$$

then this can be obtained also for the nonsymmetric case from the definition (3.4). □

For the convergence of the splitting method (3.55), we have

Theorem 8.21. *Assume that the system matrix A has the representation (3.54) for symmetric positive definite matrices P and Q. Then for any $\sigma > 0$ and any starting vector $u^0 \in \mathbb{R}^N$, the successive iterates of the method (3.55) converge to the solution u of (3.53).*

Proof: We estimate the spectral radius of the iteration matrix $T(\sigma)$ that was defined in (3.57). Because of the underlying similarity transformation, this matrix has the same spectrum as the matrix

$$(\sigma I - P)(\sigma I + P)^{-1}(\sigma I - Q)(\sigma I + Q)^{-1}.$$

Now Lemma 8.19 implies that

$$\rho(T(\sigma)) \leq \|(\sigma I - P)(\sigma I + P)^{-1}\| \, \|(\sigma I - Q)(\sigma I + Q)^{-1}\| < 1.$$

in the Euclidean norm. By Lemma 8.9 the successive iterates converge and the result then follows from (3.55). ■

Finally we mention that under the additional hypothesis that the matrices P and Q commute, the spectral radius of the iteration matrix $\rho(T(\sigma))$ can be minimized via the estimate (3.60) by an appropriately chosen parameter σ (see [HY81]).

Exercise 8.22. Express the SOR method (3.48) in the form

$$u^{k+1} = Tu^k + t, \qquad k = 1, 2, \ldots$$

and describe the matrix T. Write down T in the particular case of the symmetric Gauss-Seidel scheme (3.35).

Exercise 8.23. Apply Lemma 8.17 to find the behaviour of the spectral radius $\rho(T(\omega))$ as $\omega \to \omega_{opt}$. What recommendation follows from this for an approximate evaluation of ω_{opt}?

Exercise 8.24. Determine the value of the parameter σ that minimizes

$$\max_{1 \leq i \leq 2} \left| \frac{\sigma - \mu_i}{\sigma + \mu_i} \right|$$

for any given positive μ_1 and μ_2.

8.4 The Conjugate Gradient Method

8.4.1 The Basic Idea, Convergence Properties

In this section we examine the linear system

$$A u = b \tag{4.1}$$

where the system matrix A is positive definite. Let $(\cdot, \cdot)$ denote a scalar product in $\mathbb{R}^N$. We assume that A is self-adjoint with respect to this scalar product, i.e., that

$$(Ay, z) = (y, Az) \qquad \text{for all } y, z \in \mathbb{R}^N. \tag{4.2}$$

In the special case of the Euclidean scalar product $(x, y) = x^T y$, the matrix A is self-adjoint if and only if A is symmetric, i.e., $A = A^T$.

The computational effort required for the numerical solution of the linear system (4.1) will be significantly reduced if we can find a basis $\{p^j\}_{j=1}^N$ in $\mathbb{R}^N$ that is related to the matrix A in a special way: suppose that this basis $\{p^j\}_{j=1}^N$ has the properties

$$(Ap^i, p^j) = 0 \quad \text{if } i \neq j \tag{4.3a}$$

and

$$(Ap^i, p^i) \neq 0 \quad \text{for } i = 1, \ldots, N. \tag{4.3b}$$

Vectors $\{p^j\}$ with the properties (4.3) are said to be *conjugate* or *A-orthogonal*. Expanding the solution u of (4.1) in terms of the basis $\{p^j\}_{j=1}^N$, we have

$$u = \sum_{j=1}^N \eta_j \, p^j \tag{4.4}$$

for some $\{\eta_j\} \subset \mathbb{R}$. Then (4.3) implies that the coefficients η_j, $j = 1, \ldots, N$, are given by

$$\eta_j = \frac{(b, p^j)}{(Ap^j, p^j)}, \qquad j = 1, \ldots, N.$$

The conjugate directions p_j are not usually known a priori except in special cases such as those considered in Section 8.2. The basic idea of the conjugate gradient (CG) method is to apply Gram-Schmidt orthogonalization to construct the p_j iteratively: when the iterate u^{k+1} is found, its defect is $d^{k+1} := b - Au^{k+1}$, and the new direction p^{k+1} is

$$p^{k+1} = d^{k+1} + \sum_{j=1}^k \beta_{kj} \, p^j,$$

where the unknown coefficients $\beta_{kj} \in \mathbb{R}$ are specified by the general orthogonality condition $(Ap^{k+1}, p^j) = 0$, $j = 1, \ldots, k$. This is Step 2 in the basic CG algorithm described below.

The defect d^{k+1} points in the direction opposite to the gradient of the function

$$F(u) = \frac{1}{2}(Au, u) - (b, u)$$

at the point u^{k+1}. Thus the method is called the *conjugate gradient (CG) method*. In terms of F, the problem (4.1) is equivalent to

$$\operatorname{grad} F(u) = 0.$$

The Basic CG Method

Step 0: Select some starting vector $u^1 \in \mathbb{R}^N$. Set $k = 1$ and

$$p^1 = d^1 = b - Au^1. \tag{4.5}$$

Step 1: Find

$$v^k \in V_k := \operatorname{span}\left\{p^j\right\}_{j=1}^k \tag{4.6}$$

such that

$$(Av^k, v) = (d^k, v) \qquad \text{for all } v \in V_k, \tag{4.7}$$

then set

$$u^{k+1} = u^k + v^k, \tag{4.8}$$

$$d^{k+1} = b - Au^{k+1}. \tag{4.9}$$

Step 2: If $d^{k+1} = 0$, then stop. Otherwise find $q^k \in V_k$ such that

$$(Aq^k, v) = -(Ad^{k+1}, v) \qquad \text{for all } v \in V_k. \tag{4.10}$$

Set

$$p^{k+1} = d^{k+1} + q^k. \tag{4.11}$$

Increment k to $k + 1$ then return to Step 1.

Remark 8.25. The positive definiteness of A and the Lax-Milgram Lemma guarantee that the subproblems (4.7) and (4.10) each have a unique solution $v^k \in V_k$ and $q^k \in V_k$ respectively. □

Remark 8.26. The basic approach of the CG method can in principle also be applied to symmetric elliptic problems without any discretization. This infinite-dimensional version of the CG method has been analysed in [55]. □

In our discussions we shall exclude the trivial case $d^1 = 0$, i.e., we assume that the initial guess u^1 is not a solution of the given system (4.1).

Lemma 8.27. *The $\{p^j\}_{j=1}^k$ generated by the basic CG method are conjugate directions, i.e., one has*

$$(Ap^i, p^j) = 0, \; i, j = 1, \ldots, k, \; i \neq j,$$
$$(Ap^i, p^i) \neq 0, \; i = 1, \ldots, k.$$

(4.12)

Proof: We use induction on k. When $k = 1$, the assumption $p^1 = d^1 \neq 0$ and the positive definiteness of A imply immediately that (4.12) holds.

Assume that the result is true for the value $k \geq 1$; we wish to deduce its truth for the value $k + 1$. By (4.10) and (4.11) we have

$$(Ap^{k+1}, v) = 0 \qquad \text{for all } v \in V_k.$$

In particular this implies that

$$(Ap^{k+1}, p^j) = 0, \qquad j = 1, \ldots, k.$$

The self-adjointness of A with respect to $(\cdot, \cdot)$ then gives

$$(Ap^j, p^{k+1}) = 0, \qquad j = 1, \ldots, k.$$

Combining this result with the inductive hypothesis that the $\{p^j\}_{j=1}^k$ are already adjoint directions yields

$$(Ap^i, p^j) = 0, \qquad i, j = 1, \ldots, k+1, \; i \neq j.$$

It remains only to show that

$$(Ap^{k+1}, p^{k+1}) \neq 0. \tag{4.13}$$

Suppose that (4.13) is false. Then the positive definiteness of A implies that $p^{k+1} = 0$. From (4.11) we get

$$d^{k+1} = -q^k \in V_k. \tag{4.14}$$

On the other hand (4.7)–(4.9) yield

$$(d^{k+1}, v) = (b - Au^k - Av^k, v) = (d^k - Av^k, v) = 0 \qquad \text{for all } v \in V_k. \tag{4.15}$$

By (4.14) we can choose $v = d^{k+1}$ in (4.15), which implies $d^{k+1} = 0$. This contradicts Step 2 of the basic CG method. Hence the supposition that (4.13) is false must be incorrect; that is, (4.13) holds. ∎

Lemma 8.28. *The linear subspaces $V_k \subset \mathbb{R}^N$ generated by the basic CG method satisfy*

$$V_k = span\{d^j\}_{j=1}^k \qquad and \qquad dim\, V_k = k. \tag{4.16}$$

Proof: We use induction on k. The identity $p^1 = d^1$ and the definition (4.6) of V_k imply that (4.16) holds trivially when $k = 1$.

Assume that the result is true for the value $k \geq 1$; we wish to deduce its truth for the value $k + 1$. By (4.10), (4.11) and the definition (4.6) of V_{k+1} we obtain

$$V_{k+1} \subset \text{span} \left\{ d^j \right\}_{j=1}^{k+1}. \tag{4.17}$$

Now Lemma 8.27 implies in particular that the $\{p^j\}_{j=1}^{k+1}$ are linearly independent. Hence

$$\dim V_{k+1} = k + 1, \tag{4.18}$$

and (4.17) then gives

$$V_{k+1} = \text{span} \left\{ d^j \right\}_{j=1}^{k+1}. \qquad \blacksquare$$

Next, we show that the sub-steps (4.7) and (4.10) of the basic CG method can be simplified by using the fact that the p^j, $j = 1, \ldots, k$, are conjugate.

Lemma 8.29. *Write the vectors $v^k \in V_k$ and $q^k \in V_k$ of (4.7) and (4.10) respectively in terms of the basis $\{p^j\}_{j=1}^k$ as*

$$v^k = \sum_{j=1}^k \alpha_{kj}\, p^j, \qquad q^k = \sum_{j=1}^k \beta_{kj}\, p^j. \tag{4.19}$$

Then

$$\alpha_{kj} = \beta_{kj} = 0 \quad \text{for } j = 1, \ldots, k - 1$$

and

$$\alpha_{kk} = \frac{(d^k, d^k)}{(Ap^k, p^k)}, \qquad \beta_{kk} = \frac{(d^{k+1}, d^{k+1})}{(d^k, d^k)}. \tag{4.20}$$

Proof: The relation (4.15) shown in the proof of Lemma 8.27 implies that

$$(d^{k+1}, p^j) = 0, \qquad j = 1, \ldots, k, \tag{4.21}$$

and, also appealing to Lemma 8.28, we have

$$(d^{k+1}, d^j) = 0, \qquad j = 1, \ldots, k. \tag{4.22}$$

By (4.19) and (4.7) we get

$$\sum_{i=1}^k \alpha_{ki}\, (Ap^i, v) = (d^k, v) \qquad \text{for all } v \in V_k. \tag{4.23}$$

Choosing $v = p^j$ here, then (4.12) and (4.21) yield

$$\alpha_{kj} = 0, \qquad j = 1, \ldots, k - 1, \tag{4.24}$$

and

$$\alpha_{kk} = \frac{(d^k, p^k)}{(Ap^k, p^k)}.$$ (4.25)

Replacing $k+1$ by k in (4.11) and (4.21) shows that $(d^k, p^k) = (d^k, d^k)$. Thus

$$\alpha_{kk} = \frac{(d^k, d^k)}{(Ap^k, p^k)},$$ (4.26)

as desired. Observe that $\alpha_{kk} \neq 0$ since $d^k \neq 0$.

Consider now the coefficients β_{kj}. By (4.11) and (4.19) we have

$$p^{k+1} = d^{k+1} + \sum_{j=1}^{k} \beta_{kj} p^j.$$

This identity and the conjugacy of the directions $\{p^j\}_{j=1}^{k+1}$ give

$$0 = (Ap^i, d^{k+1}) + \sum_{j=1}^{k} \beta_{kj} (Ap^i, p^j), \qquad i = 1, \dots, k,$$

and hence

$$\beta_{kj} = \frac{(Ap^j, d^{k+1})}{(Ap^j, p^j)}, \qquad j = 1, \dots, k.$$ (4.27)

The first part of the proof and (4.8) show that

$$u^{j+1} = u^j + \alpha_{jj} p^j, \qquad j = 1, \dots, k,$$

with coefficients $\alpha_{jj} \neq 0$. Consequently

$$Ap^j = \frac{1}{\alpha_{jj}} A(u^{j+1} - u^j) = \frac{1}{\alpha_{jj}} (d^j - d^{j+1}).$$

Substituting this into (4.27) we obtain

$$\beta_{kj} = \frac{1}{\alpha_{jj}} \cdot \frac{(d^{k+1}, d^{j+1} - d^j)}{(Ap^j, p^j)}, \qquad j = 1, \dots, k.$$

Appealing now to (4.21) and Lemma 8.28, we see that $\beta_{kj} = 0$, $j = 1, \dots, k-1$, and

$$\beta_{kk} = \frac{(d^{k+1}, d^{k+1})}{(d^k, p^k)}.$$

When (4.11) and (4.21) are taken into account this gives the formula for β_{kk} in (4.20). ∎

Lemma 8.29 means that we can simplify (4.19) to $v^k = \alpha_k\, p^k$, $q^k = \beta_k\, p^k$, where (4.20) gives explicit formulas for $\alpha_k := \alpha_{kk}$ and $\beta_k := \beta_{kk}$.

These results enable us to simplify the basic CG algorithm as follows.

CG method

Select some starting vector $u^1 \in \mathbb{R}^N$. Set

$$p^1 = d^1 = b - Au^1. \tag{4.28}$$

For $k = 1, 2, \ldots$, while $d^k \neq 0$ define iteratively the coefficients and vectors

$$\alpha_k := \frac{(d^k, d^k)}{(Ap^k, p^k)}, \tag{4.29}$$

$$u^{k+1} := u^k + \alpha_k\, p^k, \tag{4.30}$$

$$d^{k+1} := b - Au^{k+1}, \tag{4.31}$$

$$\beta_k := \frac{(d^{k+1}, d^{k+1})}{(d^k, d^k)}, \tag{4.32}$$

$$p^{k+1} := d^{k+1} + \beta_k\, p^k. \tag{4.33}$$

By Lemma 8.28, (4.6) and (4.7) we obtain

Theorem 8.30. *After at most N steps, the CG method finds the solution u of the given problem (4.1) provided that all calculations are exact, i.e., are carried out without rounding errors.*

The discretization of partial differential equations leads generally to linear systems of extremely high dimension. In this context CG methods, despite the finite termination property of Theorem 8.30, are often used as iterative methods that are terminated before the exact solution is obtained. Furthermore, in the case of high dimensions, rounding errors in the calculation cannot be neglected and consequently only approximate solutions of (4.1) are generated. Thus the iterative convergence behaviour of the CG method must be investigated. For this we have:

Theorem 8.31. *The iterates u^k generated by the CG method converge to the solution u of the system (4.1) subject to the bound*

$$\|u^{k+1} - u\|_A^2 \le 2 \left(\frac{\sqrt{\mu} - \sqrt{\nu}}{\sqrt{\mu} + \sqrt{\nu}} \right)^k \|u^1 - u\|_A^2, \tag{4.34}$$

where $\mu \ge \nu > 0$ are real numbers such that

$$\nu\,(y, y) \le (Ay, y) \le \mu\,(y, y) \qquad \text{for all } y \in \mathbb{R}^N. \tag{4.35}$$

Proof: Set $e^j = u^j - u$ for $j = 1, 2, \ldots$ First we show by induction on j that there exist polynomials $r_j(\cdot)$ of degree j, with $r_j(0) = 1$, such that

$$e^{j+1} = r_j(A)\, e^1, \qquad j = 0, 1, \ldots, k. \tag{4.36}$$

This is trivially true for $j = 0$ on taking $r_0 \equiv 1$. For fixed but arbitrary $m \geq 1$ assume that (4.36) is true, r_j has degree j, and $r_j(0) = 1$ for $j = 0, 1, \ldots, m-1$ ("strong induction"); we wish to deduce (4.36) and the other two properties for the value $j = m$.

The definition of e^{m+1} and (4.30) together imply that

$$e^{m+1} = u^{m+1} - u = u^{m+1} - u^m + u^m - u = \alpha_m\, p^m + e^m. \tag{4.37}$$

By (4.33) we get, setting $\beta_0 = 0$,

$$p^m = d^m + \beta_{m-1} p^{m-1} = b - Au^m + \frac{\beta_{m-1}}{\alpha_{m-1}} (u^m - u^{m-1})$$

$$= -A(u^m - u) + \frac{\beta_{m-1}}{\alpha_{m-1}} (u^m - u + u - u^{m-1}).$$

That is,

$$p^m = \left(\frac{\beta_{m-1}}{\alpha_{m-1}} I - A \right) e^m - \frac{\beta_{m-1}}{\alpha_{m-1}} e^{m-1}.$$

The inductive hypothesis then yields

$$p^m = \left(\frac{\beta_{m-1}}{\alpha_{m-1}} I - A \right) r_{m-1}(A) e^1 - \frac{\beta_{m-1}}{\alpha_{m-1}} r_{m-2}(A) e^1.$$

Substituting this into (4.37) leads to the intermediate result

$$e^{m+1} = r_m(A)\, e^1$$

with

$$r_m(A) := \left[\left(1 + \frac{\alpha_m \beta_{m-1}}{\alpha_{m-1}} \right) I - \alpha_m A \right] r_{m-1}(A) - \frac{\alpha_m \beta_{m-1}}{\alpha_{m-1}} r_{m-2}(A),$$

completing our induction argument—it is easy to deduce from this formula and the inductive hypothesis that r_m has degree m and $r_m(0) = 1$.

Next, (e^{k+1}, Ae^{k+1}) will be studied in detail. Now $d^j = -Ae^j$ for each j and by (4.22) we have

$$(e^{k+1}, Ae^{k+1}) = -(e^{k+1}, d^{k+1}) = -\left(e^{k+1} + \sum_{j=1}^{k} \sigma_j d^j, d^{k+1}\right)$$

for any $\sigma_1, \ldots, \sigma_k \in \mathbb{R}$. Recalling (4.36) and again appealing to $d^j = -Ae^j$, this yields

$$(e^{k+1}, Ae^{k+1}) = \left((r_k(A) + \sum_{j=1}^{k} \sigma_j \, A \, r_{j-1}(A)) e^1, Ae^{k+1} \right).$$

Let $q_k(\cdot)$ be an arbitrary polynomial of degree k that satisfies $q_k(0) = 1$. Then the properties that we have proved above for r_k imply that there exist uniquely defined real numbers $\tilde{\sigma}_1, \ldots, \tilde{\sigma}_k$ such that

$$q_k(\xi) = r_k(\xi) + \sum_{j=1}^{k} \tilde{\sigma}_j \, \xi \, r_{j-1}(\xi) \qquad \text{for all } \xi \in \mathbb{R}.$$

Combining this with the previous identity, we see that

$$(e^{k+1}, Ae^{k+1}) = (q_k(A)e^1, Ae^{k+1}) \tag{4.38}$$

for any polynomial q_k of degree k for which $q_k(0) = 1$. As A is self-adjoint and positive definite, one can define a scalar product $\langle \cdot, \cdot \rangle$ on $\mathbb{R}^N \times \mathbb{R}^N$ by

$$\langle y, z \rangle := (y, Az) \qquad \text{for all } y, \, z \in \mathbb{R}^N.$$

The Cauchy-Schwarz inequality associated with $\langle \cdot, \cdot \rangle$, when applied to (4.38), yields the estimate

$$(e^{k+1}, Ae^{k+1}) \leq (q_k(A)e^1, A \, q_k(A)e^1)^{1/2} (e^{k+1}, Ae^{k+1})^{1/2}.$$

That is,

$$(e^{k+1}, Ae^{k+1}) \leq (q_k(A)e^1, A \, q_k(A)e^1). \tag{4.39}$$

Since A is self-adjoint with respect to the scalar product $(\cdot, \cdot)$, there exists an orthonormal system $\{w^j\}_{j=1}^N$ of eigenvectors of A such that

$$(Aw^j, v) = \lambda_j \, (w^j, v) \qquad \text{for all } v \in \mathbb{R}^N.$$

Now (4.35) implies that

$$\lambda_j \in [\nu, \mu], \qquad j = 1, \ldots, N. \tag{4.40}$$

Expanding in terms of the eigensystem $\{w^j\}$, by (4.40) we get

$$(q_k(A)e^1, A \, q_k(A)e^1) \leq \max_{\lambda \in [\nu, \mu]} |q_k(\lambda)|^2 \, (e^1, Ae^1). \tag{4.41}$$

We minimize the multiplier $\max_{\lambda \in [\nu, \mu]} |q_k(\lambda)|^2$ by an appropriate choice of the polynomial $q_k(\cdot)$. Using Tschebyscheff polynomials, it can be shown (see [Str04]) that there exists a polynomial $\tilde{q}_k(\cdot)$ of degree k with $\tilde{q}_k(0) = 1$ such that

$$\max_{\lambda \in [\nu, \mu]} |\tilde{q}_k(\lambda)| \leq 2 \left(\frac{\sqrt{\mu} - \sqrt{\nu}}{\sqrt{\mu} + \sqrt{\nu}} \right)^k.$$

Recalling (4.39) and (4.41), we get the convergence estimate (4.34). ∎

Remark 8.32. In the CG method it is essential that the system matrix A be self-adjoint. This property implies that $(A\cdot,\cdot)$ defines a scalar product and the problem $Au = b$ can be interpreted as a projection problem. When dealing with matrices that are not self-adjoint, one can use the *Krylov spaces* defined by

$$U_k(A, u^1) := \mathrm{span}\left\{d^1, Ad^1, a^2 d^1, \ldots, A^{k-1}d^1\right\} \tag{4.42}$$

(see Exercise 8.39). A common numerical technique that is based on Krylov spaces is the *GMRES method.* In the kth step of this method a correction $v^k \in U_k(A, u^1)$ is chosen such that

$$\|d^k - Av^k\| = \min_{v\in U_k(A,u^1)} \|d^k - Av\|$$

and the update is defined by $u^{k+1} := u^k + v^k$. Unlike the CG method, each iteration in GMRES cannot be performed using only the last two directions: instead one must use the complete Krylov space generated so far, i.e., all spanning vectors are needed. For a detailed discussion of the GMRES method we refer, e.g., to [Axe96].   □

8.4.2 Preconditioned CG Methods

In the case of the Euclidean scalar product, self-adjointness is equivalent to $A = A^T$ and the constants ν and μ in Theorem 8.31 are determined by the minimal and maximal eigenvalues of A. The estimate (4.34) of that Theorem then reads

$$\|u^{k+1} - u\|_A^2 \le 2 \left(\frac{\sqrt{\lambda_{max}} - \sqrt{\lambda_{min}}}{\sqrt{\lambda_{max}} + \sqrt{\lambda_{min}}}\right)^k \|u^1 - u\|_A^2, \quad k = 1, 2, \ldots$$

For the discrete Poisson-Problem (2.25) the eigenvalues of A are explicitly known and $\lambda_{max}/\lambda_{min} = O(h^{-2})$. This behaviour is also typical in other cases (see Section 4.9.1). The CG method has therefore a contraction factor that is $1 - O(h)$, which is unsatisfactory. Every preconditioning of the method aims to improve this situation.

Let A, B be two positive definite matrices that are self-adjoint with respect to the scalar product $(\cdot,\cdot)$. Instead of the given problem (4.1) we consider the equivalent problem

$$\tilde{A} u = \tilde{b} \quad \text{with} \quad \tilde{A} := B^{-1}A, \ \tilde{b} := B^{-1}b. \tag{4.43}$$

The properties of B imply that

$$(x, y)_B := (Bx, y) \quad \text{for all} \quad x, y \in \mathbb{R}^N \tag{4.44}$$

defines a new scalar product $(\cdot,\cdot)_B$ on $\mathbb{R}^N$. The matrix $\tilde{A}$ is self-adjoint with respect to the new scalar product because

$$(\tilde{A}x, y)_B = (B\tilde{A}x, y) = (BB^{-1}Ax, y) = (Ax, y) = (x, Ay)$$
$$= (x, B\tilde{A}y) = (Bx, \tilde{A}y) = (x, \tilde{A}y)_B \qquad \text{for all} \quad x, y \in \mathbb{R}^N.$$

We now apply the CG method on the transformed problem (4.43), using the scalar product $(\cdot, \cdot)_B$. Then one can write the steps (4.29)–(4.33) in the form (using the original scalar product $(\cdot, \cdot)$ and the non-transformed variables)

$$\tilde{\alpha}_k := \frac{(B^{-1}d^k, d^k)}{(Ap^k, p^k)}, \qquad (4.45)$$

$$u^{k+1} := u^k + \tilde{\alpha}_k p^k, \qquad (4.46)$$

$$d^{k+1} := b - Au^{k+1}, \qquad (4.47)$$

$$\tilde{\beta}_k := \frac{(B^{-1}d^{k+1}, d^{k+1})}{(B^{-1}d^k, d^k)}, \qquad (4.48)$$

$$p^{k+1} := B^{-1}d^{k+1} + \tilde{\beta}_k p^k. \qquad (4.49)$$

The method (4.45)–(4.49), which uses the unspecified matrix B to solve (4.1), is called the *preconditioned CG method* or *PCG method*. Of course, the vectors $s^k := B^{-1}d^k$ needed in (4.45) and (4.48) are computed by solving the linear system

$$B s^k = d^k. \qquad (4.50)$$

Consequently B must be chosen in such a way that the efficient solution of (4.50) is possible and this choice decides the effectiveness of the PCG method. The matrix B in (4.50) is called a *preconditioner*.

We are interested only in problems that arise from the discretization of differential equations. Both the given problem (4.1) and the auxiliary problem (4.50) depend on this discretization. We therefore write A_h, B_h and ν_h, μ_h instead of A, B and ν, μ. In the preconditioned setting the estimate (4.35), written in terms of the scalar product $(\cdot, \cdot)_B$, becomes

$$\nu_h (y, y)_{B_h} \leq (B_h^{-1}A_h y, y)_{B_h} \leq \mu_h (y, y)_{B_h} \qquad \text{for all } y \in \mathbb{R}^N.$$

This is equivalent to the condition

$$\nu_h (B_h y, y) \leq (A_h y, y) \leq \mu_h (B_h y, y) \qquad \text{for all } y \in \mathbb{R}^N \qquad (4.51)$$

if the original scalar product $(\cdot, \cdot)$ is used. If (4.51) holds then A_h, B_h are called *spectrally equivalent*.

When two-sided spectral bounds are available for both A_h and B_h, one can derive estimates of the type (4.51):

Lemma 8.33. *Assume that there are constants $\mu_A \geq \nu_A > 0$ and $\mu_B \geq \nu_B > 0$ such that the matrices A_h and B_h satisfy the bounds*

$$\nu_A(y, y) \leq (A_h y, y) \leq \mu_A(y, y)$$

$$\text{and} \quad \nu_B(y, y) \leq (B_h y, y) \leq \mu_B(y, y) \qquad \text{for all } y \in \mathbb{R}^N.$$

Then (4.51) is valid with $\nu_h = \dfrac{\nu_A}{\mu_B}$ *and* $\mu_h = \dfrac{\mu_A}{\nu_B}$.

Proof: This is immediate from the hypotheses. ∎

It should be noticed that the bounds given by Lemma 8.33 are often not optimal. Better is a direct estimate.

To treat discretized elliptic variational equations, special preconditioners have been developed to guarantee that the quotient μ_h/ν_h of (4.51) is either uniformly bounded with respect to h (then PCG works with optimal complexity) or increases very slowly as $h \to 0$. We now discuss some of the preconditioners in current use.

Assume that A can be written in the form

$$A = LL^T + R, \tag{4.52}$$

where the matrix L is lower triangular and the matrix R has a small norm. If $R = 0$, then (4.52) is the Cholesky decomposition of A, and in this case Theorem 8.31 tells us that we get the solution of (4.1) in one step. When one computes a Cholesky decomposition, the lower triangular matrix L has, in general, many more nonzero elements than the lower triangular part of A.

The Cholesky decomposition is often modified in the following way: if the absolute value of an entry of L is smaller than some user-specified threshold δ, it is simply set equal to zero. This modification is known as an *incomplete* Cholesky decomposition. If $\delta = 0$ we obtain the exact Cholesky decomposition, and on the other hand for sufficiently large δ we generate a decomposition whose matrix L has the same pattern of nonzero entries as A. The preconditioned conjugate method based on incomplete Cholesky decomposition is called the *ICCG method*.

An alternative way of constructing preconditioners for (4.1) uses discretizations on adjacent meshes. Let us denote the underlying discrete spaces of two discretizations of some given problem by U_h and U_H. We write the corresponding linear systems as

$$A_h\, u_h = b_h \tag{4.53}$$

and

$$A_H\, u_H = b_H. \tag{4.54}$$

To transform between U_h and U_H, assume that we have linear continuous mappings

$$I_h^H : U_h \to U_H \qquad \text{and} \qquad I_H^h : U_H \to U_h. \tag{4.55}$$

For instance, these mappings could be defined using interpolation.

If the given problem (4.1) coincides with (4.53), then under certain conditions the quantity

$$I_H^h A_H^{-1} I_h^H b_h \tag{4.56}$$

is an approximate solution of (4.53). Consequently the matrix

$$B_h^{-1} := I_H^h A_H^{-1} I_h^H$$

is a reasonable candidate as a preconditioner of (4.53). Preconditioners related to multigrid methods (see Section 8.5 for a detailed discussion) are carefully investigated in [Osw94], [125]. One can prove that these preconditioners have an almost optimal asymptotic convergence behaviour.

As an example we study a preconditioner proposed by Bramble, Pasciak and Xu [25] that is based on the use of a hierarchical mesh structure. We consider this *BPX preconditioner* in the context of the weak formulation of the model problem

$$-\Delta u = f \quad \text{in } \Omega, \qquad u = 0 \quad \text{on } \Gamma := \partial\Omega, \tag{4.57}$$

and analyze the spectral property (4.51). Let $f \in H^{-1}(\Omega)$ and let $\Omega \subset \mathbb{R}^2$ be a polygonal domain. Denote by $\mathcal{Z}_0$ a first (coarse) admissible triangulation of Ω and refine the mesh iteratively by the decomposition of every triangle into 4 congruent triangles as in Figure 8.9. Thus, starting from $\mathcal{Z}_0$, we get a *dyadic refinement* which we denote by $\mathcal{Z}_1, \mathcal{Z}_2, \ldots, \mathcal{Z}_m$.

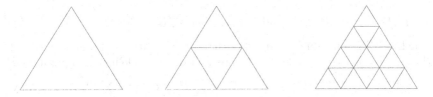

Figure 8.9 Dyadic refinement of a triangulation

Let h_l denote the mesh diameter of $\mathcal{Z}_l$. Then the dyadic refinement implies that

$$h_{l-1} = 2h_l, \quad l = 1, \ldots, m. \tag{4.58}$$

On each triangulation $\mathcal{Z}_l$ let U_l denote the space of piecewise linear elements in $C(\Omega)$. These spaces have the nested property

$$U_0 \subset U_1 \subset \cdots \subset U_{m-1} \subset U_m \subset H_0^1(\Omega).$$

We denote the Lagrange basis functions of U_l by $\varphi_{l,j}$, $j \in J_l$, $l = 0, 1, \ldots, m$, where J_l is the corresponding index set. Then

$$U_l = \text{span}\{\varphi_{l,j}\}_{j \in J_l}, \quad l = 0, 1, \ldots, m.$$

Moreover we let

$$U_{l,j} = \text{span}\{\varphi_{l,j}\}, \quad j \in J_l, \; l = 0, 1, \ldots, m,$$

denote the one-dimensional subspace spanned by the function $\varphi_{l,j}$. The discrete problem that we wish to solve is formulated on the finest triangulation $\mathcal{Z}_m$, i.e., U_m corresponds to the finite element space V_h with $h = h_m$. This discrete problem is, as usual,

find $u_h \in V_h$ such that $\quad a(u_h, v_h) = f(v_h) \quad$ for all $v_h \in V_h$, (4.59)

where

$$a(u,v) := \int_\Omega \nabla u \cdot \nabla v \, dx \quad \text{for all } u, v \in H_0^1(\Omega).$$

BPX preconditioners can be interpreted as *additive Schwarz method* (see [TW05], [125, 54]). We follow [54] in the presentation and analysis of this technique.

To simplify the notation and to avoid introducing a double index we put $V = V_h$ and write $\{V_i\}_{i=0}^M$ for the subspaces $U_0, U_{l,j}, j \in J_l, l = 1, \ldots, m$. Here $M := \sum_{l=1}^m |J_l|$ denotes the number of one-dimensional subspaces V_i that occur.

For $i = 0, 1, \ldots, M$ we assume that we have symmetric, continuous, V_i-elliptic bilinear forms $b_i : V_i \times V_i \to \mathbb{R}$. Using these forms we define projectors $P_i : V \to V_i$ and $Q_i : H^{-1}(\Omega) \to V_i$, $i = 0, 1, \ldots, M$, by

$$b_i(P_i v, v_i) = a(v, v_i) \quad \text{and} \quad b_i(Q_i f, v_i) = f(v_i) \quad \text{for all } v_i \in V_i. \quad (4.60)$$

The Lax-Milgram lemma guarantees that for arbitrary $v \in V$ and $f \in H^{-1}(\Omega)$ the projections $P_i v$ and $Q_i f \in V_i$ are well defined. Define the operators $P : V \to V$ and $Q : H^{-1}(\Omega) \to V$ by $P := \sum_{i=0}^M P_i$ and $Q := \sum_{i=0}^M Q_i$ respectively.

Lemma 8.34. *The operator P is positive definite and self-adjoint with respect to the scalar product $a(\cdot, \cdot)$. Moreover, $u \in V$ is a solution of the discrete problem (4.59) if and only if $u \in V$ is a solution of the equation*

$$Pu = Qf. \quad (4.61)$$

Proof: For arbitrary $u, v \in V$ one has

$$a(Pu, v) = a\left(\sum_{i=0}^M P_i u, v\right) = \sum_{i=0}^M a(P_i u, v) = \sum_{i=0}^M a(v, P_i u)$$

$$= \sum_{i=0}^M b_i(P_i v, P_i u) = \sum_{i=0}^M b_i(P_i u, P_i v) = \sum_{i=0}^M a(u, P_i v)$$

$$= a\left(u, \sum_{i=0}^M P_i v\right) = a(u, Pv).$$

Therefore P is self-adjoint with respect to $a(\cdot, \cdot)$.

Next we show that P is positive definite. The V_i-ellipticity of the bilinear forms b_i and the definition of the projectors P_i imply that

$$a(Pu, u) = \sum_{i=0}^{M} a(P_i u, u) = \sum_{i=0}^{M} b_i(P_i u, P_i u) \geq 0 \qquad \text{for all } u \in V.$$

Furthermore, $b_i(P_i u, P_i u) \geq 0$ for $i = 0, \ldots, M$ and V_i-ellipticity yield

$$a(Pu, u) = 0 \qquad \Longleftrightarrow \qquad P_i u = 0, \quad i = 0, \ldots, M.$$

In other words, u is orthogonal to all subspaces V_i, $i = 0, \ldots, M$ and consequently $u = 0$ if $a(Pu, u) = 0$. Thus we have shown that $a(Pu, u) > 0$ for arbitrary $u \neq 0$.

Suppose that $u \in V$ is a solution of (4.59). For arbitrary $v_i \in V_i$, $i = 0, 1, \ldots, M$, set $v = \sum_{i=0}^{M} v_i$. Then the bilinearity of $a(\cdot, \cdot)$ and the definition of the projectors P_i lead to

$$a(u, v) = a\left(u, \sum_{i=0}^{M} v_i\right) = \sum_{i=0}^{M} a(u, v_i) = \sum_{i=0}^{M} b_i(P_i u, v_i).$$

Analogously

$$a(u, v) = f(v) = \sum_{i=0}^{M} f(v_i) = \sum_{i=0}^{M} b_i(Q_i f, v_i).$$

Hence u is a solution of (4.61). Conversely, because P is positive definite the solution of (4.61) is uniquely determined. ∎

To solve the problem (4.59), we now apply the CG method to the equivalent problem (4.61). The underlying scalar product in this CG method is chosen to be $a(\cdot, \cdot)$ so that P is symmetric and positive definite, as promised by Lemma 8.34.

The dominant effort in each iteration step is the evaluation of the residuals $d^k = Qf - Pu^k$. One therefore needs an effective way of computing $Q_i f$ and $P_i u^k$ for $i = 0, 1, \ldots, M$. The definitions (4.60) tell us that we have to solve two discrete variational equations on each subspace V_i. But each space $V_i = \text{span}\{\varphi_i\}$ (say) is one-dimensional! Thus the solutions are simply

$$P_i u^k = \frac{a(u^k, \varphi_i)}{b_i(\varphi_i, \varphi_i)} \varphi_i \quad \text{and} \quad Q_i f = \frac{f(\varphi_i)}{b_i(\varphi_i, \varphi_i)} \varphi_i, \qquad i = 1, \ldots, M. \quad (4.62)$$

We assume that on the coarsest mesh $\mathcal{Z}_0$, the solution of the Galerkin equations

$$\sum_{j \in J_0} b_0(\varphi_j, \varphi_i)\, \zeta_j^k = f(\varphi_i) - \sum_{j \in J_0} a(\varphi_j, \varphi_i) u_j^k \qquad (4.63)$$

is available. Consequently we have

$$Q_0\, f - P_0\, u^k = \sum_{j \in J_0} \zeta_j^k\, \varphi_j\,.$$

Hence

$$d^k = Qf - Pu^k = \sum_{j \in J_0} \zeta_j^k \varphi_j + \sum_{l=1}^{m} \sum_{j \in J_l} \frac{f(\varphi_j) - a(u^k, \varphi_j)}{b_{l,j}(\varphi_j, \varphi_j)}\, \varphi_j, \qquad (4.64)$$

where the $b_{l,j}$ are those b_i for which V_i is associated with $U_{l,j}$. If one chooses

$$b_{l,j}(u,v) := h_l^{-2}\,(u,v)_{0,\Omega} \qquad \text{for } u,\, v \in U_{l,j}, \qquad (4.65)$$

this yields the BPX preconditioner (see [25])

$$d^k = Qf - Pu^k = \sum_{j \in J_0} \zeta_j^k \varphi_j + \sum_{l=1}^{m} \sum_{j \in J_l} 2^{-2l}\, \frac{f(\varphi_j) - a(u^k, \varphi_j)}{(\varphi_j, \varphi_j)_{0,\Omega}}\, \varphi_j\,. \qquad (4.66)$$

Alternatively, the choice $b_i(\cdot, \cdot) = a(\cdot, \cdot)$ leads to the following version of (4.64):

$$d^k = Qf - Pu^k = \sum_{j \in J_0} \zeta_j^k \varphi_j + \sum_{l=1}^{m} \sum_{j \in J_l} \frac{f(\varphi_j) - a(u^k, \varphi_j)}{a(\varphi_j, \varphi_j)}\, \varphi_j\,. \qquad (4.67)$$

Recall that ζ^k is the solution of (4.63) on the coarsest mesh $\mathcal{Z}_0$. In (4.66) and (4.67) the residual d^k is represented as an element of the finite-dimensional space U_h. The scalar products required in the CG method can be directly evaluated—see [125]—the storage of intermediate values employs representations that use the basis $\{\varphi_j\}_{j \in J_m}$ of the space U_h.

The two preconditioners (4.67) and (4.66) are spectrally equivalent with constants independent of h, as we now explain. Note first that for our piecewise linear finite elements one has

$$c\, h_l^{-2}\,(\varphi_j, \varphi_j)_{0,\Omega} \le a(\varphi_j, \varphi_j) \le C\, h_l^{-2}\,(\varphi_j, \varphi_j)_{0,\Omega}$$
$$\text{for all } j \in J_l,\, l = 1, \dots, m. \qquad (4.68)$$

The right-hand inequality here follows from an inverse inequality, while the left-hand inequality is a consequence of Friedrich's inequality where one takes into account the diameter of the support of each Lagrange basis function on the corresponding refinement level. The dyadic refinement guarantees that the constants c and C in (4.68) are independent of the refinement level. Consequently (4.58) implies the spectral equivalence of (4.67) and (4.66).

Next we study the convergence behaviour of the CG method with the scalar product $a(\cdot, \cdot)$ applied to problem (4.61), which is equivalent to the original problem (4.59). We shall use Theorem 8.31.

By Lemma 8.34, the operator P is self-adjoint and positive definite. We aim to find constants $\mu \geq \nu > 0$ such that

$$\nu\,(v,v) \leq (Pv,v) \leq \mu\,(v,v) \quad \text{for all } v \in V. \tag{4.69}$$

Towards this end, let us introduce the new norm

$$|||v||| := \min \left\{ \left(\sum_{i=0}^{M} b_i(v_i, v_i) \right)^{1/2} : v = \sum_{i=0}^{M} v_i,\ v_i \in V_i \right\}.$$

Lemma 8.35. *One has*

$$a(P^{-1}v, v) = |||v|||^2 \quad \text{for all } v \in V. \tag{4.70}$$

Proof: Let $v \in V$ be arbitrary with $v = \sum_{i=0}^{M} v_i$ for $v_i \in V_i$, $i = 0, 1, \ldots, M$. The bilinearity of $a(\cdot, \cdot)$ and the definition of the projectors P_i give

$$a(P^{-1}v, v) = \sum_{i=0}^{M} a(P^{-1}v, v_i) = \sum_{i=0}^{M} b_i(P_i P^{-1}v, v_i) = \sum_{i=0}^{M} b_i(w_i, v_i) \tag{4.71}$$

with $w_i := P_i P^{-1}v$, $i = 0, 1, \ldots, M$. Our previous assumptions ensure that $b_i(\cdot, \cdot)$ defines a scalar product in V_i. Then a Cauchy-Schwarz inequality yields

$$|b_i(w_i, v_i)| \leq b_i(w_i, w_i)^{1/2} b_i(v_i, v_i)^{1/2}.$$

A further application of the Cauchy-Schwarz inequality to the Euclidean scalar product in $\mathbb{R}^{M+1}$ and (4.71) gives

$$\begin{aligned}
a(P^{-1}v, v) = \sum_{i=0}^{M} b_i(w_i, v_i) &\leq \sum_{i=0}^{M} b_i(w_i, w_i)^{1/2} b_i(v_i, v_i)^{1/2} \\
&\leq \left(\sum_{i=0}^{M} b_i(w_i, w_i) \right)^{1/2} \left(\sum_{i=0}^{M} b_i(v_i, v_i) \right)^{1/2}.
\end{aligned} \tag{4.72}$$

On the other hand the definition of w_i and of the projectors yields

$$\sum_{i=0}^{M} b_i(w_i, w_i) = \sum_{i=0}^{M} b_i(P_i P^{-1}v, w_i) = \sum_{i=0}^{M} a(P^{-1}v, w_i) = a\left(P^{-1}v, \sum_{i=0}^{M} w_i \right)$$

$$= a\left(P^{-1}v, \sum_{i=0}^{M} P_i P^{-1}v \right) = a(P^{-1}v, v);$$

now invoking (4.72) we get $\sum_{i=0}^{M} b_i(w_i, w_i) \leq \sum_{i=0}^{M} b_i(v_i, v_i)$, i.e.,

$$a(P^{-1}v, v) \leq \sum_{i=0}^{M} b_i(v_i, v_i) \quad \text{for every representation} \quad v = \sum_{i=0}^{M} v_i, \; v_i \in V_i.$$

But if we choose $v_i = P_i P^{-1} v$, i.e., $v_i = w_i$, $i = 0, 1, \ldots, M$, then, as we saw above, our final inequality becomes an equation. This proves (4.70). ∎

To clarify the relationship between the norm $||| \cdot |||$, the norm $\| \cdot \|$ induced by $a(\cdot, \cdot)$, and (4.69) we have the following result:

Lemma 8.36. *Assume that there exist constants* $\bar{c} \geq \underline{c} > 0$ *such that*

$$\underline{c}\|v\| \leq |||v||| \leq \bar{c}\|v\| \quad \text{for all } v \in V. \tag{4.73}$$

Then the estimate (4.69) is valid with $\nu = \bar{c}^{-2}$, $\mu = \underline{c}^{-2}$.

Proof: By Lemma 8.35, the inequality (4.73) is equivalent to

$$\underline{c}^2 \|v\|^2 \leq a(P^{-1}v, v) \leq \bar{c}^2 \|v\|^2 \quad \text{for all } v \in V. \tag{4.74}$$

As P is positive definite and symmetric with respect to $a(\cdot, \cdot)$, the operator P^{-1} has the same properties. Let $\lambda_{max}(P)$ and $\lambda_{min}(P)$ be the maximal and minimal eigenvalues for the eigenvalue problem

$$a(Pv, w) = \lambda(v, w) \quad \text{for all } w \in V.$$

Then
$$\lambda_{min}(v, v) \leq a(Pv, v) \leq \lambda_{max}(P)(v, v),$$
$$\lambda_{max}^{-1}(v, v) \leq a(P^{-1}v, v) \leq \lambda_{min}^{-1}(P)(v, v), \quad \text{for all } v \in V.$$

Hence (4.74) implies that $\underline{c}^2 \leq \lambda_{max}^{-1}$ and $\bar{c}^2 \geq \lambda_{min}^{-1}$, and the conclusion of the lemma follows. ∎

To complete our study of the convergence behaviour of the CG method combined with the BPX preconditioner, we state the following result: For the BPX preconditioner, the constants in (4.73) are independent of the refinement level and of the mesh diameter (see [Osw94, Theorem 19]).

Summary: in the CG method combined with the BPX preconditioner, the contraction factor at each iteration of CG is independent of the mesh size, and the simple structure (4.66) of the preconditioner allows its efficient implementation.

For a detailed discussion of effective preconditioners see [Axe96, AB84, Bra01, Osw94].

Remark 8.37. (Wavelets)
Consider the problem

$$a(u,v) = f(v) \quad \text{for all } v \in V,$$

which satisfies the standard assumptions of the Lax-Milgram Lemma. Using wavelets, Dahmen [40] (see also [Urb02]) proposes to transform the problem to an (infinite) matrix equation over the space l_2. Let us suppose that we know a wavelet basis $\{\psi_\lambda, \lambda \in J\}$ for the space V with

$$c\|(v_\lambda)\|_{l_2} \leq \left\|\sum_\lambda v_\lambda \psi_\lambda\right\|_V \leq C\|(v_\lambda)\|_{l_2} \quad \text{for all } (v_\lambda) \in l_2.$$

Set

$$\tilde{A} = (a(\psi_\nu, \psi_\lambda))_{\nu,\lambda \in J} \quad \text{and} \quad \tilde{f} := (f(\psi_\lambda)_{\lambda \in J}).$$

Then the matrix equation

$$\tilde{A}w = \tilde{f}$$

is equivalent to the given problem and is well-conditioned: There exist constants c_1, c_2 such that

$$c_1\|w\|_{l_2} \leq \|\tilde{A}w\|_{l_2} \leq c_2\|w\|_{l_2} \quad \text{for all } w \in l_2.$$

This observation is also the basis for adaptive wavelet methods; see the references cited above for the practical details.   □

Exercise 8.38. Verify that the iterates u^{k+1} generated by the CG method are solutions of the variational problem

$$\min_{u \in V_k} F(u) = \frac{1}{2}(u, Au) - (b, u).$$

Exercise 8.39. Prove that the CG method is an example of a *Krylov-space method*, viz., one has $u^k \in u^1 + U_{k-1}(A, u^1)$, $k = 2, 3, \ldots$, where the Krylov space $U_k(A, u^1)$ was defined in (4.42). See [DH03].

Exercise 8.40. Verify that the parameters β_k of the CG method can also be written as

$$\beta_k = \frac{(d^{k+1}, d^{k+1} - d^k)}{(d^k, d^k)}.$$

Remark: this representation, which is due to Polak and Ribiére, allows the generalization of the CG method to nonlinear problems.

Exercise 8.41. Consider the two-point boundary value problem

$$-((1+x)u')' = f \quad \text{in } (0,1), \qquad u(0) = u(1) = 0$$

and its discretization using piecewise linear finite elements on an equidistant mesh.

Use the CG method to solve the discrete problem (i) directly (ii) with preconditioning based on the matrix

$$
B_h = \begin{pmatrix}
2 & -1 & 0 & \cdot & \cdot & \cdot \\
-1 & 2 & -1 & \cdot & \cdot & \cdot \\
\cdot & -1 & 2 & -1 & \cdot & \cdot \\
\cdot & \cdot & \cdot & \cdot & \cdot & \cdot \\
\cdot & \cdot & \cdot & -1 & 2 & -1 \\
\cdot & \cdot & \cdot & \cdot & -1 & 2
\end{pmatrix}.
$$

Estimate the convergence behaviour of the method in each case.

Exercise 8.42. Consider the Dirichlet problem

$$
\begin{aligned}
-\Delta u &= f \quad \text{in } \Omega := (0,1) \times (0,1), \\
u|_\Gamma &= 0,
\end{aligned}
$$

and its discretization by linear finite elements on an arbitrary nonuniform triangular mesh. Assume that a uniform mesh is available that approximates in some sense the given nonuniform mesh. Describe how in practice one would use the exact solution on the uniform mesh with the fast Fourier transform as a preconditioner for the discrete problem on the nonuniform mesh.

8.5 Multigrid Methods

When a discretized elliptic boundary value problem is solved using a classical iterative method, the convergence of the iterates u_h^k to the discrete solution u_h is typically governed by the contractive property that

$$
\| u_h^{k+1} - u_h \| \leq \| T_h \| \, \| u_h^k - u_h \|, \qquad k = 1, 2, \dots,
$$

but (unfortunately) one also has

$$
\lim_{h \to 0} \| T_h \| = 1.
$$

Now if one has two mesh sizes h and H and considers the associated discrete problems

$$
A_h \, u_h = b_h \quad \text{and} \quad A_H \, u_H = b_H,
$$

then the solution of either of these problems is a good approximation to the solution of the other problem. Multigrid methods exploit this attribute and also certain useful properties of classical iterative methods such as Gauss-Seidel; they aim to achieve contractivity with a contractivity constant that is less than one and is *independent of h*. That is, they aspire to estimates of the form

$$\|u_h^{k+1} - u_h\| \leq \gamma \|u_h^k - u_h\|, \qquad k = 1, 2, \ldots \tag{5.1}$$

with a constant $\gamma \in (0, 1)$ that is independent of the mesh.

Let us first consider a *two-grid method*. Suppose that some finite element method (or finite difference or finite volume method) generates the discrete problem

$$A_h u_h = b_h \tag{5.2}$$

on a mesh of diameter h, and that there is a second discretization

$$A_H u_H = b_H \tag{5.3}$$

available that is based on a mesh of diameter H. Assume that $h < H$. We denote the discrete finite-dimensional spaces associated with (5.2) and (5.3) by U_h and U_H. Assume that there are linear continuous mappings

$$I_h^H : U_h \to U_H \quad \text{and} \quad I_H^h : U_H \to U_h$$

that allow us to move from a mesh function on one mesh to one on the other mesh. These mapping I_h^H is called a *restriction operator* and I_H^h a *prolongation operator*. The choice of restriction and prolongation operators can influence significantly the convergence rate of the method.

Rewrite (5.2) in the equivalent fixed-point form

$$u_h = T_h u_h + t_h, \tag{5.4}$$

for which one can use a classical fixed-point iterative method such as Gauss-Seidel or SOR (see Sections 5.3 and 5.4). As was already mentioned, one typically has $\|T_h\| \leq \gamma_h$ with $\gamma_h \in (0, 1)$.

We first apply n_1 steps of our classical iterative method on the fine mesh. Then the restriction operator is used to transfer the defect to the coarse mesh and one *solves* for the correction on the coarse mesh. We then map the discrete approximation back to the fine mesh and apply n_2 steps of the iterative method. To summarize, a two-grid method has the following structure:

Two-grid method

$$z_h^{k,l+1} := T_h z_h^{k,l} + t_h, \quad l = 0, 1, \ldots, n_1 - 1 \quad \text{with} \quad z_h^{k,0} := u_h^k, \tag{5.5}$$

$$d_h^k := b_h - A_h z_h^{k,n_1}, \tag{5.6}$$

$$d_H^k := I_h^H d_h^k. \tag{5.7}$$

Determine $v_H^k \in U_H$, the exact solution of

$$A_H v_H^k = d_H^k. \tag{5.8}$$

Set

$$w_h^{k,0} := z_h^{k,n_1} + I_H^h v_h^k, \tag{5.9}$$

$$w_h^{k,l+1} := T_h\, w_h^{k,l} + t_h, \qquad l = 0, 1, \ldots, n_2 - 1, \tag{5.10}$$

$$u_h^{k+1} := w_h^{k,n_2}. \tag{5.11}$$

The steps (5.5) and (5.10) are known as *pre-smoothing* and *post-smoothing* , while (5.6)–(5.9) is called the *coarse-grid correction*.

Coalescing all the intermediate steps of the two-grid method, the algorithm defined by (5.5)–(5.11) can be written in the form

$$u_h^{k+1} := S_h\, u_h^k + s_h, \qquad k = 1, 2, \ldots \tag{5.12}$$

with

$$S_h := T_h^{n_2}\,(I - I_H^h A_H^{-1} I_h^H A_h)\, T_h^{n_1} \tag{5.13}$$

and some $s_h \in U_h$.

If, as well as the restriction and prolongation operators, the pre-smoothing and post-smoothing steps are adequately chosen, then in many concrete cases one can prove (see, for instance, [Hac03b]) that there exists a constant $\gamma \in (0,1)$, which is independent of the mesh width h, such that $\|S_h\| \leq \gamma$. At the end of this section we sketch two underlying ideas of the many proofs of this result.

Before doing this we describe the principle of the *multigrid method*. In the two-grid method we observe that on the coarse mesh one has to solve a problem having the same structure as the original problem (with, of course, a smaller number of unknowns). It is therefore natural to apply again the two-grid method when solving the coarse mesh problem (5.8). Moreover, zero is a possible starting vector because $\|d_H^k\|$ vanishes asymptotically as $H \to 0$. Thus from a two-grid method we have derived a three-grid method. Proceeding recursively, one obtains finally the multigrid method. It is always assumed that on the coarsest mesh the discrete equations generated are solved exactly.

In practice there are several variants of this basic idea. If on each level, for the approximate solution of the coarse grid correction just one multigrid step on the remaining grids is used, then we call the strategy a *V-cycle* multigrid; otherwise it is a *W-cycle* multigrid. This terminology is derived from the graphical representation of the transfer from grid to grid—see Figure 8.10.

Let

$$h_0 > h_1 > \cdots > h_{l-1} > h_l > 0$$

denote the mesh sizes of l grids used in an l-grid method. The corresponding discrete spaces and the restriction and prolongation operators are denoted by

$$U_j := U_{h_j}, \quad j = 1, \ldots, l, \qquad I_{j+1}^j : U_{j+1} \to U_j \quad \text{and} \quad I_j^{j+1} : U_j \to U_{j+1}.$$

In the figure we see a three-grid method and the different use of the underlying meshes in a V-cycle and a W-cycle.

In the first case we use one two-grid step (based on the spaces U_2 and U_1) to

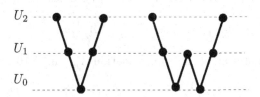

Figure 8.10   V- and W-cycle

compute the coarse mesh correction on U_2. In the second case the two-grid
method is applied twice, once on U_2 and once on U_1.

Now we consider the convergence behaviour of a multigrid method. For
this we first need some information about the structure of the method. We
assume that the multigrid method is based on a two-grid method that uses
the spaces U_j, U_{j-1} to solve the problem

$$A_j \, w_j \; = \; q_j.$$

For $q_j \in U_j$ let the two-grid method be given by

$$w_j^{i+1} \; = \; S_{j,j-1} \, w_j^i \, + \, C_{j,j-1} \, q_j, \qquad i = 0,1,\dots, \tag{5.14}$$

with, as in (5.13), an operator defined by

$$S_{j,j-1} \; := \; T_j^{n_2} \, (I - I_{j-1}^j A_{j-1}^{-1} I_j^{j-1} A_j) \, T_j^{n_1} \tag{5.15}$$

and a related matrix $C_{j,j-1}$.

Next we describe a three-grid method on U_j, U_{j-1}, U_{j-2}. Assume that
in solving the coarse mesh equations on U_{j-1} we apply σ steps of the two-
grid method on U_{j-1}, U_{j-2} with zero as initial vector. Then the three-grid
operator can be written as

$$S_{j,j-2} \; := \; T_j^{n_2} \, (I - I_{j-1}^j \tilde{A}_{j-1}^{-1} I_j^{j-1} A_j) \, T_j^{n_1}, \tag{5.16}$$

where $\tilde{A}_{j-1}^{-1}$ denotes the two-grid approximation of A_{j-1}^{-1} on U_{j-1}, U_{j-2}. Iteratively applying (5.14) we see that

$$w_{j-1}^\sigma \; = \; S_{j-1,j-2}^\sigma \, w_{j-1}^0 + p_{j-1} \tag{5.17}$$

for $\sigma = 1,2,\dots$ and some $p_{j-1} \in U_{j-1}$. But the solution w_{j-1} of

$$A_{j-1} \, w_{j-1} \; = \; q_{j-1}$$

is a fixed point of (5.17), so

$$A_{j-1}^{-1} q_{j-1} \; = \; S_{j-1,j-2}^\sigma \, A_{j-1}^{-1} q_{j-1} + p_{j-1}$$

and consequently

$$p_{j-1} = (I - S^\sigma_{j-1,j-1}) A^{-1}_{j-1} q_{j-1}.$$

As we take $w^0_{j-1} = 0$ in the algorithm, (5.17) becomes

$$w^\sigma_{j-1} = (I - S^\sigma_{j-1,j-1}) A^{-1}_{j-1} q_{j-1}.$$

Using this representation of the approximation $\tilde{A}^{-1}_{j-1}$ in (5.16), one gets finally

$$S_{j,j-2} = T^{n_2}_j (I - I^j_{j-1}(I - S^\sigma_{j-1,j-2})A^{-1}_{j-1}I^{j-1}_j A_j) T^{n_1}_j. \tag{5.18}$$

We shall use this formula to analyse the convergence behaviour of the three-grid method. Knowing the form of the two-grid and three-grid operators we conclude, recursively, that the $(m+1)$-grid operator has the form

$$S_{j,j-m} = T^{n_2}_j (I - I^j_{j-1}(I - S^\sigma_{j-1,j-m})A^{-1}_{j-1}I^{j-1}_j A_j) T^{n_1}_j. \tag{5.19}$$

Next we prove that the mesh-independent convergence of the two-grid method implies (under certain conditions) the mesh-independent convergence of a multigrid method.

Theorem 8.43. *Assume that the two-grid operators $S_{j,j-1}$ defined in (5.15) satisfy the inequality*

$$\|S_{j,j-1}\| \leq c_1, \qquad j = 2, \ldots, l$$

for some constant $c_1 \in (0,1)$. Furthermore, assume that

$$\|T^{n_2}_j I^j_{j-1}\| \, \|A^{-1}_{j-1} I^{j-1}_j A_j T^{n_1}_j\| \leq c_2, \qquad j = 2, \ldots, l$$

for some constant $c_2 > 0$. Then there exists a positive integer σ such that the multigrid operator defined by (5.19) satisfies the estimate

$$\|S_{j,1}\| \leq c, \qquad j = 2, \ldots, l$$

with a constant $c \in (0,1)$ that is independent of l.

Proof: From (5.15) and (5.19) we obtain

$$S_{j,j-m} = S_{j,j-1} + T^{n_2}_j I^j_{j-1} S^\sigma_{j-1,j-m} A^{-1}_{j-1} I^{j-1}_j A_j T^{n_1}_j.$$

Hence

$$\|S_{j,j-m}\| \leq \|S_{j,j-1}\| + \|T^{n_2}_j I^j_{j-1}\| \, \|S_{j-1,j-m}\|^\sigma \, \|A^{-1}_{j-1} I^{j-1}_j A_j T^{n_1}_j\|$$
$$\leq c_1 + c_2 \|S_{j-1,j-m}\|^\sigma. \tag{5.20}$$

But $c_1 \in (0,1)$, so there exists a positive integer σ such that the equation

$$c_1 + c_2 \alpha^\sigma = \alpha$$

has a solution $\alpha \in (0,1)$. Clearly $c_1 \leq \alpha$. From (5.20) and $\|S_{j,j-1}\| \leq c_1 \leq \alpha$ it follows inductively that $\|S_{j,1}\| \leq \alpha$. $\blacksquare$

Theorem 8.43 demonstrates the existence of a mesh-independent contraction number $\gamma \in (0,1)$ such that the estimate (5.1) is valid. For simplicity, we studied only the W-cycle with a adequately chosen number of inner steps. For the analysis of the V-cycle, which corresponds to the choice $\sigma = 1$, we refer the reader to [24].

We now analyse the convergence behaviour of a specific two-grid method. Let us consider the boundary value problem

$$-\Delta u = f \qquad \text{in } \Omega := (0,1) \times (0,1),$$
$$u|_\Gamma = 0. \tag{5.21}$$

As usual, $a(\cdot,\cdot)$ denotes the bilinear form associated with (5.21), i.e.,

$$a(u,v) = \int_\Omega \nabla u \, \nabla v \, dx,$$

and

$$a(u,v) = (f,v) \qquad \text{for all } v \in U := H_0^1(\Omega) \tag{5.22}$$

is the weak formulation of (5.21).

We discretize this problem using piecewise linear finite elements on the uniform mesh of Figure 8.11, which is our "fine" mesh.

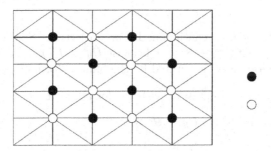

Figure 8.11 The mesh of the two-grid method

The space U_h on the fine mesh is spanned by the set of basis functions associated with the interior mesh points. As indicated on Figure 8.11, there are two classes of basis functions, each characterized by its different support.

The basis functions of these two classes span the subspaces $U_{1,h}$ (generated by the basis functions associated with the black vertices) and $U_{2,h}$ (from the basis functions of the white vertices). Then the space U_h is a direct sum:

$$U_h = U_{1,h} \oplus U_{2,h}.$$

In particular $U_{1,h} \cap U_{2,h} = \{0\}$. The splitting of the space U_h considered here does not directly correspond to the classical geometric concept of a coarse and a fine mesh.

In [22] a two-grid method was defined by alternating the solution of the variational equation between the two subspaces:

$$u_h^{k+1/2} := u_h^k + v_{1,h}^k \qquad \text{with } v_{1,h}^k \in U_{1,h} \text{ which solves} \tag{5.23}$$

$$a(v_{1,h}^k, v) = (f, v) - a(u_h^k, v) \qquad \text{for all } v \in U_{1,h}. \tag{5.24}$$

$$u_h^{k+1} := u_h^{k+1/2} + v_{2,h}^k \qquad \text{with } v_{2,h}^k \in U_{2,h} \text{ which solves} \tag{5.25}$$

$$a(v_{2,h}^k, v) = (f, v) - a(u_h^{k+1/2}, v) \qquad \text{for all } v \in U_{2,h}. \tag{5.26}$$

Remark 8.44. Let $\{\varphi_i : i \in I_{1,h}\}$ be a list of the basis functions in $U_{1,h}$. From Figure 8.11 and the definition of $U_{1,h}$, we see immediately that

$$int\ supp\ \varphi_i \cap int\ supp\ \varphi_j = \emptyset, \qquad i, j \in I_{1,h},\ i \neq j. \tag{5.27}$$

Consequently the Ritz-Galerkin equations (5.24) decouple into a simple system of scalar equations. The solution is equivalent to a half step of a red-black iteration. On the other hand, the variational equation (5.26) represents the discretized problem on the coarse mesh.

To summarize, the method (5.23)–(5.26) is a particular example of a two-grid method.   □

There exist well-defined continuous linear projections $S_1 : U_h \to U_{1,h}$ and $S_2 : U_h \to U_{2,h}$ with $S_1 + S_2 = I$ because $U_h = U_{1,h} \oplus U_{2,h}$. Let us write $\|\cdot\|$ for the energy norm generated by the bilinear form $a(\cdot, \cdot)$, i.e.,

$$\|u\|^2 = a(u, u) \qquad \text{for all } u \in U.$$

Define a *hypernorm* $[|\cdot|] : U_h \to \mathbb{R}^2$ by

$$[|u|] := \begin{pmatrix} \|S_1 u\| \\ \|S_2 u\| \end{pmatrix} \qquad \text{for all } u \in U_h. \tag{5.28}$$

This induces a hypernorm on the space of bounded linear operators $Q : U_h \to U_h$, which is defined by

$$[|Q|] := \begin{pmatrix} \|Q\|_{11} & \|Q\|_{12} \\ \|Q\|_{21} & \|Q\|_{22} \end{pmatrix} \tag{5.29a}$$

with

$$\|Q\|_{ij} := \sup_{v \in U_{j,h},\, v \neq 0} \frac{\|S_i Q v\|}{\|v\|}, \quad i,j = 1,2. \tag{5.29b}$$

From the definition one sees immediately that

$$[|Qu|] \leq [|Q|][|u|],$$

where the product on the right-hand side is matrix-vector multiplication. Similarly, one can easily verify the property

$$[|Q_1 Q_2|] \leq [|Q_1|]\,[|Q_2|] \tag{5.30}$$

for the hypernorms of two arbitrary bounded linear operators $Q_1, Q_2 : U_h \to U_h$.

In [22] (see also [Her87]) the following statement is proved.

Lemma 8.45. *The strengthened Cauchy-Schwarz inequality*

$$|a(u,v)| \leq \frac{1}{\sqrt{2}} \|u\|\, \|v\| \qquad \text{for all } u \in U_{1,h},\ v \in U_{2,h}$$

is valid.

For $j = 1,2$, define Ritz projectors $P_j : U_h \to U_{j,h}$ by

$$P_j u \in U_{j,h} \quad \text{satisfies} \quad a(P_j u, v) = a(u,v) \qquad \text{for all } v \in U_{j,h}. \tag{5.31}$$

Then we have

Lemma 8.46. *The hypernorms of the mappings $I - P_j$, $j = 1,2$, satisfy*

$$[|I - P_1|] \leq \begin{pmatrix} 0 & \frac{1}{\sqrt{2}} \\ 0 & 1 \end{pmatrix}, \qquad [|I - P_2|] \leq \begin{pmatrix} 1 & 0 \\ \frac{1}{\sqrt{2}} & 0 \end{pmatrix},$$

where the inequalities hold entry by entry in the 2×2 matrices.

Proof: From (5.31) we get

$$P_1 u = u \qquad \text{for all } u \in U_{1,h}, \tag{5.32}$$

and consequently

$$(I - P_1) u = 0 \qquad \text{for all } u \in U_{1,h}. \tag{5.33}$$

These two identities imply that

$$P_1^2 = P_1 \qquad \text{and} \qquad (I - P_1)^2 = I - P_1. \tag{5.34}$$

Suppose that $u \in U_{2,h}$ satisfies $(I - P_1)u = 0$. That is, $u = P_1 u$. But $U_{1,h} \cap U_{2,h} = \{0\}$, so

$$(I - P_1)u = 0, \quad u \in U_{2,h} \quad \Longrightarrow \quad u = 0. \tag{5.35}$$

Consider the hypernorm of $(I - P_1)$. The identity (5.33) and the linearity of the S_j show that

$$S_j (I - P_1) u = 0 \qquad \text{for all } u \in U_{1,h}, \quad j = 1, 2. \tag{5.36}$$

Suppose instead that $u \in U_{2,h}$. Now

$$(I - P_1) u = S_1 (I - P_1) u + S_2 (I - P_1) u \tag{5.37}$$

and appealing to (5.33) and (5.34) we get

$$(I - P_1) u = (I - P_1)^2 u = (I - P_1) S_2 (I - P_1) u.$$

That is,

$$(I - P_1) (I - S_2(I - P_1)) u = 0 \qquad \text{for all } u \in U_{2,h}.$$

Then (5.35) forces

$$S_2 (I - P_1) u = u \qquad \text{for all } u \in U_{2,h}. \tag{5.38}$$

Hence (5.37) simplifies to

$$S_1 (I - P_1) u = -P_1 u \qquad \text{for all } u \in U_{2,h}. \tag{5.39}$$

Lemma 8.45 and (5.31) yield the estimate

$$\|P_1 u\|^2 = a(P_1 u, P_1 u) = a(u, P_1 u) \leq \frac{1}{\sqrt{2}} \|u\| \, \|P_1 u\| \quad \text{for all } u \in U_{2,h},$$

and from (5.39) we then conclude that

$$\|S_1(I - P_1)u\| \leq \frac{1}{\sqrt{2}} \|u\| \qquad \text{for all } u \in U_{2,h}. \tag{5.40}$$

The definition (5.29) of the hypernorm and the results (5.36), (5.38) and (5.40) give immediately the estimate

$$[|I - P_1|] \leq \begin{pmatrix} 0 & \frac{1}{\sqrt{2}} \\ 0 & 1 \end{pmatrix}.$$

The bound on $(I - P_2)$ is proved similarly. ∎

As well as the hypernorm (5.28), one can define a new norm $||| \cdot |||$

$$|||u||| := \max \{ \|S_1 u\|, \|S_2 u\| \} \tag{5.41}$$

which is equivalent to the energy norm. We prove the mesh-independent convergence of the two-grid method with respect to the norm $||| \cdot |||$:

Theorem 8.47. *For any starting value $u_h^1 \in U_h$, the two-grid method (5.23)–(5.26) generates a sequence $\{u_h^k\}$ that converges as $k \to \infty$ to the solution $u_h \in U_h$ of the discrete problem*

$$a(u_h, v_h) = (f, v_h) \qquad \text{for all } v_h \in U_h, \tag{5.42}$$

with (independently of the mesh size h)

$$|||u_h^{k+1} - u_h||| \leq \frac{1}{\sqrt{2}} |||u_h^k - u_h|||, \qquad k = 1, 2, \dots . \tag{5.43}$$

Proof: Recall from (5.24) that $v_{1,h}^k \in U_{1,h}$ is defined by

$$a(v_{1,h}^k, v) = a(u_h - u_h^k, v) \qquad \text{for all } v \in U_{1,h}.$$

Consequently,

$$v_{1,h}^k = -P_1(u_h^k - u_h).$$

Hence (5.23) implies that

$$u_h^{k+1/2} - u_h = (I - P_1)(u_h^k - u_h).$$

Analogously, one has

$$u_h^{k+1} - u_h = (I - P_2)(u_h^{k+1/2} - u_h).$$

Lemma 8.46 then yields the estimate

$$[|u_h^{k+1} - u_h|] \leq \begin{pmatrix} 1 & 0 \\ \frac{1}{\sqrt{2}} & 0 \end{pmatrix} \begin{pmatrix} 0 & \frac{1}{\sqrt{2}} \\ 0 & 1 \end{pmatrix} [|u_h^k - u_h|] = \begin{pmatrix} 0 & \frac{1}{\sqrt{2}} \\ 0 & \frac{1}{2} \end{pmatrix} [|u_h^k - u_h|],$$

which gives (5.43). The convergence of $\{u_h^k\}$ to u_h follows. ∎

Remark 8.48. This method can formally be augmented by an additional half-step (see [22]), i.e., by setting

$$\tilde{u}_h^{k+1} - u_h := (I - P_1)(I - P_2)(I - P_1)(\tilde{u}_h^k - u_h),$$

but in practice it is unnecessary to apply the red-black step $(I - P_1)$ twice because $(I - P_1)^2 = (I - P_1)$. □

Remark 8.49. The method (5.23)–(5.26) can be generalized to other types of meshes and to general elliptic differential operators: see [Her87]. Nevertheless the mesh decomposition must still satisfy condition (5.27) and ensure the validity of the strengthened Cauchy-Schwarz inequality

$$|a(u, v)| \leq \mu_{ij} \|u\| \|v\| \qquad \text{for all } u \in U_{i,h},\ v \in U_{j,h}$$

with $\mu_{12}\,\mu_{21} < 1$. Condition (5.27) again allows us to solve the subproblem (5.24) in a trivial way. □

Remark 8.50. Alternatively, we sketch the basic idea of a general framework due to Hackbusch which brakes the convergence proof of a two-grid method into two separate parts.

Let us assume to consider a finite element space V_h on the fine mesh and a second space V_{2h} on the coarse mesh. For some starting value $u_h^k \in V_h$ to solve the Galerkin equation on the fine mesh we first realize ν smoothing steps:

$$u_h^{k,1} = S^\nu u_h^k .$$

Then, we compute on the coarse mesh $\hat{v}_{2h} \in V_{2h}$ which solves

$$a(\hat{v}_{2h}, w) = (f, w) - a(u_{k,1}, w) \quad \text{for all } w \in V_{2h}$$

and set

$$u_h^{k+1} = u_h^{k,1} + \hat{v}_{2h} .$$

With some adequately chosen norm $||| \cdot |||$ we assume for a convergence proof with respect to the L_2 norm the following:
It holds the

$$\text{smoothing property} \qquad |||S^\nu v_h||| \le ch^{-2}\frac{1}{\nu^\gamma}\|v_h\|_0 \qquad (5.44)$$

and the

$$\text{approximation property} \qquad \|u_h^* - Ru_h^*\| \le ch^2|||u_h^*|||. \qquad (5.45)$$

Here for given $u_h^* \in V_h$ its Ritz projection in V_{2h} is denoted by Ru_h^*.

Smoothing property and approximation property are combined in the following way. First, with the solution u_h on the fine mesh it holds

$$|||u_h^{k,1} - u_h||| \le ch^{-2}\frac{1}{\nu^\gamma}\|u_h^k - u_h\|_0 . \qquad (5.46)$$

Further we have

$$a(u_h^{k,1} - u_h + \hat{v}_{2h}, w) = 0 \quad \text{for all } w \in V_{2h}.$$

That means, $\hat{v}_{2h}$ is the Ritz projection of $u_h - u_h^{k,1}$ and the approximation property yields

$$\|u_h^{k+1} - u_h\|_0 = \|u_h - u_h^{k,1} - \hat{v}_{2h}\|_0 \le ch^2|||u_h - u_h^{k,1}|||. \qquad (5.47)$$

Combining (5.46) and (5.47) we obtain

$$\|u_h^{k+1} - u_h\|_0 \le \frac{c}{\nu^\gamma}\|u_h^k - u_h\|_0 .$$

Thus we have contractivity uniformly with respect to h if the number of smoothing steps is sufficiently large.

Remark that the verification of the approximation property is closely related to Nitsche's technique of proving L_2 error estimates.   □

As well as classical multigrid methods, *cascadic multigrid methods* have been developed and theoretically analysed in the last decade. See, for instance, [21, 109]. Like classical multigrid, cascadic methods use a nested sequence of discrete spaces, for example,

$$U_0 \subset U_1 \subset U_2 \subset \cdots \subset U_l \subset H_0^1(\Omega).$$

But unlike classical multigrid, where coarse and fine meshes are used alternately in the iteration, cascadic methods start on the coarse mesh and in the iteration process use only finer meshes. Let us denote by $B_j : V_j \to V_j$ the underlying standard iteration scheme (for instance, CG) used to solve the discrete problem

$$A_j \, w_j \; = \; q_j \, .$$

Then the fundamental approach of a cascadic method is: Apply m_j iterations of the iteration scheme on U_j and use the result as the starting point $\tilde{u}_j$ for m_{j+1} iterations on the finer mesh, i.e., on the space U_{j+1}. Thus on the space $U_h = U_l$ that corresponds to the finest mesh, the approximate solution $\tilde{u}_h$ is given by

$$\tilde{u}_h \; = \; B_l^{m_l} \, B_{l-1}^{m_{l-1}} \, \cdots B_1^{m_1} \, B_0^{m_0} \, \tilde{u}_0.$$

Here $\tilde{u}_0$ is the starting point in the discrete space U_0 defined on the coarsest mesh.

The convergence of cascadic multigrid requires that the given meshes are already sufficiently fine, and in addition the approximation of the solution u_0 on the coarsest mesh by the iteration $B_0^{m_0} \tilde{u}_0$ in U_0 has to be sufficiently good. See [21] and [109] for concrete variants of cascadic methods and its convergence analysis.

Exercise 8.51. Verify that one can choose restriction and prolongation operators for symmetric problems in such a way that

$$I_h^H \; = \; (I_H^h)^T.$$

Exercise 8.52. Consider the two-point boundary value problem

$$-(\alpha(x)y')' \; = \; f \quad \text{in } \Omega := (0,1), \qquad y(0) = y(1) = 0. \qquad (5.48)$$

Assume that $\alpha \in C^1(\overline{\Omega})$ and $\min\limits_{x \in \Omega} \alpha(x) > 0$. This problem is discretized using piecewise linear finite elements on an equidistant mesh

$$x_i := ih, \quad i = 0, 1, \ldots, N,$$

where $h = 1/N$ and $N > 0$ is an even integer. The standard basis functions are

$$\varphi_i^h(x) := \begin{cases} 1 - |x - x_i|/h & \text{if } x \in (x_i - h, x_i + h), \\ 0 & \text{otherwise.} \end{cases}$$

a) Verify that the space $U_h := \text{span}\{\varphi_i^h\}_{i=1,\dots,N-1}$ can be written as a direct sum

$$U_h = U_{1,h} \oplus U_{2,h} \quad \text{with}$$

$$U_{1,h} := \text{span}\{\varphi_i^h\}_{i=1,3,5,\dots,N-1}, \quad \text{and} \quad U_{2,h} := \text{span}\{\varphi_i^{2h}\}_{i=2,4,\dots,N-2}.$$

b) Compute the stiffness matrix for the discretization of problem (5.48) using the representation $U_h = U_{1,h} \oplus U_{2,h}$.

c) Set

$$a(u, v) := \int_0^1 \alpha u' v' \, dx.$$

For the related energy norm $\| \cdot \|$, prove the strengthened Cauchy-Schwarz inequality

$$|a(u, v)| \leq \mu_h \|u\| \|v\| \qquad \text{for all } u \in U_{1,h}, \ v \in U_{2,h} \tag{5.49}$$

with some $\mu_h \in [0, 1)$. What happens to μ_h as $h \to 0$?

d) Compute μ_h in (5.49) for the special case $\alpha(x) \equiv 1$ and interpret the result (hint: consider the Green's function of problem (5.48)).

8.6 Domain Decomposition, Parallel Algorithms

Suppose that we are given a boundary value problem for a partial differential equation that is posed on a geometrically complicated domain. When computing a numerical solution the domain is often decomposed into subdomains, each of which is separately discretized. Of course one then needs some means of recovering the properties of the global problem, typically by transmission conditions where the subdomains overlap (or on their common boundary when they do not overlap).

This technique is called *domain decomposition*. It allows us to split a discretized problem of very high dimension into subproblems, and often within the solution process one has to solve independently the discrete problems (of lower dimension) on the subdomains. This facilitates the parallelization of a large part of the computational effort needed to solve numerically the given problem; thus the problem can be solved effectively in a parallel environment. A second important attribute of the domain decomposition technique is its usefulness as a preconditioner for some iterative method that is used to solve the full discrete problem.

In this section we shall present some basic ideas of domain decomposition. As a model problem we consider the discretization of the Poisson problem with piecewise linear finite elements. For a detailed description of the theory and applications of domain decomposition methods we refer to [TW05] and [QV99].

Let us consider the problem

$$-\Delta u = f \quad \text{in } \Omega, \qquad u = 0 \quad \text{on } \Gamma.$$

As usual, $\Omega \subset \mathbb{R}^2$ is a bounded polygonal domain with boundary Γ. We use the weak formulation

$$\text{find } u \in V := H_0^1(\Omega) \quad \text{such that} \quad a(u,v) = (f,v) \qquad \text{for all } v \in V \quad (6.1)$$

and discretize with piecewise linear $C(\Omega)$ elements on a quasi-uniform mesh $\mathcal{Z}_h$ of mesh width h. Denote the triangles of the triangulation by T_j, $j \in J$ for some index set J, i.e., $\mathcal{Z}_h = \{T_l\}_{l \in J}$. Let the points $x_i \in \Omega$, $i \in I$ and $x_i \in \Gamma$, $i \in \hat{I}$ be the interior mesh points and the mesh points on the boundary, respectively. Denote the Lagrange basis functions of the finite element space V_h by φ_i, so that

$$V_h = \text{span}\, \{\varphi_i\}_{i \in I}\,.$$

The finite element discretization of (6.1) is:

$$\text{find } u_h \in V_h \quad \text{such that} \quad a(u_h, v_h) = (f, v_h) \qquad \text{for all } v_h \in V_h. \quad (6.2)$$

Now we choose m subdomains $\Omega_j := \text{int}\,(\bigcup_{l \in J_j} T_l)$ of Ω such that

$$\Omega_j \cap \Omega_l = \emptyset \text{ if } j \neq l, \qquad \text{and} \qquad \bigcup_{j=1}^{m} \overline{\Omega}_j = \overline{\Omega}.$$

Let $\Gamma_j = \partial \Omega_j$ denote the boundary of the jth subdomain. Each subdomain Ω_j is characterized by an index set $J_j \subset J$. We denote by I_j the set of all indices related to mesh points that are interior to Ω_j, i.e., $I_j = \{\, i \in I \,:\, x_i \in \Omega_j \,\}$, $j = 1, \ldots, m$.

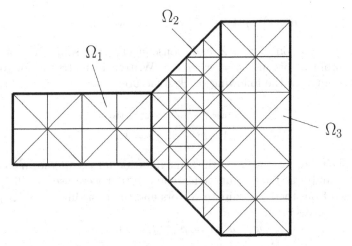

Figure 8.12 Decomposition into subdomains

To simplify the notation we omit h in what follows.

Analogously to the decomposition of Ω into subdomains Ω_j, we decompose the discrete space V as the direct sum

$$V = W \oplus Z \quad \text{with} \quad W = \bigoplus_{j=1}^{m} W_j, \tag{6.3}$$

where the spaces W_j and Z are defined by

$$W_j = \text{span}\left\{\varphi_i\right\}_{i \in I_j}, \ j = 1, \ldots, m \quad \text{and} \quad Z = \text{span}\left\{\varphi_i\right\}_{i \in I \setminus \left(\bigcup_{j=1}^{m} I_j\right)}. \tag{6.4}$$

Then the discrete problem (6.2) is equivalent to (remember that we omit h): Find $u = w + z$, $w \in W$, $z \in Z$ such that

$$a(w + z, \omega + \zeta) = (f, \omega + \zeta) \quad \text{for all } \omega \in W, \ \zeta \in Z. \tag{6.5}$$

Because $a(\cdot, \cdot)$ is bilinear and $(f, \cdot)$ is linear, this is equivalent to

$$a(w, \omega) + a(z, \omega) = (f, \omega) \quad \text{for all } \omega \in W, \tag{6.6a}$$

$$a(w, \zeta) + a(z, \zeta) = (f, \zeta) \quad \text{for all } \zeta \in Z. \tag{6.6b}$$

Consider (6.6a). For each $z \in Z$ there exists a unique $w = w(z) \in W$ such that

$$a(w(z), \omega) + a(z, \omega) = (f, \omega) \quad \text{for all } \omega \in W. \tag{6.7a}$$

Then substituting this into (6.6b) yields the reduced problem

$$a(w(z), \zeta) + a(z, \zeta) = (f, \zeta) \quad \text{for all } \zeta \in Z \tag{6.7b}$$

for the computation of $z \in Z$ and, consequently, the solution u of the full discrete problem because $u = w(z) + z$. Written in matrix form, (6.7) is a linear system with the following block structure:

$$A_{ww} w + A_{wz} z = f_w, \tag{6.8a}$$

$$A_{zw} w + A_{zz} z = f_z. \tag{6.8b}$$

For simplicity, in (6.8) we used the same notation for the unknown coordinate vectors w and z as for the functions w and v that were used in (6.7), which are formed from these coordinate vectors and the basis functions $\{\varphi_i\}$. Block elimination gives

$$w = w(z) = A_{ww}^{-1}(f_w - A_{wz} z),$$

and we get the reduced problem

$$S z = q \quad \text{with} \quad S = A_{zz} - A_{zw} A_{ww}^{-1} A_{wz}, \quad q = f_z - A_{zw} A_{ww}^{-1} f_w. \quad (6.9)$$

The matrix S is the *Schur complement* of the block A_{ww}.

In general it is not possible to compute the Schur complement explicitly for practical problems because of the presence of the term A_{ww}^{-1}. Nevertheless, the matrix-vector product Sz for a given $z \in Z$ can be computed effectively due to the structure the sub-matrices. The main ingredient in this is the determination of $w(z)$ from (6.7a) and (6.8a). Let us introduce the notation

$$a_j(u, v) = \int_{\Omega_j} \nabla u \cdot \nabla v \, dx, \qquad (f, v)_j = \int_{\Omega_j} f v \, dx, \qquad \langle \lambda, v \rangle_j = \int_{\Gamma_j} \lambda v \, dx.$$

Now $\text{int}(\text{supp } w_j) \subset \Omega_j$ for each $w_j \in W_j$, so (6.7a) is equivalent to $w(z) = \sum_{j=1}^{m} w_j(z)$ with

$$w_j(z) \in W_j \quad \text{and} \quad a_j(w_j(z), \omega_j) = (f, \omega_j)_j - a_j(z, \omega_j)$$
$$\text{for all } \omega_j \in W_j, \ j = 1, \dots, m.$$

These m sub-problems are independent discrete elliptic problems which can be solved in parallel. For an implementation of this strategy and a general discussion of parallel algorithms applied to partial differential equations, see [Haa99].

Next we return to the second sub-system (6.7b). In the continuous problem, Green's formula gives

$$-\int_{\Omega} \Delta u \, v \, dx = -\sum_{j=1}^{m} \int_{\Omega_j} \Delta u \, v \, dx = -\sum_{j=1}^{m} \int_{\Gamma_j} \frac{\partial u}{\partial n} v \, ds + \sum_{j=1}^{m} \int_{\Omega_j} \nabla u \cdot \nabla v \, dx,$$

where n denotes a unit normal on each Γ_j that points out from Ω_j. Analogously to this formula, one can in the discrete case define for each $v \in V$ its outward-pointing normal derivative λ_j on each Γ_j by the equation

$$\langle \lambda_j, \zeta \rangle_j = a_j(w_j, \zeta) + a_j(z, \zeta) - (f, \zeta)_j \quad \text{for all } \zeta \in Z, \ j = 1, \dots, m, \quad (6.10)$$

where $v = w + z$ with $z \in Z$ and $w = \sum_{j=1}^{m} w_j$ with $w_j \in W_j, \ j = 1, \dots, m$. In terms of the normal derivatives or fluxes λ_j we can write (6.7b) equivalently as

$$\sum_{j=1}^{m} \langle \lambda_j, \zeta \rangle_j = 0 \quad \text{for all } \zeta \in Z. \quad (6.11)$$

The *interior boundary* of each subdomain Ω_j is defined to be $\Gamma_j \setminus \Gamma$. Thus (6.7b) and (6.11) are discrete analogues of the continuity across interior boundaries property that the fluxes of the solution of the original problem clearly enjoy. Note that the continuity of the discrete solution on all of Ω is automatically guaranteed by our choice of the discrete space.

Consider again the reduced system (6.9), viz.,

$$S z = q. \tag{6.12}$$

We have already seen that the direct solution and computation of $w \in W$ from (6.7a) (or equivalently from (6.8a)) corresponds to a block elimination procedure. In domain decomposition however, one seeks to solve (6.12) iteratively while taking into account its underlying structure.

Let us apply a block Gauss-Seidel method to (6.7). Given $z^k \in Z$, we compute $w^{k+1} = \sum_{j=1}^{m} w_j^{k+1}$ with $w_j^{k+1} \in W_j$, $j = 1, \ldots, m$, by solving the m independent subproblems

$$a_j(w_j^{k+1}, \omega_j) + a_j(z^k, \omega_j) = (f, \omega_j)_j \quad \text{for all } \omega_j \in W_j, \, j = 1, \ldots, m, \tag{6.13a}$$

then on the interior boundaries we solve the flux continuity problem

$$a(w^{k+1}, \zeta) + a(z^{k+1}, \zeta) = (f, \zeta) \quad \text{for all } \zeta \in Z. \tag{6.13b}$$

The method (6.13) generalizes the red-black iteration of Section 8.3 (which can be interpreted as a particular example of a domain decomposition technique) to more than two subdomains while using finite elements as the underlying method of discretization.

Let us mention that the Schur complement S is symmetric (Exercise 8.56). It is therefore possible to use the CG method (in combination with an efficient method of computation of Sz for any given $z \in Z$) to solve the discrete problem (6.12). In general S is badly conditioned and this must be accommodated by using suitable preconditioners to accelerate the convergence of the CG method.

Like solution strategies for domain decomposition techniques on the continuous level (see [QV99]), the preconditioner constructed will depend on the choice of the discrete boundary values for the local subproblems and on the updating of the coupling variables. We now describe, as an example of this process, the basic approach of the Neumann-Neumann iteration.

The Neumann-Neumann algorithm modifies the block Gauss-Seidel iteration (6.13) by replacing (6.13b) by an approximating subproblem that, like (6.13a), can be described on each subdomain Ω_j independently and consequently can be solved in parallel.

The computation of the new approximation $z^{k+1} \in Z$ to the solution of (6.12), given $z^k \in Z$, is carried out in two steps. First, let $w_j^{k+1/2} \in W_j$ be the solution of the discrete Dirichlet problems

$$a_j(w_j^{k+1/2}, \omega_j) + a_j(z^k, \omega_j) = (f, \omega_j)_j \quad \text{for all } \omega_j \in W_j, \, j = 1, \ldots, m. \tag{6.14}$$

Then corrections $v^k \in W$ and $y^k \in Z$ are computed from

$$a_j(v_j^k, \omega_j) + a(y_j^k, \omega_j) = 0 \qquad \text{for all } \omega_j \in W, \tag{6.15a}$$

$$a_j(v_j^k, \zeta_j) + a_j(y_j^k, \zeta_j) = (f, \zeta_j) - a(w^{k+1/2}, \zeta_j) - a(z^k, \zeta_j) \tag{6.15b}$$

$$\text{for all } \zeta_j \in Z_j.$$

Here

$$Z_j = \text{span}\{\varphi_i\}_{i \in \hat{I}_j} \qquad \text{with} \qquad \hat{I}_j = \{i \in I : x_i \in \Gamma_j\}, \quad j = 1, \ldots, m,$$

denotes the subspace of $C(\Omega_j)$ that contributes to Z. That is, the index set $\hat{I}_j$ characterizes mesh points that lie on on the interior boundary of Ω_j. Finally, the new iterate is computed from

$$z^{k+1} = z^k + \theta \sum_{j=1}^{m} y_j^k \tag{6.16}$$

with some relaxation parameter $\theta > 0$.

The problems (6.15) are Neumann problems on the subdomains Ω_j when $\Gamma_j \cap \Gamma = \emptyset$. In this case the solution of (6.15) is defined using, e.g., the pseudo-inverse of the related coefficient matrix, which gives the least-squares solution with minimal norm. For more details and a proof of the convergence of the Neumann-Neumann algorithm see [TW05].

Remark 8.53. The problems (6.15) approximate the correction equations

$$a(w^{k+1/2} + \tilde{v}^k, \omega) + a(z^k + \tilde{y}^k, \omega) = (f, \omega) \qquad \text{for all } \omega \in W, \tag{6.17a}$$

$$a(w^{k+1/2} + \tilde{v}^k, \zeta) + a(z^k + \tilde{y}^k, \zeta) = (f, \zeta) \qquad \text{for all } \zeta \in Z, \tag{6.17b}$$

which, setting $z = z^k + \tilde{v}^k$, yield the solution of (6.12). Using (6.10) and (6.17b) the evalation of the remaining correction is often written in the form

$$(f, \zeta_j) - a(w^{k+1/2}, \zeta_j) - a(z^k, \zeta_j) = -\sum_{j=1}^{m} \langle \lambda^{k+1/2}, \zeta_j \rangle_j$$

with the fluxes $\lambda^{k+1/2}$ to be determined; see [QV99, TW05]. □

Within the Neumann-Neumann algorithm, the choice of the discrete spaces automatically guarantees the continuity of the discrete solution but not the continuity of its fluxes. Finite element tearing and integrating (FETI) algorithms, on the other hand, relax the continuity requirement; then a local space can be chosen in each subdomain independently, and continuity of the discrete solution is ensured by an additional constraint.

Let us introduce the notation

$$V_j = W_j + \text{span } \{\varphi_{jl}\}_{l \in \tilde{I}_j}$$

with

$$\varphi_{jl}(x) = \begin{cases} \varphi_l(x) & \text{if } x \in \Omega_j, \\ 0 & \text{otherwise,} \end{cases} \quad \text{and} \quad \tilde{I}_j := \{l \in I : x_l \in \Gamma_j\}.$$

For the broken space $\tilde{V}_h := \sum_{j=1}^{m} V_j$ (compare the discontinuous Galerkin methods of Section 4.8) one has

$$\tilde{V}_h \subset L_2(\Omega) \quad \text{and} \quad V_h \subset \tilde{V}_h.$$

If v_h is a function from $\tilde{V}_h$, then

$$v_h \in V_h \quad \Longleftrightarrow \quad v_h \in C(\overline{\Omega}).$$

The discrete problem (6.2) is equivalent to

$$\min_{v_h \in \tilde{V}_h \cap C(\overline{\Omega})} J(v_h) := \frac{1}{2} \sum_{j=1}^{m} \Big(a_j(v_j, v_j) - (f, v_j)_j \Big). \tag{6.18}$$

Of course, the condition $v_h \in C(\overline{\Omega})$ can be reformulated as a system of linear restrictions; for instance, by

$$\int_{\Gamma_{ij}} v_i\, z = \int_{\Gamma_{ji}} v_j\, z \quad \text{for all } z \in Z, \text{ and for all } \Gamma_{ij} := \Gamma_i \cap \Gamma_j, \quad \Gamma_{ij} \neq \emptyset. \tag{6.19}$$

Remark 8.54. The domain decomposition techniques considered here can be considered as additive Schwarz methods and analysed within this framework. See [TW05].   □

Remark 8.55. In the literature (see [TW05]), instead of (6.19) one usually requires continuity of v_h at the interior mesh points that lie on interior boundaries, i.e., at mesh points $x_i \in \Big(\bigcup_{j=1}^{m} \Gamma_j \Big) \backslash \Gamma$. With this requirement at all such points, one obtains redundant restrictions at any mesh point where more than two subdomains meet.   □

Let us define the forms $\tilde{a} : \tilde{V} \times \tilde{V} \to \mathbb{R}$ and $\tilde{b} : Z \times \tilde{V} \to \mathbb{R}$ by

$$\tilde{a}(v_h, \nu_h) := \sum_{j=1}^{m} a_j(v_j, \nu_j) \quad \text{and} \quad \tilde{b}(z_h, \nu_h) := \sum_{i,j=1}^{m} \int_{\Gamma_{ij}} z \cdot (\nu_i - \nu_j),$$

where $\nu_h = \sum_{j=1}^{m} v_j$ with $v_j \in V_j$ for each j. Then $v_h \in \tilde{V}$ is a solution of the problem (6.18) if and only if there exists a $z_h \in Z$ such that

$$\tilde{a}(v_h, \nu_h) + \tilde{b}(z_h, \nu_h) = (f, \nu_h) \qquad \text{for all } \nu_h \in \tilde{V},$$

$$\tilde{b}(\zeta_h, v_h) \qquad\qquad\; = 0 \qquad\qquad \text{for all } \zeta_h \in Z.$$

Thus we have a mixed formulation which has the advantage that it is applicable even if the meshes used do not coincide on any interior boundary. In this case the space Z of Lagrange multipliers is not automatically defined by the decomposition of the subdomains as in the conforming case where the meshes do coincide on interior boundaries. Instead, as is usual in mixed formulations, one has to choose carefully the discrete space Z; in particular, the choice of Z has to be compatible with the space $\tilde{V}$ insofar as the Babuška-Brezzi conditions must be satisfied.

A standard approach to defining Z is the *mortar technique*; see [TW05], [23, 123]). The technique prescribes the discrete values on one side of the interior boundary—the non-mortar side—and defines the basis functions on the other (mortar) side by a balance that is applied in a weak sense. The precise choice of basis functions on the mortar side depends on the discrete space used for the Lagrange multipliers; see [122].

We illustrate the mortar technique with a simple example. Decompose the domain $\Omega := (-1, 1) \times (0, 1) \subset \mathbb{R}^2$ into the subdomains (see Figure 8.13)

$$\Omega_1 := (-1, 0) \times (0, 1), \qquad \Omega_2 := (0, 1) \times (0, 1).$$

These subdomains are independently triangulated (with a view to applying a finite element method with piecewise linear elements). As a result, the mesh points of Ω_1 and Ω_2 do not coincide on the interior boundary.

At the interior boundary $\Gamma_1 \cap \Gamma_2$, we choose the Ω_2 side as the non-mortar side and write Γ_{21} for $\Gamma_1 \cap \Gamma_2$ regarded as part of Γ_2. On the mortar side, we write Γ_{12} for $\Gamma_1 \cap \Gamma_2$ regarded as part of Γ_1.

$$\Omega_1 \qquad\qquad\qquad\qquad \Omega_2$$

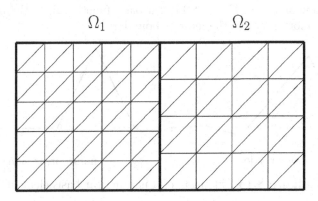

Figure 8.13 Non-matching grids

Let $x_i \in \Omega$, $i \in I := \{1, \dots, N\}$ be the interior grid points, and define

$$I_1 := \{i \in I : x_i \in \Omega_1\}, \qquad \bar{I}_1 := \{i \in I : x_i \in \Gamma_{1,2}\},$$
$$I_2 := \{i \in I : x_i \in \Omega_2\}, \qquad \bar{I}_2 := \{i \in I : x_i \in \Gamma_{2,1}\}.$$

Let $\tilde{\varphi}_i \in C(\Omega_j)$, $i \in I_j \cup \bar{I}_j$, $j = 1, 2$, be the Lagrange basis functions for piecewise linears on the corresponding subdomains. For $i \in I_1 \cup I_2$ it is easy to extend each function φ_i to the other subdomain by

$$\varphi_i(x) := \begin{cases} \tilde{\varphi}_i(x) & \text{if } x \in \Omega_j, \\ 0 & \text{if } x \notin \Omega_j, \end{cases} \qquad i \in I_j, \; j = 1, 2.$$

The functions $\tilde{\varphi}_i$, $i \in \bar{I}_2$, of the non-mortar side are extended to Ω_1 by using the mortar-side functions $\tilde{\varphi}_i$, $i \in \bar{I}_1$ in the following way:

$$\varphi_i(x) := \begin{cases} \tilde{\varphi}_i(x) & \text{if } x \in \Omega_2, \\ \displaystyle\sum_{j \in \bar{I}_1} \sigma_{ij} \, \tilde{\varphi}_j(x) & \text{if } x \in \Omega_1, \end{cases} \qquad i \in \bar{I}_2, \tag{6.20}$$

where the coefficients $\sigma_{ij} \in \mathbb{R}$, $i \in \bar{I}_2$, $j \in \bar{I}_1$, are computed from a balance equation in a weak form, i.e.,

$$\sum_{j \in \bar{I}_1} \sigma_{ij} \, \tilde{b}(\tilde{\varphi}_j, z) = \tilde{b}(\tilde{\varphi}_i, z) \quad \text{for all } z \in Z \text{ and all } i \in \bar{I}_2.$$

Choosing $Z = \text{span}\{\tilde{\varphi}_j\}_{j \in \bar{I}_1}$, one obtains the linear system

$$\sum_{j \in \bar{I}_1} \sigma_{ij} \, \tilde{b}(\tilde{\varphi}_j, \tilde{\varphi}_k) = \tilde{b}(\tilde{\varphi}_i, \tilde{\varphi}_k), \qquad k \in \bar{I}_1, \; i \in \bar{I}_2, \tag{6.21}$$

for the computation of the unknown coefficients σ_{ij}. This gives, for instance, the extension shown in Figure 8.14 of a basis function from the non-mortar side to the mortar side of the interior boundary.

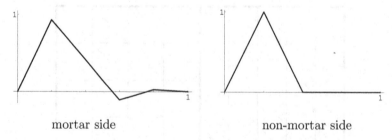

mortar side   non-mortar side

Figure 8.14 Extension of a non-mortar basis element

In this example of an extension, oscillations appear because of the properties of the matrix of the system (6.21). It is possible both to avoid such oscillations and to minimize the support of the basis functions generated by a suitable choice of test functions (Lagrange multipliers), e.g., by the use of a bi-orthogonal basis. Thus for linear elements, if one chooses

$$\psi_i(\eta) = \begin{cases} -\frac{1}{3} & \text{if } \eta \in (y_{i-1}, y_{i-1/2}) \cup (y_{i+1/2}, y_{i+1}), \\ 1 & \text{if } \eta \in [y_{i-1/2}, y_{i+1/2}], \\ 0 & \text{otherwise}, \end{cases}$$

then the resulting coefficient matrix is diagonal.

There are several other ways of choosing Z and of determining the coefficients σ_{ij} ; see [122].

Exercise 8.56. Verify that for the discrete problem (6.2) the Schur complement S is symmetric.

Exercise 8.57. Show that the block Gauss-Seidel iteration (6.13) is a special case of the method (5.23)–(5.26). How then should one choose the subspaces $U_{1,h}$ and $U_{2,h}$?

Exercise 8.58. For the decomposition of Figure 8.13, determine all mortar extensions (6.20) if piecewise constant test functions (Lagrange multipliers) are used.

Bibliography: Textbooks and Monographs

[AB84]   O. Axelsson and V.A. Barker. *Finite element solution of boundary value problems.* Academic Press, New York, 1984.

[Ada75]   R.A. Adams. *Sobolev spaces.* Academic Press, New York, 1975.

[AO00]   M. Ainsworth and J.T. Oden. *A posteriori error estimation in finite element analysis.* Wiley, 2000.

[Ape99]   T. Apel. *Anisotropic finite elements: Local estimates and applications.* Teubner, 1999.

[Axe96]   O. Axelsson. *Iterative solution methods.* Cambridge Univ. Press, Cambridge, 1996.

[Ban98]   R.E. Bank. *PLTMG user's guide (edition 8.0).* SIAM Publ., Philadelphia, 1998.

[Bel90]   A.I. Beltzer. *Variational and finite element methods: a symbolic computation approach.* Springer, Berlin, 1990.

[Bey98]   J. Bey. *Finite-Volumen- und Mehrgitterverfahren für elliptische Randwertprobleme.* Teubner, Stuttgart, 1998.

[BF91]   F. Brezzi and M. Fortin. *Mixed and hybrid finite element methods.* Springer, Berlin, 1991.

[BR03]   W. Bangerth and R. Rannacher. *Adaptive finite element methods for differential equations.* Birkhäuser, Basel, 2003.

[Bra01]   D. Braess. *Finite elements.* Cambridge University Press, 2001.

[BS94]   S.C. Brenner and L.R. Scott. *The mathematical theory of finite element methods.* Springer, 1994.

[CHQZ06]   C. Canuto, M.Y. Hussaini, A. Quarteroni, and T. Zang. *Spectral methods: fundamentals in single domains.* Springer, Berlin, 2006.

[Cia78]   P. Ciarlet. *The finite element method for elliptic problems.* North-Holland, Amsterdam, 1978.

[CL91]   P.G. Ciarlet and J.L. Lions. *Handbook of numerical analysis 2.* North Holland, 1991.

[CL00]   P.G. Ciarlet and J.L. Lions. *Handbook of numerical analysis 7.* North Holland, 2000.

[Dau88]   M. Dauge. *Elliptic boundary alue problems in corner domains.* Springer, Berlin, 1988.

[Dav04]   J.H. Davis. *Methods of applied mathematics with a MATLAB overview.* Birkhäuser, Boston, 2004.

[DH03]   P. Deuflhard and A. Hohmann. *Numerical analysis in modern scientific computing. An introduction. 2nd revised ed.* Springer, Berlin, 2003.

[dJF96]   E.M. de Jager and J. Furu. *The theory of singular perturbations.* North Holland, 1996.

[DL72]   G. Duvant and J.-L. Lions. *Les inéquations en mécanique et en physique.* Dunod, Paris, 1972.

[Dob78]   M. Dobrowolski. *Optimale Konvergenz der Methode der finiten Elemente bei parabolischen Anfangs-Randwertaufgaben.* PhD thesis, Universität, Bonn, 1978.

[Dör98]   W. Dörfler. *Uniformly convergent finite element methods for singularly perturbed convection-diffusion equations.* PhD thesis, Habilitation, Univ. Freiburg, 1998.

[EG04]   A. Ern and J.-L. Guermond. *Theory and practice of finite elements.* Springer, 2004.

[Eng80]   H. Engels. *Numerical quadrature and cubature.* Academic Press, New York, 1980.

[ET76]   I. Ekeland and R. Temam. *Convex analysis and variational problems.* North Holland, Amsterdam, 1976.

[Fri82]   A. Friedman. *Variational principles and free boundary value problems.* Wiley, New York, 1982.

[GFL+83]   H. Goering, A. Felgenhauer, G. Lube, H.-G. Roos, and L. Tobiska. *Singularly perturbed differential equations.* Akademie-Verlag, Berlin, 1983.

[GGZ74]   H. Gajewski, K. Gröger, and K. Zacharias. *Nichtlineare Operatorgleichungen.* Akademie Verlag, Berlin, 1974.

[GLT81]   R. Glowinski, J.L. Lions, and R. Trémoliere. *Numerical analysis of variational inequalities (2nd ed.).* North Holland, Amsterdam, 1981.

[Goh02]   X. Q. Goh, C. J.; Yang. *Duality in optimization and variational inequalities.* Taylor and Francis, London, 2002.

[GR89]   V. Girault and P.A. Raviart. *Finite element for Navier-Stokes equations (Theory and algorithms).* Springer, Berlin, 1989.

[GR94]   C. Großmann and H.-G. Roos. *Numerik partieller Differentialgleichungen. 2.Aufl.* Teubner, Stuttgart, 1994.

[Gri85]   P. Grisvard. *Elliptic problems in nonsmooth domains.* Pitman, Boston, 1985.

[Gri92]   P. Grisvard. *Singularities in boundary value problems.* Springer and Masson, Paris, 1992.

[GRT93]   H. Goering, H.-G. Roos, and L. Tobiska. *Finite-Element-Methoden, 3. Aufl.* Akademie-Verlag, Berlin, 1993.

[GT83]   D. Gilbarg and N.S. Trudinger. *Elliptic partial differential equations of second order.* Springer, Berlin, 1983.

[Haa99]   G. Haase. *Parallelisierung numerischer Algorithmen für partielle Differentialgleichungen.* Teubner, Stuttgart, 1999.

[Hac95]   W. Hackbusch. *Integral equations: theory and numerical treatment.* Birkhäuser, Basel, 1995.

[Hac03a]   W. Hackbusch. *Elliptic differential equations: theory and numerical treatment.* Springer, 2003.

[Hac03b]   W. Hackbusch. *Multi-grid methods and applications.* Springer, Berlin, 2003.

[Hei87]   B. Heinrich. *Finite difference methods on irregular networks.* Akademie-Verlag, Berlin, 1987.

[Her87]   R. Herter. *Konvergenzanalyse von Zweigitterverfahren für nichtselbstad-jungierte elliptische Probleme 2.Ordnung mittels Raumsplittung.* PhD thesis, TU Dresden, 1987.

[HHNL88] I. Hlaváček, J. Haslinger, J. Nečas, and J. Lovíšek. *Numerical solution of variational inequalities.* Springer, Berlin, 1988.

[HNW87]   E. Hairer, S.P. Norsett, and G. Wanner. *Solving ordinary differential equations I.Nonstiff problems.* Springer-Verlag, Berlin, 1987.

[HW91]   E. Hairer and G. Wanner. *Solving ordinary differential equations II.* Springer, Berlin, 1991.

[HY81]   L.A. Hagemann and D.M. Young. *Applied Iterative Methods.* Academic Press, New York, 1981.

[Ike83]   T. Ikeda. *Maximum principle in finite element models for convection-diffusion phenomena.* North-Holland, Amsterdam, 1983.

[JL01]   M. Jung and U. Langer. *Methode der finiten Elemente für Ingenieure.* Teubner, Stuttgart, 2001.

[Joh88]   C. Johnson. *Numerical solutions of partial differential equations by the finite element method.* Cambride University Press, 1988.

[KA03]   P. Knabner and L. Angermann. *Numerical methods for elliptic and parabolic partial differential equations.* Springer, Berlin, 2003.

[Kač85]   J. Kačur. *Method of Rothe in evolution equations.* Teubner, Leipzig, 1985.

[KN96]   M. Křižek and P. Neittanmäki. *Mathematical and numerical modelling in electrical engineering: theory and application.* Kluwer, Dordrecht, 1996.

[KNS98]   M. Křižek, P. Neittanmäki, and R. Sternberg. *Finite element methods: superconvergence, post-processing and a posteriori estimates.* Marcel Dekker, 1998.

[Krö97]   D. Kröner. *Numerical schemes for conservation laws.* Wiley-Teubner, Chichester: Wiley. Stuttgart, 1997.

[KS80]   D. Kinderlehrer and G. Stampacchia. *An introduction to variational inequalities and their applications.* Academic Press, New York, 1980.

[KT94]   A. Kaplan and R. Tichatschke. *Stable methods for ill-posed variational problems: Prox-regularization of elliptic variational inequalities and semi-infinite problems.* Akademie Verlag, Berlin, 1994.

[Lan01]   J. Lang. *Adaptive multilevel solution of nonlinear parabolic PDE systems.* Springer, 2001.

[Lax72]   P.D. Lax. *Hyperbolic systems of conservation laws and the mathematical theory of shock waves.* SIAM, Philadelphia, 1972.

[Lin07]   T. Linss. *Layer-adapted meshes for convection-diffusion problems.* PhD thesis, Habilitation, TU Dresden, 2007.

[Lio71]   J.L. Lions. *Optimal control of systems governed by partial differential equations.* Springer, Berlin-Heidelberg, 1971.

[LL06]   Q. Lin and J. Lin. *Finite element methods: accuracy and improvement.* Science Press, Beijing, 2006.

[LLV85]   G.S. Ladde, V. Lakshmikantam, and A.S. Vatsala. *Monotone iterative techniques for nonlinear differential equations.* Pitman, 1985.

[LT03]   S. Larsson and V. Thomée. *Partial differential equations with numerical methods.* Springer, Berlin, 2003.

[Mar70]   R. Martinet. Regularisation d'inequations variationelles par approximation successive. *RAIRO*, 4: 154–159, 1970.

[Mel02]   J.M. Melenk. *h-p-finite element methods for singular perturbations.* Springer, 2002.

574   Bibliography: Textbooks and Monographs

[Mic78]   S. G. Michlin. *Partielle Differentialgleichungen der mathematischen Physik.* Akademie-Verlag, Berlin, 1978.

[Mik86]   S. G. Mikhlin. *Constants in some inequalities of analysis.* Wiley, Chichester, 1986.

[MS83]   G.I. Marchuk and V.V. Shaidurov. *Difference methods and their extrapolations.* Springer, New York, 1983.

[MW77]   A.R. Mitchell and R. Wait. *The finite element method in partial differential equations.* Wiley, New York, 1977.

[NAG05]   NAG Fortran Library Manual, mark 21. NAG Ltd., Oxford, 2005.

[Näv82]   U. Nävert. *A finite element method for convection-diffusion problems.* PhD thesis, Göteborg, 1982.

[Neč83]   J. Nečas. *Introduction to the theory of nonlinear elliptic equations.* Teubner, Leipzig, 1983.

[NH81]   J. Nečas and I. Hlaváček. *Mathematical theory of elastic and elasto-plastic bodies: an introduction.* Elsevier, Amsterdam, 1981.

[NT94]   P. Neittaanmäki and D. Tiba. *Optimal control of nonlinear parabolic systems: theory, algorithms, and applications.* Marcel Dekker, New York, 1994.

[OR70]   J.M. Ortega and W.C. Rheinboldt. *Iterative solution of nonlinear equations in several variables.* Academic Press, New York, 1970.

[Osw94]   P. Oswald. *Multilevel finite element approximation.* Teubner, Stuttgart, 1994.

[PW67]   M.H. Protter and H.F. Weinberger. *Maximum principles in differential equations.* Prentice-Hall, Englewood Cliffs, 1967.

[QV94]   A. Quarteroni and A. Valli. *Numerical approximation of partial differential equations.* Springer, 1994.

[QV99]   A. Quarteroni and A. Valli. *Domain decomposition methods for partial differential equations.* Clarendon Press, Oxford, 1999.

[Ran04]   R. Rannacher. Numerische Mathematik 2 (Numerik partieller Differentialgleichungen). Technical report, Univ. Heidelberg, 2004.

[Rek80]   K. Rektorys. *Variational methods in mathematics, science and engineering.* Reidel Publ.Co., Dordrecht, 1980.

[Rek82]   K. Rektorys. *The method of discretization in time and partial differential equations.* Dordrecht, Boston, 1982.

[RST96]   H.-G. Roos, M. Stynes, and L. Tobiska. *Numerical methods for singularly perturbed differential equations.* Springer, 1996.

[Sam01]   A.A. Samarskij. *Theory of difference schemes.* Marcel Dekker, New York, 2001.

[Sch78]   H. Schwetlick. *Numerische Lösung nichtlinearer Gleichungen.* Verlag der Wissenschaften, Berlin, 1978.

[Sch88]   H.R. Schwarz. *Finite element methods.* Academic press, 1988.

[Sch97]   F. Schieweck. *Parallele Lösung der inkompressiblen Navier-Stokes Gleichungen.* PhD thesis, Univ. Magdeburg, 1997.

[Shi92]   G. I. Shishkin. *Grid approximation of singularly perturbed elliptic and parabolic problems.* Ural Russian Academy of Science (russisch), 1992.

[SN89]   A.A. Samarskii and E.S. Nikolaev. *Numerical methods for grid equations. Vol.II: Iterative methods.* Birkhäuser, Basel, 1989.

[SS05]   A. Schmidt and K.G. Siebert. *Design of adaptive finite element software. The finite element toolbox ALBERTA.* Springer, 2005.

[Str04]   J. C. Strikwerda. *Finite difference schemes and partial differential equations. 2nd ed.* SIAM Publ., Philadelphia, 2004.

[Tem79]   R. Temam. *Navier-Stokes equations. Theory and numerical analysis.* North-Holland, Amsterdam, 1979.

[Tho97]   V. Thomée. *Galerkin finite element methods for parabolic problems.* Springer, 1997.

[Tri92]   H. Triebel. *Higher analysis.* Johann Ambrosius Barth, Leipzig, 1992.

[Trö05]   F. Tröltzsch. *Optimale Steuerung partieller Differentialgleichungen.* Vieweg, Wiesbaden, 2005.

[TW05]   A. Toselli and O. Widlund. *Domain decomposition methods – algorithms and theory.* Springer, Berlin, 2005.

[Urb02]   K. Urban. *Wavelets in numerical simulation.* Springer, 2002.

[Ver96]   V. Verfürth. *A review of a posteriori error estimation and adaptive mesh-refinement techniques.* Wiley/Teubner, Stuttgart, 1996.

[VL92]   C.F. Van Loan. *Computational frameworks for the Fast Fourier Transform.* SIAM, Philadelphia, 1992.

[Wah95]   L.S. Wahlbin. *Superconvergence in Galerkin finite element methods.* Springer, 1995.

[Wlo87]   J. Wloka. *Partial differential equations.* Cambridge Univ. Press, 1987.

[Xu89]   J. Xu. *Theory of multilevel methods.* PhD thesis, Cornell University, 1989.

[Yos66]   K. Yosida. *Functional analysis.* Springer, Berlin, 1966.

[Zei90]   E. Zeidler. *Nonlinear functional analysis and its applications I-IV.* Springer, Berlin, 1985-90.

Bibliography: Original Papers

1. S. Agmon, A. Douglis, and N. Nirenberg. Estimates near the boundary for solutions of elliptic partial differential equations satisfying general boundary conditions. *Comm. Pure Appl. Math.*, 12:623–727, 1959.

2. M. Ainsworth and I. Babuska. Reliable and robust a posteriori error estimation for singularly perturbed reaction-diffusion problems. *SIAM Journal Numerical Analysis*, 36:331–353, 1999.

3. J. Alberty, C. Carstensen, and S. Funken. Remarks around 50 lines of MATLAB. *Numerical Algorithms*, 20:117–137, 1999.

4. E.L. Allgower and K. Böhmer. Application of the independence principle to mesh refinement strategies. *SIAM J.Numer.Anal.*, 24:1335–1351, 1987.

5. L. Angermann. Numerical solution of second order elliptic equations on plane domains. M^2AN, 25(2), 1991.

6. T. Apel and M. Dobrowolski. Anisotropic interpolation with applications to the finite element method. *Computing*, 47:277–293, 1992.

7. D. Arnold, F. Brezzi, B. Cockburn, and D. Marini. Unified analysis of discontinuous Galerkin methods for elliptic problems. *SIAM J. Numer. Anal.*, 39:1749–1779, 2002.

8. I. Babuška and K.A. Aziz. Survey lectures on the mathematical foundations of the finite element method. In K.A. Aziz, editor, *The mathematics of the finite element method with applications to partial differential equations*, pages 3–359. Academic press, New York, 1972.

9. I. Babuška, T. Strouboulis, and C.S. Upadhyag. A model study of the quality of a posteriori error estimators for linear elliptic problems in the interior of patchwise uniform grids of triangles. *Comput. Meth. Appl. Mech. Engngr.*, 114:307–378, 1994.

10. E. Bänsch. Local mesh refinement in 2 and 3 dimensions. *Impact of Comp. in Sci. and Engrg.*, 3:181–191, 1991.

11. C. Baumann and J. Oden. A discontinuous hp-finite element method for convection-diffusion problems. *Comput. Meth. Appl. Mech. Engrg.*, 175:311–341, 1999.

12. M. Bause and P. Knabner. Uniform error analysis for Lagrangian-Galerkin approximations of convection-dominated problems. *SIAM Journal Numerical Analysis*, 39:1954–1984, 2002.

13. M. Bergounioux, M. Haddou, M. Hintermüller, and K. Kunisch. A comparison of a Moreau-Yosida-based active set strategy and interior point methods for constrained optimal control problems. *SIAM J. Optim.*, 11:495–521, 2000.

14. C. Bernardi. Optimal finite element interpolation on curved domains. *SIAM Journal Numerical Analysis*, 23:1212–1240, 1989.

15. C. Bernardi and E. Süli. Time and space adaptivity for the second-order wave equation. *Math. Models Methods Appl. Sci.*, 15:199–225, 2005.

16. P. Binev, W. Dahmen, and R. de Vore. Adaptive finite element methods with convergence rates. *Numer. Math.*, 97:219–268, 2004.

17. P. Bochev and R. B. Lehoucq. On the finite element solution of the pure Neumann problem. *SIAM Review*, 47:50–66, 2005.

18. F.A. Bornemann. An adaptive multilevel approach to parabolic equations I. *Impact Comput. in Sci. and Engnrg.*, 2:279–317, 1990.

19. F.A. Bornemann. An adaptive multilevel approach to parabolic equations II. *Impact Comput. in Sci. and Engnrg.*, 3:93–122, 1991.

20. F.A. Bornemann. An adaptive multilevel approach to parabolic equations III. *Impact Comput. in Sci. and Engnrg.*, 4:1–45, 1992.

21. F.A. Bornemann and P. Deuflhard. The cascadic multigrid method for elliptic problems. *Numer. Math.*, 75:135–152, 1996.

22. D. Braess. The contraction number of a multigrid method for solving the Poisson equation. *Numer. Math.*, 37:387–404, 1981.

23. D. Braess and W. Dahmen. Stability estimates of the mortar finite element method for 3-dimensional problems. *East-West J. Numer. Math.*, 6:249–263, 1998.

24. D. Braess and W. Hackbusch. A new convergence proof for the multigrid method including the V-cycle. *SIAM J.Numer.Anal.*, 20:967–975, 1983.

25. J.H. Bramble, J.E. Pasciak, and J. Xu. Parallel multilevel preconditioners. *Math. Comput.*, 55:1–22, 1990.

26. F. Brezzi, W.W. Hager, and P.A. Raviart. Error estimates for the finite element solution of variational inequalities. Part I: Primal theory. *Numer. Math.*, 28:431–443, 1977.

27. F. Brezzi, D. Marini, and A. Süli. Discontinuous Galerkin methods for first order hyperbolic problems. *Math. Models a. Meth. in Appl. Sci.*, 14:1893–1903, 2004.

28. E. Burman and P. Hansbo. Edge stabilization for Galerkin approximations of convection-diffusion-reaction problems. *Comput. Methods Appl. Mech. Eng.*, 193:1437–1453, 2004.

29. C. Carstensen. Merging the Bramble-Pasciak-Steinbach and Crouzeix-Thomee criterion for H^1-stability of the L_2-projection onto finite element spaces. *Mathematics of Computation*, 71:157–163, 2002.

30. C. Carstensen. Some remarks on the history and future of averaging techniques in a posteriori finite element error analysis. *ZAMM*, 84:3–21, 2004.

31. C. Carstensen and S. Bartels. Each averaging technique yields reliable a posteriori error control in FEM on unstructured grids i. *Mathematics of Computation*, 71:945–969, 2002.

32. C. Carstensen and S. Bartels. Each averaging technique yields reliable a posteriori error control in FEM on unstructured grids ii. *Mathematics of Computation*, 71:971–994, 2002.

33. C. Carstensen and S.A. Funken. Constants in Clements interpolation error and residual based a posteriori estimates in finite element methods. *East West J. Numer. Anal.*, 8: 153–175, 2002.

34. C. Carstensen and R. Verfürth. Edge residuals dominate a posteriori error estimates for low-order finite element methods. *SIAM Journal Numerical Analysis*, 36: 1571–1587, 1999.

35. E. Casas and F. Tröltzsch. Error estimates for linear-quadratic elliptic control problems. In V. Barbu, editor, *Analysis and optimization of differential systems. IFIP TC7/WG 7.2 international working conference.*, pages 89–100, Constanta, Romania, 2003. Kluwer, Boston.

36. P. Castillo. Performance of discontinuous Galerkin methods for elliptic PDE's. *J. Sci. Comput.*, 24:524–547, 2002.

37. K.A. Cliffe, I.S. Duff, and J.A. Scott. Performance issues for frontal schemes on a cache-based high-performance computer. *Int. J. Numer. Methods Eng.*, 42:127–143, 1998.

38. B. Cockburn. Discontinuous Galerkin methods. *ZAMM*, 83:731–754, 2003.

39. M.C. Crandall and A. Majda. Monotone difference approximations for scalar conservation laws. *Math. Comput.*, 34: 1–21, 1980.

40. W. Dahmen. Wavelets and multiscale methods for operator equations. *Acta Numerica*, 6: 55–228, 1997.

41. W. Dahmen, B. Faermann, I.B. Graham, W. Hackbusch, and S.A. Sauter. Inverse inequalities on non-quasi-uniform meshes and application to the mortar element method. *Mathematics of Computation*, 73:1107–1138, 2004.

42. W. Dahmen, A. Kunoth, and J. Vorloeper. Convergence of adaptive wavelet methods for goal-oriented error estimation. Technical report, IGPM Report, RWTH-Aachen, 2006.

43. K. Deckelnick and M. Hinze. Semidiscretization and error estimates for distributed control of the instationary navier-stokes equations. *Numer. Math.*, 97:297–320, 2004.

44. V. Dolejsi, M. Feistauer, and C. Schwab. A finite volume discontinuous Galerkin scheme for nonlinear convection-diffusion problems. *Calcolo*, 39:1–40, 2002.

45. S. Domschke and W. Weinelt. Optimale Konvergenzaussagen für ein Differenzenverfahren zur Lösung einer Klasse von Variationsungleichungen. *Beitr.Numer.Math.*, 10: 37–45, 1981.

46. W. Dörfler. A convergent adaptive algorithm for Poisson's equation. *SIAM Journal Numerical Analysis*, 33: 1106–1124, 1996.

47. J.K. Duff, I.S.; Reid. The multifrontal solution of indefinite sparse symmetric linear equations. *ACM Trans. Math. Softw.*, 9:302–325, 1983.

48. J.S. Duff and J.A. Scott. A frontal code for the solution of sparse positive-definite symmetric systems arising from finite-element applications. *ACM Trans. Math. Softw.*, 25:404–424, 1999.

49. M. Feistauer and K. Svadlenka. Space-time discontinuous Galerkin method for solving nonstationary convection-diffusion-reaction problems. *SIAM Journal Numerical Analysis*, to appear.

50. J. Frehse and R. Rannacher. Eine l^1-Fehlerabschätzung für diskrete Grundlösungen in der Methode der finiten Elemente. *Bonner Math. Schriften*, 89: 92–114, 1976.

580   Bibliography: Original Papers

51. A.V. Fursikov, M.D. Gunzburger, and L.S. Hou. Boundary value problems and optimal boundary control for the Navier-Stokes system: The two-dimensional case. *SIAM J. Control Optimization*, 36:852–894, 1998.

52. E. H. Georgoulis. hp-version interior penalty discontinuous Galerkin finite element methods on anisotropic meshes. *Int. J. Num. Anal. Model.,*, 3:52–79, 2006.

53. J. Gopalakrishnan and G. Kanschat. A multilevel discontinuous Galerkin method. *Numerische Mathematik*, 95:527–550, 2003.

54. M. Griebel and P. Oswald. On the abstract theory of additive and multiplicative Schwarz algorithms. *Numer. Math.*, 70:163–180, 1995.

55. A. Griewank. The local convergence of Broyden-like methods on Lipschitzian problems in Hilbert spaces. *SIAM Journal Numerical Analysis*, 24:684–705, 1987.

56. C. Grossmann. Dualität und Strafmethoden bei elliptischen Differentialgleichungen. *ZAMM*, 64, 1984.

57. C. Grossmann and A.A. Kaplan. On the solution of discretized obstacle problems by an adapted penalty method. *Computing*, 35, 1985.

58. O. Gueler. On the convergence of the proximal point algorithm for convex minimization. *SIAM J. Control Optim.*, 29:403–419, 1991.

59. W. Hackbusch. On first and second order box schemes. *Computing*, 41:277–296, 1989.

60. K. Harriman, P. Houston, Bill Senior, and Endre Süli. hp-version discontinuous Galerkin methods with interior penalty for partial differential equations with nonnegative characteristic form. *Contemporary Mathematics*, 330:89–119, 2003.

61. A. Harten. High resolution schemes for hyperbolic conservation laws. *J. Comput. Phys.*, 49:357–393, 1983.

62. A. Harten, J.N. Hyman, and P.D. Lax. On finite-difference approximations and entropy conditions for shocks. *Comm.Pure Appl.Math.*, 19:297–322, 1976.

63. J. Haslinger. Mixed formulation of elliptic variational inequalities and its approximation. *Applikace Mat.*, 26:462–475, 1981.

64. J. Haslinger and P. Neittaanmäki. Shape optimization in contact problems. approximation and numerical realization. *RAIRO, Modlisation Math. Anal. Numr.*, 21:269–291, 1987.

65. R. Haverkamp. Zur genauen Ordnung der gleichmäßigen Konvergenz von H_0^1-Projektionen. Workshop, Bad Honnef, 1983.

66. Alan F. Hegarty, Eugene O'Riordan, and Martin Stynes. A comparison of uniformly convergent difference schemes for two-dimensional convection-diffusion problems. *J. Comput. Phys.*, 105:24–32, 1993.

67. B. Heinrich. The Fourier-finite-element method for Poisson's equation in axisymmetric domains with edges. *SIAM Journal Numerical Analysis*, 33:1885–1911, 1996.

68. M. Hintermüller. A primal-dual active set algorithm for bilaterally control constrained optimal control problems. *Q. Appl. Math.*, 61:131–160, 2003.

69. M. Hinze. A variational discretization concept in control constrained optimization: The linear-quadratic case. *Comput. Optim. Appl.*, 30:45–61, 2005.

70. M. Hinze and R. Pinnau. An optimal control approach to semiconductor design. *Math. Models Methods Appl. Sci.*, 12:89–107, 2002.

71. P. Houston, C. Schwab, and E. Süli. Discontinuous hp-finite element methods for advection-diffusion problems. *SIAM Journal Numerical Analysis*, 39: 2133–2163, 2002.

72. K. Ito and K. Kunisch. Optimal control of elliptic variational inequalities. *Appl. Math. Optimization.*, 41:343–364, 2000.

73. J.W. Jerome. On n-widths in Sobolev spaces and applications to elliptic boundary value problems. *Journal of Mathematical Analysis and Applications*, 29: 201–215, 1970.

74. C. Johnson, U. Nävert, and J. Pitkäranta. Finite element methods for linear hyperbolic problems. *Comp. Meth. Appl. Mech. Engrg.*, 45:285–312, 1984.

75. G. Kanschat and R. Rannacher. Local error analysis of the interior penalty discontinuous Galerkin method for second order elliptic problems. *J. Numer. Math.*, 10: 249–274, 2002.

76. R.B. Kellogg and M. Stynes. Corner singularities and boundary layers in a simple convection-diffusion problem. *J. Diff. Equations*, 213: 81–120, 2005.

77. R.B. Kellogg and A. Tsan. Analysis of some difference approximations for a singularly perturbed problem without turning points. *Math. Comput.*, 32(144): 1025–1039, 1978.

78. N. Kopteva. On the uniform in a small parameter convergence of weighted schemes for the one-dimensional time-dependent convection-diffusion equation. *Comput. Math. Math. Phys.*, 37: 1173–1180, 1997.

79. N. Kopteva. The two-dimensional Sobolev inequality in the case of an arbitrary grid. *Comput. Math. and Math. Phys.*, 38: 574–577, 1998.

80. S.N. Kruzkov. First order quasilinear equations with several independent variables. *Mat. Sbornik*, 81: 228–255, 1970.

81. A.Y. Le Roux. A numerical concept of entropy for quasi-linear equations. *Math. Comput.*, 31: 848–872, 1977.

82. R. Lehmann. Computable error bounds in the finite element method. *IMA J.Numer.Anal.*, 6: 265–271, 1986.

83. M. Lenoir. Optimal isoparametric finite elements and error estimates for domains involving curved boundaries. *SIAM Journal Numerical Analysis*, 23: 562–580, 1986.

84. J. Li. Full order convergence of a mixed finite element method for fourth order elliptic equations. *Journal of Mathematical Analysis and Applications*, 230: 329–349, 1999.

85. T. Linss. The necessity of Shishkin-decompositions. *Appl. Math. Lett.*, 14:891–896, 2001.

86. T. Linss. Layer-adapted meshes for convection-diffusion problems. *Comput. Meth. Appl. Mech. Engrg.*, 192: 1061–1105, 2003.

87. T. Linss and M. Stynes. Asymptotic analysis and Shishkin-type decomposition for an elliptic convection-diffusion problem. *J. Math. Anal. Appl.*, 31: 255–270, 2001.

88. J. M. Melenk. On approximation in meshless methods. In J. Blowey and A. Craig, editors, *EPSRC-LMS summer school*. Oxford university press, 2004.

89. J.M. Melenk. On condition numbers in hp-FEM with Gauss-Lobatto based shape functions. *J. Comput. Appl. Math.*, 139: 21–48, 2002.

90. C. Meyer and A. Rösch. Superconvergence properties of optimal control problems. *SIAM J. Control Optimization*, 43:970–985, 2004.

91. J.H. Michael. A general theory for linear elliptic partial differential equations. *J. Diff. Equ.*, 23: 1–29, 1977.

582   Bibliography: Original Papers

92. V.J. Mizel and T.I. Seidman. An abstract bang-bang principle and time-optimal boundary control of the heat equation. *SIAM J. Control Optimization*, 35:1204–1216, 1997.

93. P. Morin, R. Nochetto, and K. Siebert. Data oscillation and convergence of adaptive FEM. *SIAM Journal Numerical Analysis*, 38: 466–488, 2000.

94. E. O'Riordan and M. Stynes. A uniformly convergent finite element method for a singularly perturbed elliptic problem in two dimensions. *Math. Comput.*, 1991.

95. A. Ostermann and M. Roche. Rosenbrock methods for partial differential equations and fractional orders of convergence. *SIAM Journal Numerical Analysis*, 30: 1094–1098, 1993.

96. A. Ostermann and M. Roche. Runge-Kutta methods for partial differential equations and fractional orders of convergence. *SIAM Journal Numerical Analysis*, 30: 1084–1098, 1993.

97. J. Outrata and J. Zowe. A numerical approach to optimization problems with variational inequality constraints. *Math. Program.*, 68:105–130, 1995.

98. R. Rannacher and S. Turek. Simple nonconforming quadrilateral Stokes element. *Num. Meth. Part. Diff. Equ.*, 8: 97–111, 1992.

99. W. H. Reed and T. R. Hill. Triangular mesh methods for the neutron transport equation. Technical Report LA-UR-73-479. Los Alamos, 1973.

100. U. Risch. An upwind finite element method for singularly perturbed problems and local estimates in the l_∞-norm. *Math. Modell. Anal. Num.*, 24: 235–264, 1990.

101. H.-G. Roos. Ten ways to generate the Iljin and related schemes. *J. Comput. Appl. Math.*, 51: 43–59, 1994.

102. H.-G. Roos, D. Adam, and A. Felgenhauer. A novel nonconforming uniformly convergent finite element method in two dimensions. *J. Math. Anal. Appl.*, 201: 715–755, 1996.

103. A. Rösch. Error estimates for parabolic optimal control problems with control constraints. *Z. Anal. Anwend.*, 23:353–376, 2004.

104. R. Sacco. Exponentially fitted shape functions for advection-dominated flow problems in two dimensions. *J. Comput. Appl. Math.*, 67:161–165, 1996.

105. R. Scholz. Numerical solution of the obstacle problem by the penalty method. *Computing*, 32: 297–306, 1984.

106. Ch. Schwab. *p- and hp-finite element methods*. Clarendon press, Oxford, 1998.

107. H.R. Schwarz. The fast Fourier transform for general order. *Computing*, 19: 341–350, 1978.

108. R. Scott. Optimal L_∞ estimates for the finite element method on irregular meshes. *Math.Comput.*, 30: 681–697, 1976.

109. V. Shaidurov and L. Tobiska. The convergence of the cascadic conjugate-gradient method applied to elliptic problems in domains with re-entrant corners. *Math. Comput.*, 69: 501–520, 2000.

110. J. Shi. The F-E-M test for convergence of nonconforming finite elements. *Mathematics of Computation*, 49: 391–405, 1987.

111. G. I. Shishkin. On finite difference fitted schemes for singularly perturbed boundary value problems with a parabolic boundary layer. *J. Math. Anal. Appl.*, 208: 181–204, 1997.

112. R. Stevenson. Optimality of a standard adaptive finite element method. *Foundations of Comput. Math.*, to appear.

113. F. Stummel. The generalized patch test. *SIAM J.Numer.Anal.*, 16:449–471, 1979.

114. E. Trefftz. Ein Gegenstück zum Ritzschen Verfahren. In *Verhandl. 2. Intern. Kongreß tech.Mech.*, pages 131–138, 1927.

115. F. Tröltzsch. An SQP method for the optimal control of a nonlinear heat equation. *Control Cybern.*, 23:267–288, 1994.

116. M. Ulbrich. On a nonsmooth Newton method for nonlinear complementarity problems in function space with applications to optimal control. In M. et al. Ferris, editor, *Complementarity: applications, algorithms and extensions.*, pages 341–360, Madison, USA, 2001. Kluwer, Dordrecht.

117. R. Vanselow and H.-P. Scheffler. Convergence analysis of a finite volume method via a new nonconforming finite element method. *Numer. Methods Partial Differ. Equations*, 14:213–231, 1998.

118. V. Verfürth. Robust a posteriori error estimators for a singularly perturbed reaction-diffusion equation. *Numer. Math.*, 78:479–493, 1998.

119. V. Verfürth. A posteriori error estimates for finite element discretizations of the heat equation. *Calcolo*, 40:195–212, 2003.

120. S. Wang. A novel exponentially fitted triangular finite element method. *J. Comput. Physics*, 134:253–260, 1997.

121. N.M. Wigley. Mixed boundary value problems in plane domains with corners. *Math. Z.*, 115:33–52, 1970.

122. B. Wohlmuth. A comparison of dual Lagrange multiplier spaces for mortar finite element discretizations. *Math. Model. Numer. Anal.*, 36:995–1012, 2002.

123. B. Wohlmuth and R.H. Krause. A multigrid method based on the unconstrained product space for mortar finite element discretizations. *SIAM Journal Numerical Analysis*, 39:192–213, 2001.

124. J. Xu and L. Zikatanov. Some observations on Babuška and Brezzi theories. *Numer. Mathematik*, 94:195–202, 2003.

125. H. Yserentant. Two preconditioners based on the multilevel splitting of finite element spaces. *Numer.Math.*, 58:163–184, 1990.

126. A. Ženišek. Interpolation polynomials on the triangle. *Numer. Math.*, 15:238–296, 1970.

127. Z. Zhang. Polynomial preserving gradient recovery and a posteriori error estimation for bilinear elements on irregular quadrilaterals. *Int. J. Num. Anal. and Modelling*, 1:1–24, 2004.

128. M. Zlamal. On the finite element method. *Numer. Math.*, 12:394–409, 1968.

Index

Universitext